Geomorphic Analysis of River Systems

Geomorphic Analysis of River Systems: An Approach to Reading the Landscape

Kirstie A. Fryirs
Gary J. Brierley

Kirstie A. Fryirs
Department of Environment and Geography
Macquarie University
North Ryde, NSW 2109
Australia

Gary J. Brierley
School of Environment
University of Auckland
Auckland
New Zealand

⊕WILEY-BLACKWELL
A John Wiley & Sons, Ltd., Publication

Blackwell Publishing was acquired by John Wiley & Sons in February 2007. Blackwell's publishing program has been merged with Wiley's global Scientific, Technical and Medical business to form Wiley-Blackwell.

Registered office: John Wiley & Sons, Ltd, The Atrium, Southern Gate, Chichester, West Sussex, PO19 8SQ, UK

Editorial offices: 9600 Garsington Road, Oxford, OX4 2DQ, UK
The Atrium, Southern Gate, Chichester, West Sussex, PO19 8SQ, UK
111 River Street, Hoboken, NJ 07030-5774, USA

For details of our global editorial offices, for customer services and for information about how to apply for permission to reuse the copyright material in this book please see our website at www.wiley.com/wiley-blackwell.

Library of Congress Cataloging-in-Publication Data

Fryirs, Kirstie A.
  Geomorphic analysis of river systems : an approach to reading the landscape / Kirstie A. Fryirs, Gary J. Brierley.
    p. cm.
  Includes bibliographical references and index.
  ISBN 978-1-4051-9275-0 (cloth) – ISBN 978-1-4051-9274-3 (pbk.)   1. Watersheds.   2. Fluvial geomorphology.   I. Brierley, Gary J.   II. Title.
  GB561.F79 2013
  551.48'3011–dc23

                      2012016112

A catalogue record for this book is available from the British Library.

Wiley also publishes its books in a variety of electronic formats. Some content that appears in print may not be available in electronic books.

Set in 10/12.5 pt Minion by Toppan Best-set Premedia Limited

Front cover:
'Playful Platypus', Les Elvin, NAIDOC Artist of the Year 2008. © Glenelg Art Gallery, South Australia. Reproduced with permission of the artist.

Back cover:
Huang He at Dari, China. © Brendon Blue.

Cover design by:
Design Deluxe

1   2013

To those who have guided and supported us on our river wanderings

Contents

The color plate section can be found between pages 194 and 195

COMPANION WEBSITE:

This book has a companion website:
www.wiley.com/go/fryirs/riversystems
with Figures and Tables from the book

Preface

Purpose and aims of this book

Geomorphology is the science concerned with understanding the form of the Earth's land surface and the processes by which it is shaped, both at the present day as well as in the past.

British Society for Geomorphology; sourced at: http://www.geomorphology.org.uk/pages/geomorphology/

The most engaging and interesting intellectual work on geomorphic forms such as river channels has not come from computer specialists and theoretical models but from field measurements and observations.

Luna B. Leopold (2004: 10)

The scientific study of geomorphology has very divergent origins and approaches in different parts of the world. Perhaps inevitably, our perspectives and thinking are deeply rooted in our training, culture and the landscapes in which we live and work. In other words, scientific understandings have a genealogy, wherein cultural and experiential underpinnings fashion our way of thinking.

The approach to *reading the landscape* that is outlined in this book provides a way to interpret rivers across the range of environmental and climatic settings. It builds upon a solid understanding of the science of fluvial geomorphology. Field-based understandings are tied to theoretical and conceptual principles to generate catchment-specific analyses of river character, behaviour and evolution. This approach to landscape analysis views geomorphology as an interpretative and analytical science rather than a descriptive one.

Reading the landscape entails identification of river landforms and appraisal of their relationships to adjacent features. Primary controls upon contemporary dynamics are interpreted, framing analyses in relation to their landscape and catchment context, and the imprint from the past.

This book has been constructed as an introductory text on river landscapes, providing a bridge to more advanced principles outlined in Brierley and Fryirs (2005). Chapters 1–9 present foundational understandings that underpin the approach to reading the landscape that is documented in Chapters 10–14. The target audience is second- and third-year undergraduate students, as well as river practitioners who use geomorphic understandings in scientific and/or management applications.

Inevitably, no book can cover all geomorphic principles and practices, and much material has been 'left out' of the foundation chapters. For example, those interested in the details of bedload transport modelling, of hydraulic analyses of bank erosion processes and channel geometry models or of Quaternary science are encouraged to develop more advanced understanding from other geomorphology, engineering or earth science textbooks.

This book does not provide a fully grounded and comprehensive background to geomorphic analysis. Emphasis is placed upon documentation of the approach to reading the landscape. Selected readings outlined at the end of the book provide additional background information on material covered in the various chapters. The book does not deal with specific case-studies in the body of the text. Rather, many of the figures use case-studies or real examples drawn from the literature and our own sources to complement the use of principles and forms of analysis. These are accompanied by comprehensive captions that stand alone from the body of the text. These visual guides reflect the age-old saying: 'a picture is worth a thousand words'. We encourage the reader to 'ponder' each figure and consider what is embedded within it to gain a more complete understanding of the approach to river analysis. Similarly, there is minimal referencing in the text. This is an attempt to keep it clean and easy to read on one hand, and to not overemphasise some literature at the expense of other literature. Instead, an extensive reading list is provided at the back of the book.

The approach to reading the landscape that is outlined here is complementary to a plethora of other approaches available to the geomorphologist. Scientific enquiry is multifaceted. The material in this book complements, and can be used in parallel with, modelling applications, geographic information science and remote-sensing applications, quantitative process measurements, Quaternary research and sedimentology and case-study applications. Many of these fields incorporate significant technological advances. This book is based on the premise that applications of these techniques must be appropriately contextualised through field-based, landscape-scale analyses and interpretations. Such 'geographic' knowledge is integral to geomorphic applications. Technological applications cannot replace our ability to interpret a landscape. Generic information must be framed in its place-specific context. Hopefully, emphasis upon foundation principles in fluvial geomorphology provides an appropriate platform with which to ask the right questions and make interpretations of landscape forms, processes and evolution.

The approach to reading the landscape outlined in Chapter 1 has been carefully structured to scaffold the presentation of the book. However, this is not a 'how to' book, framed around prescriptive step-by-step instructions on how to interpret fluvial forms and processes. Given space limitations, we do not provide guidance on the specific tools and techniques that can be used to support such investigations (e.g. remote sensing, process-based field measurements, modelling applications, sedimentology). Instead, the book is about interpreting forms and processes and piecing them together at the landscape scale.

We hope that the contribution provided by this book is appraised in relation to these aspirations.

Kirstie A. Fryirs and Gary J. Brierley

Structure of the book

The structure of the book is shown in Figure 1. Chapters 2–9 scaffold information to provide the relevant foundations for reading the landscape. Chapter 1 sets the context for why fluvial geomorphology is important and useful in science and management. Chapter 2 documents key spatial and temporal concepts that underpin enquiry in fluvial geomorphology. Chapter 3 overviews catchment-scale relationships in river systems, describing downstream relationships along longitudinal profiles and catchment morphometrics. Chapter 4 focuses on hydrologic relationships in river systems. Chapter 5 documents impelling and resisting forces that drive river adjustment. Chapter 6 explores sediment transport in rivers in relation to entrainment, transport and deposition processes. Chapter 7 describes the range of bed and bank erosion and deposition processes that determine channel shape and size. Chapter 8 analyses process–form associations of instream geomorphic units, documenting the spectrum of features from sculpted bedrock forms to mid-channel and bank-attached bars and finally fine-grained sculpted features. Chapter 9 analyses process–form associations of floodplain geomorphic units, outlining the role of formative and reworking processes. The influence of valley confinement as a control upon floodplain forms is outlined.

The approach to reading the landscape is documented in Chapters 10–14. Tips for reading the landscape are presented at the ends of these chapters. Chapter 10 combines analyses of channels, sediment transport and geomorphic units with channel planform to assess the spectrum of river diversity from bedrock-confined, to partly-confined to alluvial river forms. This is framed around a contructivist approach to analysis of river form. Chapter 11 interprets river behaviour, outlining forms of adjustment for different types of river and the range of river behaviour at different flow stages. Chapter 12 examines river evolution and river change. The imprint of geologic and climatic controls on contemporary forms and processes is discussed. Chapter 13 explores direct and indirect human impacts on rivers. Chapter 14 brings together analyses of sediment budgets and connectivity to present a framework for examining catchment-scale sediment flux and how this can be used to predict likely future river adjustments. The final chapter draws together these threads, summarising the approach to reading the landscape under three banners: Respect diversity, Understand system dynamics and evolution, and Know your catchment.

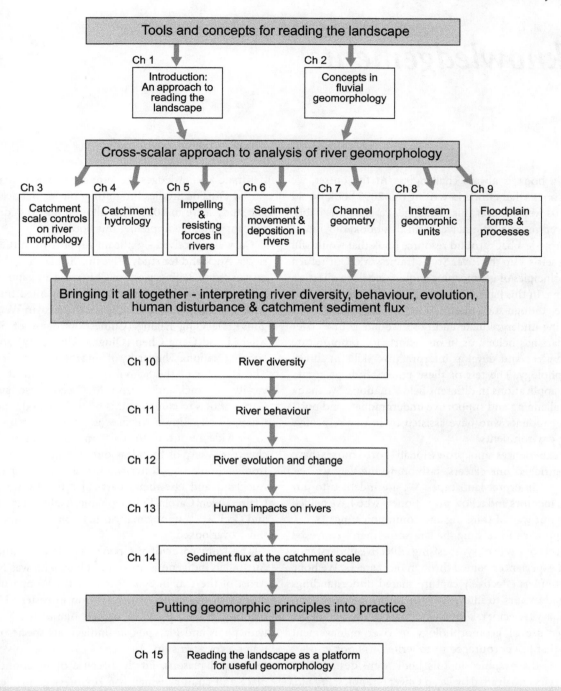

Figure 1 Structure of this book.

Acknowledgements

Writing a book is always challenging. At the outset, we thought our earlier efforts in writing our 2005 book would prepare us well to meet this challenge. Perhaps inevitably, this proved to be a little naive. With this book, our intent was to provide a background resource book that would 'fill a gap' in setting up the River Styles framework. Our return to first principles of geomorphic enquiry led us to question everything. In this light, it is amazing that this book is now complete (though such matters are never finished).

Teaching undergraduate and postgraduate courses over many years has helped us in our efforts to communicate complex ideas and develop interpretative skills in fluvial geomorphology. The test of these understandings comes through applications in different field situations. We thank those challenging and supportive undergraduate and postgraduate students who have assisted us in our respective teaching environments.

Field experiences and professional short courses have greatly enriched our careers, fashioning the way we see, analyse and interpret landscapes. We are indebted to our teachers, mentors and fellow practitioners who have helped to frame our way of thinking and communicating.

The approach to reading the landscape that is conveyed here reflects our way of synthesising collective understandings and experiences gained through our careers. We hope that our efforts effectively capture shared understandings in our endeavours to interpret river forms, processes and evolutionary trajectory. Ultimately, it was our shared desire for better use of geomorphology in river management practice that has encouraged us to write this book.

Most of the graphics in this book were designed by Kirstie Fryirs and drafted by Dean Oliver Graphics, Pty Ltd. We thank Dean for his commitment to this project. Alan Cheung also drafted several figures for the book. Comments by two anonymous reviewers substantially improved the book.

We also thank colleagues in the Department of Environment and Geography, Macquarie University, and the School of Environment, University of Auckland, for their support.

Kirstie acknowledges the support of a Macquarie University Outside Studies Program grant for finance towards her study leave in 2010. This allowed her to dedicate significant time to completing this book.

Gary also received significant support from the University of Auckland for study leave in 2009. During this period he worked in Beijing, western China and Singapore. Stimulating intellectual conversations accompanied his writing efforts at this time. He is indebted to Zhaoyin Wang (Tsinghua), He Qing Huang (Chinese Academy of Sciences), Xilai Li and Gang Chen (Qinghai University) and David Higgitt (National University of Singapore) for their support. The Director of the School of Environment at the University of Auckland, Glenn McGregor, also supported Gary's efforts to complete this book. Megan de Luca, Petra Chappell and Simon Aiken worked as research assistants to provide resources to assist in the writing of several chapters. Many of the ideas outlined here have benefited from stimulating conversations at the University of Auckland and elsewhere; particular thanks are given to Claire, Helen, Carola, Kes, Stephanie, Ashlee, Marc, Brendon and Cecilia . . . among many others, and with apologies to those overlooked.

The front cover of the book depicts a painting by an Australian indigenous artist, Les Elvin, who was NAIDOC Artist of the Year in 2008. Les is of the Wonnaruah community of the Upper Hunter region in eastern NSW. His painting 'Playful Platypus' depicts one of the local rivers with pools and Platypus, an indigenous species of freshwater ecosystems in Australia. We chose this painting for several reasons. Firstly, because of its connection to place and country which is a key message in the reading the landscape approach advocated in this book. Also, the Upper Hunter is a place where we have both spent considerable time undertaking fieldwork. This landscape, amongst many others, has shaped the reading the landscape approach.

As always, our families are our strength. Again, we thank them for their unwavering support.

CHAPTER ONE

Geomorphic analysis of river systems: an approach to reading the landscape

Introduction

Landscapes have been a source of fascination and inspiration for humans for thousands of years. Sensory responses to landscapes vary markedly from person to person. To many, spiritual associations evoke a sense of belonging, perhaps tinged with nostalgic sentiments. To others, a sense of awe may be accompanied by alienation or innate fear. Artists strive to capture the essence of landscapes through paintings, prose, poetry or other media. Our experiences in life are often fashioned by the landscapes in which we live and play. Relationships and associations vary from place to place and over time. New experiences may generate new understandings, wherein observations are compared with experiences elsewhere. These collective associations not only reflect the bewildering range of landscapes in the natural world, they also reflect the individual consciousness with which we relate to landscapes, and the influences/ experience that fashion our way of thinking, whether taught or intuitive. No two landscapes are exactly the same. Each landscape is, in its own way, 'perfect'. Different sets of controls interact in different ways in different settings, bringing about unique outcomes in any particular landscape. Just as importantly, interactions change over time, such that you cannot step in the same river twice (Heraclitus, 535–c. 475 BCE). Sometimes it seems a shame to formalise our understandings of landscapes within the jargonistic language of scientific discourse, but that is what geomorphologists do!

In simple terms, geomorphology is the scientific study of the characteristics, origin and evolution of landscapes. Geomorphic enquiry entails the description and explanation of landscape forms, processes and genesis. Implicitly, therefore, it requires both a generic understanding of the physics and mechanics of process and an appreciation of the dynamic behaviour of landscapes as they evolve through time. The key to effective use of geomorphic knowledge is the capacity to place site-specific insights and relationships in their broader landscape context, framing contemporary process–form linkages in relation to historical imprints. Theoretical and modelling advances are pivotal in the development and testing of our understanding. However, the ultimate test of geomorphological knowledge lies in field interpretation of real-world examples.

This book outlines general principles with which to interpret river character, behaviour and evolution in any given system. Emphasis is placed upon the development of field-based skills with which to read the landscape. Field-based detective-style investigations appraise the relative influence of a multitude of factors that affect landscape-forming processes, resulting patterns of features and evolutionary adjustments. Interactions among these factors change over time. Inevitably, such investigations are undertaken with incomplete information. Information at hand has variable and uncertain accuracy. Some facets of insight may be contradictory. Individual strands of enquiry must be brought together to convey a coherent story. Significant inference may be required, drawing parallels with records elsewhere. Unravelling the inherent complexities that fashion the diversity of the natural world, the assemblages of features that make up any given landscape and the set of historical events that have shaped that place is the essence of geomorphic enquiry. Just as importantly, it is great fun!

Although this book emphasises process–form relationships on valley floors, it is implicitly understood that rivers must be viewed in their landscape and catchment context. Rivers are largely products of their valleys, which, in turn, are created by a range of geologic and climatic controls. Hillslope and other processes exert a primary control upon what happens on valley floors. Sediment delivery from river systems, in turn, exerts a major influence upon coastal-zone processes. Source-to-sink relationships are a function of catchment-scale controls on sediment supply, transport and delivery. Efforts to read the landscape place site-specific observations, measurements and analyses in an appropriate spatial and temporal context. Understanding of this dynamic landscape template provides a coherent platform for a wide range of management applications.

How is geomorphology useful?

Geomorphologists have a long tradition of applying their science in environmental management. Geomorphic insights provide a physical platform with which to develop cross-disciplinary practices and applications that build upon an understanding of how the natural world looks and behaves. Landscapes determine the template upon which a range of biophysical processes interacts (Figure 1.1). For example, insights from fluvial geomorphology provide an understanding of physical processes that create, maintain, enhance or destroy riverine habitat (i.e. the physical space that flora and fauna inhabit). Habitat availability in the channel and riparian zone (and floodplain) of a river is a function of the diversity of landforms on the valley floor. Marked differences are evident; for example, along perennial and ephemeral streams or in a gorge relative to a swamp. Distinct vegetation patterns are found on differing channel and floodplain surfaces, reflecting access to water (and inundation frequency), substrate conditions and morphodynamic interactions between flow and vegetation. Vegetation may have a negligible influence upon some rivers; elsewhere, it may be a primary determinant of process–form relationships. Concerns for ecohydraulics and ecohydrology have major implications for the management of flow, sediment and nutrient fluxes. Water chemistry and turbidity are largely a function of catchment lithology, and the nature/amount of sediment that can be readily entrained by a river.

Alterations to the geomorphic structure of rivers have enormous implications for the operation of biophysical fluxes that affect the movement of water, sediment, nutrients, etc. Hence, a geomorphic template provides a basis for 'whole of system' thinking, aiding the development of coherent plans and strategies for environmental management, guiding decision-making for concerns relating to global change, natural resource management, natural hazards or conservation and rehabilitation issues. End users of geomorphological research are typically land or resource managers who address societal concerns for issues such as erosion and sedimentation problems, channel instability, hazard mitigation, pollution and contamination of water and sediments, ecosystem management, water supply and quality, and so on.

Fluvial geomorphologists have long recognised the nested, hierarchical nature of physical processes that structure river systems across various scales (Chapter 2). Geomorphic relationships vary markedly in differing ecoregions, as climatic controls upon ground cover affect runoff and sediment movement through landscapes, among many considerations. Understanding of source-to-sink relationships at the catchment scale provides a critical platform with which to develop and apply management plans and actions. If geomorphologists are to explain complex landscape behaviour and provide appropriate tools for effective management practice, process knowledge must be related to the configuration of landscape components within any given catchment and the changing nature of process linkages over time. Such understandings are required to convey a coherent view of landscape forms, processes and their evolution. These are innately geographic considerations.

Landscapes are linked and dynamic systems. Disturbance responses or management activities at one place and time may have off-site consequences over various timeframes. Although these are typically scale-dependent relationships (small impacts have minimal consequences that are restricted to closer (proximal) areas), this is not always the case (e.g. local disturbance may induce off-site responses that breach threshold conditions). Often, these relationships are predictable. Gravitationally induced flow and sediment flux is the key driver of upstream–downstream linkages in river systems. Sometimes, however, surprising outcomes may occur. For example, headcut activity and bed incision may cut back through valley floor deposits, impacting upon the river upstream. The effectiveness and efficiency of linkages vary markedly from catchment to catchment. Understanding of imprints from past disturbance events, and associated lagged and off-site responses, is critical in the development of proactive planning appli-

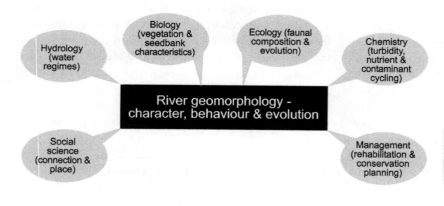

Figure 1.1 Geomorphology as a physical template atop which other interactions occur.

cations. These various considerations underpin visioning exercises that determine 'what is biophysically achievable' in the management of any given catchment.

Geomorphic analysis of river systems: our approach to reading the landscape

Analysis of geomorphic systems cannot be meaningfully formalised using a prescriptive check-list, tick-box set of procedures. Such rigidity belies the inherent diversity of landscapes, and the overwhelming range of factors, process-relationships and controls that combine to generate the pattern of features formed (and reworked) at any given place. This is not to say that all landscapes are necessarily complex; indeed, some may be extremely simple or even near featureless! An open-minded approach to enquiry recognises implicitly the potential for unique outcomes (manifest as assemblages of features and their interactions) in any given setting.

The pattern/configuration of a landscape is derived from its composition (the kinds of elements it contains), its structure (how they are arranged in space) and its behaviour (how it adjusts over time to various impulses for change). Analysis of relationships between landforms can be used to provide insight into the history of formative and reworking events, and the evolutionary history of that system. Ultimately, these space–time interactions can only be unravelled through appraisal of source-to-sink relationships at the catchment scale.

Reading the landscape is an approach by which practitioners use their knowledge and experience to *identify* the assemblage of landforms or features that make up rivers, *develop hypotheses* to *interpret* the processes responsible for those landforms, *determine* how those features have/will adjust and change over time and, finally, place this understanding in its *spatial* and *temporal context*. Successful interpretations draw on existing theory, questioning and testing its relevance to the system under investigation.

All observations and interpretations in geomorphology must be appropriately framed in their spatial and temporal context. This requires appraisal of geologic, climatic and anthropogenic controls upon landscapes at any given locality. Topographic and geologic maps, aerial photographs and satellite images, and Google Earth® provide a simple basis with which to frame analyses in their landscape context, enabling meaningful comparisons with other places. Stark contrasts can be drawn between uplifting terrains at the margins of tectonic plates and relatively stable plate-centre locations, glaciated and non-glaciated landscapes, desert and rainforest areas, or rural and urban streams. Flow–sediment relationships which fashion process–form interactions along valley floors vary mark-

edly in these different settings. It is also important to consider position within a catchment, and the scale of the system under consideration. These insights provide the contextual information within which the approach to reading the landscape is applied.

The constructivist (building block) approach to reading the landscape that is developed in this book assesses how each part of a system relates to its whole in both spatial and temporal terms (Figure 1.2). This 'bottom-up' approach synthesises the behaviour and evolution of landscapes through systematic analysis of fluvial landforms (termed geomorphic units). These features are generated by certain process–form interactions at particular positions in landscapes, and are comprised of differing material properties. Reaches are comprised of differing assemblages of landforms that are formed and reworked under a particular behavioural regime. Catchments are comprised of downstream patterns of reaches that are (dis)connected and through which fluxes of water, sediment and vegetation drive river behaviour, evolution and responses to human disturbance.

Although remotely sensed or modelled data provide critical guidance in our efforts to interpret landscapes, it is contended here that genuine understanding is derived from field-based analyses.

Reading the landscape entails four steps, for which different generic skills are required (Figures 1.3 and 1.4).

1. *Identify individual landforms (geomorphic units) and the process–form relationships that determine their process regime.*
   Landforms (or geomorphic units) are the component parts of a landscape. In general terms, they form under a given set of energy conditions at particular locations in a landscape. They are produced by a particular set of processes that fashion and rework the size and shape of the characteristic form. Geomorphologists have a good understanding of these process–form (morphodynamic) relationships, whereby the process affects the form and vice versa. Individual landforms have certain material and sedimentologic properties with a characteristic geometry and bounding surfaces (i.e. erosional or depositional contacts). Geomorphic units commonly have characteristic vegetation associations reflecting hydrologic and substrate conditions (among many considerations). Combinations of erosive and depositional processes that sculpt, create and rework the feature define the range of behaviour of each particular unit. From this, magnitude–frequency relations of formation and reworking can be inferred. This allows interpretation of the sensitivity/resilience of that feature when it is subjected to disturbance events (i.e. whether the feature will simply have additional

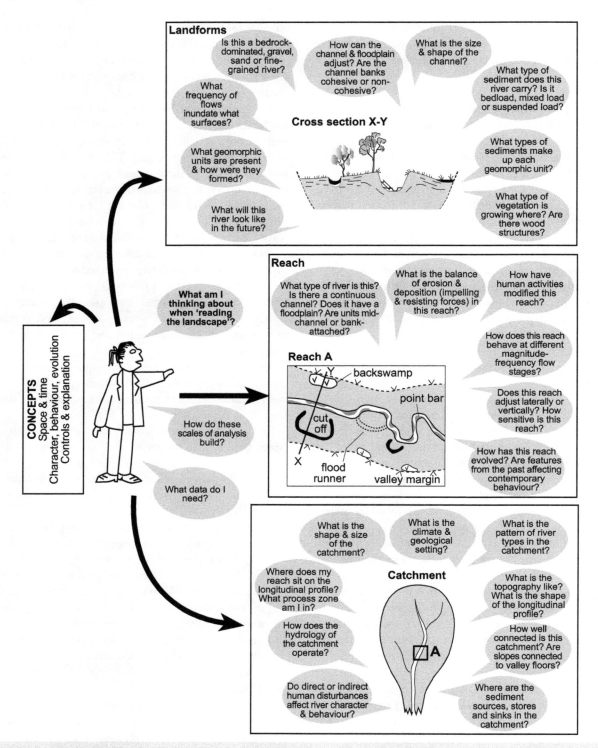

Figure 1.2 Questions you should ask when reading the landscape at the landform, reach and catchment scales.

deposits added to it, whether it will be partially reworked or whether it will be destroyed (eroded and removed)).

2. *Analyse and interpret the package and assemblage of landforms at the reach scale and how they adjust over time.*

Sections of river with a distinct assemblage of geomorphic units that reflect particular combinations of erosional and depositional processes are referred to as a reach. By definition, reaches upstream and downstream are characterised by different packages of landforms. Reading the landscape at the reach scale entails

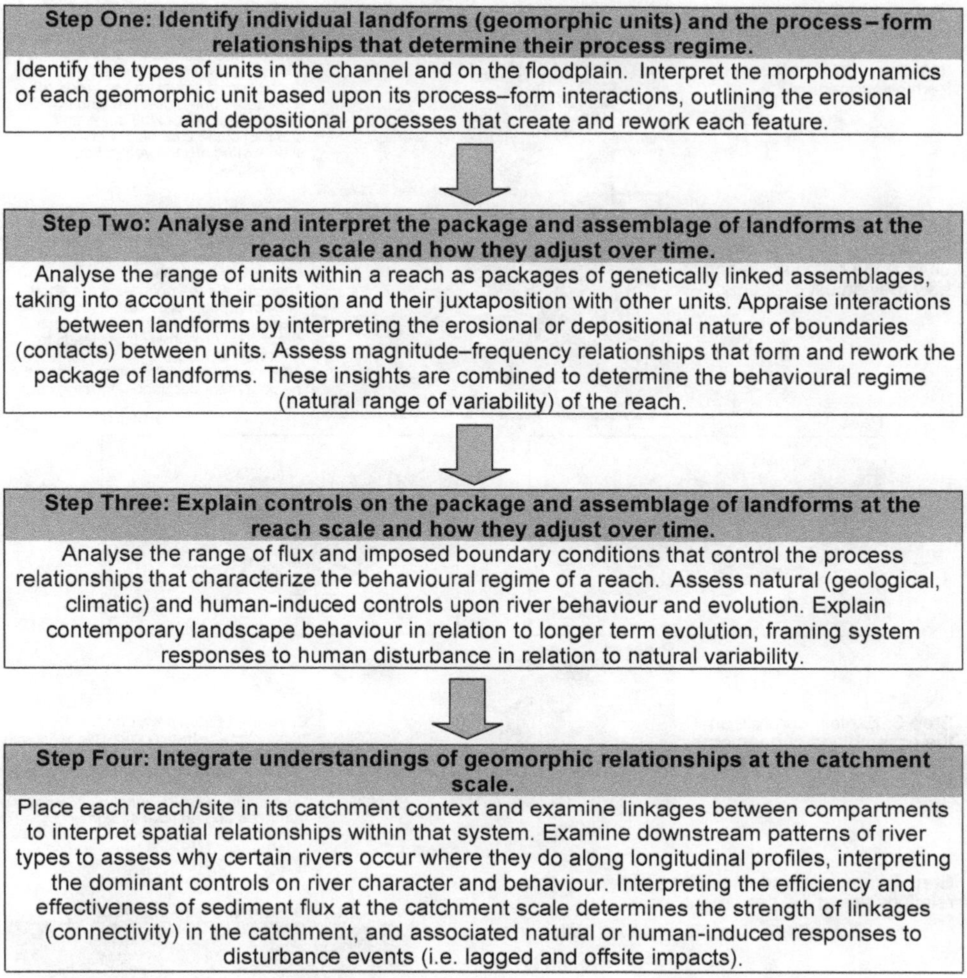

> **Step One: Identify individual landforms (geomorphic units) and the process–form relationships that determine their process regime.**
> Identify the types of units in the channel and on the floodplain. Interpret the morphodynamics of each geomorphic unit based upon its process–form interactions, outlining the erosional and depositional processes that create and rework each feature.

> **Step Two: Analyse and interpret the package and assemblage of landforms at the reach scale and how they adjust over time.**
> Analyse the range of units within a reach as packages of genetically linked assemblages, taking into account their position and their juxtaposition with other units. Appraise interactions between landforms by interpreting the erosional or depositional nature of boundaries (contacts) between units. Assess magnitude–frequency relationships that form and rework the package of landforms. These insights are combined to determine the behavioural regime (natural range of variability) of the reach.

> **Step Three: Explain controls on the package and assemblage of landforms at the reach scale and how they adjust over time.**
> Analyse the range of flux and imposed boundary conditions that control the process relationships that characterize the behavioural regime of a reach. Assess natural (geological, climatic) and human-induced controls upon river behaviour and evolution. Explain contemporary landscape behaviour in relation to longer term evolution, framing system responses to human disturbance in relation to natural variability.

> **Step Four: Integrate understandings of geomorphic relationships at the catchment scale.**
> Place each reach/site in its catchment context and examine linkages between compartments to interpret spatial relationships within that system. Examine downstream patterns of river types to assess why certain rivers occur where they do along longitudinal profiles, interpreting the dominant controls on river character and behaviour. Interpreting the efficiency and effectiveness of sediment flux at the catchment scale determines the strength of linkages (connectivity) in the catchment, and associated natural or human-induced responses to disturbance events (i.e. lagged and offsite impacts).

Figure 1.3   An approach to reading the landscape.

assessment of which types of geomorphic units are present (or absent), what types of sediments they are made of, and whether the units are formed and reworked by genetically linked contemporary processes or they reflect former conditions (Figures 1.2 and 1.4). Interpretation of the array of process–form relationships for the range of geomorphic units along a reach, and associated channel–floodplain interactions (if present), is used to determine the character and behaviour of a river. Adjustments around a characteristic state over geomorphic timeframes determine the *range of behaviour* of a river, as systems respond to disturbance events (Chapter 2). Inevitably, the magnitude–frequency domains with which these features are generated and interact may vary from system to system.

Significant insights into landscape history can be gained through analysis of whether adjacent features in a landscape are genetically linked or not (i.e. whether they formed contemporaneously, or whether they formed over differing periods of time). This provides guidance into the evolutionary history of a landscape, highlighting erosional events that rework landscape features (i.e. a temporal discontinuity). For example, terraces are older than adjacent floodplain and channel features, and they were often formed by quite different processes under differing environmental conditions.

3. *Explain controls on the package and assemblage of landforms at the reach scale and how they adjust over time.*

All landscapes adjust and evolve. Among the many inherent complexities of analysis of landscape systems is determination of the timeframe over which differing features are created and/or reworked and appraisal of the ways in which adjustments to one part of a system affect responses elsewhere in that system. The true value of geomorphic understanding lies in being

Step 1: Identify individual instream and floodplain geomorphic units and determine their process-form relationships.

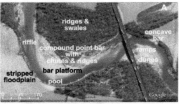

Example of interpretation
ridges & swales formed by helicoidal flow forming scroll bars, concave bank erosion and channel migration
compound point bar formed by deposition of gravel on the inside of a meander bend, flow realignment over the bar at bankfull stage and scour of chute channels. Sediment is deposited around vegetation to form ridges.

Step 2: Analyse and interpret the assemblage of geomorphic units at the reach scale and how they adjust over time.

Example of interpretation
low flow stage - flow aligned around bars and over riffles. Undercutting of concave bank occurs.
bankfull stage - point bars are short-circuited, pools are scoured, riffles are deposited, concave bank erosion and deposition on convex bank leads to channel migration.
overbank stage - flow aligned over the neck of meander bends, forming cutoffs

Past

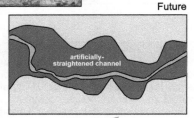

Future

Present

Step 3: Explain controls on the assemblage of geomorphic units, and 'natural' and human-induced impacts.

Example of interpretation
flux controls - transport-limited river with high sediment supply. Flashy flow regime.
imposed controls - tectonic activity resets base level.
human impacts - devegetation, cutoff formation and artificial straightening of channel.

Step 4: Integrate understandings of geomorphic relationships at the catchment scale.

Longitudinal profile and river patterns

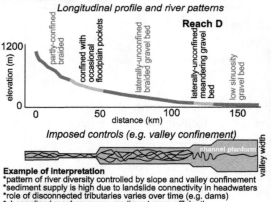

Imposed controls (e.g. valley confinement)

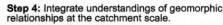

Example of interpretation
*pattern of river diversity controlled by slope and valley confinement
*sediment supply is high due to landslide connectivity in headwaters
*role of disconnected tributaries varies over time (e.g. dams)
*channelised reaches convey sediment more efficiently

Buffers, barriers and connectivity

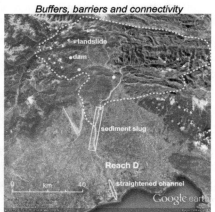

Figure 1.4   An example of how to apply the reading the landscape approach to a real river system (See Colour Plate 1). The example used here is the Tagliamento River in Italy. The approach begins by interpreting process-form relationships for individual, then assemblages of geomorphic units along different reaches of river type. River behaviour is interpreted for a range of flow stages. The role of natural and human induced disturbance on river adjustments over time is considered when analysing river evolution. Finally, each reach is placed in its catchment context, analysing flux and imposed controls on river diversity along longitudinal profiles and how reaches to fit together in a catchment. Interpreting the efficiency of sediment flux at the catchment scale determines the (dis)connectivity of the catchment and associated off-site responses to disturbance. Maps constructed using Google Earth © 2012 images. Based on information in Bertoldi et al. 2009, Gurnell et al. 2000, Surian et al. 2009 and Tockner et al. 2003. The interpretation of river types, river evolution and connectivity are our own.

able to explain the controls that drive process interactions and how they have changed/adjusted over time and interpreting what has triggered these changes/adjustments.

Differing controls upon landscape behaviour operate over variable spatial and temporal scales. By definition, the package of geomorphic units at the reach scale is fashioned by a consistent set of controlling factors. Valley setting (slope and width) is the primary determinant of *imposed boundary conditions* that are set over timeframes of thousands of years or longer (Chapter 2). In contrast, flow and sediment transfer relationships that recurrently adjust over much shorter timeframes set the *flux boundary conditions*. Primary differences in geomorphic setting (and associated behavioural regime) can be attributed to patterns of geologic (imposed) and climatic (flux) controls. Geologic factors such as tectonic setting, lithology and resulting topography affect the erodibility and erosivity of a landscape. Climatic factors influence the nature and rate of process activity (e.g. geomorphic effectiveness of flood events).

Effective integration of process-based insights through appraisals of the ways in which landscape compartments interact and evolve over time provides the basis to explain why certain behavioural adjustments have occurred. Analysis of landscape evolution enables determination of whether the contemporary system adjusts around a characteristic state, adjusts among differing states or has a different evolutionary pathway. These interpretations can be used to relate landscape responses to human disturbance to the *natural range of variability* of a system.

4. *Integrate understandings of geomorphic relationships at the catchment scale.*

Drainage basins are comprised of relatively self-contained, gravitationally induced sets of biophysical relationships. The balance of erosional and depositional processes varies markedly in source, transfer and accumulation zones of a catchment. Erosion is dominant in source zones, deposition is dominant in accumulation zones and an approximate balance of erosional and depositional processes is maintained in transfer zones (Chapter 3). Analysis of source-to-sink relationships at the catchment scale provides the most logical basis to consider the linked nature of spatial and temporal adjustments in landscapes, enabling meaningful interpretation of lagged and off-site responses to disturbance events. The unique configuration and temporal sequence of drivers, disturbances and responses of each landscape, along with the historical imprint, result in system-specific behavioural and evolutionary traits.

Catchment-scale investigations frame analyses of river character, behaviour and evolution in relation to the size and shape of the catchment, the drainage network pattern and density, and topographic relationships (especially relief, longitudinal profile shape and valley morphology) (Figures 1.2 and 1.4). Each site/reach must be viewed in its catchment context, assessing relationships to upstream and downstream considerations. Flow–sediment linkages between reaches and tributary–trunk stream relationships in differing landscape compartments (or process domains) are captured by the term *landscape connectivity* (Chapter 2). In some landscapes, hillslope and valley-floor processes are inherently coupled or connected; elsewhere they are not. Valley floors may be disconnected from adjacent hillslopes, but directly linked to sediment supply from upstream. Analysis of downstream patterns of rivers, and associated implications for flow and sediment flux, determines how adjustments to one feature (or reach) affect adjacent or other forms. The way in which disturbance responses in one part of a catchment affect river adjustments elsewhere within that system is termed a response gradient. Understanding of these catchment-scale considerations provides critical guidance in interpreting the behavioural regime and evolutionary trajectory of a river.

In summary, reading the landscape is an open-ended, interpretative, field-based approach to geomorphic analysis of river systems. Efforts to read the landscape can be summarised as follows: identify features and assess their formative processes, appraise how these features fit together in a landscape (reaches and catchments) and assess how these features adjust and evolve over time. Meaningful *identification* and *description* underpins effective *explanation*, providing a platform with which to make realistic *predictions* about likely future states. Landscape relationships are analysed through appreciation of system dynamics, recognising the variable imprint/memory of influences from the past. Behavioural regimes are differentiated from river changes as landscapes evolve. Human impacts upon rivers are differentiated from natural variability. Chapters 2–9 of this book outline contextual principles and theories with which to ground these analyses, which are explained more fully in Chapters 10–14.

Key messages from this chapter

- Geomorphology is the science concerned with understanding the form of the Earth's surface and the processes by which it is shaped, both at the present day and in the past.

- Rivers are a product of their landscape. As rivers are spatially linked systems, they are best studied at the catchment scale. Catchments synthesise process–form relationships over a range of spatial and temporal scales.
- No two landscapes (and associated river systems) are exactly the same. Reading the landscape presents a grounded basis to examine the character, behaviour and evolution of any given river system.
- Reading the landscape is a thinking and interpretative exercise. Detective-style investigations are required to differentiate among the myriad of factors that affect river character, behaviour and evolution. The approach

to reading the landscape outlined in this book has four steps:

1. Identify and interpret landforms and their process–form relationships.
2. Analyse assemblages of landforms at the reach scale to interpret behaviour.
3. Explain controls on process–form interactions at the reach scale and how they adjust over time.
4. Integrate spatial and temporal considerations through catchment-specific investigations to explain patterns of river types and their evolutionary adjustment, framing system responses to human disturbance in relation to the natural variability of the system.

CHAPTER TWO

Key concepts in river geomorphology

Introduction

This chapter outlines a range of concepts and theories about how a river landscape looks, adjusts and evolves. These spatial and temporal concepts build upon each other helping us to frame catchment-scale, system-specific applications that assess geomorphic responses to human disturbance in relation to natural variability. These concepts aid our efforts to read the landscape.

This chapter is structured as follows. First, spatial considerations are reviewed. This starts with an overview of nested hierarchical approaches to analysis of river systems. Imposed and flux boundary conditions that control the range of river character and behaviour are defined and differentiated. Then the complexity of river structure is differentiated in terms of landscape heterogeneity and homogeneity. The final spatial concept outlined here is a summary of landscape (dis)connectivity.

Second, temporal concepts that are used to characterise river systems are appraised. This starts with a synthesis of geologic (cyclic), geomorphic (graded) and engineering (steady-state) timescales. Equilibrium notions developed via negative feedback mechanisms are used to describe geomorphic adjustments around a mean (*characteristic*) state. River behaviour is differentiated from river change; the latter records a shift to a different type of river with a different behavioural regime. These transitions may be brought about by positive feedback mechanisms that breach threshold conditions. Press, pulse and ramp disturbance events are differentiated. Responses to disturbance are assessed in terms of their reaction and relaxation times. These notions are used to discuss prospects for river recovery. Magnitude–frequency relations highlight how geomorphic work and geomorphic effectiveness vary for disturbance events of differing size and recurrence. Variability in landscape sensitivity, among many factors, results in complex responses to disturbance, as landscapes preserve a variable record of past events (termed memory or persistence). Lagged and off-site responses emphasise the need to explain patterns and rates of geomorphic adjustment at the catchment scale. The principle of equifinality highlights how similar-looking forms may result from different sets of processes.

These various spatial and temporal concepts are pulled together in the final section of this chapter. The system-specific configuration of any given catchment, along with its unique history of responses to disturbance, is characterised in relation to non-linear dynamics. Principles of emergence, contingency and path dependency are outlined. System responses to human disturbance are appraised relative to natural variability. Collectively, these considerations frame the evolutionary trajectory of any given system.

Spatial considerations in reading the landscape

Catchments as nested hierarchies: the spatial configuration of landscapes

Nested hierarchical models of catchment organisation frame small-scale (and short-term) river features and processes in relation to larger scale (and longer term) factors (Figure 2.1, Table 2.1). Smaller spatial scales are nested within higher level scales. Each nested level within a hierarchical view of catchments is controlled by the conditions set by higher level scales. This allows interpretation of higher level controls on physical processes that operate at smaller scales.

Different scalar units in the nested hierarchy are commonly not discrete physical entities. Rather, they are part of a complex continuum in which the dimensions of units at each scale may overlap significantly. Interaction between units, at each scale and between scales, determines the character and behaviour of the system under investigation. When used effectively, nested hierarchical frameworks provide an elegant tool with which to organise

Geomorphic Analysis of River Systems: An Approach to Reading the Landscape, First Edition. Kirstie A. Fryirs and Gary J. Brierley.
© 2013 Kirstie A. Fryirs and Gary J. Brierley. Published 2013 by Blackwell Publishing Ltd.

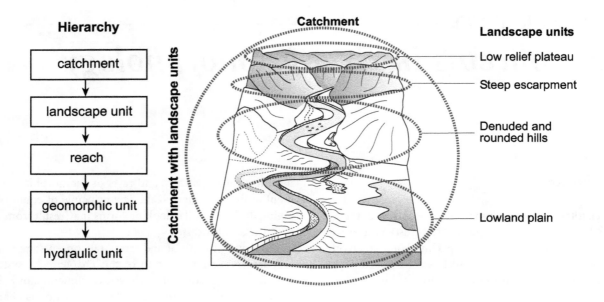

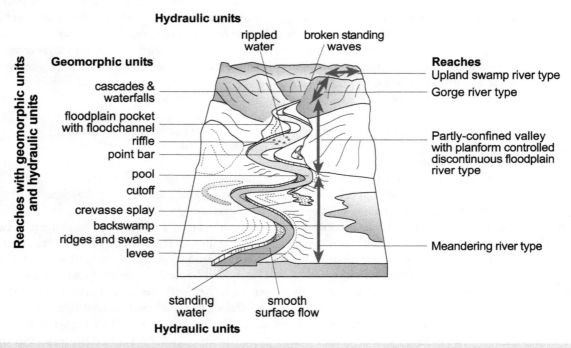

Figure 2.1   Scales of river structure. A hierarchical approach to landscape analysis frames catchments as assemblages of landscape units, reaches, geomorphic units and hydraulic units. Adjustments in bed material size operate at the hydraulic and geomorphic unit scales, channels and river planform adjust at the reach scale and slope adjustments along longitudinal profiles occur at the landscape unit and catchment scales.

information, thereby presenting a coherent platform for management applications.

Catchment (also called watershed or drainage basin)

A catchment is a single fluvial system that is linked internally by a network of channels. Regional geology and climate, among other factors, determine topography, valley width, sediment transport regime and the discharge regime. Catchment-scale factors are relatively insensitive to change, but if disturbed will take considerable time to recover. Within any catchment, individual subcatchments may have quite different physical attributes, with differing types and proportions of landscape units and associated variability

Table 2.1 A nested hierarchy of geomorphic scales

Scale	Evolutionary adjustment timeframe (yr)	Disturbance event frequency (months)	Geomorphic influence
Catchment	$10^5–10^6$	10^3	Tectonic influences on relief, slope and valley width combined with lithologic and climatic controls on substrate, flow and vegetation cover (among other factors) determine the imposed boundary conditions within which rivers operate. The drainage pattern and stream network influence the nature, rate and pattern of biophysical fluxes (these relationships are also fashioned by catchment geology, shape, drainage density, tributary–trunk stream interactions, etc.). Vegetation cover and land use indirectly influence river character and behaviour through impacts upon flow and sediment delivery.
Landscape unit	$10^3–10^4$	10^2	Landscape units are readily identifiable topographic features with a characteristic pattern of landforms. The nature, rate and pattern of biophysical fluxes are influenced by landscape configuration (i.e. the pattern of landscape units and how they fit together in any given catchment) and the connectivity of reaches. At this scale, the channel, riparian zone, floodplain and alluvial aquifer represent an integrated fluvial corridor that is distinct from, but interacts with, the remaining catchment.
Reach	$10^1–10^2$	10^1	Geomorphic river structure and function are relatively uniform at the reach scale. Morphological attributes such as channel planform and geometry are fashioned primarily by flow regime, sediment transport regime, floodplain character and vegetation and groundwater–surface-water exchange. Distinct assemblages of channel and/or floodplain landforms characterise reaches.
Geomorphic unit	$10^0–10^1$	10^0	These landform-scale features reflect formative erosional and depositional processes that determine river structure and function. Distinct features are evident in channel and floodplain compartments. Morphodynamic relationships fashion these landforms, where process influences form and vice versa.
Hydraulic unit	$10^{-1}–10^0$	10^{-1}	This scale of feature is determined by (and shapes) flow–sediment interactions that reflect the energy distribution along a river course. Relationships vary markedly with flow stage. Pronounced local-scale variability in surface roughness, flow hydraulics or sediment availability and movement may be evident around basal materials, logs and organic debris. Surface–subsurface flow linkages fashion hyporheic zone processes.

in geomorphic process zones. As such, interpretation of controls on river character and behaviour is best framed in terms of subcatchment-specific attributes such as the shape of the longitudinal profile, lithology, etc. (see Chapter 3).

Landscape units (also called land systems)

Just as drainage basins comprise a series of subcatchments, so each subcatchment can be differentiated into topographic compartments based on relief variability and landscape position. Landscape units are areas of similar topography that have a characteristic pattern of landforms. Key factors used to identify landscape units include measures of relief, slope, elevation, topography, geology and position (e.g. upland versus lowland settings). As landscape units are a function of slope, valley confinement and lithology, they not only determine the calibre and volume of sediment made available to a reach, they also impose major constraints on the distribution of flow energy that mobilises sediments and shapes river morphology. Catchment-to-catchment variability in river character and behaviour, and the operation of biophysical fluxes, are largely determined by the type and configuration of landscape units. Different landscape units tend to be associated with differing land uses.

River reaches

Topographic constraints on river forms and processes result in differing ranges of river character and behaviour in differing valley settings along a longitudinal profile (see Chapter 3). Reaches are differentiated within each landscape unit. They are defined as sections of river along which controlling conditions are sufficiently uniform (i.e. there is no change in the imposed flow or sediment load) such that the river maintains a near-consistent structure. Alternating reaches made up of different river types are referred to as *segments*. Reaches are made up of distinct assemblages of geomorphic units.

Geomorphic units (landform-scale features)

The availability of material and the potential for it to be reworked in any given reach determine the distribution of geomorphic units, and hence river structure (see Chapters 8 and 9). Some rivers comprise erosional forms that are sculpted into bedrock (e.g. cascades, falls, pools), while others comprise depositional forms in channel and floodplain compartments that reflect sediment accumulation in short- or long-term depositional environments (e.g. mid-channel bars versus a backswamp). Geomorphic units are discrete morphodynamic entities. Certain processes produce the form and the form, in turn, affects the nature and effectiveness of the process. Adjacent geomorphic units

may be genetically linked. For example, pools are functionally connected to adjacent riffles.

Hydraulic units (also called habitats or patches)

Hydraulic units are spatially distinct patches of relatively homogeneous surface flow and substrate character. These range from fast-flowing variants over a range of coarse substrates to standing-water environments on fine-grained substrates. Flow–substrate interactions vary at differing flow stages. Several hydraulic units may comprise a single geomorphic unit. For example, distinct zones or patches may be evident within individual riffles, characterised by differing substrate, the height and spacing of roughness elements, flow depth, flow velocity and hydraulic parameters such as Froude and Reynolds numbers (see Chapter 5). Some hydraulic units tend to be very sensitive to change, adjusting on an event-by-event basis, but they generally have considerable capacity to recover following disturbance.

Imposed and flux boundary conditions

Catchment boundary conditions determine the range of processes and resulting assemblages of landscape forms in any given system. *Imposed boundary conditions* do not change over geomorphic timeframes (centuries to thousands of years). These controls reflect the landscape and/or environmental setting in which rivers operate. They determine the relief, slope and valley morphology (width and shape) within which rivers adjust (Figure 2.2). For example, geologic controls such as tectonic setting and lithology influence landscape elevation and relief (i.e. slope), and the type and amount of sediment made available to be moved by the river. Long-term landscape evolution fashions the drainage pattern and stream network, along with the width and alignment of valleys within which rivers are set. Although these geologic controls are not static in their own right, they are considered here as consistent controls upon the river (i.e. factors that do not change over geomorphic timeframes). These considerations determine how much potential energy is available to be used by the river. At the same time, they impose some constraints upon the way in which energy can be used (e.g. use of kinetic energy is influenced markedly by slope and valley width, and any factors that concentrate or dissipate flow energy). Imposed boundary conditions effectively dictate the pattern of landscape units, thereby determining the valley setting within which a river behaves and/or changes.

Flux boundary conditions are essentially inset within the imposed boundary conditions (Figure 2.2). Dynamic interactions that fashion the flow and sediment regime, and vegetation associations along a reach, exert a key influence

Imposed boundary conditions set by:
Ⓐ valley confinement and slope
Ⓑ base level at bedrock riffles and downstream gorges
Ⓒ low relief topography

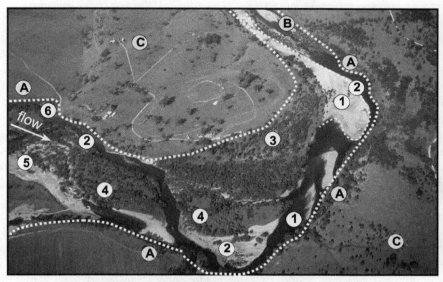

Flux boundary conditions set by interactions between water, sediment, and vegetation:

① flow and sediment interaction dictate zones of sediment transfer and deposition

② flow and sediment interaction forming different types of instream geomorphic units (e.g. bars, riffles, pools)

③ flow and sediment interaction forming and reworking floodplains

④ sediment and vegetation interaction forming different geomorphic units (e.g. ridges)

⑤ flow and vegetation interactions dictating variability in roughness

⑥ flow inundates different surfaces during discharge events of varying magnitude and frequency

Figure 2.2 Imposed and flux boundary conditions. A mix of controls affects the morphology and behaviour of any river (see Chapter 11). Imposed boundary conditions or controls remain consistent over short timeframes. Valley confinement, base level set by bedrock (i.e. slope), topography and geology influence the type of river that can form in a given setting. Flux boundary conditions fashion the flow, sediment and vegetation interactions along a river, thereby determining its character and behaviour. This figure shows how imposed and flux boundary conditions have created a partly confined valley with bedrock-controlled discontinuous floodplain river in the Clarence catchment, NSW, Australia. From Brierley and Fryirs (2008). © Island Press, Washington, DC. Reproduced with permission.

on river character and behaviour. Catchment-scale controls on the flow regime are determined largely by the climate setting. Rivers are forever adjusting to variability in flow and sediment, over timescales ranging from short pulsed events (individual floods) through to sequences of floods, through to seasonal/interannual variability and longer term trends. Stark contrasts in discharge regime are evident in arid, humid–temperate, tropical, Mediterranean, monsoonal and other climate settings, marking the differentiation of perennial and ephemeral systems, among many things. Climate also imposes critical constraints on the amount and variability of runoff, the magnitude–frequency relationships of flood events and the effectiveness of extreme events. Secondary controls exerted by climatic influences at the catchment scale are manifest through effects on vegetation cover and associated rates of runoff and sediment yield.

Flux boundary conditions determine the energy conditions under which rivers behave. The balance of erosional and depositional processes on valley floors reflects the flow–sediment balance. Any given reach develops a characteristic behavioural regime over timeframes in which flux boundary conditions remain relatively consistent. However, flux boundary conditions may change over management timeframes. For example, alterations to ground cover may change surface erodibility and rainfall–runoff relationships. Alternatively, flow regulation disrupts flow and sediment transfer within river systems. Alterations to flux boundary conditions may induce river change whereby the river adopts a new behavioural regime that adjusts to the new flux boundary conditions.

Evolutionary adjustments in river systems are brought about by alterations to the imposed and flux boundary conditions, whether as a result of 'natural' trends or human induced impacts. Typically, these events suppress or expand the manner and/or rate of adjustment of the system.

Heterogeneity and homogeneity of landscapes

Some rivers are characterised by a wide range of morphological features, while others have a remarkably simple structure. The natural diversity in geomorphic and hydraulic units is an important determinant of the range of habitat availability along a river. The more complex and diverse the array of landforms is along a reach, the broader the range of available physical habitat. For example, there is significant diversity along a meandering sand-bed river, with riffles, runs and pools in the channel zone, differing bank forms, point bars on the inside of bends (possibly dissected by chute channels), while the floodplain may comprise features such as a levee, cut-off channels (oxbow lakes or billabongs), backswamps and floodplain ponds (Figure 2.3a). In stark contrast, the physical structure of a discontinuous watercourse may be remarkably homogeneous, essentially comprising a valley fill wetland and a discontinuous channel (Figure 2.3b).

Catchment linkages and (dis)connectivity

Landscape connectivity is a primary control upon catchment-scale fluxes of water and sediment. The nature and continuity of longitudinal, lateral and vertical linkages are controlled by different sets of processes at different positions in a catchment. Fluxes may be connected (coupled) or disconnected (decoupled). Patterns and/or phases of discontinuity in these linkages vary both spatially and temporally. These are catchment-specific relationships.

Longitudinal linkages refer to upstream–downstream and tributary–trunk stream relationships in the channel network. The strength of longitudinal linkages reflects the character and distribution of landscape units in a catchment. The cascading nature of these interactions is influenced by the pattern and extent of coupling in each subcatchment. Appraisal of these linkages is required to assess how off-site impacts such as sediment release and/or decreased water supply affect reaches elsewhere in a catchment, and associated lag times. The distribution of river types in each subcatchment, and how these subcatchments fit together in the catchment as a whole, provides a physical basis to interpret these linkages.

Lateral linkages include hillslope–channel and channel–floodplain relationships. Hillslope–channel connectivity records the frequency with which channel processes rework materials derived from hillslopes. In coupled systems, water and sediment are transferred directly from hillslopes to the channel network. Conversely, in decoupled systems, materials are stored for differing intervals of time in various features between the hillslope and the channel. Floodplains are the most common sediment storage feature in this location, preventing sediment transfer directly to the channel. In many landscapes, alluvial fans are also common in this location. Channel–floodplain connectivity reflects the two-way transfer of water and sediment between channel and floodplain compartments. The magnitude, frequency and duration of overbank events are primary determinants of the periodicity of inundation that drive channel–floodplain linkages.

Vertical linkages entail surface–subsurface interactions of water and sediment. Examples include inundation levels of differing geomorphic units and the connectivity of surface and subsurface flow pathways. Within-channel linkages are controlled by the texture of the bed material and the transport regime of the channel. In a broader sense, these relationships are affected by soil/regolith characteristics that control slope hydrology and relationships between surface flow, subsurface flow and groundwater. Hyporheic and parafluvial zones are areas of subsurface flow that occur through the substrate of channel beds, bars and floodplains. Hydrologic exchange and nutrient transformation between surface waters and alluvial groundwaters may extend a considerable distance beyond the channel margin beneath the floodplain.

Various forms of physical linkage inferred for an idealised catchment are shown in Figure 2.4. In confined headwater reaches, hillslopes and channels are coupled, such

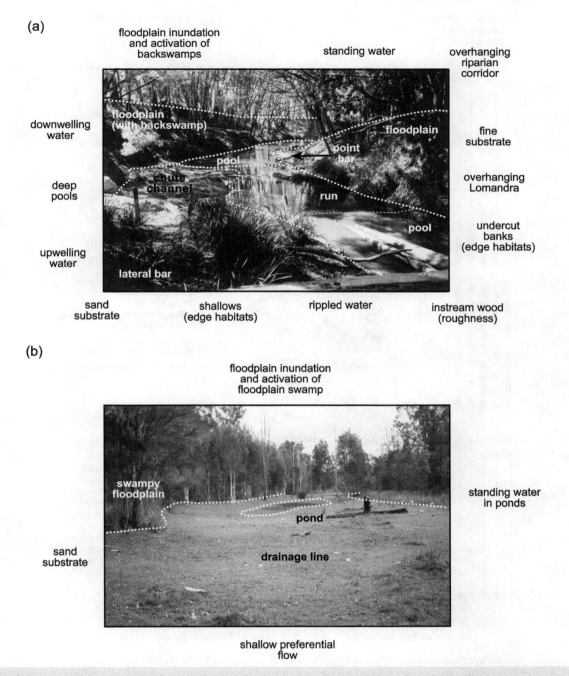

Figure 2.3 Heterogeneity and homogeneity in the physical structure of rivers. (a) This well vegetated meandering sand-bed river has a complex geomorphic structure with significant variability in the range of geomorphic units. Bucca Bucca Creek, NSW (source: K. Fryirs). (b) The discontinuous sand-bed river has a simple geomorphic structure. Six Mile Creek, NSW (source: K. Fryirs).

that water and sediment are readily conveyed from the surrounding catchment. These zones act as sediment sources. Floodplain pockets are virtually non-existent. Hyporheic zone functioning is limited by the imposed bedrock nature of these settings. Transfer reaches with strong longitudinal connectivity characterise mid-catchment locations. Dis-

continuous pockets of floodplain produce irregularities in hillslope–channel and channel–floodplain connectivity. Tributaries may be locally trapped behind floodplain pockets, disconnecting some lower order drainage lines from the trunk stream. As bedrock is prominent on the channel bed, water and nutrient exchange in the hyporheic

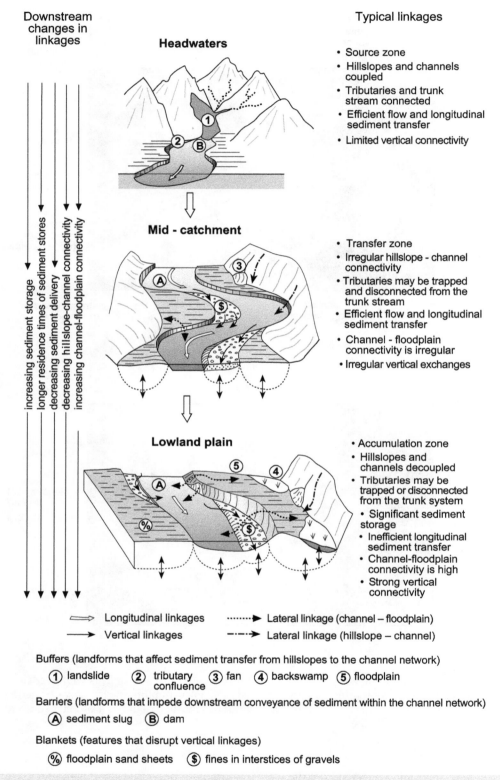

Downstream
changes in
linkages

Typical linkages

Headwaters

- Source zone
- Hillslopes and channels coupled
- Tributaries and trunk stream connected
- Efficient flow and longitudinal sediment transfer
- Limited vertical connectivity

increasing sediment storage
longer residence times of sediment stores
decreasing sediment delivery
decreasing hillslope-channel connectivity
increasing channel-floodplain connectivity

Mid - catchment

- Transfer zone
- Irregular hillslope - channel connectivity
- Tributaries may be trapped and disconnected from the trunk stream
- Efficient flow and longitudinal sediment transfer
- Channel - floodplain connectivity is irregular
- Irregular vertical exchanges

Lowland plain

- Accumulation zone
- Hillslopes and channels decoupled
- Tributaries may be trapped or disconnected from the trunk system
- Significant sediment storage
- Inefficient longitudinal sediment transfer
- Channel-floodplain connectivity is high
- Strong vertical connectivity

⇒ Longitudinal linkages ┄┄▶ Lateral linkage (channel – floodplain)

→ Vertical linkages ─·─·▶ Lateral linkage (hillslope – channel)

Buffers (landforms that affect sediment transfer from hillslopes to the channel network)
① landslide ② tributary confluence ③ fan ④ backswamp ⑤ floodplain

Barriers (landforms that impede downstream conveyance of sediment within the channel network)
Ⓐ sediment slug Ⓑ dam

Blankets (features that disrupt vertical linkages)
⅗ floodplain sand sheets Ⓢ fines in interstices of gravels

Figure 2.4 Variability in landscape connectivity for an idealised catchment. Moving in a downstream direction from the headwaters, through the mid-catchment to the lowland plain, sediment storage in the channel and on floodplains tends to increase and resides in the landscape for longer, sediment delivery is reduced, hillslopes become increasingly disconnected from channels by floodplains, and channels become more connected to their floodplains. Landforms that disrupt longitudinal, lateral and vertical linkages in river systems are called buffers, barriers and blankets (see Chapter 14).

zone is restricted to areas of alluvial sediment storage. Lowland plains are typically characterised by extensive sediment accumulation both instream and on the floodplain. Channel–floodplain connectivity is high, with ongoing exchange of water (surface or subsurface) and sediment. Sediments can reside in floodplains for considerable periods of time. Hillslopes and channels tend to be decoupled, as materials supplied from low-slope hillslopes are stored for extended periods at valley margins. These sediments only reach the channel network if floodplain reworking occurs. Some lower order tributaries may be disconnected from the trunk stream, effectively trapped behind levees. Vertical linkages of physical processes are more pronounced in these zones, as permeable alluvial materials stored along the valley floor promote surface–subsurface exchange of water.

The effectiveness of biophysical linkages varies markedly in both space and time (i.e. *spatial* and *temporal connectivity*). In some settings, the transfer of flow, sediment and nutrient may be disconnected or decoupled, whether within an individual landscape compartment or between landscape compartments. Changes to the pattern and operation of longitudinal, lateral or vertical linkages through the formation of blockages may exert a significant impact upon flow and sediment conveyance through catchments (see Chapter 14).

Conceptualisation of time

Temporal context is a key consideration when reading the landscape. Typical questions include:

- How often and how far do individual clasts move over a channel bed during floods of differing magnitude?
- What processes shape the channel and over what timeframe?
- How has the floodplain formed and over what timeframe?
- How does the channel adjust its position on the valley floor and over what timeframe?
- How/when did the valley form, and what factors influenced it shape and size?

Timeframes of river analysis

Just as nested hierarchical arrangements can be used to interpret differing scales of river analysis, river behaviour, change and evolution can be considered over timeframes that are nested within each other. Differing sets of controls act as key determinants of the process relationships that fashion landscape evolution over differing timeframes. The role of any given factor may vary dependent upon the

timeframe over which it is analysed. For example, the nature and extent of ground cover are major controls upon rainfall–runoff relationships and resistance factors, influencing spatial variability in the effectiveness of erosion and deposition processes over the short term (days, weeks, months). Over decades this may influence the aggradational/degradational balance of a reach. However, over tens of thousands of years, this imprint upon landscape evolution is overridden by landscape responses to climatic and geologic events (such as glacial/interglacial cycles or earthquake events).

River adjustments over geologic, geomorphic and engineering timescales are conceptualised in Figure 2.5. Geologists are primarily concerned with long-term evolution of the Earth system, typically viewed over timescales of millions of years. In this context, evolution is viewed as cyclical phases of uplift and longer term downwearing (Figure 2.5a). Erosion and transfer of sediments from headwater areas to the lowland zone result in gradual levelling of the landscape, lowering relief over time. Subsequent phases of uplift reinitiate this cycle, bringing about landscape rejuvenation (making the landscape young again). The cycle was initially conceptualised by William Morris Davis, and is referred to as the Davisian cycle of erosion (see below). Inevitably, these process relationships vary markedly in differing tectonic settings. Many other geologic processes may disrupt river evolution (e.g. folding, faulting, subsidence). Resulting differences in topographic setting bring about pronounced spatial variability in patterns of river evolution. Differing stages of adjustment may be observed. Some landscapes are at the uplift and incision stage, others are experiencing a phase of relative stability (although progressive downwearing is ongoing). Processes that operate over geologic timescales determine the nature and distribution of landscape units, thereby shaping relief, slope and valley confinement, and associated patterns of aggradation and degradation along a river.

Geomorphologists are primarily concerned with process–form relationships that shape the Earth's surface over timeframes of hundreds and thousands of years. Over these timescales, river adjustments primarily occur in response to flow variability and sediment flux brought about by climatically induced events. Flow regimes vary markedly in different climatomorphogenetic regions, resulting in marked differences in river behaviour in, say, arid versus temperate versus tropical settings. Regardless of the specific setting, the notion of a dynamic equilibrium, whereby landscapes adjust around a characteristic form, is the dominant conceptualisation of time that underpins geomorphological thinking (Figure 2.5b). This has been referred to as graded time. Any reach has a natural range of variability as it adjusts to disturbance events, but morphological adjustments retain a characteristic (near-equivalent) form

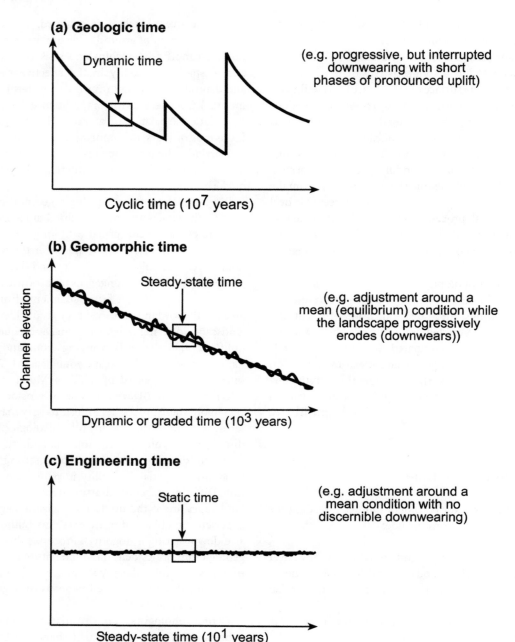

(a) Geologic time

Dynamic time

(e.g. progressive, but interrupted downwearing with short phases of pronounced uplift)

Cyclic time (10^7 years)

(b) Geomorphic time

Steady-state time

Channel elevation

(e.g. adjustment around a mean (equilibrium) condition while the landscape progressively erodes (downwears))

Dynamic or graded time (10^3 years)

(c) Engineering time

Static time

(e.g. adjustment around a mean condition with no discernible downwearing)

Steady-state time (10^1 years)

Figure 2.5 Conceptualisation of landscape adjustments over (a) geologic (cyclic), (b) geomorphic (dynamic equilibrium, graded) and (c) engineering (steady state, static) timeframes.

over geomorphic timeframes. Much of the theory that has been developed in fluvial geomorphology is based on the premise that alluvial channels self-regulate their form via negative feedback mechanisms under conditions of dynamic equilibria. Impacts of disturbance events are damped out or self-corrected via internal adjustments such that the assemblage of geomorphic features that make up a reach in an equilibrium condition remains uniform over

hundreds or thousands of years. The inherent resilience of a system determines its capacity to return to this characteristic state following disturbance.

Finally, engineers frame conceptualisations of landscape adjustment in terms of notional stability (Figure 2.5c). Small perturbations around an average condition over timeframes of years or decades maintain a steady-state equilibrium. Engineering solutions to river problems assume that the governing conditions at the geomorphic timescale are constant. Based on this assumption, principles of fluid mechanics provide quantitative insights into flow fields, the capacity/rate of sediment transport, and the associated nature and rate of channel bed and channel form adjustments at finer scales of resolution. Concern for design solutions (stability) are usually framed in terms of event-driven changes in response to extreme events (e.g. the 1:100 yr event). Using regime theory principles (see Chapter 7), erosion and deposition are kept in balance within a reach, whereby sediment flux reflects oscillations around a steady state. Over even shorter timeframes, typically viewed as weeks or less, a reach may be viewed to be unchanging (hence the label static time).

Davisian cycle of landscape erosion

The Davisian cycle of erosion has three stages: youth, maturity and old age (Figure 2.6). After uplift, a *youthful* landscape is characterised by incising V-shaped valleys separated by broad, flat drainage divides that are undissected by erosion. *Rejuvenation* occurs when the landscape is uplifted or adjustments to base level occur. Confined, bedrock-controlled rivers are characterised by waterfalls and rapids, with no floodplains. Headward erosion is prominent. During *early maturity*, incision continues and the V-shaped valleys become more accentuated. At this stage, relief is at its maximum. Drainage divides become sharp as sidewall erosion widens valleys, creating opportunities for

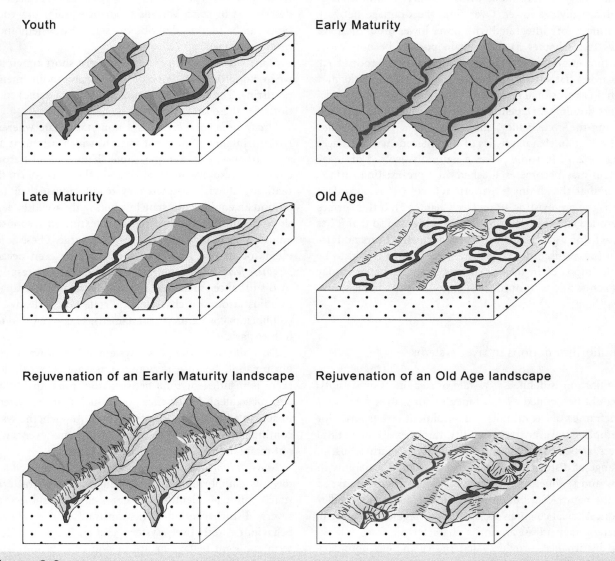

Figure 2.6   The Davisian cycle of erosion depicting youthful, mature and old age landscapes as a landscape evolves. Rejuvenation can occur at any stage if a disruption to the cycle occurs (e.g. responses to uplift, volcanic activity and earthquakes).

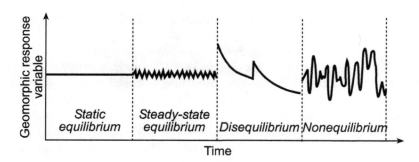

Figure 2.7 Different types of equilibria. Various types of equilibrium conditions can be used to describe the state of a system, its behaviour and evolution.

floodplain pockets to develop. In *late maturity*, incision ceases as the channel has cut down to the reframed base level (i.e. rejuvenation is complete). At this stage, the valley cannot incise further, but valley sidewall retreat produces wider valleys and rounded divides. Relief is diminished as drainage divides lower. Over time the drainage network becomes more integrated with fewer lower order drainage lines (see Chapter 3). Floodplain pockets become more continuous. In *old age*, the landscape has been eroded back towards a flat surface (peneplain), but resistant rocks may remain as erosional remnants. Valleys are wide and alluvial rivers dominate. Depending on when rejuvenation occurs during the cycle, valleys may return to a youthful stage, but maintain characteristics of the antecedent older stages (Figure 2.6). In some places a naturally evolved drainage system may become established on a pre-existing surface, such that the drainage pattern reflects (or is accordant with) the pre-existing strata (see Chapter 3). If this surface is eroded and the drainage pattern is lowered so that it lies across a new geological structure to which it bears no relation (i.e. is discordant), the drainage pattern is said to be 'superimposed'. In these instances, drainage patterns may cut across fold belts or rocks of various types and resistance (see Chapter 12).

Equilibrium notions in river systems

Equilibrium is defined as a state of steadiness or stability. It could be defined as constancy of form over a relevant timeframe or continuity of sediment transport, for example. However, true stability rarely exists in natural rivers, as bed materials and channels recurrently adjust to a range of disturbance events and associated variability in flow and sediment flux. These relationships are mediated by the nature/distribution of resisting elements along valley floors. Alluvial channels are able to self-regulate their form, whereby they adjust around a characteristic form following disturbance, provided that the catchment boundary conditions remain relatively constant. Interpretations of the 'equilibrium state' of a river require insight into how long it takes for a river to develop its characteristic form,

the nature of that form/alignment and the time period over which this form is likely to persist. Although equilibrium notions are important and useful, it is difficult to confirm their application to any specific site. Rather, these ideas and concepts are best applied as hypotheses or descriptors of state. Various types of equilibrium conditions can be used to describe the state of a system and its behaviour (Figure 2.7).

Static equilibrium operates over very short timeframes when no process activity or geomorphic adjustment is occurring. The system maintains a constant (stable) condition over these short timeframes.

Steady-state equilibrium is a non-static state wherein a river maintains a consistent form to which it returns after a disturbance. For example, some channels adjust around a norm in response to seasonal and other short-term fluctuations. Alluvial channels may readily adjust their form around an equilibrium state by altering their width, depth, slope (sinuosity) and bed material texture in response to adjustments in flow and sediment discharge. These adjustments occur over timeframes of years or even decades. *Negative feedbacks* counteract, inhibit and absorb responses to disturbance, such that the system retains its capacity to adjust around a characteristic state (i.e. a steady-state equilibrium condition is maintained). This is referred to as homeostasis.

Disequilibrium occurs as a system adjusts towards equilibrium but, because response times are relatively long, there is insufficient time between the initial disturbance and subsequent disturbance events such that the system is unable to adjust towards a steady state. As such, the system continually responds to recurrent disturbance events without maintaining an equilibrium condition.

Non-equilibrium occurs when there is no tendency towards equilibrium, such that an average characteristic state cannot be identified. Systems that are subjected to severe disturbance often demonstrate non-equilibrium behaviour. Alternatively, systems that are in a state of long-term transient behaviour are in non-equilibrium. In the latter cases, the contemporary character and behaviour of a system retains a *memory* of conditions that were set at some stage in the past.

Differentiating behaviour from change

In this book, adjustments around a characteristic state define the behavioural regime of different types of rivers. Reach behaviour is appraised over timeframes in which flux boundary conditions have remained relatively uniform, such that flow and sediment load inputs and outputs are near consistent, and a characteristic set of attributes is maintained. Ongoing adjustments occur as flow/sediment fluxes respond to alterations in impelling and/or resisting forces, but the reach-scale configuration of geomorphic attributes is maintained. Assessments of the *natural capacity for adjustment* for the reach under investigation are framed in relation to the forms of adjustment that occur for that type of river and the ease (recurrence) with which those adjustments take place (i.e. reach

sensitivity). System responses to differing forms of disturbance are appraised.

If a reach is subjected to a significant change in physical fluxes or other catchment boundary conditions, such that a wholesale shift in the capacity for adjustment of a river brings about a different set of form–process relationships, *river change* is said to have occurred. This records a shift in the natural capacity for adjustment of the river and the transition to a different type of river with a different behavioural regime. River behaviour is differentiated from river change in Figure 2.8.

Assessments of river behaviour versus river change must be framed in relation to the morphological adjustments that are able to occur in a given setting. The likelihood that adjustments will take place reflects the *degrees of freedom* within which a river operates. Each degree of freedom (bed

River behaviour refers to 'natural' adjustments that occur for a particular river type (e.g. lateral migration for meandering sand bed rivers).

River change occurs when a wholesale change in river type occurs (e.g. from an intact valley fill to a channelised fill)

Figure 2.8 River behaviour differs from river change. River behaviour describes adjustments that occur for a certain type of river. River change is a wholesale change in river type and associated behavioural regime. Photographs: British Columbia (G. Brierley), Barbers Swamp and Wolumla Creek, NSW (K. Fryirs).

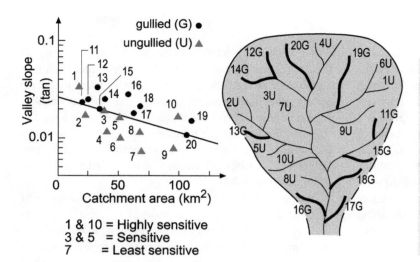

Figure 2.9 Thresholds define transitions in geomorphic state. In this example, unincised valley floors generally lie below the discriminating function line with low slopes for a given catchment area, while incised valley floors generally lie above the discriminating (slope–catchment area) function. Disturbance events are likely to exceed threshold conditions, thereby inducing gullying for sites 1 and 10 while sites 3 and 5 are also prone to incision given their closeness to the threshold (modified from Patton and Schumm (1975)).

character, geomorphic units, channel morphology and channel planform) records the ability of a certain component of the river system to adjust or change. Analysis of river behaviour and change is discussed more fully in Chapters 11 and 12.

Change in state may reflect breaching of *threshold* conditions. A threshold is defined as a point at which a stimulus (e.g. a disturbance) of increasing strength induces a specific response and a transition from one state to another. The simplest example of a threshold differentiates the application of energy to the channel bed to determine whether bed material moves or is stationary. An *extrinsic threshold* is induced from outside the system. For example, land use or climate changes may push a system towards a new state. An *intrinsic threshold* is induced from within the system, whereby adjustments in sediment supply, slope or flow regime can push a system towards a new state. For example, progressive increases in the slope of a valley fill may ultimately instigate incision as an intrinsic threshold is breached. Similar intrinsic threshold relationships describe the susceptibility of alluvial fans to fan-head incision and the susceptibility of a reach to a change in channel planform based on a relationship between sinuosity and valley slope. If a river is close to a threshold condition, relatively small events may induce a disproportionately large response. *Positive feedback* mechanisms strengthen or reinforce responses to disturbance. Snowball effects or a chain reaction may magnify and self-perpetuate impacts such that threshold conditions are breached and the system adopts a new state.

Proximity to threshold analysis (or threshold spotting) seeks to identify characteristics of relatively sensitive and insensitive landforms of the same generic type and determine the threshold conditions under which change is inferred to occur. Data are collected from a range of locali-

ties in which the same set of processes operate (i.e. landscape setting is equivalent), such that a relationship can be established to differentiate among states of system behaviour (i.e. condition). Proximity to threshold analysis is valuable in assessing the relative sensitivity of a landscape to change. For example, the distribution of gullied and ungullied tributary stream lines can be predicted using discriminant analysis based on valley slope for a given catchment area. In the schematic example presented in Figure 2.9, gullied tributaries plot above the discriminating function and ungullied tributaries plot beneath it. However, this is not an entirely consistent relationship, as some systems are yet to become incised. Hence, while tributaries 1 and 10 are highly sensitive, as they sit above the threshold of gullying, tributaries 3 and 5 also lie very close to this threshold condition, while tributary 7 is most distant from the threshold and, therefore, is considered to be least sensitive to change. From this perspective, if a major storm was to impact upon this catchment, tributaries 1 and 10 are most likely to be subjected to dramatic change.

Disturbance events

A disturbance is defined as a change in process intensity. Disturbance events in river systems refer to any factor that affects the boundary conditions under which rivers operate. This may reflect changes to geological controls upon imposed boundary conditions, such as tectonic events. More typical and frequent disturbance is induced by flood events. Patterns and rates of river adjustment or change reflect the nature of disturbance events.

Useful differentiation can be made between *pulse* and *press* events based on the intensity and duration of the disturbance. *Pulsed* disturbance events are episodic events of low frequency, high magnitude and limited duration whose effects tend to be localised (e.g. a seasonal flood). Extreme events may produce a lasting effect, especially if a threshold condition is breached.

During a *press* type of disturbance, controlling variables are sustained at a new level as a result of more permanent shifts in input/flux conditions. The impact of flow regulation following construction of a dam is a typical example. Press disturbance events typically affect much larger areas than pulsed events. Responses are not spatially uniform and they tend to be more permanent, prospectively altering the evolutionary pathway of a reach. Knock-on effects can induce geomorphic changes in reaches that were not directly impacted by the initial disturbance, often a considerable period after the initial disturbance (i.e. there is a notable *lag time*). This reflects the connectivity of the system and the sensitivity of the reaches under consideration.

In *ramp* disturbance events, the strength of the disturbance steadily increases over time and space. Examples of ramp disturbances include drought or increasing sedimentation of a channel bed following clearance of forest cover.

System responses to press, pulse and ramp disturbance events vary markedly. Among many factors, this reflects the condition of the landscape at the time of any given event (i.e. how close to a threshold the system sits) and the connectivity of the system. In strongly coupled catchments, disturbance effects are often conveyed efficiently through the landscape. In contrast, responses to disturbance are inefficiently propagated through decoupled or disconnected landscapes, as barriers or buffers inhibit conveyance of water and sediment, absorbing or damping the impacts of disturbance (see Figure 2.4; Chapter 14).

Reaction time refers to the time taken for the system to respond to a change in conditions or a change in the intensity of a process (Figure 2.10). *Relaxation time* refers to the time taken for the system to attain a characteristic form (whether a previous state or a new state). *River recovery time* is a measure of the system's overall ability to return to a previous state or attain a characteristic form between disturbance events. As such it is a measure of time taken for a system to adjust to a disturbance (including both reaction and relaxation times). Depending on the sensitivity of the system to disturbance and time between disturbance events, various forms of recovery can occur. Rapid recovery will occur in self-adjusting systems. The system is able to adjust quite readily, and quickly, following disturbance. Reaction and relaxation times are quick. Delayed recovery to a disturbance occurs after a *lag time*. Reaction time may be quick, but the system takes some time to return to its previous state. In this book, these recovery responses are referred to as *restoration* trajectories, as they reflect a return towards a pre-disturbance state. Some disturbances may push a system beyond its capacity to retain

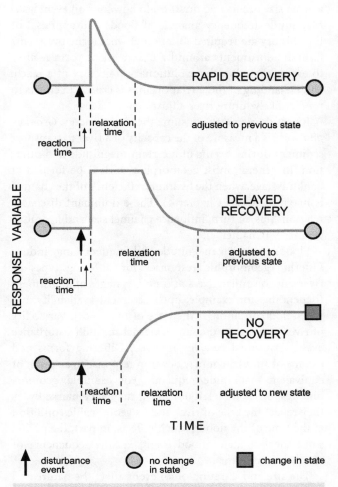

Figure 2.10 System responses to disturbance events. Depending on system sensitivity to adjustment, responses can be rapid and recovery quick, they can be delayed (or lagged) or they can induce a shift in state such that the system adapts to a new state.

or regain its previous form (i.e. threshold conditions are exceeded) or the time between disturbance is so short that the system cannot adjust in time. In these cases, the system adjusts to an altered set of flux conditions and a new characteristic state is formed. In this book, this trajectory is referred to as *creation*. In the latter case, individual events may leave a persistent imprint upon the landscape.

Magnitude–frequency relationships in river systems

Floods of varying magnitude, frequency and duration bring about differing forms and rates of geomorphic adjustment for different types of rivers. Magnitude refers to the size of an event, frequency to how often an event of a given size occurs and duration to how long an event lasts. Magnitude–frequency analysis of floods and appraisal of flow history are required to interpret what formative events fashion adjustments around a characteristic form relative to events that alter the evolutionary trajectory of a reach. Understanding of these relationships is critical in efforts to predict likely future river adjustments.

Rivers do 'work' to consume their own energy. *Geomorphic work* is a measure of the capacity of a river to transport sediment during events of a certain magnitude and duration. In general, most geomorphic work is performed at bankfull stage, when the hydraulic efficiency of the channel is maximised (see Chapter 4). These dominant discharge relationships, in turn, influence channel size and hydraulic geometry (Chapter 7).

Floods of similar magnitude and frequency may induce differing geomorphic responses for differing types of rivers, or to similar rivers at differing stages of adjustment (depending, for example, upon sediment availability and vegetation condition at the time of the event). Variability of river response to the same external stimuli (disturbance event) is termed *complex response*. Differing forms and extents of adjustment may occur in response to an event of equivalent magnitude/frequency. For example, geomorphic responses to a 1:100 yr event may vary markedly, as they reflect the type of river and its geomorphic condition at the time of the flood (which reflects, in part, the period since the last major flood event; i.e. the sequencing of events is important).

The size and duration of an event affect the nature and extent of geomorphic adjustment and the timeframe over which impacts linger (or *persist*). *Geomorphic effectiveness* refers to the capacity of an event to shape landscapes through erosional and/or depositional processes that rework materials on valley floors. It can be measured by the stream power (energy) of the flow (see Chapter 4) and its duration. For example, Figure 2.11 outlines variable geomorphic effectiveness of three flood events. Flood A has a long duration and low stream power which is unable to exceed the threshold for alluvial sediment transport. As a result, no geomorphic work is done, even though the event is relatively long-lived. Flood B has a large peak instantaneous stream power per unit area, but a short duration. Although the alluvial erosion threshold is exceeded, geomorphic adjustments are likely to be limited as the period of time exceeding the erosion threshold is short. Flood C has a large peak instantaneous stream power per unit area and long duration. This geomorphically effective event has substantive time beyond the alluvial and bedrock erosion thresholds, such that flows are able to rework channel boundaries, whether alluvial or bedrock.

Some rivers are fashioned primarily by recurrent and persistent processes that generate low forces (termed *grad-*

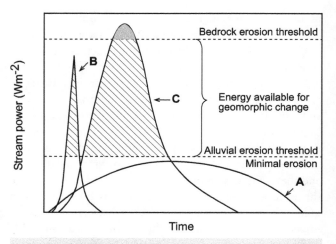

Figure 2.11   A conceptual approach to analysis of the geomorphic effectiveness of floods. Flood C is the most geomorphically effective flood because the energy available to do work is high, the duration of the event long and the thresholds of alluvial and bedrock erosion exceeded. Flood B is less effective as it has short duration beyond the alluvial threshold, and is unable to rework bedrock. Flood A does not extend beyond the alluvial threshold, so it is geomorphically ineffective. Modified from Costa and O'Connor (1995). An edited version of this paper was published by AGU. © 1995 American Geophysical Union. Reproduced with permission.

ualism). Others are shaped primarily by responses to large rare floods that generate large forces (termed *catastrophism*); in these instances, these may be the only events that are able to mould channel boundaries. In reality, most rivers are not the product of a single formative discharge. Rather, they are produced and reworked by a range of flow events. The sequencing/timing of events and the condition of the system at the time of any given event affect patterns, rates and persistence of geomorphic adjustments.

Many attributes of landscapes are inherited from process–form interactions in the past. Conditions that continue to influence contemporary river character and behaviour are termed *antecedent controls*. For example, terrace-controlled rivers may reflect a history of geologic, climatic and anthropogenic controls upon river adjustments. The terraces that remain in the landscape continue to influence the contemporary process–form relationship along the river. An understanding of river evolution is required to determine the degree to which disturbances or events that occurred in the past retain an imprint in the current system, and whether the contemporary system continues to adjust to that disturbance after a considerable lag time. Each river has a *memory* of the consequences of past events that may shape current morphology and the capacity of the system to adjust to subsequent flood events. In contrast, events that erode landscapes and remove materials may leave no record of their impact; indeed, they may remove the imprint of past events. This is referred to as *erasure*.

The history of flood events affects the frequency of reworking of deposits on valley floors. *Residence time* refers to the length of time that sediment is stored in the landscape before it is reworked. Landforms with long residence time provide the fluvial archives that are used to analyse river evolution. For example, the residence time of sediment stored in a distal floodplain is much longer than the residence time of sediment stored in a mid-channel bar. Closely spaced high-magnitude–low-frequency events are likely to have a greater geomorphic impact than widely spaced high–magnitude–low-frequency events because geomorphic recovery may occur between events. In contrast, a sequence of small–moderate floods may induce greater change than a single high-magnitude–low-frequency flood.

To interpret the effect of any given flood on a system accurately, each event must be viewed in relation to flood history and the condition of the catchment at the time of any given event. The timing and sequence of events influences the extent and persistence of change. *Persistence* is defined as the length of time the impact of a disturbance lingers in the system (i.e. the length of time before this imprint is reworked and/or erased). The persistence of landforms reflects their scale and ease of reworking. For example, bed configuration persists in a channel for a short period of time, whereas channel planform and floodplain morphology persist for much longer periods of time, recording events that occurred at various times in the past. Timeframes of adjustment are notably shorter for alluvial (self-adjusting) rivers than for rivers in confined or partly confined valley settings, where imposed conditions such as bedrock restrict the ease of adjustment and extend the timeframe over which these changes occur.

The concept of *equifinality* or convergence postulates that the end state (post disturbance) is one of similarity even though the initial conditions are different (i.e. similar outcomes may result from different pathways and/or formative processes; Figure 2.12). Detailed field and sedimentological investigations may be required to interpret former states in an evolutionary sequence. Understanding of these adjustments is critical in appraisals of the likely future trajectory of change.

In some instances, various stages of geomorphic adjustment can be identified at different positions in a landscape. From this, likely evolutionary pathways can be inferred. An example of this conceptual approach to landscape analysis is shown for the Davisian cycle in Figure 2.6, where landscapes at youthful stage will adjust subsequently into mature and old age states. This is dependent upon the sequence of disturbance events, and in this instance adjust-

ments will take millions of years. Alternatively, ungullied states outlined in Figure 2.9 may be transformed into gullied states almost overnight. This interpretative process of space for time substitution is referred to as ergodic reasoning (or invoking the ergodic hypothesis).

River sensitivity and resilience

Geomorphic responses to disturbance events reflect the nature of the impact on the one hand and the sensitivity/resilience of the system on the other hand. Landscape sensitivity is a measure of the likelihood that a given disturbance will produce a sensible, recognisable and persistent response. Measures of *river sensitivity* reflect the capacity for system adjustment (the ease with which adjustments can take place for that type of river), proximity to a threshold condition and the pre-conditioning of the system (i.e. the state of the river at the time of the disturbance event, reflecting, among many factors, how the system has responded to recent events and whether this has made the system more or less vulnerable to adjustment). These factors provide a measure of the response potential of a river to external influences. *Sensitive rivers* are readily able to adjust to perturbations but are prone to dramatic adjustment or change. Conversely, *resilient rivers* have an inbuilt capacity to respond to disturbance via mutual adjustments that operate as negative feedback mechanisms. In this scenario, the self-regulating nature of the system mediates external impacts such that long-term stability is retained. These rivers are readily able to adjust to perturbations without dramatic adjustment or change in process–form associations.

The ability of a system to absorb perturbations, such that disturbance events do not elicit a morphological response, is referred to as the *buffering capacity*. In systems with large buffering capacity and/or with large thresholds to overcome, there may be considerable time lags between perturbation and morphological response. Morphological responses to disturbance events are likely to be more pronounced along more sensitive reaches.

River sensitivity is also influenced by within-catchment position and patterns/rates of geomorphic linkages (i.e. connectivity). In sensitive landscapes, lag times are short as geomorphic responses occur relatively quickly after the disturbance event and are readily conveyed through highly connected landscapes. In resilient landscapes, lag time may be long such that geomorphic responses are not manifest in the system for some time. These landscapes respond slowly (if at all) to disturbance. Geomorphic threshold conditions are rarely exceeded, and there may be significant decoupling in the system such that change in one part of the system is absorbed and not manifest elsewhere.

Intact valley fill   Meandering sand bed

Low sinuosity sand bed

Figure 2.12 Equifinality refers to attainment of a similar endpoint (morphologic response) from different evolutionary pathways and starting points. In the case shown, an intact valley fill river and a meandering sand-bed river have been transformed into a low sinuosity sand-bed river following human disturbance. Removal of riparian vegetation triggered incision and channel expansion in both instances. Unravelling controls and triggers on river change, whether natural or human induced, is an important consideration in determining the evolutionary trajectory of a river. Photographs: Budderoo Swamp (K. Fryirs), Thurra River (A. Brooks) and Cann River (K. Fryirs).

Catchment-specific analysis of river systems: combining spatial and temporal concepts

The pattern of a landscape is derived from its composition (the kinds of elements it contains), its structure (how they are arranged in space) and its behaviour (how it adjusts over time to various impulses for change). Process–response relationships are fashioned by the way in which a system is put together, i.e. its *configuration*, defined by the spatial distribution of various components and their topological relationships (i.e. connectivity). Each system has its own memory of natural and human-induced disturbance events. These forcing factors fashion the evolutionary trajectory of

the system. Given this situation, each catchment has its own *response gradient*, whereby patterns of sensitivity to disturbance and the way in which these disturbance responses are conveyed through systems (i.e. reach-to-reach connectivity) reflect catchment-specific considerations (Chapter 14). Sometimes responses are predictable, other times they are not.

Catchments are complex systems within which the operation of a range of factors may change over varying timeframes. Variability in lagged and offsite responses to disturbance reflect considerations such as threshold conditions, complex response, sensitivity/resilience and feedbacks. The evolutionary trajectory of some systems can be characterised by adjustment around an equilibrium state, whereas other rivers are subjected to recurrent threshold breaches and chaotic patterns of adjustment. Differing systems retain a variable memory of past events (i.e. these are catchment-specific relationships).

Process interpretations must always be placed in an evolutionary context for the system under consideration, recognising that there is no unique solution for a given set of controlling variables. Viewed in this light, the evolutionary nature of river systems is better viewed in relation to *non-linear dynamics* rather than equilibrium notions.

Linear systems are characterised by self-organising processes which maintain an orderly sequence of system adjustments around an equilibrium state. On the one hand, this is a question of timescale (see Figures 2.5 and 2.7); on the other hand, it reflects the state of the system and the extent to which the river has fully adjusted to prevailing conditions. Many rivers are not in equilibrium. They may be adjusting to antecedent controls. They may be highly disturbed or are set 'out of kilter' by particular events. Non-linear systems exhibit discontinuities in system evolution. These bifurcations may send a system onto a range of potential evolutionary trajectories. Bifurcations often occur when extrinsic or intrinsic thresholds are breached. Systems evolve slowly between bifurcations, but rapidly and potentially chaotically whenever thresholds are breached. As such, river evolution may entail rapid and gradual changes. Discontinuities may record transitions from regular to chaotic behaviour.

Landscape behaviour – both process operation and the suite of landforms that result – is *contingent* on a wide range of factors. As a consequence, deterministic predictions of system behaviour will rarely, if ever, be possible. This explicitly requires an acceptance of probabilistic approaches to landscape analysis. Although we can make statements about the likelihood of occurrence of an event of given magnitude within a given spatio-temporal context, and under certain conditions, we cannot make absolute predictions about time, location or magnitude of process occurrence, or indeed of anything more than its immediate consequences. Hence, it is the statistics of frequency–magnitude spectra of process behaviour that are important. System-specific responses to alterations in imposed/flux boundary conditions determine pathways of river behaviour and change. Response to disturbance reflects its *effectiveness* in inducing change and how landscape *configuration* and *connectivity* influence off-site and lagged responses.

Each catchment operates under its own set of imposed and flux boundary conditions. Landscapes are open, complex systems that are comprised of mosaics of landforms of differing sizes and longevity. Different components of landscapes can be primed to change in response to a diverse array of triggers. Impacts in one place may dampen or buffer effects elsewhere. Responses to external stimuli reflect initial conditions, the history of change and perturbation, proximity to threshold conditions and the degree to which forcing processes and geomorphic forms are in synchronicity with each other. These conditions set the *path dependency* of any given river system.

As landscapes respond to perturbing impulses in complex, non-linear ways, process and form cannot be simply linked together in terms of frequency, magnitude and effectiveness of formative events. Simply bolting together models of individual processes cannot explain whole of system changes over time. Disrupted evolutionary pathways, variable historical imprints and system-specific configuration and connectivity ensure that there may be pronounced variability in behavioural responses to disturbance events. Landscapes are *emergent*. Their evolutionary trajectory and responses to disturbance cannot always be predicted through linear cause-and-effect reasoning. They are also *contingent*. Responses to a given event depend upon the condition of the system at the time of the event (the principle of complex response). Unravelling these complex, systematic interactions is a core attribute of reading the landscape.

Conclusion

River geomorphology is not a deterministic cause-and-effect science. Surprising outcomes should be expected. Such is the way of the natural world. General trends help us to guide our interpretations, but their prescriptive application may impact negatively upon the inherent range of geodiversity of river systems. The challenge now is to synthesise and extend understanding beyond the mechanics of process, providing more holistic appreciation of the diversity and dynamics of riverscapes as individual systems. This entails analysis of landscape configuration and connectivity, alongside appraisals of the process–form relationships for any individual component. Such a synthesis embraces flux and change as defining characteristics of geomorphic systems. Non-linear thinking has become the norm! Each individual landscape is indeed 'perfect' (Phillips, 2007), with its own configuration and history. Reading the landscape provides a practical tool with which such system-specific applications can be performed.

Key messages from this chapter

- A range of spatial and temporal concepts can be used to read the landscape, assessing how a landscape looks, adjusts and evolves.

- Catchments can be considered as nested hierarchies, incorporating the following scales: catchment, landscape unit, reach, geomorphic unit and hydraulic unit. Interactions at coarser scales control processes and interactions at finer scales. A building-block approach to river analysis 'fits together' smaller scale features to 'construct' broader scale landscapes.
- Imposed boundary conditions determine the coarser scale controls on river character and behaviour. Geological factors determine relief, slope and valley morphology.
- Flux boundary conditions drive flow, sediment and vegetation interactions on valley floors. These relationships primarily reflect climatic factors.
- Some rivers have a simple (homogenous) geomorphic structure, while others are heterogeneous.
- Landscape connectivity is a primary control upon fluxes of water and sediment. It is characterised by longitudinal, lateral and vertical relationships. The strength of linkages varies at differing positions in catchments, and over time. Blockages may disrupt water and sediment transfer.
- Timeframes of geomorphic adjustment are key considerations in reading the landscape. River adjustments occur over geologic, geomorphic and engineering timescales.
- Rivers retain a variable memory of past disturbance events. That is, contemporary process-form interactions may be influenced to a variable degree by past conditions.
- Various types of equilibrium conditions describe river behaviour and adjustment. Some rivers operate as non-linear systems.
- River behaviour can be differentiated from river change. Behaviour refers to the natural capacity for adjustment. River change refers to a wholesale shift in river character and behaviour.
- Adjustment and change in some rivers is driven by threshold exceedance. Some rivers gradually adjust, while others are fashioned primarily by catastrophic events.
- Feedbacks and lagged responses can produce complex responses, whereby events of a given magnitude and frequency induce differing responses from system to system.
- Pulse, press and ramp disturbance events drive river adjustment.
- Reaction, relaxation and recovery times that record system responses to disturbance events reflect the sensitivity/resilience of the system.
- Floods of variable magnitude and frequency drive geomorphic adjustment. Geomorphic work is a measure of how much sediment is transported by a river during events of differing magnitude and frequency. Geomorphic effectiveness is a measure of landscape-forming events (erosion and deposition).
- Contemporary process–from relationships are often influenced by geomorphic memory (antecedent controls).
- Differing river types have variable sensitivity to adjust.
- Spatial and temporal relationships must be interlinked through system-specific applications that read the landscape at the catchment scale. Landscapes are contingent, non-linear and emergent. Geomorphology is not a deterministic cause-and-effect science.

CHAPTER THREE

Catchment-scale controls on river geomorphology

Introduction: what is a catchment?

A *catchment* (also called a drainage basin or a watershed) is a single fluvial system that is linked internally by a network of channels (Figure 3.1). These self-contained topographic and hydrologic systems are the fundamental spatial unit of landscapes. The catchment boundary defines the separation of surface flow from one hydrologic system to another. It is typically demarcated by a ridge line. Catchments are comprised of subcatchments, such that tributary–trunk stream relationships are primary determinants of patterns and rates of river processes and forms at the catchment scale. While hillslope–channel linkages exert a key control on water and sediment transfer through catchments (see Chapter 4), emphasis in this book lies with geomorphic process–form relationships on valley floors. *Drainage network composition* refers to the internal structure of a channel network within a catchment.

Catchment-scale morphometrics and the development of stream profiles are fashioned over geologic timeframes, framing the imposed boundary condition controls within which rivers operate (see Chapters 1 and 2). These controls constrain the range of river behaviour and associated morphological attributes within a catchment. For instance, regional geology and climate, among other factors, determine topography, sediment transport and the discharge regime. These considerations, in turn, influence patterns and rates of flow–sediment interactions through controls on the distribution and use of available energy.

In this chapter, we first examine how the balance of erosion and deposition along river courses determines the distribution of sediment process zones in catchments and resulting patterns of river morphology. Analysis of catchment-scale relationships along longitudinal profiles (the downstream gradation in elevation along a river course from its source to sink) sets the foundations for examination of how water and sediment are transferred through catchments. We then discuss how various morphometric measures (i.e. catchment area, shape, relief, drainage density and stream order) can be used to interpret sediment and water flux in catchments.

Process zones in catchments: sediment source, transfer and accumulation zones

Rivers are slope-induced systems which convey water and sediment from their headwaters to the mouth. Their catchments are typically comprised of steep headwaters, moderate-slope mid-catchments and low-lying plains. Sediment transfer relationships in catchments can be differentiated into source, transfer and accumulation zones (Figure 3.1). Although sediments are eroded, transported and deposited in each zone, the dominance of each process varies spatially and temporally in these differing landscape compartments. Assessment of whether you are in a source, transfer or accumulation zone is a critical starting point in efforts to read the landscape.

Source zones are dominated by erosion processes such that sediment is liberated and supplied downstream (Figure 3.1). These zones are net exporters of sediment. Source zones classically occur in headwater regions of catchments. Elevated areas such as mountain ranges, tablelands and escarpments primarily comprise erosional landforms cut into bedrock (Table 3.1). Vertical downcutting is the dominant fluvial process in mountain zones. This produces steep and narrow (i.e. confined) valleys. Sediments supplied from hillslopes are fed directly to the valley floor, where narrow and deep bedrock-controlled channels transport most of the available sediment, such that there is limited sediment storage on the channel bed. As a result, sediment supply greatly exceeds deposition in these zones. Coarse-grained (boulder and cobble; see Chapter 6) bedload transport mechanisms are dominant as finer grained materials are flushed through these reaches.

A *transfer zone* occurs where there is a balance between sediment supply and sediment export such that sediment

---

Geomorphic Analysis of River Systems: An Approach to Reading the Landscape, First Edition. Kirstie A. Fryirs and Gary J. Brierley.
© 2013 Kirstie A. Fryirs and Gary J. Brierley. Published 2013 by Blackwell Publishing Ltd.

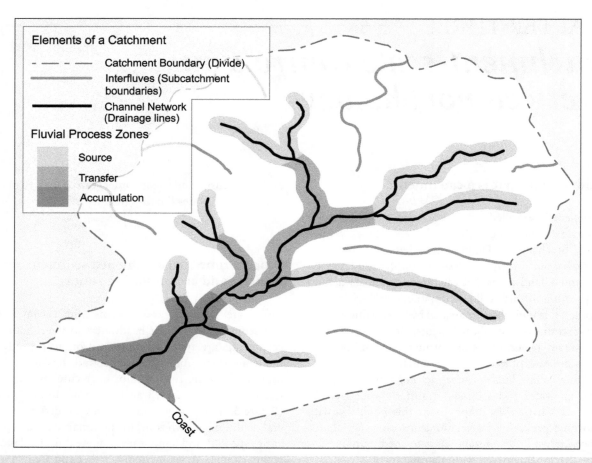

Figure 3.1 Elements of a catchment. Sediment source zones tend to occur in the headwaters, transfer zones in the middle reaches and accumulation zones in lowland reaches. Based on ideas in Schumm (1977).

Table 3.1 Relationship between geomorphic process zones and landscape units

	Examples of landscape units	Dominant fluvial process	Valley setting
Source	Mountain ranges, escarpment zones	Erosion via vertical cutting; minimal sediment storage	Confined or partly confined
Transfer	Tablelands, foothills, erosional piedmonts	Erosion via lateral cutting; fluctuating sediment storage	Partly confined of laterally unconfined with bedrock base
Accumulation	Coastal plains, alluvial plains, depositional piedmonts, playas	Deposition and net sediment accumulation	Laterally unconfined with fully alluvial channel boundaries

is being conveyed downstream but is being supplemented by a roughly equivalent supply from upstream. These reaches tend to occur in the middle parts of catchments, downstream of the headwater sediment source zones in landscape units such as rounded foothills or piedmont zones (Figure 3.1; Table 3.1). In these areas, there is sufficient energy to sustain bedload transport along channels, but a balance between net input and output is approached such that sediment supply is approximately equal to deposition. Moving downstream, the grain size of sediments decreases and channel size increases. Initially, instream sediment storage units are primarily short-term sediment stores with limited residence times (e.g. mid-channel bars or fans at tributary confluences), but a progressive downstream transition in sediment storage units occurs as valley slope is reduced and valley width increases. Considerable energy is expended eroding the base of confining hillslopes. These processes, combined with vertical incision, create the space in which floodplain pockets are able to form in partly confined valley settings (i.e. spatial segregation of channel- and floodplain-forming processes occurs). The character of the valley trough, in combination with slope and bed/bank material, exerts considerable control on river morphology. These reaches have sufficient stream power to rework sediment stores on the bed and banks.

Accumulation zones are dominated by depositional processes and associated sediment stores (Figure 3.1). These zones are net importers of sediment. The accumulation zone or sediment sink is marked by alluviation, aggradation and long-term sediment storage. Accumulation zones typically occur in lowland regions of catchments. Materials eroded and transported from upstream areas are deposited in flanking sedimentary basins (aggradational zones), such as lowland plains or broad alluvial plains in endorheic (inland draining) basins (Figure 3.1; Table 3.1). Typically, these zones are comprised of large, alluvial channels with continuous floodplains along each bank. Flow energy is dissipated across broad alluvial surfaces. In these low slope settings, long-term valley widening and denudation have produced a broad valley trough in which the channel infrequently abuts the bedrock valley margins (i.e. hillslopes and channels are decoupled). As a consequence, sediments are delivered to the channel almost entirely from upstream sources. Deposition rates greatly exceed erosion rates in these parts of catchments. Relatively low stream power conditions reflect low slopes as base level is approached. The decline in stream power is marked by a decrease in bed material texture (silt and clay materials are dominant, especially on floodplains and deltas).

While the classic view is that catchments comprise a downstream sequence of source, transfer and accumulation zones, there are many instances where this is not the case and process zones are repeated along the river course. In escarpment-dominated catchments, for example, extensive plateaux and tablelands may occur above the escarpment. These settings often contain alluvial sediment stores (such as swamps) of considerable age that act as accumulation zones in the headwaters of the catchment. These are often transitional to source zones in the escarpment area, which in turn are transitional to transfer and accumulation zones. In other catchments, mid-catchment or lowland gorges may be evident, disrupting the typical downstream sequence of source, transfer and accumulation zones. The configuration of the catchment and the sequence of sediment process zones reflect the nature and rate of erosional and depositional processes in differing landscape settings. These geomorphic transitions in catchments are best depicted through analysis of process relationships along longitudinal profiles.

Longitudinal profiles of rivers

Longitudinal profiles provide a powerful platform with which to analyse interactions among many attributes of river systems. The *longitudinal profile* of a river depicts the change in elevation of a channel from its headwaters to its mouth, thereby showing the rate of change of slope (or gradient) with distance downstream. *Valley slope (or gradient)* is a measure of the relative fall in elevation (rise) of the valley floor over a certain distance (run). *Channel slope*, for the same section of valley, has the same elevation change (rise), but the distance (run) is longer, as this records the length of the channel within the valley (i.e. the channel is not straight, so channel length is greater than valley length; adjustments to channel length, as such, is one of the primary mechanisms by which a river can consume its own energy).

The shape of a longitudinal profile reflects the imprint of large-scale, long-term environmental changes and landscape evolution in any given catchment. The level to which a channel naturally cuts is referred to as the *base level*. At the mouth of exorheic basins (i.e. systems that drain to the ocean), this is sea level. Differing segments of a catchment may cut to differing base levels, imposed by features such as lakes or bedrock-steps (waterfalls). Tributary streams cut to the base level that is set at the confluence with the trunk stream. Base level may change over time. For example, climate change induces adjustments to sea level over glacial–interglacial cycles. The base level in inland-draining (endorheic) basins is set by the lowest elevation in the catchment (typically demarcated by a lake, though these are often ephemeral).

Most rivers adjust their form to create a smooth, concave-upwards longitudinal profile. A *graded stream* can be defined as one in which channel slope adjusts over annual timeframes to enable available discharge and prevailing channel characteristics to have sufficient energy to transport the load supplied from the drainage basin. The graded stream is a system in equilibrium; its diagnostic characteristic is that alteration any of the controlling factors will cause a displacement of the equilibrium in a direction that will tend to absorb the effect of the change. Equilibrium profiles are characterised by orderly downstream patterns of increasing flow (discharge) and channel size, and smooth downstream decrease in slope and bed material size (discussed below).

The longitudinal profiles of many of the world's largest rivers are smooth; however, many longitudinal profiles have convex sections or reaches characterised by abrupt changes in slope. For example, resistant bedrock may form steps or waterfalls that act as local base levels along longitudinal profiles. This represents a *knickpoint* (i.e. a discontinuity in bed elevation and slope that progressively moves upstream. Erosion processes are accentuated at

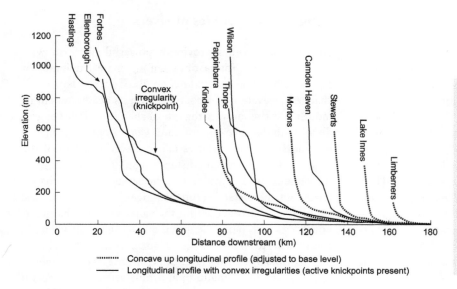

Figure 3.2   Longitudinal profiles and knickpoints of rivers. Longitudinal profiles of rivers adjust to create smooth, concave-up forms. Irregularities such as bedrock steps and waterfalls (e.g. knickpoints) are common. The example shown is for the Hastings River and its tributaries, North Coast, NSW, Australia. Irregularities in the concavity and form of these longitudinal profiles reflects the extent of escarpment retreat through this catchment.

these breaks of slope. Elsewhere, tectonic uplift may induce convex irregularities along longitudinal profiles if rates of bed incision are less than rates of uplift. Irregularities along longitudinal profiles can also be a function of antecedent controls associated with long-term landscape evolution. For example, rates of uplift, incision and knickpoint retreat determine the positions of escarpments along longitudinal profiles and the resulting proportions of the profile that contains plateau, escarpment and lowland plain sections (see Figure 3.2). Finally, changes in base level can induce irregularities along profiles. For example, construction of a dam/reservoir alters the base level of the trunk and tributary streams.

Analysis of longitudinal profile form provides a powerful tool for interpreting downstream changes to the erosion–deposition balance and the position of process zones along the profile. Profile concavity can be defined as:

$$\text{Concavity} = \frac{2A}{H}$$

where A (m) is the height difference between the profile at mid-distance and a straight line joining the end points of the profile and H (m) is the total fall of the longitudinal profile.

Slope s and distance L (or catchment area) are related by a power function:

$$s = kL^n$$

where k and n are steepness and concavity variables respectively.

Irregularities in profile form, and associated interpretation of sediment process zones, can be appraised through analysis of segments of longitudinal profiles that have different values of k, n or both.

Other important transitions along longitudinal profiles include changes in valley confinement, especially the transition from bedrock to alluvial rivers that reflects marked increases in instream sediment storage and the emergence of floodplains.

Geomorphic transitions along river longitudinal profiles

Relationships between valley slope and confinement determine the imposed boundary conditions within which rivers operate. Inset within these controls, interactions between discharge and sediment calibre/volume fashion the distribution of erosion and deposition (i.e. sediment process zones), and resulting transitions in river character and behaviour along longitudinal profiles. A set of interactions along a classic concave-upward longitudinal profile are depicted on Figure 3.3. A series of fundamental transitions in river geomorphology occur along this profile.

Relief variability, manifest primarily through the slope and confinement of the valley floor, is a key determinant of the *valley setting* in which rivers form. Valley setting is a function of the rate and extent of bedrock incision relative to valley widening. Tectonic setting and lithology are primary controls on this relationship. Three broad classes of valley setting are differentiated, namely confined, partly confined and laterally unconfined (Figure 3.3; see Chapter 10). Confined valleys occur where rates of bedrock incision are greater than rates of sidewall retreat and valley widening. In these valleys, erosion processes dominate and bedrock-confined rivers occur. Confined valleys tend to occur in the source zones of catchments. Partly confined valleys occur where the rate of bedrock incision roughly equals rates of valley sidewall retreat. Sediment transfer

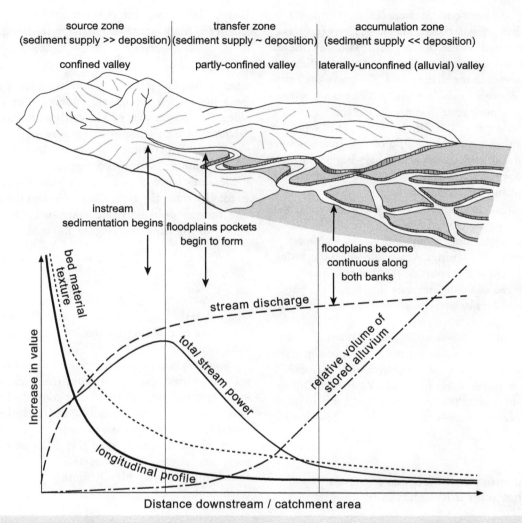

Figure 3.3 Schematic representation of the relationship between downstream changes in slope, discharge, bed material texture, total stream power and stored alluvium along a typical concave-up longitudinal profile, and associated transitions in sediment process zones and valley-setting pattern. Based on Church (1992).

zones are most commonly found in this valley setting where sediment supply and output are roughly balanced. Floodplains represent discontinuous sediment stores outside the channel zone. Laterally unconfined valley settings occur where rates of valley incision are low and sidewall retreat has produced wide, open valleys in which large volumes of alluvium can accumulate. Sediment supply is significantly greater than sediment output in these settings, producing alluvial streams with continuous floodplains in these accumulation zones.

Several other transitions in river character and behaviour are related to the downstream pattern of valley settings and process zones (Figure 3.3). The downstream gradation in bed material size mirrors the slope of the longitudinal profile. This reflects the combined effects of abrasion and hydraulic sorting as slope decreases (see Chapter 6). The distribution of total (gross) stream power (the product of slope and discharge) has a 'hump' shape, as steep slopes and high discharges produce a peak in total stream power around the transition from the source to the transfer zone. This reflects increases in discharge as smaller tributary networks join the trunk stream and relatively steep slopes within partly confined valley settings. Upstream of the stream power 'hump' the discharge is too low to produce high stream-power conditions, and downstream of the hump the slope is too low. As a result of this relationship, significant geomorphic work (i.e. sediment transport and transfer) occurs in these sections of a catchment. Transitions in the distribution of instream and floodplain sediment stores and sinks occur near this hump.

Montgomery *et al.* (1996) derived a fundamental set of relationships between drainage area (a surrogate for discharge) and slope along rivers that can be used to discriminate between bedrock rivers (i.e. those that effectively flush channel materials) and alluvial rivers (i.e. those that store sediment on the channel bed). This determines the downstream position at which instream sediments begin to be stored on the channel bed. The transition from fully confined rivers to rivers where sediments are stored outside the channel zone (i.e. in floodplains) typically occurs around slopes of 0.008 m m^{-1}. Floodplains typically begin to form on the receding limb of the stream power hump, where energy decreases and valleys begin to widen such that *accommodation space* (i.e. space on the valley floor) is available for sediments to accumulate. Initially, floodplains are discontinuous and alternate along the valley floor in partly confined valleys. Eventually, floodplains become continuous along both banks within laterally unconfined valley settings. As a result of this increasing trend of sediment storage outside the channel, the relative volume of stored alluvium increases significantly in partly confined valleys and in laterally unconfined accumulation zones (Figure 3.3).

The relationships shown in Figure 3.3 provide a simplified representation of downstream transitions in river character and behaviour in response to changes in imposed boundary conditions along longitudinal profiles. One of the key skills in reading the landscape is the ability to relate catchment-specific patterns to these 'classic' downstream trends.

Catchment morphometrics as controls on river character and behaviour

Landscape setting is a key determinant on catchment morphometrics. Analysis of relief (change in elevation/slope), drainage density (i.e. landscape dissection) and valley width aids in the interpretation of the distribution of erosion/deposition (process zones), and sediment and water flux in catchments, thereby guiding interpretations of controls upon patterns of river character and behaviour. Catchment morphometrics (i.e. shape, area, relief and drainage density) can be measured quickly and efficiently using digital elevation models (DEMs) and geographic information systems. *Valley width* is measured as the distance between bedrock valley margins (i.e. hillslopes). It is normally measured across the top of floodplains or terraces and perpendicular to the channel.

Catchment shape

Catchment shape is a major influence upon hydrologic relationships in landscapes (see Chapter 4). Lithology and long-term landscape evolution are key controls on catchment shape. The relationship between catchment area, stream length and resultant catchment shape, can be expressed as:

$$L = 1.4A^{0.6}$$

where L (km) is stream length measured in a straight line from the highest topographic point to the river mouth along the longest axis of the catchment and A (km^2) is catchment area. The exponent 0.6 suggests that catchments elongate with increasing size and that large catchments are relatively longer than smaller catchments.

Measures used to assess catchment shape include the circularity ratio, the elongation ratio and the form factor. The 'normal' pear-like ovoid shape of catchments can be related to circular forms (Figure 3.4) to determine the *circularity ratio*:

$$R_c = \frac{A}{A_c}$$

where R_c is the circularity ratio, A is catchment area and A_c is the area of a circle with the same circumference as the catchment.

Using this ratio, catchments with low ratios (about 0.4) are relatively elongate and are controlled primarily by geologic structure. Basins that are not controlled by structure have circularity ratios between 0.6 and 0.7 and are relatively round (i.e. the ratio is close to 1.0).

Unlike the circularity ratio that relies on the measurement of circles, the *elongation ratio* measures the catchment area to length relationship to give a measure of catchment shape:

$$E_r = \frac{A^{0.5}}{L}$$

where E_r is the elongation ratio, A (km^2) is the catchment area and L (km) is the catchment length along its axis. The closer to 1.0 the ratio is, the more round the catchment is. Catchments with elongation ratios around 0.6 are relatively elongate. In theory, the more elongate the catchment is, the slower the runoff from the basin is.

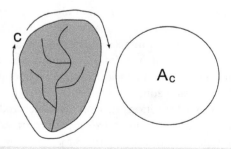

Figure 3.4   Relation of the 'normal' pear-shaped ovoid catchment to a circular form.

The *form factor* is another measure of the relationship between catchment area and length. However, unlike the elongation ratio that gives a measure of the shape of the catchment, the form factor provides a measure of the relationship between catchment area and catchment length and it's effect on hydrology:

$$R_f = \frac{A}{L^2}$$

where R_f is the form factor, A (km^2) is the catchment area and L (km) is the catchment length along its axis. Catchments with a ratio of 4 have flashy flood regimes, while catchments with ratios closer to 8 tend to have lower flood intensities (see Chapter 4).

Catchment relief

Maximum catchment relief H is defined as the difference between the elevation of the catchment mouth E_{min} and the highest peak in the catchment E_{max}:

$$H = E_{max} - E_{min}$$

However, this only measures the total fall of a catchment. To gain a clearer picture of the relative height over which water falls and the distance it travels, the maximum catchment relief is calibrated for catchment length using the *relief ratio*. This provides a measure of the average drop in elevation per unit length of river:

$$R_h = \frac{H}{L}$$

where R_h is the relief ratio, H (m) is the maximum catchment relief and L (m) is the basin length along its axis. Note: units of H and L should be the same (e.g. metres), so as to make R_h dimensionless.

The *hypsometric interval* is measured as the proportion of the catchment area that lies above and below a certain elevation (Figure 3.5). Moving upstream, elevation h progressively increases (Figure 3.5a). The area a of the catchment cumulatively increases with each incremental increase

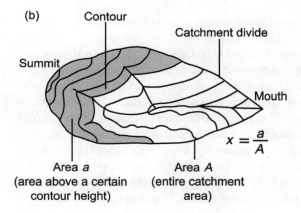

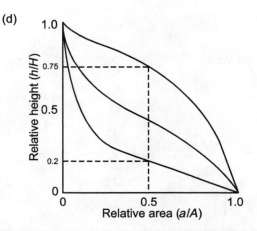

Figure 3.5 The hypsometric interval measures the relationship between elevation and area in a catchment, thereby providing a guide of catchment relief. Modified from Ritter *et al.* (1978).

in *h* (e.g. between contours) (Figure 3.5b). Relative values of *h*/*H* and *a*/*A* can be used to derive the hypsometric curve where the *x* and *y* values are dimensionless, representing proportions of the total area and height (Figure 3.5c). For *y* = 0, all heights are above the datum plane. As such, they lie within the total area (i.e. *x* = 1). The area below the curve is calculated as the hypsometric interval (HI). This measure of topographic setting varies for different tectonic zones and geologic settings. For example, the top curve in Figure 3.5d typifies a relatively steep terrain in a catchment that has a significant proportion of its catchment comprising high-relief mountains that readily transfer flow/sediment (over half the catchment area is high relief, *h*/*H* = 0.75). The bottom curve in Figure 3.5d reflects low-lying terrain in a catchment with a significant proportion of its area in rounded foothills and lowland/coastal plain plains (for half of the catchment area, *h*/*H* = 0.2).

These various measures of relief may vary markedly from terrain to terrain or for differing subcatchments within a catchment, dependent upon the nature, extent and pattern of landscape units.

Drainage density and network extension

Drainage density D_d is measured as the total length of stream channels per unit area of a catchment (e.g. km km^{-2}). It provides guidance into the degree of landscape dissection, which in turn exerts a significant influence upon flow and sediment transfer through a catchment. Higher surface areas promote greater runoff and sediment generation. Average drainage density in moderately resistant lithologies range from 8.0 to 16.0. Ratios below this range are considered low. At the other end of the spectrum, dissected badlands may have drainage densities >1000. Maximum

efficiency of flow and sediment transfer is achieved in these basins with complex bifurcating networks of small channels. These conditions promote rapid geomorphic responses to disturbance events. Vegetation cover and land use influence drainage density. Sparse vegetation cover leaves the landscape exposed to intense rainfall events that induce high rates of erosion and landscape dissection, maintaining and/or increasing drainage density.

Drainage networks evolve over time to generate a leaf-vein pattern of streams. In the evolutionary sequence shown in Figure 3.6, drainage density increases in response to channel extension and/or rejuvenation. In initial stages, incision along the trunk stream lowers the base level along tributaries, inducing headcut development along these streamlines. Drainage density of the basin is low at this stage. Drainage extension and channel expansion progressively increase drainage density. However, once the drainage network has reached its maximum extent for a given catchment area, it is no longer possible to maintain rates of incision and erosion. As a result, drainage network integration reduces channel numbers and drainage density.

Drainage pattern

Drainage patterns describe the ways in which tributary streams are connected to each other and the trunk stream (Figure 3.7). Drainage network patterns are a product of the lithology and structure of a region. *Dendritic* drainage patterns are the most common form. They develop in areas of homogeneous terrain in which there is no distinctive geologic control. A pattern analogous to veins in a leaf is produced. Tributaries join the trunk stream at acute angles, less than 90°. The lack of structurally controlled impediments ensures that this configuration promotes relatively smooth downstream conveyance of sediment. In many

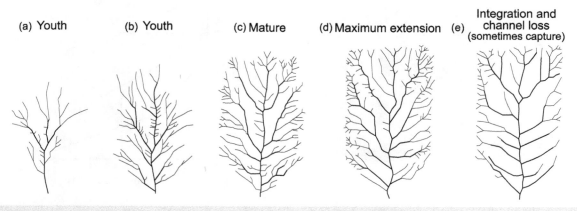

Figure 3.6 Stages of drainage network evolution and extension. Modified from Chorley *et al.* (1984).

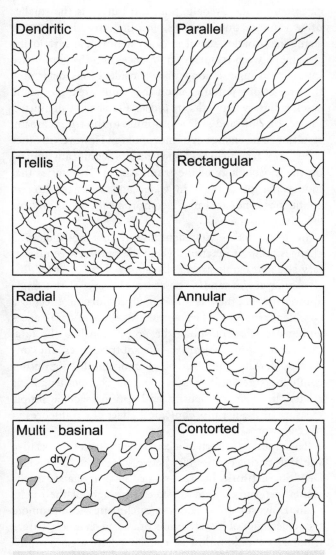

Figure 3.7 Geologic controls upon drainage pattern. From Howard (1967). © AAPG. Reprinted with permission.

and faulting, a *rectangular* pattern is commonly observed. Streamlines are concentrated where the exposed rock is weakest. Tributaries join the trunk stream at sharp angles. *Radial* and *annular* drainage patterns develop around a central elevated point. This pattern reflects differential erosion of volcanoes and eroded structural domes respectively. *Multi-basinal* (or deranged) networks occur where the pre-existing drainage pattern has been disrupted. These networks are typically observed in limestone terrains or in areas of glacially derived materials. Finally, *contorted* drainage networks occur where the drainage network has been disrupted by neotectonic and volcanic activity.

Geologic controls on drainage network form, and river character and behaviour

Geologic controls on slope and sediment calibre exert a primary influence upon river character and behaviour. *Imposed boundary conditions* determine the relief, slope and valley morphology (width and shape) within which rivers adjust. In a sense, these factors influence the maximum potential energy conditions within which a river can operate. They also constrain the way that energy is used, through their control on valley width and, hence, the concentration (or dissipation) of flow energy. Imposed boundary conditions effectively dictate the pattern of landscape units, thereby determining the valley setting within which a river behaves and/or changes. Drainage basin evolution over millions of years often provides a significant antecedent control on contemporary river forms and processes.

Lithologic controls upon sediment calibre and volume

The calibre and volume of sediment supplied to valley floors fashion the behavioural regime of rivers. Rivers can only move the sediments available to them. Lithology influences both the calibre and volume of available sediments. The mineralogical composition of any rock determines the texture and hardness of its weathering breakdown products. Hence, the regional lithology influences whether these materials are resistant to erosion. The lithology of any given place is a product of geologic history. Minerals derived from upper mantle materials make their way to the Earth's surface either directly via volcanic events or indirectly via subsurface (endogenetic) processes and subsequent removal of overlying materials. The enormous pressure and strain exerted by tectonic forces, and burial, induce metamorphic adjustment of igneous rocks and their reworked sedimentary counterparts. Weathering processes that break down parent rocks exert a significant influence upon the

other settings, however, geologic structure exerts a dominant influence on drainage pattern. For example, a *trellis* pattern is indicative of both a strong regional dip and the presence of folded sedimentary strata. Trunk streams flow along valleys created by downturned fold structures called synclines. Short tributaries enter the main channel at sharp angles approaching 90°. These tight-angle tributary junctions may induce short runout zones for debris flows. A *parallel* pattern is found in terrains with a steep regional dip or in regions where parallel, elongate outcrops of resistant rock impose a preferred drainage direction. Tributaries tend to stretch out in a parallel fashion following the slope of the landscape surface. In areas of right-angled jointing

mix of grain sizes that are available to be reworked by geomorphic processes. In river environments, many sediments along valley floors are derived from reworking of upstream sediment stores that have been derived from rocks with a completely different mineralogical composition, and associated range of weathering breakdown products.

Differing lithologic settings produce rivers with differing bed material sizes. Channels that are lined with large boulders and cobbles are not found in areas where the regional lithology generates materials that are very friable. Resistant materials, such as gneiss or marble, generate coarse-bed, bedload-dominated rivers. Rivers in granitic environments have a distinctly bimodal sediment mix, with coarse granules and sand on the bed, while floodplains are made up largely of silt–clay materials (these are mixed-load rivers with composite banks; see Chapters 6 and 7). Rivers that flow through sandstone are often remarkably clear because they lack fine-grained sediments that induce turbid flow. These streams have a uniform sediment mix of sand-sized materials and are characterised by non-cohesive banks. A stark contrast is evident along rivers in basaltic terrains, where the lack of coarse-grained materials results in turbid, muddy, suspended-load streams. Flow in many limestone (karst) terrains is ephemeral, and most of the sediment load is transported in solution. Hence, the mix of available grain sizes exerts a primary control upon whether the river operates as a bedload-dominated river, a mixed-load river, a suspended-load river, or a solution-load river (see Chapter 6).

The erodibility of bedrock also influences the volume of sediment that is supplied to a river system. Hard, resistant lithologies supply small amounts of sediment to rivers, resulting in supply-limited, bedrock-dominated landscapes (see Chapter 6). Such rocks often create steps along longitudinal profiles demarcated by waterfalls and oversteepened sections, along with narrow valleys. Rock hardness also affects the abrasive capacity of bed materials, influencing the rate of downstream decrease in grain size along a river. Weak, highly erosive rocks commonly oversupply a river with sediment, such that aggradation ensues in these transport-limited environments (see Chapter 6). Badland (gullied) environments commonly occur in such highly erosive rocks. The vast surface areas in these landscapes generate enormous volumes of sediment that result in aggradational valley floors (i.e. they are aggradational settings).

Tributary–trunk stream relationships

The spatial arrangement of tributaries in a river network exerts a primary influence upon process relationships at the catchment scale. By definition, a tributary is the smaller of two intersecting channels, and the larger is the trunk stem. The tributary–trunk stream catchment area ratio, the spacing between tributary confluences and the confluence intersection angle, among many considerations, determine the impact of tributaries upon the trunk stream. In some cases, tributary networks are too small to have a significant impact on flow and sediment inputs to the trunk stream, resulting in no change in its morphology. However, in other cases, tributary networks may have a significant impact on the morphology of the trunk stream. Tributaries that induce abrupt changes in water and sediment flux at confluence zones are called 'geomorphically significant (or effective) tributaries'.

In general terms, consistent flow-related morphological changes occur at junctions where the ratio between tributary size and trunk stream size approaches 0.6 or 0.7. Intersection angles tend to be acute. However if this angle approaches 90°, the likelihood of a geomorphic effect at a confluence increases. The cumulative effect of confluences within a catchment should be proportional to the total number of geomorphically significant tributaries. The confluence density (number of geomorphically significant confluences per unit area or per unit channel length) is related to drainage density and can provide a simple measure of the net morphological effect of confluences in river networks.

The drainage pattern of a catchment dictates the relative size and spacing of tributary networks (Figure 3.8a and b). Dendritic networks in heart-shaped or pear-shaped catchments instigate confluence effects throughout the catchment. Downstream increases in catchment width promote the coalescence of hierarchically branched channels. Larger tributaries that join downstream may have a geomorphically significant effect upon the trunk stream. In contrast, narrow, rectangular catchments with trellis networks lack larger tributaries. These networks have a small number of geomorphically significant tributaries. Also, the effectiveness of these similarly sized tributaries diminishes downstream, as their size is progressively smaller relative to the trunk stream.

Catchment configuration and network geometry influence the distance between geomorphically significant confluences. Large tributary junctions that are closely spaced may have confluence effects that overlap, particularly during large floods. In contrast, more widely spaced geomorphically significant tributaries exert a localised effect on factors such as downstream grain size (see Chapter 5). In general, as basin size increases, the channel length and area affected by individual confluence-related channel and valley morphological modifications increase. This measure can be used to determine how the degree

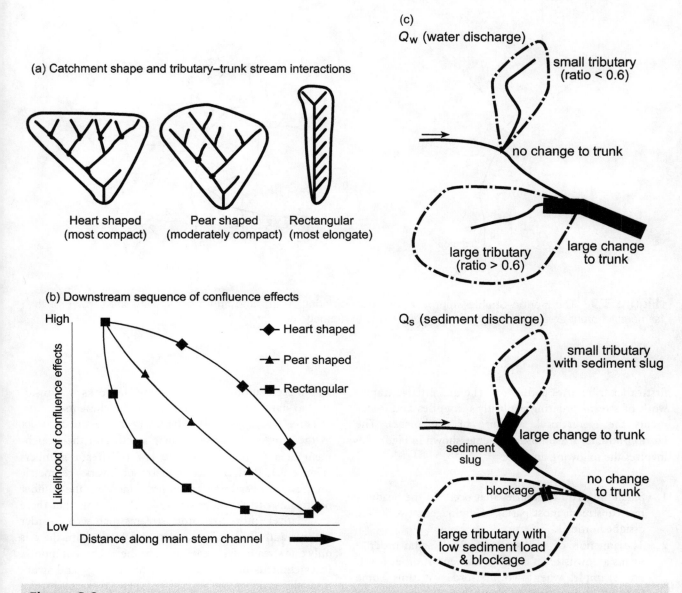

Figure 3.8   Tributary–trunk stream relationships in river systems. Geomorphically significant tributaries exert a major influence upon the flow and sediment characteristics of trunk streams. (a) Catchment shape dictates the length and subcatchment area of any tributary systems. (b) Relative catchment area of tributaries and trunk streams, and the pattern/frequency of confluence zones varies for catchments of differing shape and drainage pattern, with associated variability in tributary impacts upon the trunk stream as differing positions in a catchment. (c) Relative flow–sediment impacts, and the occurrence of blockages, affect the geomorphic effectiveness of tributary inputs. (a, b) Based on Benda *et al.* (2004). © American Geophysical Union. Reproduced with permission.

and spatial extent of disturbance events in tributaries (floods and changes to sediment supply) impacts upon trunk stream dynamics. If the catchment configuration is altered, for example by emplacement or removal of blockages such as dams, the significance of tributaries to overall flow and sediment flux can be altered considerably (Figure 3.8c).

Stream order

Stream order provides a measure of the relative size and pattern of channels within a drainage network. This exerts a significant influence upon the relative discharge of streams at any position in a drainage network. First-order streams have no tributaries, second-order streams only have

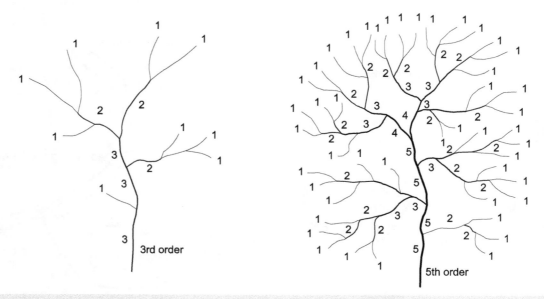

Figure 3.9   The Horton–Strahler approach to stream ordering. Third- and fifth-order streams are shown for the same catchment area, indicating a marked difference in drainage density.

first-order tributaries and so on. The quantitative framework of stream ordering explicitly recognises and documents the hierarchical structure of catchments. The Horton-Strahler stream order scheme, shown in Figure 3.9 involves the following analysis:

1.   Small, fingertip tributaries that occur in the headwaters (upstream most parts) of drainage networks are assigned order 1.
2.   The junction of two streams of the same order u forms a downstream channel segment of order $u + 1$. For example, when two first-order streams come together, the segment of channel downstream of the confluence is assigned an order of 2. If two second-order streams come together a third-order stream is formed downstream.
3.   The junction of two streams of unequal order u and v, where $v > u$, creates a downstream segment with an order equal to that of the higher order stream v. For example, if a second-order stream meets a third-order stream, no change in order results and the segment downstream of the confluence remains as a third-order stream.

This approach does not consider the relative change in channel size and discharge that occurs when smaller, lower order tributaries meet a larger order stream. Determination of stream order is highly dependent on the scale of analysis

and interpretations of where channel networks are considered to start in the headwater areas of catchments.

Three *laws of network composition* relate stream order to the number of streams, their length and their catchment area (Figure 3.10). The law of stream numbers (Figure 3.10a) is characterised by an inverse geometric progression whereby as stream order increases the number of streams of that order decrease. This means that there are relatively more first-order streams than second-order streams and third-order streams and so on until there is only one stream of a higher order at the catchment mouth. In catchments of relatively uniform lithology and structure, the ratio of the number of first- to second-order streams equals the ratio of the number of second- to third-order streams and so on. This is called the *bifurcation ratio* R_b. The higher the bifurcation ratio, the more frequently a drainage line splits into a tributary and trunk stream and the higher the drainage density. The law of stream lengths (Figure 3.10b) states that as stream order increases there is a direct increase in stream length for that order, such that first-order streams tend to be relatively short compared with streams of a higher order. The rate of increase in stream length typically lies between 1.5 and 3. Finally, the law of catchment area (Figure 3.10c) states that catchment area increases in a smooth progression with increasing stream order. The relative increase in stream length has a ratio of between 3 and 6. These various parameters provide a descriptive summary of basin network composition.

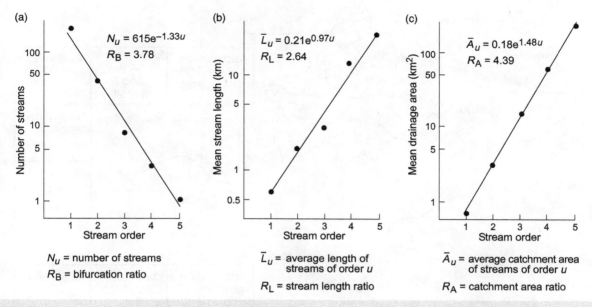

Figure 3.10 Hortonian laws of stream network composition. As stream order increases so the (a) number of streams of a given order decreases (i.e. there are many first-order streams and fewer fifth-order streams in a network), while (b) mean stream length and (c) catchment area increase.

In a sense, these measures of catchment morphometrics build upon an implicit assumption that the upstream or upslope parts of landscapes are connected to downstream or downslope areas. In many instances, however, this assumption does not hold entirely true. While many landscapes are effectively connected (or coupled), some are at best partly connected, while others may be disconnected. As noted in Chapter 2, these relationships are spatially and temporally contingent. Hence, different parts of the same system may have very different biophysical relationships.

The influence of catchment configuration upon flow and sediment flux

Figure 3.11 shows contrasting examples that demonstrate the influence of catchment morphometrics upon flow and sediment flux within the same region. The first example, the catchment drains an area of around 1800 km². The regional geology is made up of volcanic and metasedimentary rocks. The second example, the catchment drains a similar area and is made up of granitic rocks in the southern half of the catchment and metasedimentary rock to the north. Both catchments have relatively small areas of plateau landscape (tablelands) atop a 'Great Escarpment'. They are both fifth-order catchments. Longitudinal profiles have steep, concave-upward forms, with short sections atop the escarpment and gorge (steepness and concavity varia-

bles are very similar). Given their similar tectonic setting, catchment relief, relief ratios and hypsometric intervals are also very similar. In both instances, knickpoint retreat and valley expansion have been dominant long-term controls upon landscape and river evolution, However, the first example has a more elongate catchment, with a rectangular (joint controlled) drainage pattern (Figure 3.11a). In contrast, the second catchment drains an amphitheatre-shaped catchment, with a dendritic drainage pattern (Figure 3.11b). Catchment shape has a significant effect on the length of tributaries and their catchment area, and the spacing of tributary confluences along the trunk stream. The first catchment has short tributaries that join the trunk stream irregularly. The spacing between tributary confluences is long. This means that water and sediment are contributed irregularly and in small quantities. These tributaries are not geomorphically significant. As a result, the hydrograph is long and squat (see Chapter 4). The form factor is low, suggesting that flood intensity is relatively low. In contrast, the second catchment is round and contains long tributaries that drain large catchment areas. The spacing between tributaries along the trunk stream is short, such that significant inputs of water and sediment are regularly contributed. These tributaries are geomorphically significant and the resultant hydrograph is long and peaked (see Chapter 4). The form factor is high, suggesting a flashy flow regime. In summary, the second catchment has a much higher form factor and slightly higher drainage density

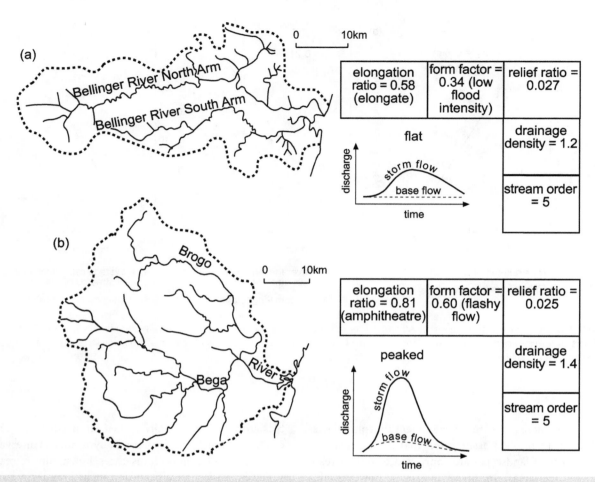

Figure 3.11   Schematic representation of morphometric impacts upon flow hydrographs for the (a) Bega and (b) Bellinger catchments, NSW, Australia (note, only major stream lines are shown on this figure).

than the first catchment. As a result, this sand-bed river is subjected to flashier (more variable) discharges than the lower flood intensity (less peaked) events that characterise the gravel-bed Bellinger River.

Conclusion

Efforts to read the landscape build upon meaningful analysis of catchment-specific morphometrics. Differentiation of source, transfer and accumulation zones provides helpful guidance in framing analysis of river systems. Appraising relations to longitudinal profiles and associated understandings of downstream changes in slope and valley width helps to explain the balance of erosion and deposition, and resulting river forms, at different positions in landscapes. These considerations, alongside catchment shape, size and tributary–trunk relationships, fashion the flux of water and sediment through a drainage network. Tectonic, lithologic

and climatic controls upon drainage density exert a primary influence upon the availability of materials to be distributed and their erodibility. Analyses of flow and sediment fluxes must consider how landscape components fit together at the catchment scale (i.e. their connectivity). Critically, site-specific investigations must be framed within their landscape and catchment context.

Key messages from this chapter

- A catchment is a hydrologically bounded landscape unit that is linked internally by a network of channels.
- Differing balances of erosion and deposition processes characterise sediment source, transfer and accumulation zones in catchments.
- Longitudinal profiles depict the change in elevation (and slope) of a river from its source zone to its mouth. Base level defines the level to which a river naturally

cuts. Rivers adjust their slope to develop a smooth concave-upward (graded) profile. Steep sections of long profiles, typically demarcated by waterfalls, are referred to as knickpoints.

- Moving down a concave-upward longitudinal profile, slope decreases while catchment area and sediment storage increase. The transition from source through transfer to accumulation zones is typically characterised by transitions from confined through partly confined to laterally unconfined valley settings. The dominance of erosion processes induces the formation of bedrock rivers with an imposed morphology in source zones. Depositional processes are dominant in accumulation zones, promoting the development of self-adjusting alluvial rivers. Erosion and deposition processes are approximately in balance in transfer reaches.

- Imposed boundary conditions determine the relief, slope and valley morphology (width and shape) within which rivers adjust. Lithologic controls exert a primary control upon sediment calibre and volume.

- Catchment shape, relief, drainage density and pattern and tributary–trunk stream relationships affect flow and sediment transfer in river systems.

- Stream order provides a measure of the relative size of channels and, therefore, the relative discharge of streams at any position within a drainage network. Laws of drainage network composition relate stream order to the number of streams, their length and their catchment area.

CHAPTER FOUR
Catchment hydrology

Introduction: what is hydrology?

Hydrology is the study of water movement through the hydrological cycle. *Hydraulics* is the study of the mechanics of water flow and the impelling forces that induce sediment movement in river systems (see Chapter 5). This chapter focuses on the hydrology of catchments. Before examining the processes by which water is supplied, transferred and stored in catchments, an introduction to the global hydrological cycle is presented. Measures of discharge, and the magnitude and frequency of river flows, are considered in the second half of the chapter.

The hydrological cycle

The hydrological cycle is a conceptual model that describes the storage and movement of water between the biosphere, atmosphere, lithosphere and the hydrosphere. Water can exist in liquid, gaseous and solid states. It is stored in various reservoirs, such as atmosphere, oceans, lakes, rivers, soils, glaciers, snowfields and groundwater. The hydrological cycle describes the movement of water between the atmosphere, oceans and land through energy and matter exchanges as it evaporates, precipitates and flows. Water moves from one reservoir to another by processes such as evaporation, condensation, precipitation, deposition, runoff, infiltration, sublimation, transpiration, melting and groundwater flow.

At the global scale the hydrological cycle is a closed system in which water is neither created nor destroyed. Much more water in the hydrological cycle is 'in storage' than is actually moving through the cycle. Oceans are by far the largest reservoirs, holding about 97% of all water on Earth (Table 4.1). The remaining 3% is freshwater, around 78% of which is stored in ice in Antarctica and Greenland, while a further 21% is groundwater, stored in sediments and rocks beneath the Earth's surface. Freshwater stored on land in rivers, streams, lakes and soil is less than 1% of the freshwater on the Earth and less than 0.02% of all the water on the Earth. River discharge comprises only 0.0001% of the world's water budget.

The pattern of water storage varies spatially and temporally. During colder climatic periods, water accumulation in snowfields, glaciers and ice-caps reduces the amounts of water in other parts of the hydrological cycle. The reverse is true during warm periods. During the last ice age (the last glacial maximum occurred 15 000–18 000 yr ago) glaciers covered almost one-third of Earth's land mass. As a result, the oceans were 120–150 m lower than today. During the last global 'warm spell' about 125 000 yr ago (i.e. interglacial conditions) the seas were about 5.5 m higher than they are today. About 3 million years ago the oceans were up to 50 m higher than current levels.

The *residence time* of a reservoir within the hydrological cycle is a measure of the average age of the water in that store (Table 4.2). Water stored in the soil remains there very briefly and is readily lost by evaporation, transpiration, throughflow or groundwater recharge. After evaporating, the average residence time of water in the atmosphere is about 9 days before it condenses and falls to the Earth as precipitation (global precipitable water vapour averages 25 mm, while global average annual precipitation is around 1000 mm; therefore, atmospheric water is completely recycled 40 times per year or every 9 days). On average, water

Table 4.1   Global distribution of water

Reservoir	Volume (10^6 km^3)	Percentage of total (%)
Oceans	1370	97.25
Ice-caps and glaciers	29	2.05
Groundwater	9.5	0.68
Lakes	0.125	0.01
Soil moisture	0.065	0.005
Atmosphere	0.013	0.001
Streams and rivers	0.0017	0.0001
Biosphere	0.0006	0.00004

Geomorphic Analysis of River Systems: An Approach to Reading the Landscape, First Edition. Kirstie A. Fryirs and Gary J. Brierley.
© 2013 Kirstie A. Fryirs and Gary J. Brierley. Published 2013 by Blackwell Publishing Ltd.

Table 4.2 Average residence times of water in differing reservoirs

Reservoir	Average residence time
Oceans	3 200 yr
Groundwater: deep	10 000 yr
Groundwater: shallow	100–200 yr
Lakes	50–100 yr
Glaciers	20–100 yr
Rivers	2–6 months
Seasonal snow cover	2–6 months
Soil moisture	1–2 months
Atmosphere	9 days

is renewed in rivers once every 16 days. Slower rates of replacement occur in large lakes, glaciers, ocean bodies and groundwater, where turnover can take from hundreds to thousands of years. Groundwater can spend over 10 000 yr beneath the Earth's surface. In many areas where humans are dependent on groundwater for domestic and agricultural purposes, extraction rates far exceed rates of renewal.

Operation of the hydrological cycle

There are four main components to the hydrological cycle: atmospheric water, precipitation, evaporation and transpiration, and surface water. *Atmospheric water* exists as water vapour, droplets and crystals in clouds. Once water vapour is in the air, it circulates within the atmosphere. When an air package rises and cools, the water vapour condenses to liquid water around particulates such as dust. These are called condensation nuclei. Initially, these condensation droplets are much smaller than raindrops and are not heavy enough to fall as precipitation, thereby creating clouds. As the droplets continue to circulate within the clouds, they collide and form larger droplets, which eventually become heavy enough to fall as rain, snow or hail. The volume of water in the atmosphere is very small. It varies with changes in temperature, pressure and geographical location. High-velocity winds in the upper atmosphere are able to move water vapour long distances in a relatively short period of time.

Precipitation is defined as the transfer of water from the atmosphere to the Earth's surface in the form of rain, freezing rain, drizzle, sleet, ice pellets, snow or hail. The distribution of precipitation on the Earth's surface is generally controlled by the presence or absence of mechanisms that are able to lift air masses, thereby causing saturation to occur. This is also influenced by the amount of water vapour held in the air, which is a function of air temperature. Some 86 % of evaporation and 78 % of all *rain* occurs

over the ocean (14 % of evaporation and 22 % precipitation falls on land). Hence, the land receives a net moisture donation from the oceans. Precipitated water over terrestrial surfaces is taken up by soil and plants. Some percolates into groundwater reservoirs and some falls on glaciers and snowfields where it may accumulate as glacial ice. Some accumulates in lakes. Ultimately, however, the deficit in precipitation between the land surface and oceans is accounted for by *global runoff* from streams and rivers.

Evaporation is the term used to describe the phase change from liquid water to water vapour. This process is driven by the energy provided by the sun. The rate of evaporation increases with increasing temperature and wind speed and decreases with increasing humidity. Evaporation rates vary for large and small bodies of water, soils and vegetation. Most evaporation occurs over oceans, which cover more than 70 % of the Earth's surface. On average, the depth of the world's oceans is about 3.9 km. However, maximum depths extend beyond 11 km. In the Southern Hemisphere there is four times more ocean than land, whereas the ratio between land and ocean is almost equal in the Northern Hemisphere. Of the evaporated water in the atmosphere that is supplied from oceans, only 91 % is returned to ocean basins by way of precipitation. The remaining 9 % is transported to areas over landmasses where climatological factors induce the formation of precipitation.

The imbalance between rates of evaporation and precipitation over land and ocean is corrected by runoff and groundwater flow to the oceans. Surface flow makes up a small proportion (<10 %) of all water that travels from the land to the oceans. Most of this surface flow returns to the ocean as overland flow and streamflow. Only 5 % returns to the ocean by means of slow-moving groundwater.

Transpiration is the loss of water by plants. This loss occurs through stomata, which are open during the day to enable the absorption of carbon dioxide needed for photosynthesis. Transpiration rates depend upon the temperature, humidity and wind speed near the leaves of plants. Since plants draw water from the soil, transpiration rates can greatly affect soil moisture content. Soil water loss results from both transpiration and evaporation and is termed *evapotranspiration*. In general, four factors control the amount of water entering the atmosphere via evapotranspiration: energy availability, the humidity gradient away from the evaporating surface, the wind speed immediately above the surface and water availability.

When rain hits the Earth, drops are either *intercepted* by vegetation or more impervious surfaces such as roofs or roads, or they *infiltrate* into the soil (Figure 4.1). Infiltrated water moves through the pores of the soil until the soil becomes saturated. Infiltration rates lessen with soil saturation, leading to surface flow. Once infiltrated, water continues to filter through soil or rock through vertical movement.

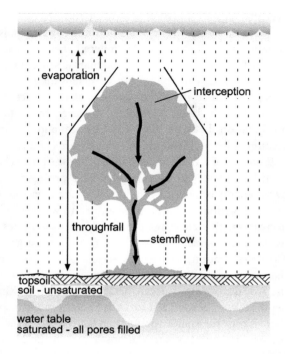

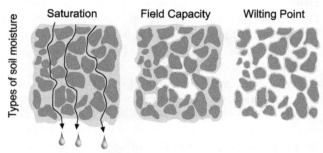

Figure 4.1　Flow interception, stem flow and through-fall, and visual representations of the three states of soil moisture.

This is called *percolation*. Percolation transfers water from the soil layer to the groundwater. This is usually seasonal. It occurs only when the soil is saturated and when roots and evaporation do not result in a net movement of soil water towards the surface.

Vegetation intercepts some of the falling rain. Some of the water is stored in the canopy as *interception storage*. Interception storage is inversely related to rainfall intensity such that the gentler the rainfall, the more storage occurs on the plant. The volume of interception storage is dependent on vegetation type and leaf density. For example, crops may only intercept 10–20% of rain, whereas forests may intercept up to 50% of rain or 100% of drizzle. Interception storage plays only a minor part in intense or long-duration storms, where the storage capacity of the vegetation decreases. Some of the water that is intercepted never makes it to the ground surface. Instead, it evaporates

from the vegetation surface directly back to the atmosphere (*interception loss*). A portion of the intercepted water can travel from the leaves to the branches and then flow down to the ground via the plant's stem. This is called *stemflow*. Another portion of the precipitation may flow along the edge of the plant canopy to cause canopy drip. Both of these processes can increase the concentration of water added to the soil at the base of the stem and around the edge of the plant's canopy (see Figure 4.1). Rain that falls through the canopy onto the soil surface, without being intercepted, is called *throughfall*. Water that falls onto the Earth's surface either directly or indirectly may result in *rain splash*. Rain splash is an important erosion process on hillslopes. Dislodged soil particles may become suspended in the surface runoff and carried into streams and rivers. Raindrop size affects the potential for surface soil erosion.

Several factors influence the amount of precipitation that soaks into the soil: the amount and intensity of precipitation, the prior condition of the soil, hillslope angle and the presence of vegetation. These factors can interact in sometimes surprising ways. For example, very intense rainfall onto very dry soil in desert environments may not soak into the ground at all because of a surface crust, creating flash-flood conditions. Water that does soak in becomes available to plants. Plants take up water through their root systems; the water is then pulled up through all parts of the plant and evaporates from the surface of the leaves via transpiration.

Infiltration is the movement of water into the soil. Once infiltrated, this water is referred to as *soil moisture*. The rate of infiltration (termed *infiltration capacity*) is a function of soil type. The rate of infiltration is lower for fine-grained soils with smaller pore sizes, relative to coarser grained materials. Sandy soils can have infiltration rates >20 mm h^{-1}, whereas heavy clay soils may have infiltration rates of only 1–5 mm h^{-1}. Additional controls on the rate of infiltration include precipitation intensity, porosity, the presence or absence of hydrophobic substances, the amount of organic matter, soil compaction, slope angle and soil moisture at the time of rain. Antecedent soil moisture is a key factor that determines when and where the infiltration capacity of the soil will be exceeded such that runoff is generated (see below). If a soil is already wet from a previous storm, the infiltration capacity of the soil is reduced. Any subsequent storm will have a lower infiltration capacity, as soil pores are filled.

Between storms, infiltration water percolates through the soil. This produces a wetting front down the soil profile. Depending on the time since the storm and the rate of infiltration, wetting fronts may take several weeks to percolate through a soil profile. *Saturation* occurs when soil is at its maximum retentive capacity and soil pores are filled with water. Saturation tends to be short-lived, as a portion of this water quickly drains away as gravitational water.

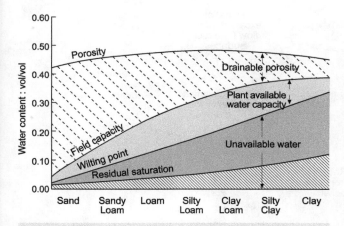

Figure 4.2 Variability in soil moisture content with textural soil types. The finer the grain size, from sand to clay, the less readily a soil is able to transfer water. For a given water content, one soil may be at field capacity (e.g. a sand) while another may be saturated (e.g. a clay).

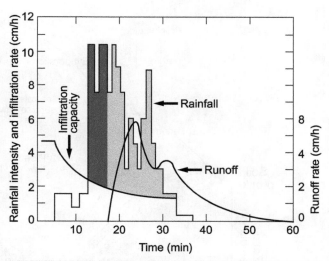

Figure 4.3 Schematic representation of the relationship between rainfall, infiltration and runoff. Rain starts at time $T = 5$ min. At this time the infiltration capacity of the soil is at its highest. At around $T = 12–18$ min the rainfall intensity is at its greatest. As water fills soil pores, the infiltration rate steadily decreases, before plateauing. The dark area represents rainfall that enters the soil via infiltration. Around this time, $T = 18$ min, all excess rain (light grey shade) is converted to runoff. The peak in runoff at $T = 22$ min is lagged behind rainfall because of the infiltration effects of the soil. Runoff may still occur up to 30 min after rain has ceased as soil water is contributed further down the hillslope/catchment. Modified from Dunne and Leopold (1978).

This condition tends to occur during heavy rainfall events or when soils are irrigated. *Field capacity* is the amount of water that is held in soil after it has been fully wetted and all gravitational water has been drained away. The field capacity of a soil is reached faster in a coarser textured soil (e.g. sand) than in a fine-textured soil (e.g. clay). This state is reached 1–2 days after heavy rainfall or irrigation ceases. At field capacity the soil holds the maximum amount of water that can be stored and used by plants. If soil moisture drops to the *wilting point*, water is not available to plants and they wilt. Water is retained in micropores and in very thin films around soil particles, but adhesive forces make this water unavailable for use by plant roots and microbes.

Saturation point, field capacity and wilting point are all influenced by soil texture. For example, the field capacity in one soil type (e.g. sand) might be the wilting point in another (e.g. loam) and saturation in another (e.g. clay) at the same water content (Figure 4.2). Ultimately soil type, textures and thickness, and the extent of groundcover on hillslopes influence the volume and rate of water transfer to river channels.

Runoff generation

Drainage basins are by definition closed to inputs of surface water (see Chapter 3). Thus, the number of inputs is minimised or essentially reduced to one, precipitation, although interbasin transfer of groundwater may occur.

The transfer of water from the land surface to the ocean is termed *runoff*. Runoff occurs when the soil can no longer absorb the water that is made available to it (Figure 4.3).

The soil is infiltrated to full capacity and excess water from rain, snowmelt or other sources flows through the soil or over the land surface. Depending on antecedent soil moisture, most runoff occurs during the most intense phase of rainstorms. The *runoff ratio*, which describes the fraction of precipitation that appears as runoff, is largely dependent upon soil moisture content. Soil moisture content must exceed a threshold before any significant runoff occurs. Runoff eventually makes its way into streams and rivers creating discharge.

There are four types of runoff (Figure 4.4): throughflow, pipeflow, infiltration-excess overland flow (Hortonian overland flow) and saturation-excess overland flow. Each of these can contribute water to a river channel.

Throughflow is defined as the rate of movement of flow through the subsurface soil matrix. Infiltrated water flows through the soil matrix in the intergranular pores and small structural voids (termed micropores). The water moves laterally downslope under gravity and can occur when the soil is saturated or unsaturated. Rates of throughflow vary with soil type, slope gradient and the concentration

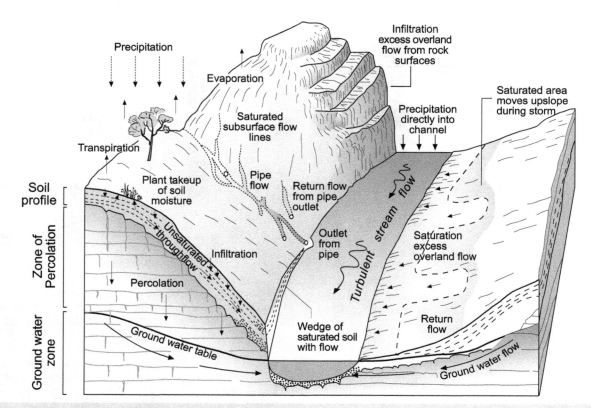

Figure 4.4 Block diagram showing how throughflow, pipeflow, infiltration-excess overland flow (Hortonian overland flow) and saturation-excess overland flow interact in a landscape. Different types of overland flow occur at different positions in a landscape. Infiltration occurs into soil profiles and flows through soil pores as throughflow. Where pipes occur in the soil, water travels faster through the soil profile and downslope. Infiltration-excess overland flow occurs in areas with thin soil profile or bedrock. This occurs when the intensity of the rain exceeds the capacity of the substrate to infiltrate water. In low-lying areas at the base of hillslopes, or in floodplain zones, or in areas of return flow, saturation-excess overland flow is common. This occurs when the soil profile becomes saturated and additional rain runs over the surface. Modified from Selby (1993). © Oxford University Press. Reproduced with permission.

of water in the soil. In permeable geologies, water may be lost through percolation and eventually recharge groundwater. The hydraulic conductivity k of a soil is a measure of how fast the soil transmits or transfers water. Typical values of saturated hydraulic conductivity vary from 0.3 to 300 mm h^{-1}. Hydraulic conductivity is commonly measured using Darcy's law:

$$v = kA\frac{h}{l}$$

where v (m s^{-1} or mm h^{-1}) is the flow rate/velocity, A (m^2) is the cross-sectional area of the aquifer and h/l (m m^{-1}) is the slope of the water table (i.e. the hydraulic gradient). Throughflow may feed streams during dry conditions as soil water is slowly released into channels. This is called the maintenance of *baseflow* (see later).

Pipeflow occurs when water flows through large voids or macropores below the surface. Macropores include pipes that are open passageways in the soil caused by decaying roots and burrowing animals. Macropores also include larger structural voids within the soil matrix that serve as preferential pathways for subsurface flow. Infiltrating water follows preferential pathways and macropores and may result in increases in moisture content at depth before saturation or similar increases in moisture content occur higher in the soil profile. The overlying surface of a pipe must be strong enough to support the roof and the walls. Hence, piping is best formed in cohesive, fine-grained sediments. Pipes may be between 0.02 m and >1 m in diameter and several metres to >1 km in length. Flow through pipes is more rapid than throughflow, occurring at rates of 0.005–500 mm h^{-1}. Rapid lateral flow through a network of macropores and the effusion of old water into stream channels is the primary mechanism for runoff generation in many humid regions where overland flow is rarely observed. Piping is often a precursor to rill and gully development, and is commonly observed in highly dissected badland environments characterised by very high drainage densities.

Infiltration-excess overland flow (also called Hortonian overland flow after R.E. Horton, or unsaturated overland flow) occurs when rainfall intensity or rate is greater than the infiltration capacity of the soil. The soil is unable to absorb water quickly enough and excess rainfall accumulates on the soil surface in small depressions. Once these depressions are filled, the water spills out and runs off over the surface as overland flow. The amount of water stored on the hillside is called *surface detention*. The transition from depression storage to surface detention and overland flow is not sharp, as some depressions may fill and contribute to overland flow before others.

Hortonian overland flow is often shallow, sheetlike and fast moving. As such, it brings about extensive erosion of soil and bedrock by interill and rill flow. This occurs more frequently in areas where rainfall intensity is high and soil infiltration capacity is low, typically as a result of soil type, vegetation and land use. Hortonian overland flow is widespread in desert landscapes, where infiltration rates are low because of thin vegetation cover, shallow soils and abundant bedrock outcrops. It is much less common in humid landscapes, where thick vegetation and deep soils favour high infiltration rates. Exceptions occur where the natural vegetation cover has been thinned or removed. Destruction of vegetation permits raindrop impact to seal the soil surface and lower infiltration rates without saturating the soil. Infiltration excess overland flow is also common in steep terrains with thin soils, in areas of compacted or impervious surfaces (e.g. urban areas) and in areas of frozen or burnt ground. In cold environments with frozen ground, infiltration capacity is reduced to zero, producing runoff during storm events. In contrast, fire results in water repellency by soils which may reduce infiltration capacity for months to years. The heat from fire also removes the thin films of water adhered to soil particles, disconnecting potential flow paths and limiting penetration of water into soil micropores.

Saturation-excess overland flow occurs where subsurface soil is saturated, the water table rises to the surface and water flows over the surface. Once 'saturation from below' occurs, all further surface water input becomes overland flow. This form of runoff occurs in two ways: by rain falling onto already saturated areas or by return flow. The extent of the saturated area in a catchment varies between storms. This results in expansion and contraction of the runoff zone as parts of the catchment become saturated and then dry out. In general, areas at the bases of hillslopes or near stream areas with flat topography are particularly susceptible to saturation excess overland flow. At the base of hillslopes, the zone of saturation moves upslope during a rainfall event as the valley bottom and footslopes become saturated (Figures 4.4 and 4.5).

Return flow occurs where throughflow is forced back to the soil surface (exfiltrated) at areas of saturation and becomes overland flow. Return flow may result in the formation of springs. As shown in Figure 4.5, return flow typically occurs at four places: (1) at hillslope concavities, where the hydraulic gradient inducing subsurface flow from upslope is greater than that inducing downslope transmission; (2) at thin/less-permeable soil boundaries, where a sudden decrease in soil water storage capacity occurs; (3) in hillslope hollows, where subsurface flow lines converge and water arrives faster than it can be transmitted downslope as subsurface flow; and (4) on lower hillslopes and at the channel–valley margin interface, where ground is already saturated, the shallow water table is close to the surface and the saturation zone expands.

Groundwater flows

Groundwater flows occur beneath the surface beyond the soil-moisture root zone. Groundwater is the zone in the ground that is permanently saturated with water. In river systems this includes the hyporheic zone beneath the channel and the parafluvial zone beneath active floodplain compartments. The top of groundwater is known as the water table. This defines the boundary between aerated rocks and soils and the saturated zone. Excess surface water moves through soil and rock until it reaches the water table. Groundwater flows occur from areas with a higher water table to areas down slope where the water table is lower. A distinction can be made between shallow groundwater flows through surficial materials and deep groundwater flows through underlying bedrock. An aquifer is a permeable layer of rock which can both store and transmit large amounts of groundwater. Some infiltration stays close to the land surface and can seep back into surface-water bodies (and the ocean) as groundwater discharge. Some groundwater finds openings in the land surface and emerges as freshwater springs. Aquifers can take thousands or millions of years to recharge naturally.

In many instances the terrestrial movement of water within the hydrological cycle is scarcely perceptible, as it is concealed from view at the Earth's surface. In terms of fluvial geomorphology, far greater account is given to the movement of water within river channels, and associated processes of erosion and sedimentation. The effectiveness of these processes is fashioned by, and in turn fashions, channel formation.

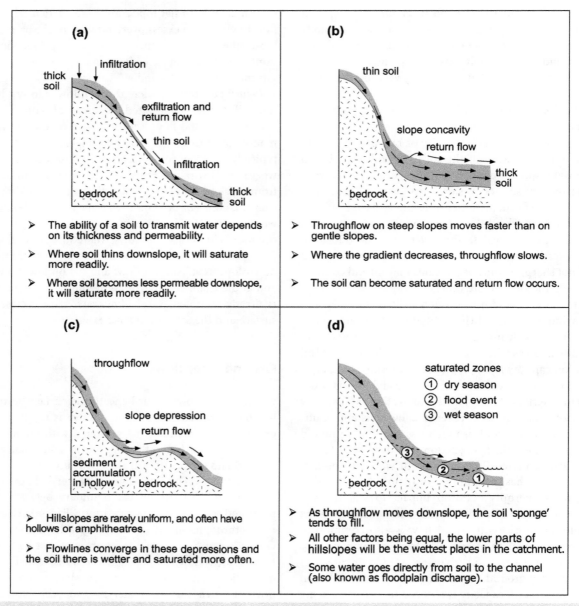

Figure 4.5   Forms of return flow. Return flow occurs where throughflow is forced back to the soil surface (exfiltrated) when it is saturated. Four scenarios are shown: return flow associated with (a) changes in soil thickness, (b) soil concavities, (c) hollows and (d) saturated zones on valley bottoms.

Catchment-scale runoff and discharge generation models

Not all forms of runoff are active at the same time or during different rainfall events. In any given catchment, runoff will be produced from *variable source areas* by a range of mechanisms that are activated at different times during a storm. This means that some areas of a catchment are more likely to generate runoff than others under certain conditions. Antecedent soil moisture is a key determinant

of the variable source area involved in the generation of saturation overland flow. In this *partial area model* of runoff generation (Figure 4.6) it is often the case that 90% of the surface runoff in a catchment is generated from 10% of the hillslope area, depending on storm size. The longer and more intense a rain event is, the larger the area of the catchment that will generate runoff and the more active certain types of runoff generation will become.

The main processes involved in runoff generation are summarised in Figure 4.7. Most rainfall–runoff models

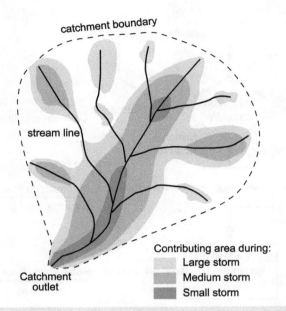

catchment boundary

stream line

Catchment
outlet

Contributing area during:
Large storm
Medium storm
Small storm

Figure 4.6 The partial area model of runoff generation. During small storms only small areas of the catchment are saturated and generate runoff. However, as storm size increases (either more rain falls or rain lasts longer), the total area that contributes water via various forms of runoff increases.

partition flow according to whether it is derived by infiltration excess, saturation excess and groundwater/baseflow pathways. Surface water inputs may generate runoff from either infiltration excess or saturation excess. Infiltration excess pathways are often the first to be activated, particularly during intense storms when rainfall inputs are greater than infiltration capacity (Figure 4.7a). Runoff over the surface can be widespread or localised (Figure 4.7b). Once the soil is saturated and further rain falls on the already saturated area, saturation excess overland flow and return flow pathways are activated. Depending on the timing of saturation in different parts of the catchment, water may be contributed from variable source areas at different rates; hence the partial area model of runoff generation. Depending on antecedent soil moisture and local conditions, both infiltration excess and saturation excess flow may occur at the same time. Infiltrated water enters the soil regolith where it activates throughflow and pipeflow pathways and percolates into deeper groundwater. These processes occur over different timeframes, contributing baseflow to channels during dry times. The character of the soil profile and the structure of the underlying topography may create preferential subsurface flow paths that also contribute runoff over variable timeframes.

Each mechanism of runoff generation has a different response to rainfall or snowmelt in the volume of runoff produced, the peak discharge rate and the timing of contributions to streamflow in the channel. The relative importance of each process is affected by climate, geology, topography, soil characteristics, vegetation and land use. The dominant process may vary between large and small storms.

Channel initiation

The initiation of channels via interrill and rill flow may occur on remarkably steep hillslopes close to drainage divides, as small channels are cut into colluvial hollows (accumulations of hillslope-derived sediments) within zero-order basins. The headcut that demarcates the initiation of channels marks a key transition from overland flow mechanisms to channelised flow.

Interrill flow, also known as sheet flow, sheet wash or slope wash, generally appears as a thin layer of water with threads of deeper, faster flow diverging and converging around surface protuberances, rocks and vegetation (Figure 4.8). As a result of these diverging and converging threads, flow depth and velocity may vary markedly over short distances, giving rise to changes in flow turbulence (see Chapter 5). Erosion by interrill flow involves soil detachment and sediment transport. Soil particles may be detached by a variety of processes, but diffusive ones, such as raindrop impact, frost action and animal activity, dominate. Raindrop impact is the most widespread of these processes. Raindrop detachment rates are positively correlated with rainfall intensity and soil erodibility and negatively correlated with plant canopy and ground cover.

Rill flow is deeper and faster than interrill flow and is typically turbulent. Rill flow occurs in small hillslope channels. These rills are commonly only a few centimetres wide and deep but grade into gullies (Figure 4.8). The boundary between rills and gullies is necessarily arbitrary, but one that has been widely adopted specifies that gullies are wider than 0.33 m and deeper than 0.67 m. Rill depth usually reflects the degree of rill development or depth to a resistant layer in the soil. Small rills exist for only short distances before they are obliterated by other hillslope processes. Large rills, on the other hand, may persist for decades. Rill flow is typically too deep for raindrop impact to have a significant effect on either soil detachment or sediment transport. As a result, sediment transport in rills is a function of sediment availability and erodibility and of the energy of the flow within the rill/gully channels. As such, these systems have characteristics of channelised flow observed in river channels.

Gully and channel formation

Once flow over a surface becomes sufficiently concentrated, rills and preferential drainage lines can become incised,

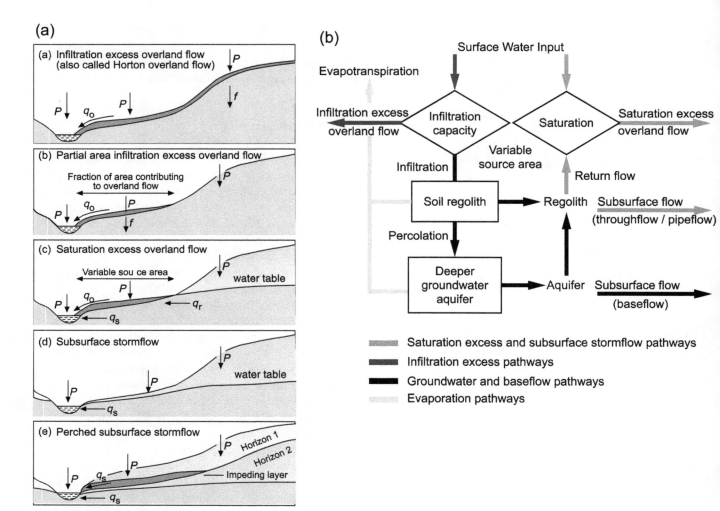

Figure 4.7 Runoff generation. (a) Mechanisms of runoff generation reflect interactions among precipitation *P*, infiltration *f*, overland flow q_o, saturated throughflow q_s and return flow q_r. From Bevan (2001). © John Wiley and Sons, Ltd. Reproduced with permission. (b) Hydrological pathways involved in runoff generation vary for different types of runoff between infiltration-excess pathways, saturation-excess and subsurface stormflow pathways and groundwater (baseflow) pathways. Modified from Tarboton (2003).

forming *gullies* when they cut into colluvial materials on hillslopes and *channels* if they cut into alluvial materials on valley floors (Figure 4.8). The latter may also be called channelised or entrenched streams. These features may be longitudinally discontinuous during the initial stages of their formation. This longitudinally disconnected channel terminates in fan-like forms (or floodouts) that sever the upstream–downstream transfer of sediments (i.e. these are disconnected landscapes; see Chapters 2 and 14).

Phases of gully and incised channel formation reflect cut-and-fill cycles or stages (Figure 4.9). In the initial incision phase, flows breach an unchannelised surface, cutting into relatively loose surficial materials and/or regolith. This may reflect an area of local oversteepening of the valley floor or an area with lower resistance. It may also occur in

response to a fall in base level. This is a threshold-driven phenomenon. Development of a primary headcut (or knickpoint in bedrock – see Chapter 3) incises and lowers the channel bed. This forms a deep, narrow channel with a steep headwall at its upstream end. Headcuts mark an abrupt change in elevation and slope on the longitudinal profile. Concentration of flow energy at the headwall produces a plunge pool that accentuates erosion at the base of the bank. As flow falls over the headwall, sediment is eroded such that upstream migration of the headcut occurs. This results in upstream extension of the channel network. As a headcut passes and the drainage network is extended upstream, the channel expands and greater volumes of sediment are transferred downstream. Secondary headcuts may develop along the channel bed as it adjusts to the new

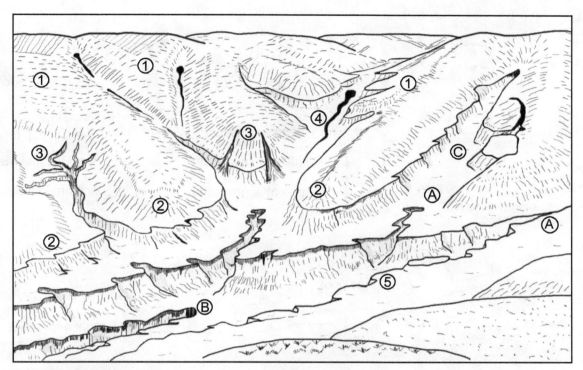

Types of drainage lines

① interrill ② rill ③ gully ④ incised channel - discontinuous

⑤ incised channel - continuous

Incised channel forms

Ⓐ primary headcut Ⓑ secondary headcut Ⓒ intact valley fill

Figure 4.8   Channelised flow. Rills and gullies develop on hillslopes, whereas incised and discontinuous channels form on valley floors. Incision is initially triggered by a primary headcut. Subsequent bed level adjustments are induced by secondary headcuts. Modified from Schumm *et al*. (1984) (Fig 6.7). © Water Resources Publications. Reproduced with permission.

base level. The channel expands via sidewall erosion processes. Mechanisms include: fluting, via rain splash producing rills in the channel bank; wall failure, as sections of bank topple and slump into the channel; seepage, where piping and tunnelling remove sediment from the bank; and overfalls, where undercutting induces bank collapse. These processes create a wider channel that is able to accommodate more flow. Eventually the channel enlarges to such a degree that it can no longer maintain sediment transport and then infilling and aggradation of the channel bed occurs. In some cases the channel may be refilled to form an intact valley floor once more. Repeated phases of these cut-and-fill cycles create distinct bands of sediment layer on valley floors.

The width and depth of a channel change markedly at different stages of cut-and-fill cycles. Once a channel incises and deepens, flow energy is increasingly concentrated and

notable widening occurs. Subsequent aggradation phases are also associated with channel expansion and an accompanying increase in the width/depth ratio. As flows are increasingly dispersed over the wider channel bed, bed level continues to aggrade and channel depth is further reduced and the width/depth ratio is increased further. Eventually, flow energy may become dissipated across an intact valley floor once more. These cut-and-fill processes occur in environments with a non-perennial flow regime.

Flow regimes of perennial, intermittent and ephemeral rivers

Perennial, intermittent and ephemeral rivers can be differentiated on the basis of their frequency of flow (Figure 4.10). Perennial streams have continuous flow throughout

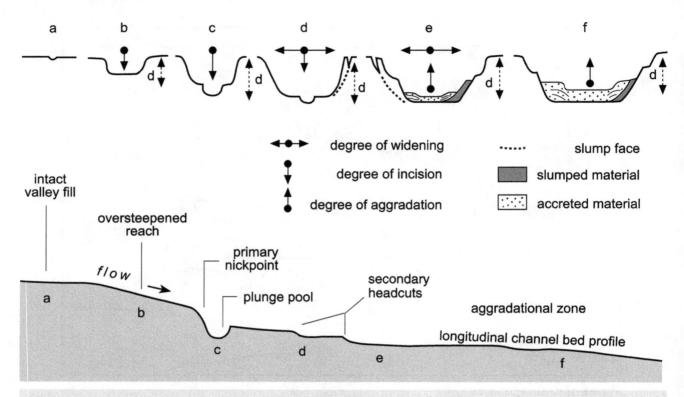

Figure 4.9 Incised channel development via incision and headward retreat. Incision into intact valley floors occurs due to factors such as lowering of base level or oversteepening of the valley floor. A preferential drainage line is developed on the surface (section a). Once a primary headcut develops, channels incise into the surrounding alluvium and become narrow and deep (sections b and c). A plunge pool develops at the base of the headwall (section c). To accommodate the increased discharge, the channel expands laterally through sidewall and bank erosion processes, producing a wide, deep channel (section d). Once the transport capacity of the channel to carry sediment is exceeded, sediment is deposited on the channel bed and aggradation occurs (section e). Eventually, channel expansion ceases and bank-attached, inset floodplains may develop (section f). Modified from Schumm *et al.* (1984). © Water Resources Publications. Reproduced with permission.

the year. An intermittent stream dries up from time to time on an irregular basis. Ephemeral streams are dry most of the year, but may have seasonal flow. As most large rivers flow through different climatic and topographic zones they have subcatchments that are both perennial and ephemeral. This is especially evident for rivers that span large areas/latitudes, and have areas of both high and low altitude (where precipitation patterns are markedly variable). Collectively, these various considerations determine the discharge regime of river systems.

Discharge and the magnitude/frequency of flow in river systems

Discharge, Q, is measured as the volume of water that passes a particular channel cross-section point every second ($m^3 s^{-1}$), as defined by the *continuity equation*:

$$Q = wdv$$

where Q ($m^3 s^{-1}$) is the discharge, w (m) is the width of the channel (or flow in stage-dependent analyses), d (m) is the depth of the channel (or flow in stage-dependent analyses) and v ($m s^{-1}$) is the velocity of the flow (i.e. its speed)

Channel cross-sections are normally presented from left to right bank looking downstream. Channel width is normally measured perpendicular to the channel bank from 'top-of-bank' to 'top-of-bank' (Figure 4.11). Depth is normally measured to the deepest part of the channel. Channel slope is measured using the rise over run method (see Chapter 2). Procedures used to measure velocity are outlined below.

The continuity equation describes how channels can adjust their size and shape in response to the available flow and prevailing sediment conditions. Flow affects channel size through a suite of mutual adjustments. Channel size,

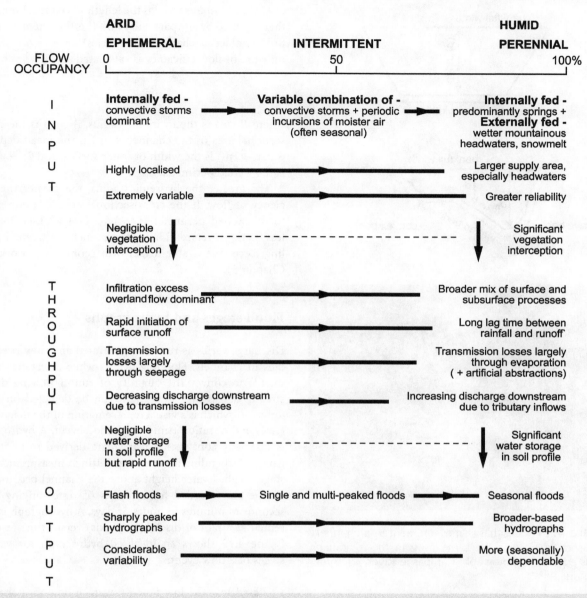

Figure 4.10   Contrasts in hydrological input, throughput and output between ephemeral, intermittent and perennial rivers. Ephemeral rivers are characterised by localised storms, rapid initiation of surface runoff and flash-flood conditions. In contrast, perennial rivers are often fed by reliable water inputs such as snowmelt or recurrent rain in the headwaters. Recurrent water inputs result in increasing discharge downstream. Sustained flows may occur over seasonal and longer timeframes. Modified from Knighton (1998).

in turn, affects the way in which flow energy is utilised and channel boundaries are altered. The extent to which a balance is reached reflects the ease with which the channel is able to adjust and the frequency, periodicity and variability of flow. Channel boundaries that are comprised of readily mobilised sediments and relatively regular flow are much more likely to demonstrate mutually adjusting equilibrium relationships as described by the continuity equation than are rivers with resistant (i.e. irregularly reworked) boundaries. In the latter instance, channel morphology is

likely to be imposed by bedrock or coarse alluvial sediments, or the resulting form is determined by infrequent high-magnitude events.

The level at which flow fills a channel to the top of the banks before flowing overbank is called the *bankfull stage* (Figure 4.11). Flow energy is concentrated at this stage. However, once flow spills onto the floodplain, energy is dissipated. Bankfull stage flow perform most geomorphic work. These are the most efficient sediment transporting events, as energy losses are minimised. In some instances

(a)

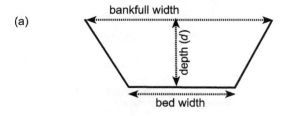

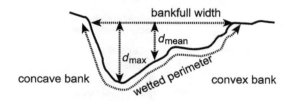

(b)

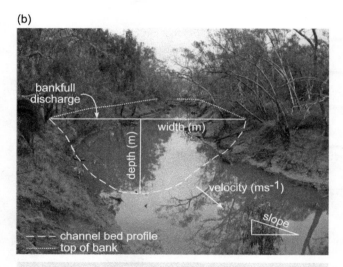

Figure 4.11 Discharge measurement in a channel cross-section d_{mean} is the same as y in the hydraulic radius equation. Photograph: Campaspe River, Victoria (K. Fryirs).

Hydraulic radius records the length of channel boundary that is wetted by any particular flow (i.e. the extent of solid–fluid contact at channel boundaries) (Figure 4.11). This measure of flow efficiency is calculated as:

$$R = \frac{A}{2y + w}$$

where R (m) is the hydraulic radius, A (m^2) is the cross-sectional area of the channel, y (m) is the mean depth of water, w (m) is the width of water surface and $2y + w$ (m) is the wetted perimeter.

The greater the hydraulic radius, the greater the efficiency of flow. If the cross-sectional area is large relative to the wetted perimeter (i.e. $A \gg 2y + w$) then there is less frictional retardation in the channel and greater impelling force that is able to perform geomorphic work (see Chapter 5).

Flood stages and hydrographs

Discharge varies as runoff is generated and flow rises and falls in channels. Flooding occurs when a watercourse is unable to convey the quantity of runoff flowing downstream. Stages of a flow event can be depicted on a *hydrograph*, which is a visual representation of the magnitude (size) and duration (time) of a flood event. A hydrograph is normally constructed using data derived from stream gauges. Depending on the gauge settings, measurements of stage height (water height above the channel bed) or discharge are recorded for intervals of time ranging from seconds to minutes to hours or days. A hydrograph is presented as a plot of discharge (y-axis) versus time (x-axis). Figure 4.12 shows an idealised hydrograph for various stages of a flow event:

- *Baseflow* – the low flow or average flow in a channel. This is derived from the slow drainage of soil water or groundwater. Baseflow is important for the maintenance of refugia for fauna and flora (e.g. in pools and waterholes).
- *Stage 1, flood pulse* – the first increase in flow above baseflow conditions. Infiltration induces a lag time between rainfall, runoff generation and the flood pulse. This lag time can be short (hours in the case of a flash-flood in steep terrain) or weeks/months (in the case of an ephemeral river that lies hundreds of kilometres from its source where a flood wave will be generated).
- *Stage 2, rising stage* – a steep increase in the volume of water being carried by a channel. Flow becomes more turbulent.
- *Stage 3, bankfull stage* – the channel is filled to the top of the banks (without spilling onto the floodplain).

it may be difficult (or impossible) to define bankfull stage (e.g. in confined valleys). For example, several levels of inset floodplain or bench may produce a compound river channel (see Chapter 7). Also, bankfull stage may change over time.

Bankfull discharge often coincides with the minimum *width/depth ratio* flow stage, which equates to the highest energy condition. The *width/depth ratio* of a channel is measured as bankfull width (top of bank to top of bank) divided by maximum channel depth. For example, if the ratio is 5, this means that the width of the channel is five times the channel depth and the channel is considered to be narrow and deep. A width/depth ratio of 20 means the channel is 20 times wider than it is deep and it is said to be wide and shallow.

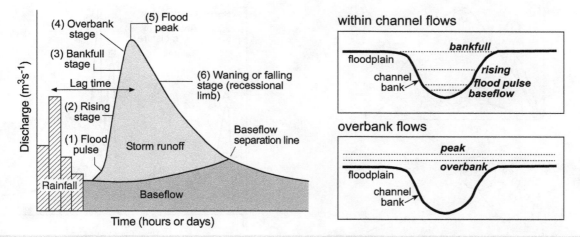

Figure 4.12 Flow stages on a hydrograph. Flow is often lagged behind the most intense phases of rainfall because of infiltration and throughflow effects. Lag time is measured as the time between peak rainfall and peak discharge. Storm runoff occurs when discharge increases above the baseflow condition. Six stages of flood flow are differentiated on the hydrograph and as flow through channel cross-sections.

Flow efficiency is maximised at this stage (i.e. energy loss is minimised).

- *Stage 4, overbank stage* – flow spills onto the floodplain, where it spreads in a sheetlike form, dissipating energy. Depressions on the floodplain, such as abandoned channels, are the first areas to be inundated.
- *Stage 5, peak stage* – the highest point of the flood. The stage at which the volume of water is at its maximum.
- *Stage 6, waning stage* (or falling stage) – the gentle receding limb of the floodwaters back towards baseflow conditions.

Hydrographs are comprised of three key components: the baseflow, the rising limb and the receding limb. Flow above baseflow is sometimes referred to as quickflow. Several factors control the shape of hydrographs. The role of catchment morphometrics, topography, rainstorm pathways and antecedent moisture conditions are discussed here.

Catchment shape dictates the length of tributary streams, and hence their catchment area, and where these tributaries join the trunk stream (Chapter 3). Catchments that have long tributary streams with relatively large catchment areas that join the trunk stream over a short distance have relatively peaked hydrographs. In more elongate catchments, small tributary streams recurrently join the trunk stream separated by relatively long distances. In these settings, relatively small amounts of discharge are supplied to the trunk stream over a greater distance, resulting in a relatively flat hydrograph. The same sets of relationships may occur in catchments with high and low *drainage densities* (see Chapter 3). The large number of channels feeding the primary trunk stream in highly dissected landscapes results in peaked flow hydrographs.

Catchment topography is probably a more significant determinant of the shape of flood hydrographs than catchment shape. For example, intense rain events in short, steep, escarpment-dominated coastal catchments may generate 'torrents of terror'. These narrow and peaked hydrographs reflect the short lag time (hours) between intense rainfall and peak discharge. In contrast, long, low-relief rivers have broad and flat hydrographs with long lag times (weeks and months) between rainfall and peak discharge.

Although event-driven responses, *rain storm pathways* also influence hydrograph shape. Four different storm types that produce four different hydrographs in the same catchment are shown in Figure 4.13. Storms that track across the headwaters of a catchment produce flash-floods with a narrow and peaked hydrograph (1). If a storm of equivalent magnitude tracks across the catchment outlet, a low (or no) flood will occur (2) and the hydrograph will be squat. At the catchment outlet, a storm tracking up the catchment will have a broader hydrograph with lower peak (3) than if the storm tracks down the catchment (4).

Finally, the peak discharge resulting from any given rainfall event may vary from flood to flood depending upon *antecedent soil moisture conditions*. For example, repetition of rainfall events of a given magnitude may trigger differing runoff responses, as shown in Figure 4.14. Peak discharge 'A' reflects relatively low antecedent moisture conditions. A similar-magnitude storm that follows after a short period of time causes a much larger peak discharge at 'B' because the rain falls on soils that are already wet. Coalescence of storm hydrographs can cause floods that are much larger

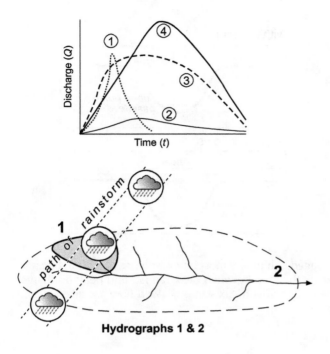

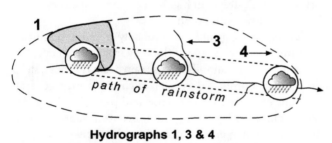

Figure 4.13   The impact of storm tracking on the shape of hydrographs. Storms that track across catchments produce peaked hydrographs in the subcatchments activated (hydrograph 1), but only broad hydrographs at the catchment outlet (hydrograph 2). Storms that track down or up a catchment produce broad and peaked hydrographs (hydrographs 3 and 4).

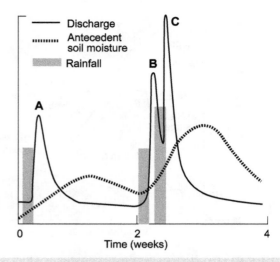

Figure 4.14   The role of flood sequencing and antecedent soil moisture on the magnitude of hydrographs. Rain onto a dry catchment (low antecedent moisture) will only produce a small-magnitude flow. However, recurrent rain or rain onto already wet and saturated catchments will produce large-magnitude flows with multiple peaks. This reflects ongoing inputs of rain and the activation of various forms of runoff from different parts of a catchment.

than those that occur in response to isolated storms (e.g. 'C' on Figure 4.14). This is an example of complex response (Chapter 2).

Analysis of hydrograph shape

Different components of the rising limb, receding limb and baseflow conditions on a hydrograph can be analysed mathematically to provide guidance on the characteristics of flows at any given gauge site and to compare flow characteristics between sites. All hydrographs tend to be skewed, with a longer tail on one side of the mean than the other.

Most hydrographs tend to be positively skewed with a steep rising limb and a gentle receding or waning limb. This occurs because flow tends to increase rapidly above baseflow as various forms of runoff become activated and become concentrated on the valley floor during the rain event. Once rain stops it takes considerable time for the flow to return to baseflow conditions as residual water and stores pass over or through the landscape. Surface runoff is depleted first and is followed by interflow. Baseflow declines slowly. The timeframe/duration over which these processes occur varies markedly from catchment to catchment. The steepness of the curve of the rising limb of the hydrograph records the rate of channel infilling and subsequent floodplain inundation. Steep rising limbs reflect quick runoff and accumulation of water on the valley floor. Gentle rising limbs reflect lagged and significant infiltration or water storage in the catchment.

Kurtosis is a measure of the peakedness of the hydrograph. A high-kurtosis distribution has a sharper peak and longer, fatter tails. This reflects a flashy flow regime with high initial rates of runoff and quick receding flows. Flashy flow regimes are common in ephemeral, arid zone rivers. A low-kurtosis hydrograph has a broad, low-amplitude peak, is more rounded and has a shorter tail. This reflects a flow regime where flows accumulate slowly over time (possibly due to infiltration and runoff, followed by cumulative tributary inputs) and then recede slowly over time.

Flow duration records the time between the rising limb and the receding limb. The width of the hydrograph can be measured to give an indication of the period of inundation at differing flow stages. The duration of a flood is most commonly measured as flow extends above baseflow, at bankfull discharge and at peak discharge. Flow duration can extend from hours to weeks, depending on landscape and climatic setting. Recession curve analysis records the rate of decrease of flow stage on the receding part of the hydrograph. This can be calculated as:

$$Q_t = Q_0 k^t$$

where Q_t ($m^3 s^{-1}$) is the discharge at time t, Q_0 ($m^3 s^{-1}$) is the discharge at the start of the recession and k is the recession constant.

The recession constant is a ratio of the flow at one time period to the flow in the previous time period calculated from daily data. For example, a recession constant of 0.9, the flow on any day will be 0.9 times the previous day. Recession constants for surface runoff may range from 0.195 to 0.799, while interflow ranges from 0.733 to 0.940 and baseflow ranges from 0.929 to 0.988. Analysis of base-flow entails separation of baseflow from quickflow. This is not a straightforward task. Following a filtering process, the volume of baseflow is calculated and divided by the total flow volume to estimate the baseflow index. Baseflow index values greater than 0.3 are usually associated with perennial streams, while values less than 0.3 reflect ephemeral conditions.

Discharge measurement

In many instances stream gauges provide a continuous read-out of discharge (typically related to flow stage measurement). Elsewhere, direct measurements can be made for ungauged cross-sections. Alternatively, predictive techniques are applied, typically based on extrapolative procedures from existing gauge data. In these cases slope–area, stage–discharge, catchment–area discharge and retrospective estimates can be made.

Direct measurements in the field

Direct measurement of discharge requires measurement of flow velocity for given segments of a surveyed channel cross-section. The continuity equation is applied to determine the discharge for each segment, which is then added together for the cross-section. A current meter measures flow velocity. The mean velocity in a vertical profile is approximated by making a few velocity observations and

using a known relation between those velocities and the mean in the vertical. Flow velocity in natural channels generally pulsates. Velocity is therefore measured for at least 40 seconds in an effort to better represent average velocity at a point. Two key methods are used to determine mean vertical velocity: the two-point method or the six-tenths depth method. The two-point method is used when velocity profiles are relatively uniform. In these instances, average velocity is adequately estimated by averaging velocities at 0.2 and 0.8 of the depth below the water surface. The vertical-velocity curve may be distorted by overhanging vegetation that is in contact with the water or by submerged objects. If those elements are close to the vertical in which velocity is being measured, an additional velocity measurement should be made at 0.6 of the depth. In the six-tenths depth method, a measure of velocity made in the vertical at 0.6 of the depth below the surface is used as the mean velocity in the vertical, as velocity at this depth most closely approximates the mean velocity.

Slope–area method

The slope–area method uses water surface slope and flow area to derive discharge based on Manning's n equation (see Chapter 5):

$$v = \frac{d^{2/3} s^{1/2}}{n}$$

where v ($m s^{-1}$) is the flow velocity, d (m) is the flow depth, s ($m m^{-1}$) is the water surface slope and n is the Manning roughness coefficient (see Chapter 5).

Once velocity is calculated it can be placed in the continuity equation along with the depth and width of flow to derive discharge. This method is particularly useful for measuring the discharge of large floods, where debris lines can be used to indicate the slope, water depth and cross-section of the flow.

Stage–discharge relationships

Some gauges only record flow stage (i.e. the water level that reflects flow height). The stage–discharge method relates flow stage to the cross-sectional area of the flow in those instances when velocity is not recorded. The stage–area method requires numerous gaugings or calculations of discharge at different flow depths to represent the change in discharge as the channel fills with water. Once these measurements have been made, a stage–discharge rating curve is constructed. From this, measures of stage can be used to estimate discharge at that site. Confidence limits of the curve are critical, as errors up to 30% may result, even in those instances where measurements are of high quality.

Catchment area–discharge relationships

Broader, regional-scale estimates of discharge can be estimated for a given channel cross-section using catchment area–discharge relationships. Ideally, areas that lie within a particular climatic and/or topographic setting, with a relatively homogeneous rainfall and climatic pattern, are selected. Discharge and catchment area data are obtained for all suitable gauges in this region. This is usually conducted for the 1:1.01, 1:2, 1:5, 1:10, 1:20, 1:50 and 1:100 recurrence interval flow events (see below). Regression analyses are used to derive discharge estimates for a given recurrence interval for a given catchment area. The quality of these estimates varies significantly depending on the number, reliability and continuity of records at the gauging stations in any given region. It is particularly important to have data records for the full range of flow conditions; this is something that is especially difficult for extreme high flows – ironically, the very events that we typically need data for! In regions with poor records, this method merely provides an indicative guide to likely discharges for different flow conditions.

Retrospective analysis of high flow stage

Retrospective analysis is commonly used to analyse flood flows, especially if direct measurement is hampered by access and safety concerns. Various methods can be used to estimate flood stage, from which stage–discharge methods can be applied to estimate flood discharge. Flood debris is the most obvious indication of flood stage (Figure 4.15). Leaf litter, wood and rubbish are often trapped by riparian vegetation. In fine-grained systems, mud drapes are commonly found on the upstream sides of trees (like a bath-tub ring). In coarse-grained systems, tree scars often occur on the upstream sides of trees, as bark is removed from the trunk as clasts or debris collide with the tree. In some instances, scar tissue within tree rings can be used to determine the year in which these flood events occurred. In other places, regolith scour lines occur where surfaces are stripped of vegetation and sediment along channel margins and hillslopes.

Palaeohydrology is the study of past or ancient flood events. It involves the analysis of flood sediments, debris or landforms produced by floods in the past. Slackwater deposits comprised of sand or silt deposits emplaced from suspension in deep, high-velocity floods can be used to reconstruct the height of palaeo-flood events (Figure 4.15). These flood sediments or debris are commonly found in sheltered locations high above the contemporary channel, where they are preserved in features such as caves, in the lee of bedrock spurs, in areas downstream of valley constrictions or in tributary systems where trunk stream backwater effects occur. Dating of these deposits enables the

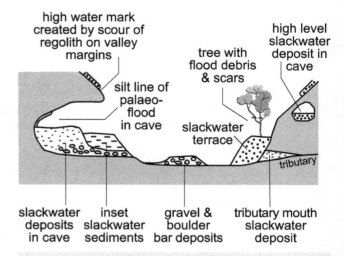

Figure 4.15 Slackwater deposits found in elevated situations that are not recurrently reworked can be used to reconstruct high-flow stages of past events. Reprinted from Journal of Hydrology, 96, Baker, V.R., Palaeoflood hydrology and extraordinary flood events, 79–99. © 1978, with permission from Elsevier.

timing of high-magnitude flood events from the past to be determined (see Chapter 12).

Flow frequency

The frequency of a flood of a given magnitude, referred to as the *average recurrence interval* (ARI) (also called return period), is often expressed as 1-in-N years. Estimates of the recurrence interval of a flow of a certain magnitude/discharge are derived primarily from historical records (mainly gauge records). In most instances, this gauged record is so short that statistical procedures must be applied to undertake analysis of extreme flood events. A statistical technique called Gumbel analysis first ranks the *maximum annual flood series*, defined as the highest peak flow in each year, from highest to lowest. Then the recurrence interval of events of a given discharge is calculated as follows:

$$r_i = \frac{m+1}{n}$$

where r_i is the recurrence interval, m is the total years of record and n is the rank of the flood event.

Gumbel analysis allows a hydrologist to determine the likely recurrence interval of flood events of differing magnitude. The recurrence interval of a flow only tells us the average period in years between floods of the same magnitude. For example, a 1-in-100-yr ARI flood is expected to occur once every 100 yr. However, this does not mean that two 1-in-100-yr events cannot occur in one year! A more meaningful way to express the frequency of floods of various magnitude is the *annual exceedence probability*.

This is a measure of the statistical probability of that flood occurring in any one year. For example, a 1-in-100-yr flow has a 1% probability of occurring in any one year (i.e. 1/100), a 1-in-10-yr flow has a 10% probability of occurring in any one year (i.e. 1/10) and a 1-in-2-yr flow has a 50% probability of occurring in any one year (i.e. 1/2).

Flood frequency curves are often used to represent discharge versus flood recurrence interval. A commonly used curve is a log Pearson III (LP3) distribution, as shown in Figure 4.16. The recurrence interval of the flow is presented as the annual exceedance probability representing a range of flows from the 1:1.01- to the 1:100-yr flows. If the discharge of the flood event is known, these plots can be used to determine the recurrence interval of the flow. When analysed in conjunction with field-based assessment of channel cross-sections, geomorphologists can produce *rating curves* for particular river systems. These plots represent the relationship between channel cross-section size and discharge. They can be used to determine the recurrence interval (or stage – i.e. water level) at which different surfaces are inundated. From this, bankfull recurrence, and the likely morphological changes that will result when dif-ferent types of rivers are subjected to events with differing magnitude–frequency relationships, can be estimated.

Statistical bankfull discharge has a recurrence interval of 1.58 yr, while the *mean annual discharge* occupies 40% of the total bankfull channel capacity with a recurrence interval of 2.33 yr and a frequency of about 25%. Mean annual flow can be calculated by averaging daily data from complete years. Many engineering applications, such as channelisation of urban streams or design of dam spillways, are framed in relation to the *probable maximum flood*. This analysis is based on maximum rainfall intensities, whereby the probable maximum rainfall is defined as the greatest depth of precipitation for a given duration that is meteorologically possible for a given storm area at a particular location at a particular time of year.

Flow variability

Understanding flow variability is critical to interpretations of the behavioural regime of rivers. Some rivers have adjusted their form to irregular (extreme) flow events,

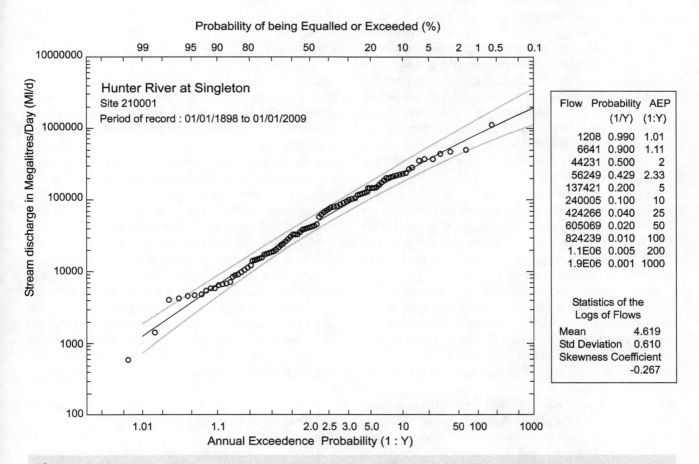

Figure 4.16 The Log Pearson III graph of flow magnitude–frequency relationships for the Hunter River at Singleton, NSW, Australia. Extracted from the NSW Office of Water Pinneena database for gauged streams.

while others adjust to perennial or seasonal flow (i.e. the effectiveness of flow events of a given magnitude and recurrence interval varies markedly from place to place and from system to system). Three commonly used approaches to represent streamflow data are presented below: time-series plots, measures of flow variation and flow duration curves.

Time-series plots are graphical representations of mean daily flow versus time (whether daily, monthly or annual flows). Subjective assessments of *seasonal variability* in flow, and maximum and minimum flows, can be used to relate the flood history of a system to geomorphic changes (see Figure 4.17). These plots can be analysed to determine the seasonality of flow, when floods of various magnitude occurred and the timing, magnitude and recurrence of the flood of record (largest flood).

Various measures are used to describe the variability of flow. The *coefficient of variation* C_v provides a measure of the variability in annual flows in a system (*interannual variability*) and is measured as:

$$C_v = \frac{S_Q}{Q}$$

where S_Q is the standard deviation of the annual discharges and Q is the mean annual discharge. A high value of C_v indicates that flows are more variable from year to year. High values of C_v are around 2 and low values are around 0.1. In general, ephemeral streams in arid and semi-arid regions and areas that are affected by infrequent tropical storms have high variability. In these areas, long dry periods are punctuated by occasional floods. The number of days of zero (or low) flow provides a useful measure of flow variability in these systems. In contrast, perennial streams in the humid tropics have low C_v, indicating limited interannual flow variability.

Graphical representations of the ratio of the mean annual or 2-yr discharge to discharges of higher recurrence intervals provide another useful measure of flow variability (see Figure 4.18). Systems that are subjected to discharge events that are 10 times greater than the mean annual flood are prone to catastrophic erosion.

Flow duration curves provide a visual summary of flow data that show the percentage of time that a given flow magnitude is exceeded. Daily, monthly or annual flow data can be used to appraise high and low flow periods over differing timeframes. Most commonly, flow duration curves are plotted using a dimensionless flow axis of daily mean flow divided by mean daily flow (e.g. see Figure 4.17). The steeper the flow duration curve, the more variable the flow (and the greater the coefficient of variation). Steep areas at the tail of the curve (low-flow end) suggest that baseflow is minimal and the channel dries up quickly once rain ceases. A steep area at the head of the curve suggests that a small number of high-flow events produce flows that are a lot higher than those that occur most of the time. Flow duration curves with a flat slope result from dampening effects of high infiltration and groundwater storage.

Considerable caution must be exercised in the use of hydrological generalisations to characterise system-specific applications. The contrast between hydrogeomorphic regimes in, say, arid, Mediterranean, tropical, temperate and polar settings, with their profoundly differing vegetation covers, could scarcely be more stark. Hydrologic and geomorphic relationships in rivers that freeze for significant parts of the year are quite different to those demonstrated by rivers subjected to high-flow stages throughout most of the year (e.g. many tropical rivers). Impacts of flow regulation and other forms of human disturbance sit atop this 'natural' variability (see Chapter 13). As a consequence, flow and sediment transfer vary markedly from region to region and from system to system. In many instances, hydrological analyses of flow duration curves are a critical part of environmental flow assessments, aiding determination of the timing and periodicity of inundation of differing geomorphic surfaces in a manner that mimics the natural flow regime.

Conclusion

Hydrological setting has enormous implications for the geomorphology of river systems. Interactions among a wide range of mechanisms within the hydrological cycle determine the manner and rate of runoff generation at any given place. This results in significant differences between the hydrological conditions under which intermittent, ephemeral and perennial streams operate. The frequency with which bankfull and overbank events occur is a key determinant of geomorphic work and the redistribution of sediments in river systems. In some instances rivers are closely attuned to regular (annual) flow events; elsewhere, extreme events are a primary determinant of contemporary channel forms. These relationships may change over time, whether in response to climatic conditions, alterations to ground cover or direct human impacts such as dam construction. Critically, however, it is not just the amount of flow that determines the geomorphic response of a river; rather, it is the way in which the river uses the flow that is made available to it.

Key messages from this chapter

- Hydrology is the study of water through the hydrological cycle. The hydrological cycle is a conceptual model

(a) Time Series Plot for the Hunter River at Singleton

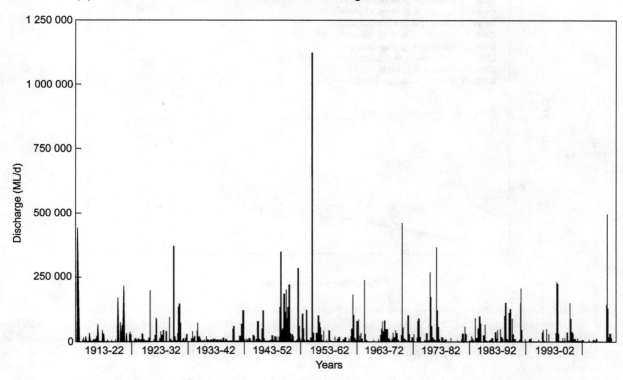

(b) Flow Duration Curve for the Hunter River at Singleton

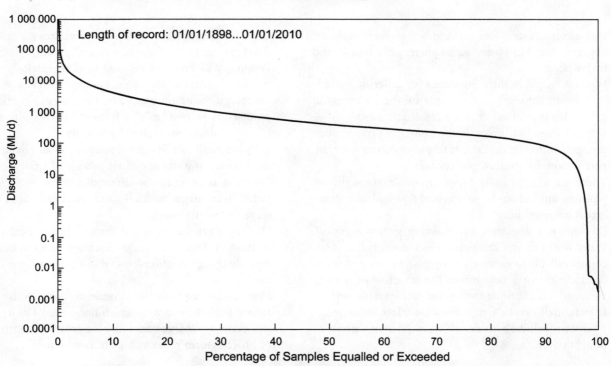

Figure 4.17   (a) Time-series plot showing the history of floods for the Hunter River at Singleton, NSW, Australia. (b) Flow duration curve used to determine the amount of time flows of different magnitude are exceeded for the Hunter River at Singleton, NSW, Australia. Extracted from the NSW Office of Water Pinneena database for gauged streams.

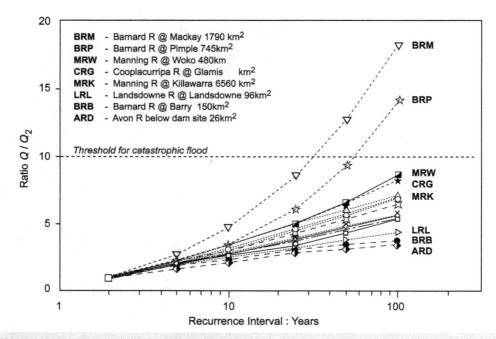

Figure 4.18 Flow variability analysis (coefficient of variation) for annual maximum flood data for a range of tributaries in the Manning Catchment, NSW, Australia (based on gauge records longer than 10 yr). Significant variability in flow regime is evident. The Barnard River has the steepest curve, reflecting a flashy flow regime and highly variable discharge that can induce catastrophic floods. This is the only stream for which the discharge of the 100-yr flood is more than 10 times greater than the 2-yr flood. The Avon River has a low slope curve and experiences less discharge variability. These patterns are largely controlled by topography and rainfall patterns across the catchment.

that describes the storage and movement of water between the biosphere, atmosphere, lithosphere and hydrosphere.

- Water is stored in differing forms for differing periods of time in differing components of the hydrological cycle. This is referred to as the residence time.
- Key processes that drive the hydrological cycle include precipitation, evaporation, transpiration, interception, percolation, infiltration and runoff.
- There are four primary types of runoff: throughflow, pipeflow, infiltration-excess overland flow and saturation-excess overland flow.
- Precipitation does not immediately generate runoff. Water works its way through various stores before it become available to channels. Progressive release of water from these stores determines the baseflow of a river. Hence, rivers continue to flow for considerable periods after rainfall events. Groundwater–surface water interactions are critical considerations in the maintenance of flow in rivers.

- Channel initiation on hillslopes commences via headcut development in the form of rills and gullies.
- Frequency of flow can be used to differentiate among perennial, intermittent and ephemeral rivers.
- Discharge, the amount of water that passes a particular point every second ($m^3 s^{-1}$), is measured as width multiplied by depth multiplied by velocity.
- A hydrograph is a visual representation of changes in the amount of water at differing stages of a flood event. Several flow stages can be differentiated: base flow, flood pulse, rising stage, bankfull stage, overbank stage, peak stage and waning stage.
- Discharge can be measured directly in the field or in relation to slope–area, stage–discharge and catchment-area–discharge relationships. Rating curves are often used.
- Magnitude–frequency relationships characterise the recurrence intervals with which flood events of a given size occur. Discharge variability is a critical determinant of process–form relationships in river systems.

CHAPTER FIVE

Impelling and resisting forces in river systems

Introduction

When water accumulates in a channel on an inclined surface it has the ability to flow. The energy of that flow is able to perform geomorphic work such as transporting sediment or deforming channel boundaries. Slope and volume of water are key determinants of the amount of energy and the way in which that energy is used. Channel boundary factors influence these relationships: they determine the amount of seepage (and hence flow continuity), the manner/rate of energy consumption in overcoming friction and the ease with which bed and banks can be deformed. As tributaries join the trunk stream, flow volume increases. However, in general terms, slope tends to decrease. Changes to these controls affect the capacity of rivers to transport materials of differing texture or induce erosion and deposition as ways of using their available energy. Adjustments balance out to generate smooth longitudinal profiles, as noted in Chapter 3.

This chapter outlines the primary forms of impelling and resisting forces in river systems. The chapter is structured as follows. Following summary comments on the mechanics of fluid flow, impelling and resisting forces are outlined. This is followed by a discussion of the way that energy is used in river systems in the context of the degradation–aggradation balance along longitudinal profiles, and the associated distribution of erosion and depositional processes.

Impelling and resisting forces and Lane's balance of erosion and deposition in channels

Rivers act to move water and sediment downslope. In doing this they expend energy and perform geomorphic work. However, a critical energy level or threshold must be reached before a river can perform this work. The potential energy of flow within a channel is measured as the mass of water entering a river at a certain height above a given base level. As water moves downstream, potential energy is converted to kinetic energy. The *conservation of energy principle* states that the potential energy plus kinetic energy must remain constant within the system (i.e. no energy is lost). Hence, any loss in potential energy is matched by an equivalent gain in kinetic energy. However, rivers are non-conservative systems and friction causes much available energy to be dissipated in the form of heat, which performs no geomorphic work. Whether geomorphic work is done is dependent on the available amount of potential energy and the balance of energy expended and energy conserved at any particular location, such that erosion thresholds are, or are not, breached. Three possibilities exist: (1) a river may have more energy than that required to move its water and sediment load, in which case it has surplus energy and will adjust in the form of erosion; (2) it may have exactly that required, in which case it is stable; and (3) it may have an energy deficit, which will result in adjustment in the form of deposition.

In physics, a *force* refers to any influence that causes a free body (object) to undergo a change. In the case of river systems, water acts as the primary force by which matter in the form of sediment is moulded and shaped as it moves downstream. This largely reflects the amount of water (discharge) acting on a given slope. The proportion of erosion and deposition that occurs along a river channel is a function of the relative balance of *impelling* and *resisting* forces. The Lane balance diagram provides a key conceptualisation of this dynamic (Figure 5.1). There are four key components to the Lane balance. The left bucket depicts the volume of bed material load Q_s, with a sliding scale of median bed material size/calibre D_{50}. The right bucket depicts the volume of water in the river channel Q_w (discharge), with a sliding scale of channel slope s. The relative sizes of the buckets and their positions along the sliding scale determine whether the balance is tipped to the left or to the right and whether aggradation (deposition) or degradation (erosion) occurs. In theory, the channel acts to maintain the balance (Figure 5.1a). If discharge increases

(a) In balance (equilibrium)

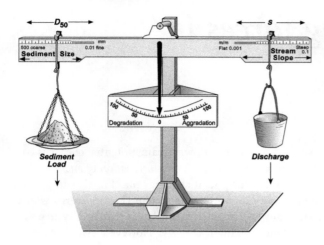

(b) Degradation (erosion)

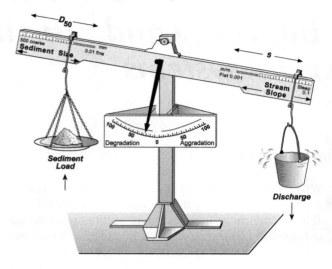

(c) Aggradation (deposition)

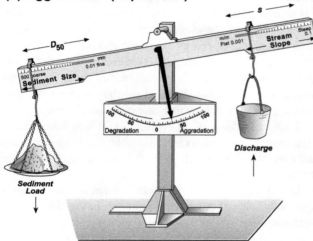

Figure 5.1 The Lane balance diagram. Flow–sediment interactions determine the aggradational–degradational balance of river courses. (a) The river maintains a balance, accommodating adjustments to the flow/sediment load. (b) Excess flow over steep slopes, or reduced sediment loads, tilts the balance towards degradation and incision occurs. (c) Excess sediment loads of a sufficiently coarse nature, or reduced flows, tilt the balance towards aggradation and deposition occurs. The arrows on (b) and (c) indicate the way in which the channel adjusts its flow/sediment regime to maintain a balance. Modified from Lane (1955).

or the channel increases its slope (e.g. a bend is cut off, so that channel length is reduced), the balance will tip to the right and degradation (erosion) results (Figure 5.1b). This means that there is excess energy in the system relative to the volume and size of sediment and that energy is consumed via incision (the channel cuts into its bed). Alternatively, the same outcome occurs if the sediment load Q_s is reduced or if the bed material size is decreased. In contrast, if excess sediment is added to the stream (i.e. Q_s increases), especially if that bed material is coarser (i.e. D_{50} increases), the available discharge is unable to move all available material within the channel and aggradation (deposition) occurs (i.e. sediments accumulate on the bed; Figure 5.1c). Once more, the same outcome arises if discharge Q_w decreases or channel slope decreases (i.e. channel length increases as the channel becomes more sinuous).

Critically, the Lane balance is used to describe how a channel is likely to adjust to maintain its balance in response to changes to flow and sediment conditions. In general, the water/discharge 'bucket' is primarily a function of climatic controls upon flow availability and variability, whereas the sediment 'bucket' is primarily a function of geological controls upon sediment availability (calibre and volume, determined primarily by weathering breakdown products, the erosivity of those materials and the erodibility of the landscape).

The Exner equation links erosion or deposition to a deficit of, or excess in, sediment flux respectively, thereby providing a means to quantify these relationships. The equation describes conservation of mass between sediment on the channel bed and sediment in transport. It states that bed elevation increases (i.e. aggradation occurs) proportional to the amount of sediment that drops out of transport, and conversely decreases (i.e. degradation occurs) proportional to the amount of sediment that becomes entrained by the flow. As such, the Exner equation can be used to predict the occurrence of erosional and depositional forms along a reach. The equation is often used in its one-dimensional form as follows:

$$\frac{\partial n}{\partial t} = -\frac{1}{\varepsilon_0}\frac{\partial q_s}{\partial x}$$

where $\partial n/\partial t$ is the change in bed elevation over time, ε_0 is the grain packing density, q_s is the sediment discharge, ∂x is the downstream direction. Values of ε_0 for natural channels range from 0.45 to 0.75. The value for randomly packed spherical grains is 0.64.

Impelling forces drive adjustments through erosion and reworking of materials as a given volume of water flows over a certain slope (i.e. discharge; see Chapter 4). This is often measured in terms of the 'energy' and 'efficiency' of flow within a channel. Flow with sufficient energy is able to perform geomorphic work. To do this, it must overcome a number of threshold conditions to entrain and transport sediment (see Chapter 6), whereby it is able to erode the channel margins (see Chapter 7). Measures of *stream power* and *shear stress* are commonly used to explain how sediment is transported along a river channel.

Resisting forces reduce flow energy via friction. They determine how a channel consumes its available energy, i.e. the ability of the river to carry sediment of a given volume and calibre. These factors resist change, limiting the extent of river activity and adjustment, striving to maintain river morphology. They are commonly measured as *flow*, *boundary* and *channel resistance*. Prior to analysing and interpreting impelling and resisting forces, the mechanics of fluid flow are briefly described.

Mechanics of fluid flow

An understanding of the mechanics of fluid flow is required to quantify flow energy and the efficiency with which channels are able to use that energy. A fluid is defined as a material that deforms continuously and permanently under the application of a shearing stress, no matter how small. The inability of fluids to resist shearing stress gives them their characteristic ability to change their shape or to flow. Overcoming friction is a key characteristic of water flow. The ability of flow to overcome friction is dependent on flow volume and the nature of the surface over which it is moving. There are two fundamentally different types of flow motion:

1. *Laminar flow* refers to smooth, orderly motion in which fluid elements or particles appear to slide over each other in layers or laminae with no large-scale mixing.
2. *Turbulent flow* refers to random or chaotic motion of individual fluid particles with rapid macroscopic mixing of particles through the flow.

The gradation from laminar to turbulent flow is presented in Figure 5.2. Flow is turbulent in natural channels. Velocity profiles represent the displacement of particles of water with respect to the bed in a given time period. The velocity gradient in turbulent flow is uniformly steep. Flow speed increases rapidly away from the boundary (channel bed), with the gradient being proportional to boundary roughness.

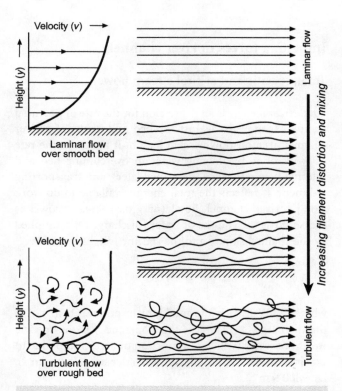

Figure 5.2 Laminar and turbulent velocity profiles. Laminar flow is sheetlike and fluid particles slide over each other in layers or laminae. This occurs in smooth channels or pipes. Turbulent flow has a more abrupt vertical velocity profile, with notable bed disruption and chaotic mixing of particles within the flow. Schematics on the right show how increasing filament distortion within the water column disrupts the water surface and flow lines are more variable as turbulence increases.

Isovels are contours of equal downstream velocity viewed in cross-section. The deepest part of the channel is referred to as the *thalweg*. Generally, the highest velocity filament of flow is located in this part of the channel. Velocity profiles, the pattern of isovels and the position of the thalweg vary for channels of differing shape and size (Figure 5.3a). In general terms, however, isovels are more closely packed near the channel bed than further away and they are less closely packed near to the banks than near to the bed (i.e. velocity increases as you move away from the rough boundary – channel bed and banks). The thalweg tends to sit just below the surface of the flow due to free-surface resistance.

Helicoidal flow is the anticlockwise, corkscrew-like motion of water in a meander bend (sinuous channel). This secondary flow is initiated by oscillation or perturbation in the flow and associated pressure gradient forces (Figure 5.3b). A number of secondary currents may be evident that diverge and converge at different positions in the channel.

Impelling forces in river channels

Total, specific and critical stream power

Total stream power is an expression for the rate of potential energy expenditure against the bed and banks of a river channel per unit downstream length. It measures the rate of work done by flowing water in overcoming bed and internal flow resistance (described later), and transporting sediment. It reflects the total energy available to do work along a river channel. Total (or gross) stream power is measured as the volume of water (discharge Q) multiplied by the channel slope s and the specific weight of water:

$$\Omega = \gamma Q s$$

where Ω (W m^{-2}) is the total stream power, Q (m^3 s^{-1}) is the discharge, s (m m^{-1}) is the slope and γ is the specific weight of water (which is a function of acceleration due to gravity (9.8 m s^{-2}) multiplied by water density (1000 kg m^{-3}), i.e. 9800 N m^{-2}).

Specific (or unit) stream power is a measure of energy expenditure per unit width of channel. It is measured as total stream power divided by the width of flow:

$$\omega = \frac{\Omega}{w}$$

where ω (W m^{-2}) is the specific stream power, Ω (W m^{-2}) is the total stream power and w (m) is the water surface width at a specific discharge.

Indicative thresholds of channel erosion and floodplain reworking have been defined in relation to critical values of unit stream power. For example, the thresholds for movement of pebbles, cobbles and boulder are around 1.5 W m^{-2}, 16 W m^{-2} and 90 W m^{-2} respectively. The threshold level of channel instability is around 35 W m^{-2}, while 300 W m^{-2} is a threshold for floodplain stripping. These threshold values are merely indicative estimates. Real-world values vary dependent upon reach- and catchment-specific conditions, reflecting topographic, climatic and vegetation factors, among many considerations.

Critical stream power is the power needed to transport the average sediment load supplied to a stream. Where critical power is greater than the total stream power generated, there is insufficient energy to entrain and transport sediment (i.e. these are transport-limited conditions; see Chapter 6). In contrast, where critical power is less than the total stream power generated there is sufficient energy available to move sediment and deposition occurs (i.e. these are supply-limited conditions).

Mean boundary shear stress

Shear stress, also referred to as tractive force, is the force applied by flowing liquid to its boundary. Put simply, shear stress describes the force of water along a channel boundary. Bedload movement and sediment transport are functions of shear stress (see Chapter 6). When the drag force of flowing water against a particle is greater than the gravitational force holding it in place the particle begins to move. Mean boundary shear stress is a measure of the force of flow per unit bed area. In other words, it is a measure of the drag exerted by the flow on the channel bed. It is computed as:

$$\tau_0 = \gamma R s$$

where τ_0 (N m^{-2}) is the shear force per unit area of the surface (alternatively, $1\,\text{N m}^{-2} = 1\,\text{kg m}^{-1}\,\text{s}^{-2} = 1$ Pa (pascal)), γ is the specific weight of water (9800 N m^{-3}), R (m) is the hydraulic radius, s (m m^{-1}) is the slope.

In many cases, channel depth d is substituted for hydraulic radius, especially for channels with a high width/depth ratio. Mean boundary shear stress is used to determine the ability of flow to perform geomorphic work, especially bedload transport. It measures the force acting on the bed and banks of a channel. In general, shear stress on the banks of a channel tends to be 0.7 to 0.8 of that acting on the bed.

The concept of *critical shear stress* can be used to determine threshold conditions required to initiate bed erosion

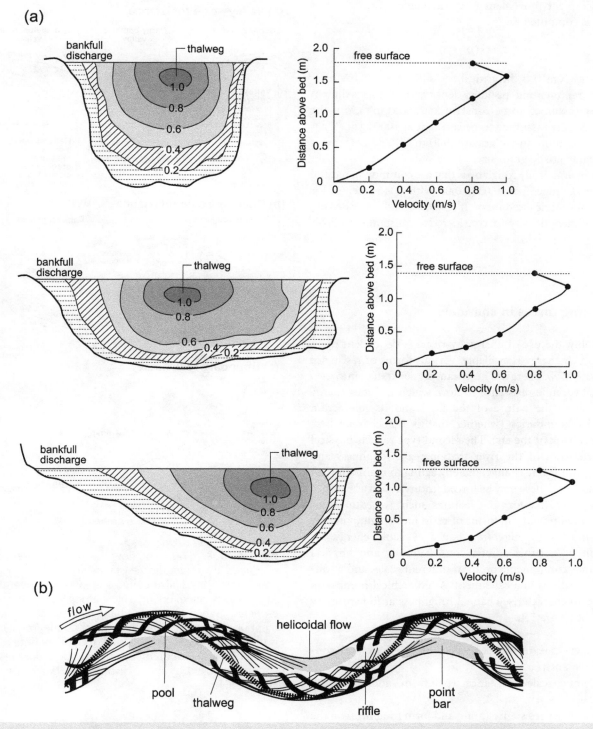

Figure 5.3 (a) Isovel patterns within channels of different shape. Isovels are areas of equal velocity. Velocity increases away from the bed and banks of the channel as the effects of roughness diminish. The thalweg lies in the deepest part of the channel just below the surface, where free-surface resistance is reduced. Vertical velocity profiles through the thalweg are shown on the right. (b) In plan view, helicoidal flow structures in meander bends produce an anticlockwise, corkscrew effect.

and sediment movement (see Chapter 6). Critical shear stress is computed as:

$$\tau_c = k(\rho_s - \rho)gD$$

where τ_c ($N\,m^{-2}$) is the critical bed shear stress, k is a coefficient representing packing density, ρ_s is the sediment density (assumed to be constant at $2650\,kg\,m^{-3}$), ρ is the water density (assumed to be constant at $1000\,kg\,m^{-3}$), g is the acceleration due to gravity ($9.81\,m\,s^{-2}$) and D (mm) is the characteristic grain size.

For hydraulically rough beds that are common in natural streams, k ranges from 0.03 to 0.06, with 0.045 accepted for uniform spherical sediment. If $k = 0.045$ and water density and sediment density are considered as constants, it follows that:

$$\tau_c = 0.73D$$

Resisting forces in channels

Inevitably, the use of energy by rivers is dependent upon how much energy is available. Various forms of resistance dissipate flow energy. These can be separated into two types: hydraulic resistance to flow, which includes energy lost due to the nature of the fluid and its interaction with its boundaries (whether that is the channel bed, channel bank or the air). The second type reflects physical characteristics of the river that increase roughness and resist geomorphic change, thereby acting to maintain river morphology. These may be forced features, such as bedrock outcrops, or free-forming features such as bedforms or bend development. All forms of resistance result in energy dissipation and, therefore, reduce available energy for erosion or channel adjustment. Hydraulic and physical resistance components occur at different scales and at different positions in a catchment. A hierarchical framework is used to characterise channel roughness at four primary scales in Figure 5.4:

1. valley-scale resistance (valley morphology and confinement);
2. channel-scale resistance (planform and bed-bank roughness);
3. boundary resistance (grain and form roughness on the channel bed);
4. fluid resistance (internal viscosity and free-surface resistance).

Valley-scale resistance

At the broadest scale, valley alignment and the position of the channel on the valley floor can induce energy loss from

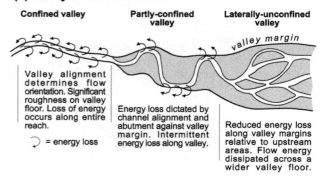

(a) Valley-scale resistance

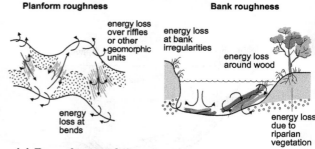

(b) Channel-scale resistance

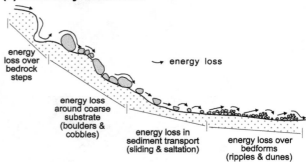

(c) Boundary resistance

Figure 5.4 Forms of flow resistance. These can produced by hydraulic resistance to flow or by the presence of physical roughness elements in rivers. (a) At the valley scale, energy is lost along valley margins, especially when there are changes to valley confinement or alignment. (b) At the channel scale, resistance is induced by planform and bank roughness (including vegetation and wood). (c) Local-scale boundary resistance is imposed by bedrock outcrops, bedforms and grain roughness.

flow (Figure 5.4a). For example, energy is dissipated and lost where channels abut valley margins or where channels enter broad open valleys after being confined. These 'forced' forms of resistance are typically dictated by the position of bedrock valley margins and bedrock spurs.

Channel-scale resistance

Channel resistance is a function of factors such as planform roughness and bed-bank roughness elements. *Planform roughness* is determined by the morphological configuration (e.g. curvature) of the channel and adjustments to channel position on the valley floor that reduce flow momentum (Figure 5.4b). For example, loss of energy occurs along the concave banks of meander bends and over riffles.

Bed-bank roughness reflects channel alignment, channel geometry and the role of vegetation and wood (Figure 5.4b). Irregularities in bed-bank morphology modify flow patterns and consume flow energy. Riparian vegetation influences channel size and shape through controls on increased bank shear strength and/or reduced boundary-layer shear stress. Instream vegetation and the loading of wood can comprise a significant proportion of channel roughness and, hence, total hydraulic resistance. Wood can impart significant hydraulic resistance either as individual pieces or through its collective influence upon the size and type of instream features such as pools, bars and steps and associated channel morphology. These factors may dramatically reduce bedload transport rates. Variations in the height, density and flexibility of aquatic vegetation influence reach-scale flow resistance. The role of instream vegetation and wood as agents of hydraulic resistance depends upon the size of the obstruction relative to the scale of the channel (see below).

The Chezy and Darcy–Weisbach equations are useful velocity-based equations for computing roughness. The Chezy equation provides a measure of resistance to flow as a function of hydraulic radius, slope and velocity, such that:

$$v = C\sqrt{Rs}$$

where C is the Chezy resistance coefficient, v ($m\,s^{-1}$) is the mean flow velocity, R (m) is the hydraulic radius and s ($m\,m^{-1}$) is the slope of the energy gradient.

The Darcy–Weisbach friction factor ff is also often used to describe friction losses in open channel flow. It is also a function of hydraulic radius, energy gradient and velocity such that:

$$ff = \frac{8gRs}{v^2}$$

where g is the acceleration due to gravity ($9.81\,m\,s^{-2}$), R (m) is the hydraulic radius, s ($m\,m^{-1}$) is the slope of the energy gradient and v ($m\,s^{-1}$) is the mean flow velocity. Values around 0.02 are considered low levels of resistance, whereas numbers over 0.2 are considered high levels of resistance.

Boundary resistance

Boundary resistance is differentiated into grain roughness and form roughness (Figure 5.4c). In general terms, *grain roughness* refers to the relationship between grain size and flow depth. It reflects the friction induced by individual grains or clasts. In gravel or coarser textured streams, grain roughness can be the dominant form of roughness, exerting a considerable drag on the flow. As depth increases with discharge at a cross-section, the effect of grain roughness is drowned out and flow resistance decreases. *Form roughness* is derived from features developed in the bed material that increase turbulence and form drag, resulting in energy dissipation. In sand-bed streams this commonly exceeds grain roughness in importance, and in streams with coarser bed material, where grain friction might be expected to dominate, it can still be a major contributor to flow resistance. Bed configuration varies with flow stage, as differing bedforms alter form roughness (see Chapter 6). The Darcy–Weisbach friction factor ff can be used to analyse channel bed resistance provided by grain size and the (de)formation of bedforms. Types of form roughness in gravel-bed rivers include pebble clusters or bar forms that induce resistance because of ponding upstream of steps, riffles or bars. Types of form roughness in sand-bed streams include ripples and dunes. Although this effect is most pronounced at low flow stage, bar resistance can still account for 50–60 % of total resistance at bankfull stage. In bedrock rivers, step–pool sequences and coarse substrate (i.e. boulders) are the key resisting bed elements that dissipate energy through hydraulic jumps and ponding.

Fluid resistance (Reynolds and Froude numbers)

Fluid resistance is produced via the dynamics of water flow. There are two key resistance elements to fluid dynamics: internal and surface resistance. *Internal resistance* occurs in all flows as a result of viscosity and/or changes in velocity. For low velocities of flow in a smooth pipe, a thin filament of dye does not diffuse but remains intact through the pipe as a straight line. This reflects laminar flow layers within the body of the fluid. However, as velocity increases and flow becomes more turbulent, the filament wavers, mixing the layers within the body of the fluid (Figure 5.2). Flow energy is consumed through this process. The critical velocity at which the shift from laminar to turbulent flow occurs is characterised by a dimensionless number called the *Reynolds number (Re)*. The Reynolds number, which measures the ratio between inertial and viscous forces within the flow, is measured as:

$$Re = \frac{uh}{v}$$

where u (m³ s⁻¹) is the mean flow velocity, h (m) is the flow depth and v (m² s⁻¹) is the kinematic viscosity (this defines the ratio of dynamic viscosity μ (N s m⁻² or kg m⁻¹ s⁻¹) to density ρ, where μ is a measure of the internal forces within a fluid which resist the forces causing flow – molecular viscosity).

Viscous forces are significant during laminar flow, where Re numbers are normally <500. In contrast, inertial forces that produce eddies, vortices and flow instabilities predominate in turbulent flow and Re numbers generally exceed 2000. Turbulent flow involves random secondary motions through which flow energy is lost. The relative intensity of turbulence is greater over rough channel beds and at high Re numbers.

Just as the internal structure of a fluid changes with increases in velocity, so does the surface morphology of the fluid. *Free-surface resistance* occurs when the water surface is distorted by standing waves and hydraulic jumps. Hydraulic jumps reflect rapid transition from fast-flowing to slow-flowing water (e.g. a steep riffle entering a pool). This transitional stage, which is referred to as the critical velocity, defines the difference between subcritical and supercritical flow. It is measured by a dimensionless number called the *Froude number* (Fr). To explain this, envisage dropping a rock into a pool of water. This results in concentric waves travelling outwards. Energy is dissipated by means of these gravity waves radiating from the point of disturbance. The velocity at which these gravity waves travel, termed celerity, is defined by:

$$V_w = \sqrt{yg}$$

where V_w (m s⁻¹) is the velocity of the wave, y (m) is the depth of fluid and g is the acceleration due to gravity (9.81 m s⁻²).

The Froude number is derived by relating the velocity of this surface wave (celerity) to the velocity of the flow in the channel, such that:

$$Fr = \frac{v}{\sqrt{yg}}$$

where Fr is the Froude number, v (m s⁻¹) is the flow velocity, y (m) is the depth of fluid and g is the acceleration due to gravity (9.81 m s⁻²).

Flow is subcritical or tranquil if Fr < 1. At this stage, gravity wave velocity exceeds flow velocity so that the ripples on the water surface are able to travel upstream. When Fr > 1, the flow is supercritical or rapid. Gravity waves cannot migrate upstream at this stage. The surface waves may become unstable and break, resulting in considerable energy loss. Free-surface instability then results in standing waves and hydraulic jumps which increase resistance to flow. Flow is critical when Fr = 1.

Froude number can be estimated by observing the surface water morphology at certain flow stages (Figure 5.5). Flows with Fr < 1 have water surfaces that undulate. When Fr > 1 the flow has standing waves and undular jumps, transitioning to hydraulic jumps for higher Froude numbers.

Visual guides of surface flow provide a relatively quick tool with which to assess and interpret the velocity and energy of flows (see Table 5.1). These flow types can be related to Froude number, velocity or resistance elements along a reach. The degree of turbulence, velocity, Froude number and stream energy decrease from the top to the bottom of Table 5.1.

Flow type varies with flow stage. For any given reach, low-flow conditions are likely to be more tranquil with a lower Froude number than the same channel at bankfull stage, where more turbulent flow has a higher Froude number. Hence, assessments of free-surface resistance should note the flow stage at the time of field observations or measurements. Thresholds of sediment entrainment, transport and deposition, and associated changes to river morphology, can be linked to changes in flow type (see Chapter 6).

Vegetation and wood as resistance elements in river systems

Morphodynamic relationships between geomorphic river structure and riparian vegetation are manifest at various spatial and temporal scales. Patterns of erosion and deposition in the channel zone, and resulting distributions of substrate types, are influenced at local scales. Reach-scale heterogeneity may be induced by vegetation interactions with geomorphic units along rivers. At the catchment scale, vegetation cover influences catchment hydrology and rates of surface erosion and sediment supply (see Chapters 4 and 14 respectively). Vegetation cover on hillslopes influences interception, evapotranspiration, throughflow, etc. Hence, a catchment with an intact forest cover has higher water infiltration capacities, longer runoff lag times and decreased flood peaks relative to areas in which forests have been cleared. Forested catchments also have higher runoff thresholds for erosion and gully initiation, leading to low rates of sediment supply to channels relative to cleared catchments. Sediment yields tend to be greatest in semi-arid environments, where there is negligible ground cover, but sufficient flow to transfer sediments intermittently. Limited ground cover may induce high sediment availability in arid areas, but sediment yields are limited by availability of flow. Greater ground cover in more temperate areas inhibits rates of erosion.

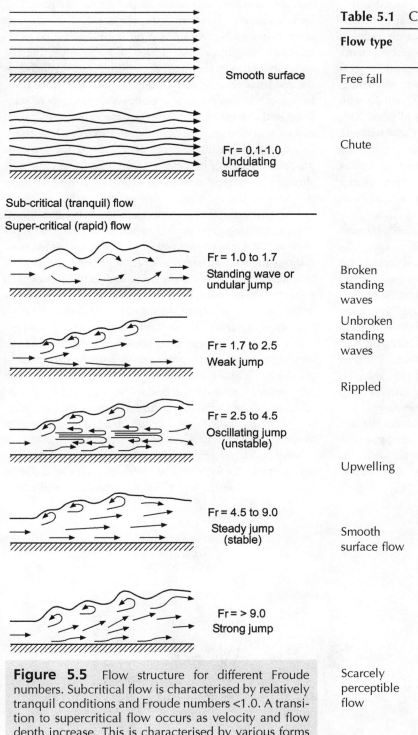

Smooth surface

Fr = 0.1-1.0
Undulating
surface

Sub-critical (tranquil) flow

Super-critical (rapid) flow

Fr = 1.0 to 1.7
Standing wave or
undular jump

Fr = 1.7 to 2.5
Weak jump

Fr = 2.5 to 4.5
Oscillating jump
(unstable)

Fr = 4.5 to 9.0
Steady jump
(stable)

Fr = > 9.0
Strong jump

Figure 5.5 Flow structure for different Froude numbers. Subcritical flow is characterised by relatively tranquil conditions and Froude numbers <1.0. A transition to supercritical flow occurs as velocity and flow depth increase. This is characterised by various forms of water surface disruption. Waves and jumps are formed at Froude numbers >1.0.

Table 5.1 Classification of surface flow types[a]

Flow type	Geomorphic unit[b]	Description
Free fall	Cascade/ waterfall	Water falling vertically without obstruction. Often associated with a bedrock or boulder step.
Chute	Chute	Fast, smooth boundary turbulent flow over boulders or bedrock. Flow is in contact with the substrate and exhibits upstream convergence and divergence.
Broken standing waves	Rapid	White-water tumbling waves with crest facing in an upstream direction.
Unbroken standing waves	Riffle	Undular standing waves in which the crest faces upstream without breaking.
Rippled	Run	Surface turbulence does not produce waves, but symmetrical ripples which move in a general downstream direction.
Upwelling	Vertical flow	Secondary flow cells visible at the surface by vertical 'boils' or circular horizontal eddies.
Smooth surface flow	Glide	Relative roughness is sufficiently low that very little surface turbulence occurs. Very small turbulent flow cells are visible, reflections are distorted and surface 'foam' moves in a downstream direction.
Scarcely perceptible flow	Pool	Surface foam appears stationary, little or no measurable velocity, reflections are not distorted.
Standing water	Backwater	Abandoned channel zone or backswamp with no flow except at flood stage.

[a] From Thomson *et al.* (2001).
[b] See Chapter 8.

At the channel/reach scale vegetation and wood resistance elements reduce flow velocity and shear stress. Near-bank velocities and shear stress are significantly reduced against rough banks. Vegetation also increases bank strength, as root masses bind sediments. As a result, vegetation and wood may reduce rates of erosion or channel migration by several orders of magnitude. Channels with dense riparian vegetation and high loadings of wood have a more irregular channel morphology than those without. Channels are also smaller; on average, they are 0.5–0.7 times the width and 1.4 times the depth of an equivalent channel that is only vegetated by grass. However, even a thin grass cover may inhibit incision and act as an efficient trap for sediments.

Vegetation acts as a blocking agent in channels, increasing channel roughness and reducing flow velocity. The hydraulic effect of vegetation is proportional to the area that is projected into the channel cross-section. It may comprise 30–50% of the total hydraulic resistance in a channel. However, hydraulic resistance varies with flow stage and the size, structure and flexibility of the vegetation (including its root structure). In general, roughness is reduced as flow stage increases and the resistance effect of vegetation is damped out. Also, as a general rule, the proportion of vegetation occupying a channel cross-section decreases downstream as the channel becomes larger.

Channel geometry (size and shape) influences the frequency and periodicity of inundation of differing geomorphic surfaces, thereby affecting the capacity for sediment reworking and resulting patterns of erosion and deposition. Different types of vegetation grow on different geomorphic surfaces, controlled largely by the frequency of inundation and substrate type (Figure 5.6). In-channel vegetation (macrophytes, water reeds, or shrubs and trees) increases roughness and reduces flow velocity, thereby increasing rates of sediment deposition. Binding increases the residence time of these sediment stores. Positive feedback mechanisms enhance deposition and storage of sediment in the channel, promoting prospects for germination and growth of vegetation on different surfaces. Once established, vegetation-induced resistance affects the geomorphic effectiveness of flow events upon differing geomorphic surfaces. Conversely, reductions to ground cover may make sediment stores more vulnerable to movement.

Different vegetation patterns are affected by, and in turn induce, differing geomorphic responses for different types of rivers. This may range from systematic successional associations induced by lateral migration of bends, to patches characterised by abrupt transitions, such as those associated with channel abandonment along wandering gravel-bed or anabranching rivers (see Chapter 10). As a result, vegetation may exert a critical control on adjustments to channel geometry and planform. For example, incursions of exotic vegetation may induce profound adjustments to channel planform. Colonisation of bars and islands by vegetation may induce a shift from a braided to a meandering river. Alternatively, avulsion may be induced, as dense vegetation or wood accumulations provide focal blockage points around which channel shift occurs (see Chapter 11).

The resistance role played by riparian vegetation varies markedly for different types of rivers in different environmental settings. For example, the geomorphic role

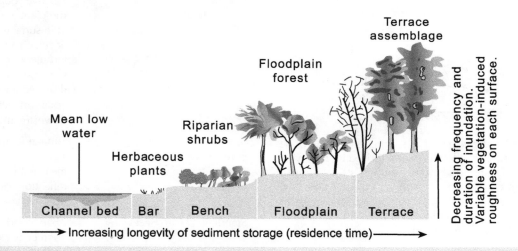

Figure 5.6   Different types of vegetation grow on different geomorphic surfaces reflect height above low flow, substrate conditions and degree of disturbance. Herbaceous plants grow close to the water in areas that are disturbed regularly. Floodplain forests are inundated and disturbed less regularly and often contain a mix of canopy and understorey species.

of riparian vegetation and wood is quite different in arid and semi-arid climates (whether hot or cold desert conditions) relative to humid–temperate or tropical settings. The presence and age of trees along riparian margins are key determinants of the recruitment potential of wood supply, exerting a longer term influence upon river morphodynamics. The preservation potential of wood influences its geomorphic role. For example, the decay rate of wood in tropical rivers may be rapid, while temperate streams may retain the same pieces of wood for extended periods of time. Markedly different patterns of riparian vegetation are evident along, say, upland swamps, gorges, meandering rivers in rainforests and alluvial plains in arid zones. Some rivers are dominated by grasses, others by shrubby scrub vegetation. Others have dense riparian vegetation but open floodplains. Some rivers have floodplains covered with closed rainforest; others contain significant wetland vegetation as valley margins. Other rivers flow through swamps or peatlands.

Given the genetic link between riparian vegetation and recruitment of wood along rivers, wood loadings vary markedly in different environmental settings. Wood may accumulate as single branches, logs, whole trees or assemblages of trees trapped within log jams. The resistance posed by wood depends on its orientation, size and density/frequency in the channel. Depending on their size relative to the channel, wood may span the channel, accumulate in clusters or be transported downstream (Figure 5.7). In low-order channels and headwater zones, wood may induce channel blockage ratios as high as 80%. Narrow channels are directly linked to hillslopes in these settings, resulting in high wood supply via debris flows, landslides and windthrow (direct fall). In these parts of catchments, wood structures tend to span the channel. In general, types of wood structures reflect vegetation type (e.g. size, root networks, wood density) and river morphology. For example, widely spreading or multiple-stemmed hardwoods are more prone to forming wood structures than accumulating as racked members of large log jams because they extend laterally as well as beyond their bole diameter. In contrast, coniferous wood tends to produce cylindrical pieces that are more readily transported through river systems, resulting in local concentrations of log jams.

In lowland zones, channels are wide and hillslopes are further away from the channel margin. Wood is supplied by local undercutting of banks or is transported from upstream. Log jams may accumulate along the channel bed/bank. Wood tends to be rotated sub-parallel to the flow, minimising the blockage ratio, but maximising its role in bar accretion and bank toe protection. Logs may also be incorporated into the channel bed, increasing bed stability. Depending on flow alignment and wood size, a range of different structures or jams can be formed

(Figure 5.7). In wider channels, wood may be transported beyond the fall point and become incorporated into log jams. This may cause local bank erosion, trigger channel avulsion or cut-off development, or promote island development.

Manning's *n* as a unifying roughness parameter

Manning's *n* is the most commonly used measure of roughness in river systems. It is related to the Chezy and Darcy–Weisbach coefficients, in that it is derived from a velocity-based equation. This coefficient integrates the effects of flow resistance caused by bed roughness, the presence and flexibility of vegetation, the amount of sediment or debris carried by the flow and other factors. Spatially, values of Manning's *n* vary for different parts of a channel cross-section such as overbank areas (whether bedrock controlled or floodplains) and the channel zone itself (see Figure 5.6). Manning's *n* also varies throughout a system depending upon which types of roughness influence flow, and the individual characteristics of roughness elements. Manning's *n* changes with flow stage. Roughness decreases as flow strength increases. Individual grains protrude less into the flow as flow depth increases, thereby reducing resistance.

Manning's *n* can be derived using a component technique or a velocity equation. The component method uses measures of various forms of roughness in channels (e.g. the shape of the cross-section, the type of vegetation on the banks, the sinuosity of the channel) to derive the overall value (Table 5.2):

$$n = (n_0 + n_1 + n_2 + n_4 + n_4)\,m_5$$

where n_0 is a value for the material of a straight uniform channel, n_1 is a factor for surface irregularities, n_2 is a factor for variations in the shape and size of the channel cross-section, n_3 is a factor for obstructions, n_4 is a factor for vegetation and m_5 is a multiplier for channel meandering.

The component method is unsuitable for large channels (with $W \gtrsim 10\,\text{m}$). Care must be taken to avoid double counting channel characteristics in the various components. In most cases, Manning's *n* is calculated using a velocity equation:

$$v = \frac{d^{2/3}s^{1/2}}{n}$$

where v (m s^{-1}) is the flow velocity, d (m) is the channel depth, s (m m^{-1}) is the channel slope and n is Manning's roughness coefficient. Visual guides have been developed

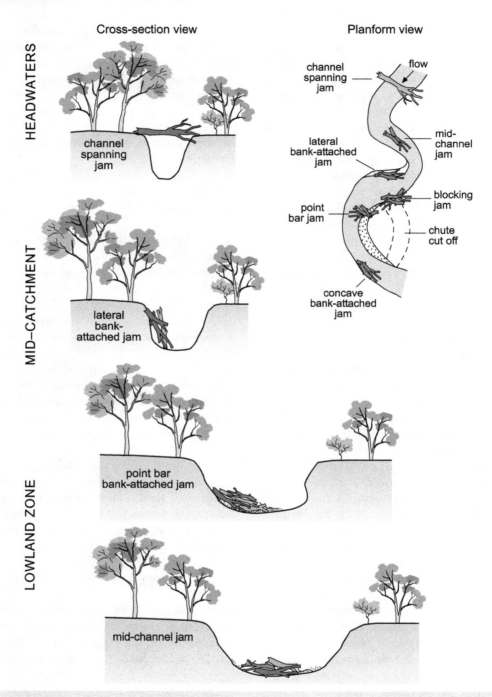

Figure 5.7 Different types of wood structures. Depending on within-catchment position and the size and configuration of the channel, wood structures can span a channel, occur along channel banks or in mid-channel locations. A range of different types can form, including point bar jams, concave bank jams, lateral bank jams, blocking jams, mid-channel jams and spanning jams. In middle and lowland rivers, the formation of log jams requires an obstruction to flow, or a key member to prevent flushing from the reach. Wood provides significant resistance to flow through its blocking effect. The blocking ratio is a measure of the percentage of the channel cross-section that contains wood. Based on Cohen and Brierley (1997).

Table 5.2 Values of various forms of roughness used in the component method of Manning's n derivation

Material, n_0		Degree of surface irregularity, n_1	
earth	0.020	smooth	0.000
rock	0.025	minor (e.g. minor slumping)	0.005
fine gravel	0.024	moderate (e.g. moderate slumping)	0.010
coarse gravel	0.028	severe (e.g. badly slumped or irregular bedrock)	0.020
Variation of channel cross-section, n_2		**Relative effect of obstructions (debris, roots, boulders etc), n_3**	
gradual	0.000	negligible	0.000
alternating occasionally	0.005	minor	0.010–0.015
alternating frequently	0.010–0.015	appreciable	0.020–0.030
		severe	0.040–0.060
Vegetation, n_4		**Degree of meandering, m_5**	
none	0.000	minor (sinuosity <1.2)	1.00
low	0.005–0.010	appreciable (sinuosity 1.2–1.5)	1.15
medium	0.010–0.025	severe (sinuosity >1.5)	1.30
high	0.025–0.050		
very high	0.050–0.100		

that outline the range of Manning's n expected for channels of certain sizes and configurations and areas of overbank flow with various forms of vegetation coverage (Table 5.3).

The balance of impelling and resisting forces along longitudinal profiles

The potential energy that is available to do work along a river (e.g. entrain/transport sediment, erode bed/banks) is a function of the balance of impelling and resisting forces at any given point in a river system. Conceptually, a range of equilibrium relationships driven by negative feedback mechanisms attempt to maintain a balance between slope, bed material size, channel size, etc. These relationships determine the consumption of energy at different positions along the longitudinal profile of a river (especially energy that is required to overcome frictional resistance). This attempt to balance impelling and resisting forces along a reach is a critical determinant of a river's behavioural regime, indicating whether there is a relative dominance of erosional or depositional tendencies at any given location along the longitudinal profile. Adjustments to the magnitude or distribution of impelling or resisting forces are primary agents of river behaviour (Chapter 11) and change (Chapter 12).

The balance of impelling and resisting forces determines the potential for threshold exceedance in any given reach. Within a given set of boundary conditions, rivers may adjust to prevailing water, sediment and vegetation conditions to generate a range of river morphologies (see Chapter 10). These different types of river use their energy in efforts to balance impelling and resisting forces in different ways.

Some rivers act to maximise resistance, while others act to maximise sediment transport. Understanding these different behavioural attributes is a key step in interpreting why a river looks and behaves in the way that it does (Chapter 11). By interpreting differences in these relationships, major transitions in process–form relationships can be detected along river systems. For example, in upstream sections of catchments, resistance is imposed by forcing elements such as bedrock and so bedrock-confined rivers dominate. Adjustments to these resistance elements occur over geological timeframes. In contrast, in downstream reaches, alluvial channels are able to create and sustain their own forms of resistance elements around which adjustments occur over geomorphic timeframes.

There are many enigmatic relationships in the study of river systems. One of the most puzzling scenarios to unravel is how rivers adjust to balance impelling and resisting forces at differing positions along smooth, concave-upwards longitudinal profiles. The fact that most river system 'hold together' over extensive periods of time attests to the effectiveness of this balancing act. Although rivers are forever subjected to disturbance events, whether natural or human induced, dramatic responses are the exception rather than the norm. Typically, systems internally adjust to minimise the impacts of external events. In other words, rivers adjust the balance of impelling and resisting forces in any given reach such that the system is able to use available water (flow) to transport available sediment. Of course there are exceptions to this generalisation, and this balance may be disrupted by extreme events. Indeed, some systems are especially sensitive to change, especially if they lie close to a threshold condition. In general terms, however, it is the ability of channels to alter the resistance that they provide

Table 5.3 Visual estimates of Manning's *n* coefficient[a]

	Manning's *n*
Channels excavated in earth	
Straight, uniform and clear	0.016–0.020
Windy with grass and some weeds	0.025–0.033
Channels excavated in rock	
Smooth, uniform	0.025–0.040
Jagged, irregular	0.035–0.050
Small natural streams (bankfull <30 m)	
Straight, uniform and clean	0.025–0.033
Clean, winding with some pools and shoals	0.033–0.045
Sluggish, weedy with deep pools	0.050–0.080
Very weedy with deep pools	0.075–0.150
Steep mountain stream with gravel, cobbles and boulders	0.030–0.070
Major natural streams (bankfull >30 m)	
Regular cross section with no boulders or brush	0.025–0.060
Irregular and rough cross-section	0.035–0.100
Overbank flow areas	
Short pasture grass with no brush	0.025–0.035
Long pasture grass with no brush	0.030–0.050
Light brush and trees	0.040–0.080
Medium-dense brush	0.070–0.160
Dense growth of willows	0.110–0.200
Channel material types	
Concrete pipe	0.012–0.014
Sand-bed channel	0.010–0.040
Gravel/cobble bed	0.020–0.070
Boulder-bed	0.030–0.200
Bedrock step–pool	0.100–0.500

[a]Modified from Chow (1959).

to flow that modifies the operation of the Lane balance in any given reach. Adjustments to bed material size and the arrangement of materials on the bed, channel geometry (size and shape) and channel planform (number of channels and their sinuosity) use flow energy in differing ways, allowing the channel to minimise the extent of response

to disturbance events. For example, channel contraction results in greater dissipation of energy in overbank flows that occur more frequently because the channel is smaller (and vice versa). Alternatively, increases in channel sinuosity decrease the slope of the channel, thereby reducing flow energy (and vice versa). A similar increase in roughness, and consumption of energy, arises from an increase in bed material size or an increase in channel multiplicity (and associated surface area, or boundary resistance). These various relationships adjust in different ways at differing positions along longitudinal profiles.

Slope and available energy are high in steepland settings of the source area headwater compartment of river systems with smooth, concave-upward longitudinal profiles. However, catchment areas are small, minimising available discharge. Typically, infrequent high-intensity storms are required to trigger formative flow events. Given steep slopes, these events have significant capacity to perform geomorphic work. However, roughness elements on the valley floor consume large amounts of energy, minimising the geomorphic effectiveness of these flows. There are two primary components to the high inherent roughness in these areas. First, these bedrock-controlled rivers have irregular channel boundaries (i.e. channel geometry), irregular channel alignment (i.e. planform), steep slopes with a large number of steps, cascades and waterfalls, and many forced roughness elements induced by features such as trapped wood (log jams). Second, most fine-grained sediments are flushed through these high-energy settings, leaving behind the coarse fraction of the bed material load. This is usually comprised of boulders, cobbles and gravels. These materials are only mobilised by the highest (extreme) flows. Indeed, they are often lag deposits from the last formative flow. Critically, these materials generate considerable flow resistance to all events other than the deepest flow. As a consequence, the valley floor is relatively stable for the vast majority of the time. Infrequent high-magnitude events may bring about localised disturbance, but to all intents and purposes the channel adjusts very little. Incrementally, when critical shear stress is exceeded and the coarsest fraction is mobilised, the channel is able to cut into its bed (i.e. these are degradational systems).

Transfer reaches along rivers with smooth, concave-upwards longitudinal profiles are typically found downstream of source zones along sections with lower slopes and wider valley settings. These are transition zones in river systems. The two primary components of impelling forces, slope and discharge are adjusting along differing trajectories. The rate of decrease in elevation decreases with distance from the headwaters, reducing slope with distance along the longitudinal profile. In contrast, discharge tends to increase as catchment area increases. In general terms, a peak in total stream power conditions occurs around the beginning of the transfer zone, typically for third- to fifth-order streams (see Chapter 3). However, total stream power decreases downstream, as the rate of decrease in slope is greater than the rate of increase in discharge. Just as importantly, however, the increase in valley width with distance downstream alters the way in which the channel uses its available energy. Rather than simply concentrating energy in efforts to incise into its bedrock bed, flow energy is dissipated in wider sections of valley in which floodplain pockets may form (i.e. the river is able to deposit and rework finer grained sediments in these reaches). As a consequence, the river uses its energy in a different way to upstream reaches, achieving a different configuration in which erosional and depositional processes are approximately in balance.

Finally, the accumulation zone of catchments has lower slope and higher discharge conditions than upstream reaches. In these areas the channel is most able to consume its own energy through self-adjustment processes, as the channel flows within its own sediments. Hence, adjustments to channel planform and geometry are able to accommodate alterations to flow and sediment conditions more readily than elsewhere in the catchment. Impelling forces are lower than upstream, as lower slopes exert a greater influence upon total stream power than the impact of higher discharge. Typically, flow is smoother in the larger channels of accumulation zones that flow within finer grained (less rough) sediments (except where large bedforms are produced), so resisting forces are also reduced. Bed materials are reworked more frequently in these areas relative to upstream. In these parts of catchments, the channel is less able to transport all sediments made available to it, so aggradation occurs on valley floors and sediments prograde into receiving basins.

In summary, flow energy is used in differing ways in source, transfer and accumulation zones, but the quest for balance is sustained throughout. Understanding of these relationships at any given site, appreciating the ways in which differing forces are applied in any given system and unravelling the ways in which one reach is connected to another are key skills in efforts to read the landscape. Geomorphologists analyse how the balance works (or does not work) along any given section of river and how the causes of that adjustment may induce on- or off-site impacts, whether locally or in upstream or downstream parts of the longitudinal profile.

Conclusion

The balance of impelling and resisting forces provides critical guidance to understanding the distribution of erosional and depositional processes in river systems. This results in pronounced downstream variability in the pattern of river types, and their associated character and behaviour. Prior to analysing these relationships, the following chapter considers how flow energy is used to transport sediments through river systems.

Key messages from this chapter

- Hydraulics is the study of the mechanics of water flow and the impelling forces that induce sediment movement.
- Laminar flow is smooth, orderly motion of flow; turbulent flow is random, chaotic motion of flow. The range of flow types reflects the relationship between flow depth, velocity and substrate conditions.
- The Lane balance diagram and Exner equation summarise primary controls upon the degradational–aggradational balance of a river system. The erosion–deposition balance primarily reflects the volume of water (discharge) flowing on a given slope and the ability of the river to carry sediment of a given volume and calibre. The 'buckets' on the Lane balance dictate whether aggradation (deposition) or degradation (erosion) occurs.
- Impelling forces perform geomorphic work through erosion and reworking of materials. They are a function of the volume of water acting on a given slope. This can be measured as stream power or shear stress.
- Total stream power is computed as the volume of water flowing over a certain slope. Unit stream power is calculated as energy expenditure per unit width of channel. This provides a measure of the way that energy is used in a river system. Critical stream power is the power needed to transport the average sediment load supplied to the stream.
- Shear stress is the tractive force applied by a flowing liquid to its boundary. It is measured as the product of fluid density, the acceleration due to gravity, flow depth and slope. Critical shear stress defines the threshold stress that is required to initiate bed erosion and sediment movement.
- Whether geomorphic work is done is dependent on the amount of available energy and the balance of energy expended and energy conserved at a particular location. Three possibilities exist: a river may have more energy than that required to move its water and sediment load, in which case surplus energy is used to erode boundaries; it may have exactly that required, in which case it is stable; it may have an energy deficit, which results in deposition.
- Resisting forces are frictional forces that reduce flow energy, providing a measure of how a river consumes

its energy. Primary resisting elements include: valley-scale resistance (valley morphology and confinement), channel-scale resistance (planform and bed-bank roughness), boundary resistance (grain and form roughness on the channel bed) and fluid resistance (internal viscosity and free-surface resistance). Manning's *n*, a unify-

ing roughness parameter, is measured by relating flow velocity to channel depth and slope.

• The balance of impelling and resisting forces along longitudinal profiles is the key determinant of the balance of erosional and depositional processes that make up source, transfer and accumulation zones (process domains).

CHAPTER SIX

Sediment movement and deposition in river systems

Introduction

Although rivers only account for a small proportion of water on Earth, they are the primary agent of sediment transfer in landscapes. Chapter 5 focused on the mix of impelling and resisting forces that fashion the flow–sediment balance in river systems. The Lane balance was used to describe how water flowing over a certain slope interacts with sediment of a given calibre to determine the aggradational–degradational balance of a river and resulting river morphology. The balance of impelling and resisting forces determines whether the river has excess energy and is scouring and transporting sediment, or whether there is insufficient energy to move available sediment and deposition occurs. Chapter 4 examined hydrological considerations that affect the Lane balance. This chapter focuses upon the 'sediment bucket' in the Lane diagram (sediment calibre and volume). Processes of sediment entrainment, transport and deposition are discussed in the context of the Hjulström diagram. Controls upon these processes are outlined and the use of sedimentological properties to aid efforts to read the landscape is discussed.

Grain size (sediment calibre) and definitions of bedload, mixed load and suspended load in rivers

Sediment calibre refers to the size of material that is available to be carried by a river. Grain size exerts an influence upon particle entrainment, modes of sediment transport, distances travelled and patterns of deposition. Rivers sort their load in longitudinal, lateral and vertical directions, giving rise to characteristic morphological traits.

The Wentworth scale is the primary framework that is used to define sediment calibre (Table 6.1). This scale differentiates among boulders, cobbles, gravel, sand, silt and clay. For materials coarser than sand-sized particles, the b-axis or intermediate axis, i.e. the axis perpendicular to the longest (orthogonal) axis, is usually measured in the

field. Of the three primary axes, the b-axis most closely reflects the weight of a particle. Point samples can be obtained by defining a grid for a particular depositional feature (locale) on the bed/bar surface and systematically measuring b-axes for 100 particles or more at each node of the grid. Alternatively, a bulk sample can be removed from the bed/bar and sieved at 0.5ø interval in the field. Each bulk sample should be sufficiently large such that the largest stone in the sample is not more than 1 % of the total sample weight. Visual-estimation grain-size charts can be used to assess mean and maximum grain sizes for sand-size particles at 0.5ø grain-size intervals. Alternatively, samples are returned to the laboratory for more detailed grain-size analysis. For silt and clay fractions, field texturing can be used to differentiate among grain-size classes (Table 6.2), or an array of laboratory procedures can be applied.

Unlike coarse sediments, cohesive sediments are not amenable to classification by grain size and distribution. Rather, complex particle bonds affect the properties of cohesive sediments, and associated behaviour in terms of erosion, deposition and resuspension. Physical, electro-chemical and biological effects interact to affect these behavioural properties. Physical factors affecting erodibility include clay content, water content, clay type, temperature, bulk density and pore pressure. Physico-chemical properties of the overlying fluid also affect erodibility. The chemistry of the eroding fluid and the pore fluid (e.g. pH, salinity, cation exchange capacity, sodium adsorption ratio) influences the valence of clay particles. As such, it plays a critical role in interparticle bonding, thereby influencing the erosion rate and the degree of flocculation of the clay–water suspension. Natural organic matter, measured as the percentage of organic carbon adsorbed to clay particles in the sediment, can increase the interparticle bonding, thereby increasing resistance to erosion.

Sediment size is often quoted as a characteristic grain diameter D, such as the median D_{50} or the D_{84} (the grain diameter at which 84 % of the material is finer). Field measurements are required to derive the grain-size distribution plot from which the D_{50}, D_{84} and other grain-size statistics

Geomorphic Analysis of River Systems: An Approach to Reading the Landscape, First Edition. Kirstie A. Fryirs and Gary J. Brierley.
© 2013 Kirstie A. Fryirs and Gary J. Brierley. Published 2013 by Blackwell Publishing Ltd.

Table 6.1   The Wentworth scale of grain size[a]

Class name	Grain size (mm)	Grain size (ø)	How measured in the field?
Boulder	≥256	≤−8	*b*-axis with ruler
Cobble	64–256	−6 to −8	*b*-axis with ruler
Gravel	2–64	−1 to −6	*b*-axis with ruler
(pebbles)	(4–64)	(−2 to −6)	
(granules)	(2–4)	(−1 to −2)	
Sand	0.063–2	4 to −1	Sand grain-size card
(very coarse sand)	(1–2)	(0 to −1)	
(coarse sand)	(0.5–1)	(1 to 0)	
(medium sand)	(0.25–0.5)	(2 to 1)	
(fine sand)	(0.125–0.25)	(3 to 2)	
(very fine sand)	(0.063–0.125)	(4 to 3)	
Silt	0.004–0.063	9 to 4	Field texture guide
Clay	≤0.004	≥−9	Field texture guide

[a]Modified from Parker (2008).

Table 6.2   Field texture grades

Texture grade	Behaviour of moist bolus	Approx. clay content (%)
Sand (S)	Coherence nil to very slight; cannot be moulded; single sand grains adhere to fingers	normally <5 always <10
Loamy sand (LS)	Slight coherence; can be sheared between thumb and forefinger to give a minimal ribbon of 6–7 mm; discolours fingers with dark organic stain	5–10
Clayey sand (CLS)	Slight coherence; sticky when wet; many sand grains stick to fingers; minimal ribbon of 6–13 mm; discolours fingers with clay (sesquioxide) stain	5–10
Sandy loam (SL)	Bolus just coherent but very sandy to touch; will form ribbon of 13–25 mm; dominant sand grains are medium size (250–500 μm)	10–15
Fine sandy loam (FSL)	Bolus coherent; fine sand (63–250 μm) can be felt and heard when manipulated; will form ribbon of 13–25 mm; sand grains clearly evident under hand lens	10–20
Loam (L)	Bolus coherent and rather spongy; feel smooth when manipulated but with no obvious sandiness or silkiness; may be somewhat greasy to touch if organic matter present; will form ribbon of about 25 mm	~20–25
Silt loam (SiL)	Coherent bolus; very smooth to silky when manipulated; forms ribbon of about 25 mm	~25+ = 25% silt
Sandy clay loam (SCL)	Strongly coherent bolus; sandy to touch; medium-size sand grains (250–500 μm) visible in matrix; will form ribbon of 25–38 mm	20–30
Clay loam (CL)	Coherent plastic bolus; smooth to manipulate; will form ribbon of 38–50 mm	30–35
Silty clay loam (SiCL)	Coherent smooth bolus; plastic and silky to touch; will form ribbon of 38–50 mm	30–35+ = 25% silt
Sandy clay (SC)	Plastic bolus; fine to medium sands can be seen, felt or heard in clayey matrix; will form ribbon of 50–75 mm	35–40
Silty clay (SiC)	Plastic bolus; smooth and silky to manipulate; will form ribbon of 50–75 mm	35–40+ = 25% silt
Light clay (LC)	Plastic bolus; smooth to touch; slight resistance to shearing between thumb and forefinger; will form ribbon of 50–75 mm	35–40
Medium clay (MC)	Smooth plastic bolus; handles like plasticine; can be moulded into rods without fracture; some resistance to ribboning shear; will form ribbon of about 75 mm or more	40–50
Heavy clay (HC)	Smooth plastic bolus; handles like stiff plasticine; can be moulded into rods without fracture; firm resistance to ribboning shear; will form ribbon of about 75 mm or more	= 50

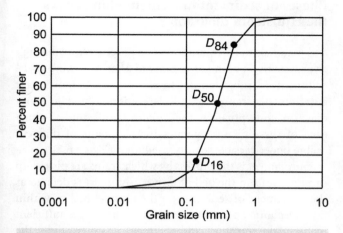

Figure 6.1 Example of a grain-size distribution (cumulative frequency) curve. The *x*-axis represents grain size ranging from clay to gravel. The *y*-axis represents the percentage of the sample that is finer than each size fraction. The D_{50} (median), D_{84} (grain size at which 84% of the sample is finer) and D_{16} (grain size at which 16% of the sample is finer) are shown. These are commonly used measures of grain size and variability within a sample.

can be determined. The geometric standard deviation of sediment sizes σ_g provides a useful guide to the mix of grain sizes (i.e. the degree of sorting). This is computed as $(D_{84}/D_{16})^{0.5}$. If $\sigma_g < 1.3$, the sediment mix is considered uniform. For more detailed investigations, cumulative frequency plots of grain-size distributions are produced (Figure 6.1). This commonly entails a combination of field and laboratory procedures to textural analysis.

Alluvial rivers can be broadly divided into two types: *sand-bed* streams have surface median size D_{50} in the range 0.0625–2 mm, while for *gravel-bed* streams $2 < D_{50} < 256$ mm. For simplicity, cobble- and boulder-bed streams are integrated into the latter category. The dividing line between sand- and gravel-bed streams is not arbitrary; streams with a characteristic size between 2 and 16 mm (pea gravel) are relatively rare. Sand and silt often move through a gravel-bed river as throughput load during floods, with little interplay with the beds beyond partial filling of the interstices of newly deposited gravels. When the concentrations of these 'fines' are too high, or when the flow velocities are too low to prevent excess accumulation within the gravel framework, the gravels can become polluted with fines. The grain-size distributions of most sand-bed streams are unimodal and can often be approximated with a normal distribution function. However, many gravel-bed rivers have bimodal grain-size distributions, with both

a gravel mode and a sand mode, but a paucity of pea-gravel size sediment (2–16 mm).

Bed material load is that part of the sediment load that exchanges with the bed (and thus contributes to morphodynamics). *Wash load* is transported without exchange with the bed. In other words, the wash load of a river consists of sediment moving in suspension that is too fine to be present in measurable fractions in the bed. Wash load is more properly termed 'floodplain material load' because it exchanges with the floodplain. In rivers, material finer than 0.0625 mm (silt and clay; i.e. mud with $D < 0.062$ mm) is often approximated as wash load. Bed material load is further subdivided into bedload and suspended load. *Bedload* refers to movement of material by sliding, rolling or saltating in a trajectory just above the channel bed (see below). Turbulence plays an indirect role in this motion. In contrast, suspended load is subjected to the direct dispersive effect of turbulent eddies within flow, such that particles may be moved high into the water column. Low-slope sand-bed rivers move their bed material load (typically sand) as both bedload and suspended load, but suspended load far dominates bedload at the flood conditions that transport most of the sediment. In most large, low-slope sand-bed streams, mud comprises the great majority of the sediment transported on a mean annual basis.

Bedload, suspended-load and mixed-load rivers are differentiated on the basis of the relative proportion of grain sizes carried within the flow or along the channel bed (Schumm, 1968). *Bedload rivers* carry more than 11% of their load as sand-sized grains or larger along the channel bed. These rivers tend to have a high width/depth ratio with non-cohesive banks and loose, coarse materials on the bed. *Suspended-load rivers* carry the vast proportion of their load as fine sand, silt and clay materials that are suspended in the body of the flow. Bedload is <3% of the total load. These rivers tend to have a low width/depth ratio with fine-grained cohesive banks and beds. *Mixed-load rivers* carry a mix of bedload and suspended load (typically 3–11% bedload). Their bed often comprises a diverse range of grain sizes. Banks are composite, with non-cohesive materials capped by cohesive fine-grained deposits. Material size and distribution are key influences upon the mobility of particles on the channel bed and banks.

Fine cohesive sediment can cause the bed's strength to be greater than the shear stress required to entrain coarser particles. Erosion rates of silts, sands and gravels are limited to the entrainment rate of the clay when more than 10% of the bed is composed of clay. If the bed shear stress is larger than the critical shear stress for the finer size classes, but smaller than that for coarser size classes, only the finer classes are eroded. Dependent upon sediment supply conditions, this will eventually armour the bed surface and prevent further erosion.

Cohesive sediments are closely linked to water quality. Many pollutants, such as heavy metals, pesticides and nutrients, preferentially adsorb to cohesive sediments. In addition to the contaminants absorbed to the sediments, the sediments themselves are sometimes a water quality concern. The turbidity caused by sediment particles can restrict the penetration of sunlight and decrease food availability, thus affecting aquatic life.

Sediment coarser than 62 μm is coarse, non-cohesive material. Sediment sizes smaller than 2 μm (clay) are generally considered cohesive sediment. Silt (2–62 μm) is considered to be between cohesive and non-cohesive sediment. Indeed, the cohesive properties of silt are considered due to the existence of clay. Thus, in practice, silt and clay are both considered to be cohesive sediment. For sediment containing more than approximately 10% clay, the clay particles control the sediment properties. Cohesive sediments consist of inorganic minerals and organic material. Inorganic minerals consist of clay minerals (e.g. silica, alumina, montmorillonite, illite and kaolinite) and non-clay minerals (e.g. quartz, carbonates, feldspar and mica, among others). The organic materials may exist as plant and animal detritus and bacteria.

Phases of sediment movement along rivers: the Hjulström diagram

Bed materials move intermittently and recurrently through river systems, in a similar manner to a 'jerky conveyor belt'. Sediment is subjected to three phases of movement: entrainment, transport and deposition. The Hjulström diagram conceptualises the circumstances under which each of these phases operates for sediments of different calibre under variable velocity conditions (Figure 6.2).

Entrainment is the process by which grains are picked up or plucked from the bed of a river. *Transport* is defined as the movement of sediment on the channel bed or within flow. *Deposition* occurs when energy is no longer sufficient to maintain transport and sediment is stored along the river. The threshold between entrainment and transport demarcates the flow velocity and sediment size conditions under which sediment is either picked up (entrained) or remains stationary. Given the dynamics of flow in natural channels, the entrainment threshold spans a range of flow velocities and is represented by a band on the Hjulström diagram. A sharply dipping threshold separates transport and deposition domains. This threshold depicts the

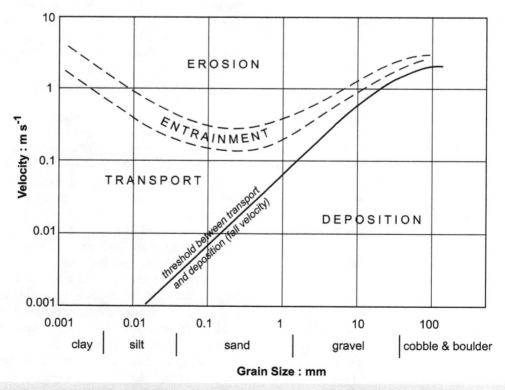

Figure 6.2 The Hjulström diagram depicts phases of sediment entrainment (erosion), transport and deposition based on the grain size of the sediment and the velocity of flow. Modified from Hjulström (1935).

grain size and velocity conditions under which sediment is removed from transport and is deposited. This is defined in relation to the fall velocity, the velocity at which sediments fall out of flow and are deposited (see below).

The low point for the threshold between entrainment and transport occurs for medium-sized sands. These non-cohesive grains are the most readily entrained sediments in river channels, with a threshold velocity of around 0.2 m s^{-1}. Other sand-sized materials are entrained at velocities between 0.2 and 0.4 m s^{-1}. Velocities greater than 1 m s^{-1} are required to entrain coarser clasts. Electrochemical cohesive properties ensure that the silt–clay fraction also requires high flow velocities to be entrained.

Once entrained, turbulence maintains silt and clay materials in transport under a wide range of flow velocities, such that they can be transported considerable distances. A significant drop in velocity (or even standing water with velocities <0.01 m s^{-1}) is required for these fine-grained particles to settle out from the water column. In contrast, coarse-grained sediments (particularly gravel clasts or coarser) spend little time in transport. This is represented by the closeness of the entrainment threshold to the deposition threshold on Figure 6.2. Once entrained, coarse sediments tend to be transported short distances as bedload. Small decreases in flow velocity (<0.2 m s^{-1}) may breach the fall velocity threshold, resulting in deposition of the grain. As a result, coarse-grained sediments tend to spend significant time in storage.

Entrainment of sediment in river channels

Entrainment processes detach grains from a surrounding surfaces making them available to be transported. Detachment occurs via a number of mechanisms. *Corrasion* is the mechanical (hydraulic and abrasive) action of water on a surface. *Cavitation* occurs when pressure differentials caused by shock waves generated by the collapse of vapour pockets detach particles from a surface. Entrainment occurs at higher flow velocities than both transport and deposition. This is because the impelling forces of the flow must overcome the resistance forces acting on the grain, including friction (the weight of particle, its roughness and interlocking), and cohesion (the electrochemical and surface tension forces of the grain) (Figure 6.3). Grain density also exerts a critical influence upon entrainment.

The two primary impelling forces that act to entrain particles are fluid drag and lift force. *Fluid drag* is exerted by an erosive agent (i.e. water) exerting a force in the direction of flow. Horizontal drag is affected by the velocity of the flow and flow density. The greater the velocity and viscosity of flow, the greater the potential for drag to exert

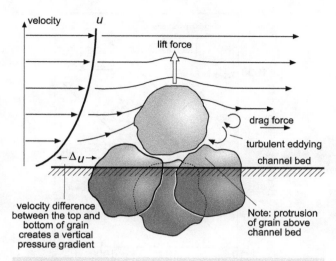

Figure 6.3 Forces acting on a grain. Flow velocity increases away from the channel bed as resistance decreases. Drag and lift forces, associated with flow turbulence, entrain grains from the channel bed. Entrainment depends on particle size and the extent to which it protrudes into the flow or is embedded against other grains. Modified from Knighton (1998).

a force on the grain. In addition, a *vertical lift* force is required. Vertical lift is affected by flow turbulence and buoyancy of the particle. The pressure gradient induced by the difference in flow velocity between the top and bottom of a grain acts to lift the particle vertically. Turbulent eddying may also induce vertical velocity components close to the bed. However, these forces decrease rapidly away from the bed as velocity and pressure gradients diminish. Therefore, grains that protrude from the channel bed have more vertical lift exerted upon them than particles which are embedded or interlocked in surrounding substrate. The downslope component of movement is affected by particle weight and slope angle. Once the combined lift and drag forces exceed the cohesion and friction forces, grains are raised off the channel bed and move into transport. The conditions under which initiation of motion occurs is called the entrainment threshold and can be measured as a critical velocity v_{cr} as depicted in the Hjulstrom diagram (Figure 6.2), or as a critical shear stress τ_c (Figure 6.4).

In general, the critical bed shear stress τ_c required to move grains increases with grain size (Figure 6.4). However, the shear stress at which the grain will actually move can differ by an order of magnitude. For example, a grain that is 100 mm in size could move under a shear stress as low as 20 N m^{-2}, but is more likely to move at shear stresses >100 N m^{-2}. This variance reflects instantaneous stresses within the flow, bed roughness and whether the grain is loose or embedded.

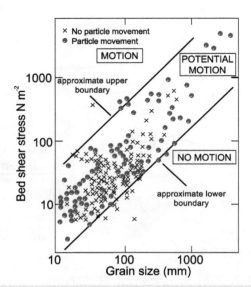

Figure 6.4 Bed shear stress as a function of grain size. In general, as grain size increases, greater shear stress is required to entrain a grain. However, this threshold of motion is not distinct. Rather, it is best viewed as a zone of motion where grains may, or may not, move depending on the shape of the grain and the looseness of the channel bed. Greater shear stresses are required to entrain grains that are embedded, hidden or armoured relative to loose particles. From Williams (1983). © John Wiley and Sons, Ltd. Reproduced with permission.

Frictional resistance influences the efficiency of entrainment in natural stream beds. For coarse sediment, grain size is the key control. Other controls include bed packing, armouring and hiding (see below). For fine-grained sediment, cohesion is the key control on the efficiency of entrainment, but sheltering within the laminar layer also occurs. Other controls include the presence of in-channel roughness elements (e.g. vegetation, wood) that reduce energy and/or increase bed strength. These additional resistance forces must be overcome before entrainment can occur. Higher velocity or shear stress is required to entrain sediments that are well protected by these elements.

The Shields number is a dimensionless parameter with which to quantify sediment mobility. The Shields equation is often used to calculate the entrainment threshold of grains of various sizes. This equation relates dimensionless critical shear stress τ_c (i.e. bed drag force acting on the flow per unit bed area; see Chapter 5) to grain size and density of grain packing (i.e. embeddedness in the channel bed or submersion in the laminar layer). For grains to move on a channel bed, the boundary shear stress τ_0 must exceed the critical shear stress τ_c for a grain of a given size. However, this relationship cannot be used for silt- and clay-sized

materials because these sediments are affected by electrostatic forces.

The Shields equation is:

$$\tau^* = \frac{\tau_c}{\rho RgD}$$

where τ^* is the Shields parameter, τ_c ($N\,m^{-2}$) is the critical bed shear stress, R is the submerged specific gravity, measured as $(\rho_s/\rho) - 1$, where ρ_s is the density of sediment (assumed to be constant at $2650\,kg\,m^{-3}$) and ρ is the density of flow ($1000\,kg\,m^{-3}$), g is the acceleration due to gravity ($9.81\,m\,s^{-2}$) and D (mm) is the characteristic grain size.

The Shields parameter can be interpreted as a ratio scaling the impelling force of flow drag acting on a particle to the force resisting motion acting on the same particle. The threshold of motion for a river bed composed of grains of characteristic size D and submerged specific gravity R and subjected to bed shear stress τ_b is quantified by the modified Shields curve:

$$\tau_c^* = 0.5(0.22Re_p^{-0.6} + 0.06 \times 10^{-7.7Re_p^{-0.6}})$$

where τ_c^* is the critical Shields number above which motion starts and Re_p is the particle Reynolds number, given as:

$$Re_p = \sqrt{RgD}\,\frac{D}{\nu}$$

where R is the submerged specific gravity (where sediment density ρ_s is assumed to be constant at $2650\,kg\,m^{-3}$ and water density $\rho = 1000\,kg\,m^{-3}$), g is the acceleration due to gravity ($9.8\,m\,s^{-2}$) and ν ($m\,s^{-1}$) is velocity. When first mobilised, the particles roll, slide or saltate close to the bed as bedload.

The Shields diagram (Figure 6.5) shows how the entrainment threshold varies for bedload, mixed-load and suspended-load rivers. On a smooth surface with a low Reynolds number (Re), small grains (i.e. suspended load particles <0.2 mm in size) are submerged in the laminar sublayer and are not entrained. Under this condition, for a given grain size a larger shear stress is needed to initiate movement. On hydraulically rough beds with higher Reynolds number, which are the norm in natural streams, the Shields number becomes constant between 0.03 and 0.06. As bed shear stress decreases over a rough bed, the maximum sediment size that can be moved decreases (from bedload to mixed load). Minimum bed shear stress occurs for grain sizes between 0.2 and 0.7 mm (medium sands), making them the most easily entrained. For grain sizes greater then 0.7 mm, higher shear stress is required to entrain larger and heavier clasts. This parallels relationships described by the Hjulström diagram (Figure 6.2).

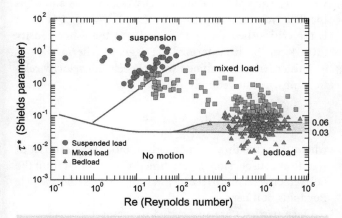

Figure 6.5 Shields diagram for bedload, mixed-load, and suspended-load rivers. Re, the Reynolds number, can be interpreted as scaled grain size. The threshold of motion at high Re is represented between upper (0.06) and lower (0.03) limits of Shields parameter τ^*. This reflects sand bedload movement. Below this threshold no motion occurs. Higher shear stresses and velocities are required to entrain silt and clay particles into the suspended load. From Miller *et al.* (1977). © John Wiley and Sons, Ltd. Reproduced with permission.

For coarser bedload fractions, grain roughness is often the dominant component of resistance on a channel bed, acting as an impediment to transport. This is particularly evident if the bed material consists of gravel (2–64 mm) or cobbles (64–256 mm). A simple measure of the ability of flow to transport sediment of a certain calibre is represented by the ratio:

$$\frac{d}{D}$$

where D (m) is the characteristic grain size and d (m) is the flow depth.

As flow depth increases, the effect of grain roughness is drowned out such that entrainment and transport can occur. In small channels $d/D < 1$, such that flow depth is less than the size of the bed material and flow occurs around large clasts. Individual cobbles and boulders protrude from the water surface and mobilisation of bed sediment is extremely unlikely. In these instances the bed material may be organised as a series of packing arrangements that further inhibit mobility (e.g. step–pool sequences and cascades – see Chapter 8). In intermediate channels $10 > d/D > 1$, such that flow depth is up to 10 times greater than bed material size. Individual clasts on the channel bed are submerged and bed material can be mobilised under a range of flows. In these settings, the bed is typically organised in a series of pools and riffles, the position and morphology of which reflect adjustments in river planform. In large channels $d/D > 10$ and the depth of flow is over 10 times greater than the bed material size. Under these conditions, bed material is mobile and a range of well-defined morphologies results. The pattern of units and the ability of flow to rework sediments reflect the calibre and volume of sediment supply and discharge.

Surface erosion of cohesive sediments occurs as individual particles or small aggregates are removed from the body of materials by hydraulic (hydrodynamic) forces such as drag and lift. The ability of a cohesive sediment to resist surface erosion is known as erosional strength. Resistance to surface erosion differs from resistance to mass erosion. Mass erosion is determined by the undrained strength of the sediment, or yield strength. Mass erosion occurs when the yield strength is exceeded. Examples of this mechanism include slip failure of a streambank (see Chapter 7) or when large chunks of sediment are eroded from the streambed. There is a difference of one to three orders of magnitude between erosional strength and yield strength.

Biological effects work alongside physical and electrochemical effects in determining the behaviour of cohesive materials. Biogenic stabilisation or biostabilisation refers to situations whereby biological action directly or indirectly induces a decrease in sediment erodibility. Discrete particles may become covered by bacteria and diatom growth, causing cohesion to increase and roughness to decrease. Further binding is caused by bacterial secretion that forms cohesive networks between the diatoms. In this way, originally non-cohesive materials may become biostabilised. Some organisms may grow on the bed surface, filling the interparticle voids and forming a microbial mat, thereby creating a smooth, protective biofilm and reducing the hydraulic roughness. This decreases the stress in the near-bed margin, thereby strengthening the bed by effectively increasing the velocity at which particles are entrained. Alternatively, some organisms rework sediments by bioturbation or create uneven surfaces with protrusions that increase hydraulic roughness.

Burrowing organisms may have positive or negative effects on the stability of surficial soils. For example, Oligochaeta (burrowing worms) may reduce the critical stress for erosion 10-fold. Chironomids (common midges with burrowing larva) also have a negative effect on sediment stability, but this influence may diminish over time as the organisms excrete mucus and develop tube houses, cementing the bed and making it less erosive. Elsewhere, burrowing organisms may strengthen the bed by locally increasing the critical shear stress of sediments. These are complex biogeochemical interactions. Site-specific variability is common and, indeed, may change over differing timeframes (e.g. seasonally).

Transport of sediment in river channels

Transport processes are responsible for the downslope movement of sediments along a river. As the Hjulström diagram shows, a critical velocity must be reached before a channel can perform transportational work. The time a particle spends in transport varies significantly for sediments of different size. As fine-grained sediments are easily held in suspension, they spend the most time in transport and can travel considerable distances. Sediment transport is maintained by turbulent flow, where chaotic patterns of currents and vortices suspend particles within the water column. In contrast, coarser grained sediments are transported intermittently along the bed, moving short distances.

Sediment transport occurs in three ways: dissolved load, suspended load and bedload.

Dissolved-load transport

Dissolved load contains material that is transported in solution (e.g. minerals, nutrients, contaminants). This may make up more than 50 % of the total sediment load of large rivers, but it tends to be negligible in mountain streams. Often, high dissolved load occurs on the rising limb of the hydrograph as pre-storm accumulations of solutes are released during erosion. Subsurface flow sources that are in contact with soluble materials also contribute to this phenomenon. Dissolved load commonly declines with increasing discharge or on the waning stage of flow events as a result of dilution. Catchment geology and human disturbance are dominant influences on the concentration of dissolved load. Climatic factors also influence rates and volumes of solute mobilisation and dissolution. Rivers in wet, humid regions and rivers in karst landscapes have high dissolved loads. Although dissolved loads often exceed solid transport and they are critical for ecosystem functioning in rivers, they rarely exert a significant control on channel morphology.

Suspended-load transport

Flow in rivers is invariably turbulent. Suspended load transport occurs when particles are maintained in the body of flow by turbulent mixing and convection. Turbulent eddies are able to waft bed sediment high up into flow if turbulence is strong enough and grains are not too heavy. Although the onset of suspension is not a sharp phenomenon, a standard rule of thumb is:

$$\frac{u^*}{V_0} = 1$$

where V_0 ($m^3 s^{-1}$) is the fall velocity (see below) and u^* is the shear velocity, defined as $u^* = (\tau_c/\rho)^{0.5}$, where τ_c ($N\,m^{-2}$) is the critical bed shear stress and ρ is the water density ($1000\,kg\,m^{-3}$). This is termed the Bagnold criterion.

This criterion, combined with particle fall velocity relationships, can be reduced to:

$$\tau_{sus}^* = f_{sus}(Re_p)$$

where τ_{sus}^* is the threshold Shields number for the onset of significant suspension, f_{sus} is a function computed from the fall velocity relation (see below) and Re_p is the particle Reynolds number.

Once particles are suspended they travel as suspended load. For most river systems, particles <0.2 mm (fine sand) are transported in suspension. Depending on flow conditions, grains up to 1 mm (coarse and medium sands) may also be carried in suspension. Flow is turbid in channels with high suspended sediment loads.

During suspended-load transport, particles move at approximately the same speed as the flow itself, such that near-continuous transport occurs. Once entrained, sediment can stay in suspension for long periods of time and can travel long distances. However, concentrations tend to be highest near the bed and decrease with distance from the bed. Suspended sediment transport rates are a function of suspended sediment concentration (mass per unit volume), flow velocity and depth such that:

$$Q_{ss} = c_s vd$$

where Q_{ss} is the suspended sediment transport rate, c_s (kg L^{-1}) is the depth-averaged sediment concentration, v ($m\,s^{-1}$) is the mean flow velocity and d (m) is the flow depth.

Sediment will only settle out of suspension when velocity is greatly reduced. Channels may carry up to 90 % of their total sediment load in suspension. Very high concentrations of suspended sediment may damp down turbulence within the flow and increase the apparent viscosity of the water. Fine sediment concentrations greater than 10 000 ppm start to significantly increase the water viscosity and, thus, increase the ability of a stream to transport sediment. This ability increases as the suspended sediment concentration increases. Extreme suspended sediment concentrations, termed hyperconcentrated flows, have loads around 1 kg L^{-1}.

Most material that forms the suspended load of rivers is sourced from surface erosion, bank erosion and localised point sources. Suspended sediment discharge tends to increase with distance downstream (Figure 6.6a). Hence, suspended-load transport is often less important in upland areas, where bed sediment is coarse and sources of suspended sediment are limited. Suspended sediment

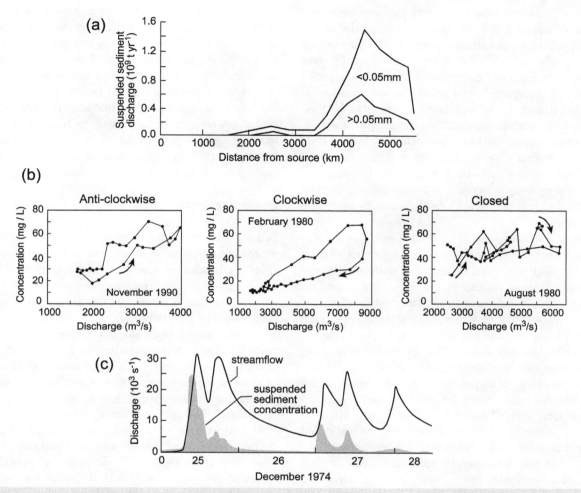

Figure 6.6 Variability in suspended sediment load in space and time. (a) Suspended sediment discharge along the Yellow River, China (from Long and Xiong (1981)). Suspended sediment discharge (or load) increases with distance downstream as the river flows through the highly erodible materials of the loess plateau, which greatly increases lateral and tributary sediment inputs. © IAHS Publication. (b) Hysteresis in suspended sediment concentration along the Rhine River. The type of hysteresis reflects the availability and contribution of suspended sediment sources, the timing of those contributions to the trunk stream and position in catchment. Clockwise hysteresis is most common, followed by closed and then anticlockwise forms. From Asselman (1999). © John Wiley and Sons, Ltd. Reproduced with permission. (c) An example of the relationship between streamflow and suspended sediment concentrations for a series of flow events. Suspended sediment discharge peaks on the rising limb of the hydrograph. From Walling and Webb (1982). © IAHS Publication.

transport is generally not a linear transport function, as suspended sediment load is controlled more by catchment variation in sediment supply and hydrology than mean or peak discharge and the transport capacity of the flow. For example, significant asynchronous behaviour (hysteresis) between discharge and suspended sediment concentration has been recorded in single storm events (Figure 6.6b). This occurs when the sediment wave is not synchronous with the water wave.

Clockwise variation in suspended sediment concentration occurs when the sediment wave precedes the water wave so sediment concentration peaks before discharge

(Figure 6.6). Higher sediment concentrations for flow of a given discharge occur on the rising limb than on the falling limb of a flood hydrograph. This form of hysteresis is more common in small catchments where sediment sources are close to the channel and sediment availability is depleted or flushed early in the flow during the rising stage. Suspended sediment supply increases dramatically as runoff and bank erosion are activated during the rising stages of the flood. Most of the sediment reaches the channel while discharge is still rising. On the falling limb of the flow, sediment supply is depleted as runoff ceases or sediment supply is exhausted.

An *anticlockwise* variation in suspended sediment concentration reflects a peak in water discharge prior to the peak in sediment concentration (Figure 6.6b). This occurs when the relative travel time of water travelling as a wave is faster than the mean flow velocity at which suspended sediment is transported. As a result, the suspended sediment flux tends to lag behind the flood wave. The lag time increases with distance downstream and as upstream sediment sources continue to supply material, and so anticlockwise hysteresis is more common in large catchments.

A *closed* response describes perturbations in cyclic hysteresis where local inputs (e.g. tributary inputs) become (de)activated, producing a more chaotic response curve (Figure 6.6c).

In some instances peaks in suspended sediment concentration occur during the rising stage of a flow event (Figure 6.6c). However, this is not necessarily a uniform relationship, and phases of increasing and decreasing suspended sediment concentrations may be observed as discharge increases. Factors that may account for this pattern include the availability of sediment sources, time since the last storm event, intensity and duration of rainfall, antecedent soil moisture and the magnitude of the previous sediment-transporting event. If suspended-load rivers operate at less than their maximum transport capacity, then the spatial and temporal variabilities in sediment supply from the catchment most likely drive the variability in suspended sediment concentration. Hysteresis effects make it difficult to predict and model suspended load transport.

Computationally, the Rouse profile provides a measure of suspended sediment concentration. It is measured as:

$$P = \frac{V_0}{ku^*}$$

where P is the Rouse number, V_0 (m s^{-1}) is the fall velocity (see later), k is the upwards velocity on a grain (von Kármán constant: 0.41) and u^* (m s^{-1}) is the shear velocity.

The Rouse number is used to define a concentration profile with flow depth. Relationships between particle density and size and the density and viscosity of the liquid determine the part of the flow in which the sediment particle is carried. In flows with a suspended load component of 50% the Rouse number will be >1.2 but <2.5, whereas for a flow that is 100% suspended load the initiation of motion will occur at Rouse numbers >0.8 but <1.2. Bedload occurs at Rouse numbers >2.5.

Assessment of total suspended solids is the most commonly used technique for measuring suspended sediment concentration. This can be determined by pouring a carefully measured sample (typically 1 L) through a pre-weighed filter of a specified pore size and then weighing the filter again after drying. The gain in weight is a dry weight measure of the particulates present in the water sample expressed in units derived or calculated from the volume of water filtered (typically milligrams per litre, mg L^{-1}). Dissolved substances or small algae/organic material may add to the weight of the filter as it is dried.

Turbidity provides an actual weight of the particulate material present in a sample. Total suspended solids (TSS) measurements can be correlated to turbidity measurements at a given site. Once established, this correlation can be used to estimate TSS from more frequently made and easily obtained turbidity measurements.

Bedload transport

Bedload transport occurs when grains or clasts are transported in a thin layer on the channel bed. Once shear stress exceeds critical shear stress, particles roll, slide or saltate along the bed in a shallow zone a few grains thick (Figure 6.7). *Sliding* occurs when particles are pushed along the channel bed by horizontal drag. *Rolling* occurs when larger grains are pushed along the channel bed but roll rather than slide. *Saltation* temporarily lifts grains into the water column. The grains essentially 'jump' along the channel bed prior to falling back to the channel bed. Saltation tends to occur at higher flow velocity than rolling and sliding, when turbulence 'picks up' grains from the channel bed. Rolling tends to dominate in gravel-bed rivers, whereas saltation is common along sand-bed streams. Bed material in sand-dominated rivers tends to be relatively well sorted, with values of geometric standard deviation of bed sediment ranging from 1.1 to 1.5, while gravel-bed rivers commonly have values >3 given the amount of sand and fine-grained materials that occurs in the interstices of the gravel substrate.

Bedload sediment transport is a sporadic process. Grains stay in transport for short periods of time and travel relatively short distances. This is because, once the entrainment threshold is reached, transport occurs at velocities less than the surrounding flow. Small changes in flow velocity around the entrainment–transport–fall-velocity threshold result in distinct pulsing or intermittent transport of bedload materials of various sizes (Figure 6.8a). This may be associated with the progressive downstream movement of bedforms or bars. As flow intensity increases, impelling forces eventually exceed resisting forces on the channel bed. This condition is usually expressed by either a critical shear stress τ_c or a critical velocity v_{cr}. Unlike dissolved- and suspended-load transport, the rate of bedload transport is a function of the transporting capacity of the stream. In all cases, the velocity at which bedload moves is less than the velocity of the transporting flow.

While bedload transport may only be a small proportion of the total sediment load of a river, it is a key determinant

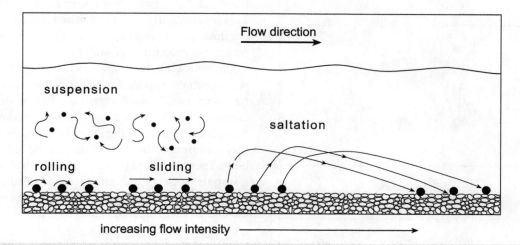

Figure 6.7 Forms of bedload sediment transport. At low flow intensities, rolling and sliding occur. As flow intensity increases, grains can be picked up and transported short distances via saltation.

of adjustments to river morphology. In general, bedload transport increases with increasing discharge, but bedload transport rate can vary by 10–100 times for a given discharge (Figure 6.8b). As such, it is very difficult to model and predict bedload-sediment transport.

The balance of impelling and resisting forces that fashion flow and sediment conditions acting within a channel on a given slope is a key control upon bedload movement (see Chapter 4). A range of sediment transport equations has been developed to estimate bedload transport by rivers. Almost all express the maximum amount of sediment (capacity) that can be transported for a given flow and sediment condition (transport capacity) per unit width of channel q_b. Essentially, bedload transport equations can be differentiated into those that relate sediment transport rate to excess shear stress ($\tau_0 - \tau_{cr}$), to excess discharge ($q - q_{cr}$) or to excess stream power ($\omega - \omega_{cr}$) as follows:

Bu Boys type $q_b = X'\tau_0(\tau_0 - \tau_{cr})$

Schoklitsch type $q_b = X''s^k(q - q_{cr})$

Bagnold type $q_b = (\omega - \omega_{cr})^{3/2}d^{-2/3}D^{-1/2}$

where X' and X'' are sediment coefficients, d (m) is the flow depth and D (mm) is the grain size. These parameters represent the flow force per unit area acting on the channel bed or the potential rate of doing geomorphic work (energy expended) in moving sediment.

The most commonly used sediment transport equations were developed from laboratory flume and/or field studies of sediment in motion. Each equation is generally applicable for a range of sand- or gravel-sized materials, or both.

As many of these equations rely upon empirically derived coefficients, significant inaccuracies may result if equations are used outside the ranges for which they were derived.

One of the most widely used bedload transport equations for gravel bed rivers is the Meyer-Peter and Müller formula:

$$q_s^* = 8(\tau^* - \tau_c^*)^{3/2}.$$

Where,
  q_s^* = nondimensional measure of bed shear stress
  τ^* = the Shields parameter
  τ_c^* = critical Shields number above which motion starts
        (experimentally derived value of 0.047)

For sand bed systems, the Engelund-Hansen formula is widely used:

$$q_s^* = \frac{0.05}{C_f}(\tau^*)^{5/2}$$

  q_s^* = nondimensional measure of bed shear stress
  C_f = is a friction factor
  τ^* = the Shields parameter

Unlike other formulae, the Engelund-Hansen formula does not relate bedload movement to a critical shear stress. This is a key consideration in predicting bedload movement for coarser-grained gravel materials. The Engelund-Hansen is particularly useful in estimating bedload transport for sand-bed systems under subcritical flow conditions of low flow regime.

More sophisticated approaches to predicting bedload transport have been developed for mixed load rivers with

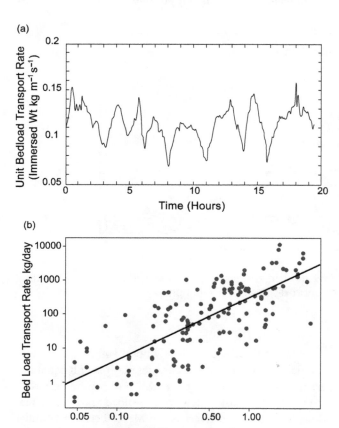

Figure 6.8 Variability in bedload transport rate. (a) Bedload transport rate is highly episodic, as sediments are entrained, transported short distances and then deposited. From Gomez *et al.* (1989). © John Wiley and Sons, Ltd. Reproduced with permission. (b) A characteristics bedload transport rating curve, showing how the bedload transport rate increases with discharge, but is spread over several orders of magnitude. From Moog and Whiting (1998). © American Geophysical Union (an edited version of this paper was published by AGU). Reproduced with permission.

threshold flow at which entrainment and motion occur. This is particularly evident in mixed bedload systems. For example, gravel transport is often enhanced if there is a reasonable proportion of sand in the mix.

Ultimately, it is questionable whether any bedload transport equations provide a satisfactory prediction of bedload transport rate, and all equations should be used with caution. This is because many of these equations use mean flow parameters rather than local near-bed parameters and they assume that the channel has infinite sediment availability. They also fail to consider other controls on sediment transport, such as bed armouring or loading of wood. Other factors that influence sediment transport capacity include water temperature, sediment concentration, sediment specific gravity, particle shape, settling velocity and bedforms. Sensitivity tests must accompany field investigations and theoretical developments to validate these applications.

Direct measurement techniques used to assess bedload transport rate include a range of samplers, loggers and traps/pits that are installed on channel beds. Samplers are used when moving sediment is collected during a flood event. Sediment baskets sit on the top of the bed and collect sediment moving along the bed into a mesh basket. Traps are pits excavated in the channel bed and into which sediment falls as it travels over the channel bed. Tracers can take a number of forms, including painted clasts and electronic chips and magnets. Electromagnetic sensors can be installed in the river bed to monitor the inception, intensity and duration of electromagnetic tracers. Other forms of tracing involve investigation of the mineralogy of sediments relative to source areas or the tracing of artificial substances such as contaminants and nutrients. Tracer techniques can provide accurate measurements, especially for small rivers. In most coarse-grained rivers the transport of sediment is difficult to measure accurately with bedload traps or samplers due to the heterogeneity of sediment sizes and complex channel topography. It is especially difficult during high flow stages, when most of the bedload transport occurs. Field deployment of most methods is difficult and involves a high level of uncertainty.

Emerging technologies have achieved considerable success using morphological methods to estimate sediment flux, wherein repeat field surveys are used to measure changes in channel position. From this, estimates of the minimum amount of bedload passing a cross-section or reach can be determined. This provides relatively accurate medium-term estimates of bedload yield, particularly on large gravel-bed rivers.

Total sediment discharge is often estimated in the absence of in situ hydraulic measurements using a sediment rating curve. Sediment rating curves are empirical relations between the total water discharge and the sedi-

multiple grain sizes across gravel and sand range. Many favour the Wilcock and Crowe sediment transport formula that requires consideration of reference shear stresses for each grain size, the fraction of the total sediment supply that falls into each grain size class, and a 'hiding function' that accounts for the trapping of finer sediments between larger grains. In these mixed grain size populations, a higher proportion of sand-sized materials may enhance the mobility of coarser grained fractions.

As a general rule, shear stress functions provide more reliable assessments of bedload transport in gravel-bed rivers and stream power functions are more reliable predictors in sand-bed streams. The primary difficulty faced when using these equations is determination of the critical

ment discharge (see Figure 6.8b). Most sediment rating curves implicitly assume capacity-limited conditions. The standard form is:

$$Q_s = aQ^b$$

where Q_s is the sediment discharge, Q is the water discharge and a and b are empirical constants. a can vary wildly, but b generally takes a value between 1.5 and 2.5. Because of this constraint on b, a value of b can be assumed and, with only one measurement, the value of a can be calculated. However, very few rivers are capacity limited all of the time. During extreme floods, even classic capacity-limited streams exhibit supply-limited behaviour. This results in hysteresis in the sediment–discharge relationship. The only way supply-limited rivers can be modelled in the absence of measurements is if an entrainment relationship is coupled with detailed routing of suspended load, along with a model for bedload. Suspended load presents the biggest problems in estimating the sediment load of supply-limited rivers.

Bed material size and organisation exert a significant influence upon sediment transport relationships in river systems. As such, they are primary considerations in differentiating among the character and behaviour of bedload, mixed-load and suspended-load rivers (see Chapters 10 and 11). Hence, it is important to give careful attention to material properties that affect sediment and bedform movement in river systems.

Material properties that affect sediment movement in river systems

As noted on the Hjulström diagram (Figure 6.2), grain size exerts a significant influence upon sediment transport relationships in river systems. However, particle size is just one of the material properties that affect entrainment, transport and depositional processes. In this section, the influence of grain-by-grain interactions on the channel bed (i.e. packing arrangements), bedform generation in sand- and gravel-bed rivers, partial and equal mobility, supply- and transport-limited rivers and the role of material cohesiveness in fine-grained (silt–clay) channels is discussed, prior to outlining controls upon depositional processes in river systems.

Grain-by-grain interactions on the channel bed

Grain-by-grain interactions affect the ease with which individual particles can be mobilised along the channel bed. This has profound implications for sediment entrainment, transport and deposition. These hydraulic relationships, in turn, determine the organisation of particles on the channel bed (i.e. these are mutual interactions). For example, bed-

form formation influences bed roughness, thereby affecting the energy that is available to entrain and transport sediments. As noted from the Hjulström diagram (Figure 6.2), medium–coarse sands are the most readily entrained sediments because they are the smallest sized particles that are cohesionless. As such, these sediments are readily moulded into bedforms. Conditions are markedly different in gravel-bed rivers, where grain-by-grain interactions induce various structures on the channel bed that work alongside coarser grain sizes to limit rates of bedload movement. This fashions the threshold-induced nature of bedload movement in gravel-bed rivers.

Grain-by-grain interactions on the channel bed can be viewed in terms of bed material organisation on the bed itself and in terms of the relationship between surface and subsurface fractions. The former category essentially refers to packing and imbrication on the channel bed, while the latter refers to armouring and paving. *Packing* refers to how tightly particles fit together on the bed of a river. In some instances, grains are relatively loose and clasts can be picked up easily. Elsewhere, they are tightly packed and the entire surface must be broken before individual grains are released and are able to be entrained. Turbulent flow juggles and shakes bedload materials across the channel bed, winnowing away finer grained particles. The channel bed becomes more tightly packed as grains vibrate, making it more difficult to remove individual grains. Typically, the grains become *imbricated*, whereby clasts stack up against each other (see Figure 6.9a). This is a dominant form of bed organisation when materials are platy or subrounded; however, it cannot occur when materials are well rounded or sub-spherical. Imbrication stabilises the bed, approximately equivalent to a threefold increase in particle size in terms of the constraint that it imposes upon initiation of

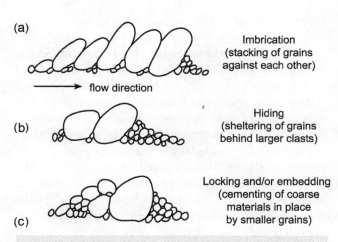

Figure 6.9 Schematic representations of (a) imbrication, (b) hiding and (c) embedding of grains in gravel-bed rivers.

sediment movement. Smaller particles that are deposited behind large clasts are sheltered from transport due to *hiding* and may become *embedded* as finer material cements coarser clasts in place (Figure 6.9b and c).

Surface–subsurface relationships refer to the grain-size population on the channel bed relative to materials that are stored beneath the surface (Figure 6.10). Winnowing processes selectively entrain finer grained particles from the bed, resulting in preferential retention of the coarser grain-size fraction. This coarse layer protects subsurface layers from entrainment and erosion, limiting the supply of fine material from the subsurface to the bedload at high flow in gravel-bed rivers. Vertical stratification of bed materials takes various forms. In *armoured* beds the surface layer is usually one grain thick, comprised of the D_{90} fraction, and has the same texture as subsurface materials (Figure 6.10a). A paved bed, or *pavement*, refers to situations in which the coarsest size at the surface is greater than in the subsurface fraction (Figure 6.10b and d). Typically, armour is almost absent whenever bedload transport rates are high. Spatial impacts of these bed-stabilising mechanisms may vary across the active channel zone. In some instances, the surface bedload fraction may effectively become static, such that it is immobile under virtually all flows. For example, drastic reduction in peak flows following dam construction may inhibit the capacity of flow to mobilise the coarsest fraction, while smaller grain sizes are progressively winnowed (Figure 6.10c). However, some gravel-bed streams show no vertical stratification in bed material sizes or composition. Hence, while perennial streams with low sediment supply and moderate floods often generate a well-defined armour layer, ephemeral streams subjected to high sediment supply and violent floods deposit chaotically mixed (unstratified) materials which are infrequently reworked.

Downstream gradation in bed material size

The size, shape and organisation of bed materials are determined by sediment supply and subsequent transport and deposition. Bed material size generally decreases downstream in conjunction with decreasing channel slope. This reflects two primary processes: abrasion and hydraulic sorting. *Abrasion* is the process by which grinding, breakage and impact rubbing of grains either in transport or in situ results in clast breakdown and reduction in size. The rate at which a grain is abraded is controlled by lithology and hydraulic conditions. For example, basalt typically abrades more rapidly than granite, which abrades less quickly than sandstone and shale. The volume of available materials also affects the abrasion rate. Erosion rates are enhanced if the rate of removal matches the rate of supply. Alternatively, if supply rate exceeds rate of removal, erosion rates may be

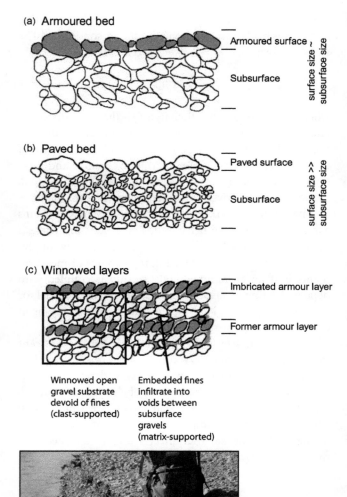

Figure 6.10 Schematic examples of armoured and paved beds. (a) Armouring occurs when the surface grain size approximately equals the subsurface grain size. (b) Paving occurs when the surface grain size is larger than the subsurface sediments. (c) Winnowing occurs when finer grained particles are selectively entrained, generating an open gravel matrix that is devoid of fine grains. (d) Paved surface along the Ngaruouro River, New Zealand. Photograph G. Brierley.

diminished as these materials protect the subsurface materials. *Hydraulic sorting* is the process by which a downstream reduction in velocity and transport capacity of flow leads to selective entrainment and transport of sediment, so that a gradation in grain size occurs from upstream to downstream. The combined effect of these two processes results in coarser, angular, mixed size and poorly sorted bed materials in upstream locations, and smaller, better rounded, more uniformly sized and well-sorted bed materials in downstream locations. Abrasion and sorting often lead to an exponential decline in grain size with distance downstream, where:

$$D = D_0\, e^{-aL}$$

where D (mm) is the particle size, D_0 (mm) is the initial grain size, L (km) is the distance downstream and a is a coefficient representing both sorting and abrasion.

The rate of change in grain size is highly variable, but tends to be lower where sediments are finer. In other words, the decrease in particle size per unit distance is proportional to particle size. So, particle size decreases significantly in headwater areas where the initial material is much coarser, and decreases at a much slower rate downstream, particularly in sand-bed situations. In general terms, the median grain size of bed material is reduced by 50% over distances that range from 10 to 100 km. Abrupt gravel–sand transitions are commonly observed. This reflects the tendency for grain sizes in the range of pea gravel to be relatively scarce in rivers. The relative importance of abrasion versus sorting as a control upon the downstream diminution of grain size varies from system to system, but selective transport is often the dominant process.

Tributary sediment inputs often disrupt the downstream pattern of grain-size gradation along a river, creating scatter along the longitudinal profile of bed sediment size (Figure 6.11a). As these input materials have travelled a shorter distance from source, they are often coarser than bed materials along the trunk stream itself. Alternatively, tributaries that drain fine-grained geologies can input sediments that are smaller in size than those occurring along the trunk stream. Inputs from geomorphically significant tributaries may create discrete 'sedimentary links' along a river course (Chapter 3 and Figure 6.11b). The grain-size pattern will only be disrupted if the calibre of sediment is sufficient to change the characteristic grain size of the sediment mix (either coarsening or fining; Figure 6.11c). The volume of the input will determine whether the effect on trunk stream grain size is long- or short-lived, and the length of river required for the downstream pattern to be 'reset'. Downstream fining processes usually dominate within sedimentary links. Slow downstream fining tends to result in longer sedimentary links (unless a new input occurs) and fast

fining tends to produce short sedimentary links (Figure 6.11c). The effect of tributary inputs tends to be more pronounced along downstream sections of the trunk stream where sediments are finer grained. Alternatively, lateral inputs from connected (coupled) hillslopes via landslides and other processes may disrupt longitudinal grain-size patterns along the trunk stream. Disruptions to downstream grain-size trends may also reflect the differing susceptibility to weathering and abrasion of different rock types along a river course. In contrast, blockages along the sediment conveyor belt may restrict downstream sediment to certain grain-size fractions (e.g. a series of dams or weirs, or sediment slugs), altering the downstream gradation in grain size in these decoupled landscapes. Identification of significant lateral sources (either hillslope or tributary derived) and their degree of connectivity in the system are key determinants of catchment-scale sediment flux (Chapter 14).

Bedform generation and its impact upon bedload movement in sand-bed rivers

The bed morphology of sand-bed streams adjusts readily to changes in flow and/or sediment supply conditions. Bedload transport generates distinct micro-scale structures called bedforms that migrate along the channel bed. Given the small size and weight of individual grains, bed material is mobile over a wide range of flows, creating instabilities in the form of ripples, dunes and antidunes (Figure 6.12). When shear stress exceeds a critical threshold, cohesionless beds are moulded into differing geometric forms dependent upon flow characteristics (especially velocity and depth, and their effect upon bed material of differing sizes). In turn, bedform geometry influences flow resistance and the nature/distribution of flow energy, as measured by the Darcy–Weisbach resistance coefficient ff (see Chapter 4). These are complex feedback relationships. Sediment transport rates vary across individual bedforms as a result of form-induced accelerations and decelerations in flow, promoting scour in troughs and deposition towards crests. Shear forces induce sorting at local scales. Selective deposition begins once flow velocity falls below the settling velocity of a particle. Settling velocity is closely related to particle size (Stokes' law; see below), but is greatly affected by the sediment mix and associated drag forces. The net result is horizontal and vertical (downstream and transverse) gradation of sediment sizes. Length and height of bedforms vary with flow depth and sorting of sediment.

Bedform development is a threshold-based hydraulic process, whereby specific forms reflect the magnitude/duration/stage of flow (especially velocity and depth) and bed material texture. Bedforms cannot be developed when there is insufficient energy to transport sediment. A

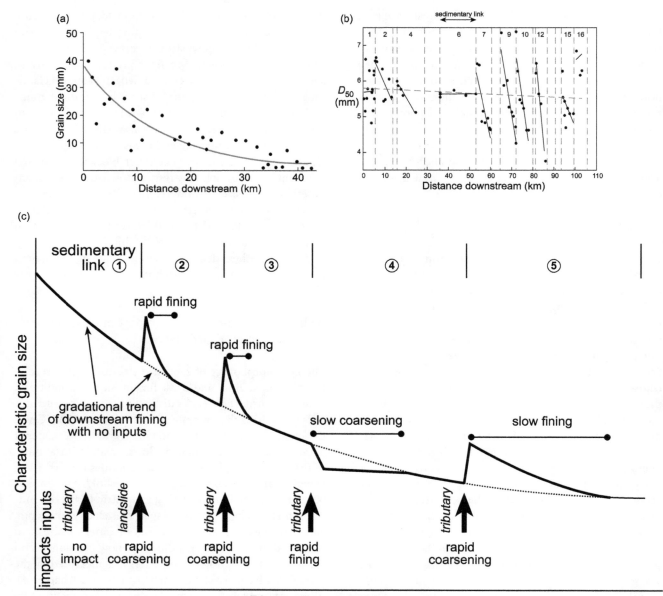

Figure 6.11   Downstream gradation in bed material size (a) Example of scatter along a downstream plot of bed material size that reflects lateral and tributary inputs of sediment. (b) Sedimentary links can be defined where the difference in grain size at the confluence is sufficiently large to alter the size distribution of the trunk stream sediments. Several sedimentary links are evident along the Pine River, British Columbia, Canada. These are defined by vertical dashed lines. Exponential relationships have been fitted to the entire longitudinal profile (long dashed line) and to individual sedimentary links. Reprinted from Rice and Church (1998). © John Wiley and Sons, Ltd. Reproduced with permission. (c) Tributary impacts upon downstream grain size trends may vary markedly. In some instances, grain-size fining can occur, while other tributaries have no effect on trunk stream dynamics. This schematic shows how, depending on the slope of the longitudinal profile and the size of sedimentary inputs, downstream fining may be a rapid or slow process.

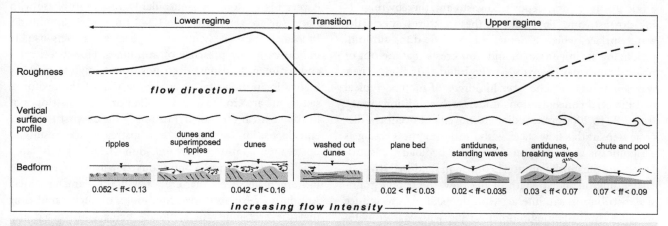

Figure 6.12 Bedform configurations in sand-bed channels. Increasing flow intensity generates a range of bedforms at lower, transitional and upper flow regime stages. These are free-forming features. Modified from Knighton (1998) and Simons and Richardson (1966).

well-defined sequence of bedforms ranging from ripples–dunes–plane bed–antidunes has been identified with increasing flow intensity in sand-bed rivers. Lower and upper flow regime forms are classified according to their shape, resistance to flow and mode of sediment transport. Lower flow regime conditions comprise plane bed with no motion, ripples or dunes. At these stages, form roughness is dominant. Upper flow regime conditions comprise plane bed with motion and antidunes. At these stages, grain roughness is dominant. Bed configuration in the transition zone between these two regimes is characterised by the washing out of dunes as the bed approaches plane bed with motion conditions.

A flat sandy bed (lower stage plane bed) is deformed at relatively low competent stresses into small wavelets instigated by the random accumulation of sediment and then into *ripples*. These bedforms are roughly triangular (asymmetrical) in profile, with gentle upstream (stoss) and steep downstream (lee) slopes, separated by a sharp crest. Slightly higher velocities are required to mobilise fine sand relative to coarse sand. Rarely occurring in sediments coarser than 0.6 mm, ripples are usually less than 0.04 m high (typically 2–100 grain diameters) and 0.6 m long. These dimensions are seemingly independent of flow depth. They can provide local flow resistance on the channel bed with friction factors ff up to 0.13. With coarser grain sizes, wavelengths tend to be longer, while ripple height is marginally greater. A reasonably well-defined viscous sublayer is required if ripples are to form. Because of this sublayer, ripples do not interact with the water surface. Ripples are initiated by the turbulent bursting process. These small bedforms translate downstream at speeds inversely proportional to their height, reflecting discontinuous movement of bed material

load. Sediment transport capacity is low. Sand moves in an intermittent manner, whereby grains are moved up the back of the ripple (stoss slope) fall over the face of the ripple and accumulate on the lee side. The lee face is formed by the avalanching of material transported up the stoss side to the crest. The entire bedform migrates downstream because of this process. Considerable segregation and sorting of bed material occurs (i.e. ripples are internally graded, dipping structures). Typical bed material concentrations under such conditions range from 10 to 20 parts per million (ppm).

As flow energy increases, ripples become superimposed on dunes. An increased rate of transport and increased turbulence are associated with this transition. Typical bed material concentrations under such conditions range from 100 to 1200 ppm. As shear stresses increase further, ripples are overtaken and eventually replaced by *dunes*. These bedforms provide significant local flow resistance on the channel bed with friction factors ff up to 0.16. Although superficially similar, dunes can be distinguished from ripples by their larger height and wavelength, attaining values in excess of 10^1 m and 10^2 m respectively in large rivers. Unlike ripples, dune height and wavelength are directly related to water depth, approximately in the form whereby height is up to one-third of flow depth and wavelength is four to eight times the flow depth. Dunes are produced by the same process that generates ripples, but flows are faster, deeper and more turbulent. They have an asymmetric form, with a gentle stoss (upstream) side (slopes 1–8°) and a steep lee (downstream) side. Flow accelerates and sediment transport increases from trough to crest. The lee face is formed by avalanching of material transported up the stoss side to the crest. Hence, dunes erode upstream and deposit downstream, thereby migrating downstream. The scale of these features affects the water surface, which is depressed over the dune summit, producing sediment ridges and flow crests that are out of phase. Dunes may be up to 10 m high and 250 m long. Their wavelength may extend over hundreds of metres. Typical bed material concentrations under such conditions range from 200 to 2000 ppm. Sediment movement occurs in discrete steps at this flow stage, with form roughness acting as a significant control upon the rate of bedload transport. Dunes are the most common bedform in sand-bed streams. Depending on the strength of the flow, the parent grain-size distribution can interact with the bedforms to induce strong vertical and longitudinal sorting, with coarser material accumulating preferentially in dune troughs.

As friction and velocity increase, dunes elongate and flatten. Eventually, they are washed out to a transitional flat-bed regime before upper regime bedforms develop at higher discharge and shear stress conditions. This *plane bed with movement* stage, which occurs at Froude numbers between 0.3 and 0.8 (see Chapter 5), is characterised by marked reduction in bed roughness with friction factors ff around 0.02. Bed friction at this stage is controlled solely by grain roughness. Once more, mutual interactions between bed roughness and sediment transport are evident. The lack of three-dimensional erosional and depositional forms reflects the intensity of sediment transport at this stage. Indeed, bed material transport occurs as continuous bedload sheets at this phase. As these rhythmic waves transfer downstream, they generate alternating zones of fine and coarse sediment, rather than altering bed elevation. *Sand waves* have straight, continuous crests. These shallow forms have high length/height ratios and typically range from 5 to 100 m in length. They have also been referred to as bars or flattened dunes. Washed-out dunes and plane bed with movement phases have bed material concentrations in the range 1000–3000 ppm and 2000–6000 ppm respectively. Bedload sheets result in strongly pulsating bedload transport in terms of both total rate and characteristic grain size.

As flow intensity increases further, standing waves develop at the water surface and the bed is remoulded into a train of sediment waves which mirror the surface forms. These *antidunes* are more transitory and much less common than dunes. They form in broad, shallow channels of relatively steep slope when the sediment transport rate and flow velocity are particularly high. A further increase in flow intensity leads to *standing wave* and *antidune* development and renewed form roughness with friction factors ff of between 0.02 and 0.035. They are formed from continuous grain movement under sediment-charged situations (i.e. high sediment transport capacities). High flow velocities are readily able to overcome resistance factors (i.e. resistance to flow is relatively low and sediment transport

is very high). Roughly sinusoidal waves are in phase with water surface waves that are 1.5–2 times their amplitude. Indeed, a train of symmetrical surface waves is usually indicative of the presence of antidunes. Flow decelerates and sediment transport decreases from trough to crest. As high-velocity flows fall over the lee side of the bedform, sediments are eroded and deposited on the stoss slopes of the downstream bedform. Erosion of the downstream face and deposition on the adjacent upstream faces results in upstream migration (hence *anti*-dunes). During this 'rapid flow' there is almost continuous downstream sediment movement in sheets that are a few grain diameters thick. Antidunes develop under conditions of such rapid flow that the probability of bedforms being constructed and preserved is very limited. Typical bed material concentrations under such conditions exceed 2000 ppm.

Rapid energy dissipation occurs when waves break. This destroys the bedform temporarily, creating a pattern of chutes and pools. *Chute and pool trains* develop under very steep flows with supercritical Froude numbers. Steps are delineated by hydraulic jumps (immediately downstream of which the flow is locally subcritical). Cyclic steps migrate upstream. This is associated with abundant sediment flux, with bed material concentrations greater than 2000 ppm.

Sediment slugs reflect an oversupply of sediment such that the channel becomes capacity limited and the valley floor aggrades. They tend to reflect broad-scale disturbance events such as volcanic eruptions, cyclone-induced hillslope failures or forest clearance. *Macroslugs* are small-scale features such as bedforms and unit bars that pass through a system on a timescale up to individual transport events. *Megaslugs* tend to be contained within the channel and are comprised of assemblages of sand/gravel sheets and complex bar structures that alter channel geometry as they pass sporadically downstream over decadal timeframes (Figure 6.13a). *Superslugs* reflect abrupt changes to sediment supply that result in major aggradation and transfer of sediments in both channel and floodplain zones (Figure 6.13b). Sediment slugs and waves may translate downstream as a coherent body of sediment or they may disperse downstream (selectively entraining materials, resulting in sorting and progressive fining over time). Channels are filled and geomorphic heterogeneity is markedly reduced. It may take hundreds of years for these features to propagate through a catchment.

Bedform generation and its impact upon bedload movement in gravel-bed rivers

As noted for sand-bed rivers, the generation of bedforms in gravel-bed rivers is a threshold-driven phenomenon. Sediment trapping is controlled by turbulence on the bed surface. Once clasts become established, they 'attract' clasts

Figure 6.13 Sediment slugs. These bodies of sediment reflect oversupply of sediment such that a channel becomes capacity limited and the valley floor aggrades. (a) Sediment slugs along the lower Bega River have formed as a result of sediment release from valley fills and channel expansion throughout the catchment (see Chapter 14). Photograph: G. Brierley. (b) Sediment slugs in the Hunter catchment deposited sand slugs that were several metres thick after the 1955 flood, the largest flood on record. Photograph reproduced with permission from the NSW Office of Water archives.

of the same size. Addition of further clasts may promote the transition from a hydraulic bedform to a storage bedform. Competent flows move the least stable particles into more stable positions, systematically sorting and configuring clasts on the bed. This essentially reflects the same set of processes noted for sand bedforms, but bigger and faster (shallow and rapid) flows are required to generate these features in gravel-bed rivers. Features produced range in size from particle clusters, arranged around a single obstacle clast, to transverse ribs (Figure 6.14). The succession of features has differing transport rates and friction

factors (i.e. composite grain and form resistance). However, in contrast to sand-bed situations, gravel bedforms are not clearly related to flow properties; rather, their form is primarily determined by grain size, sorting and shape. Bedform features tend to scale to the size of the largest clast. Flow of sufficient strength is able to generate dunes in gravel-bed streams.

Pebble clusters generally consist of a single obstacle protruding above neighbouring grains. This clast acts as a focal point for accumulation of particles upstream and downstream (Figure 6.14). Pebble/particle clusters are typically <10 to 100 m long, with their long axis parallel to flow. A large, immobile core or 'keystone' acts as an obstacle, with upstream stoss deposition and downstream wake deposition. These 'keystones' can be historical or lag gravels which barely move. The stoss is composed of coarser clasts which lodge against the core clast in an imbricated pattern. Wake deposits are generally finer grained and are protected from lift forces by the core clast. Wake length tends to be of similar size to the core clast.

Ostler lenses are similar to pebble clusters (Figure 6.14). Fine-grained sediment settles due to separation of flow on the downstream side of boulders and cobbles during the falling stage. Deposits consist of pebbles and sands that are transported during flows that are inadequate to move cobble-size materials. These are temporary storage sinks generated during capacity-limited flows.

Transverse clast dams are ridges of gravel that are elongated normal to flow (Figure 6.14). They are arranged as steps along the bed. Backfill dominantly comprises low-flow accumulation, with clast size increasing to the next ridge (or step). The dam front is comprised of loose, well-sorted gravels. Sediment finer than that comprising the ridge accumulates upstream, suggesting that ridges behave like dams by retaining finer sediment. These features may be over 1.5 m high; their length is a function of their height. Owing to their influence on turbulence, ridge position influences the position of the next ridge downstream.

Transverse ribs form as sheetlike deposits under highly sediment-charged conditions (Figure 6.14). They range in size from 1 to 10 m. Regularly spaced pebble, cobble or boulder ridges are typically one or two clasts thick and several clasts wide. They are oriented transverse to flow, with a spacing relationship that is similar to transverse clast dams (i.e. proportional to the size of the largest particle in the ridge crest). However, these features differ from transverse clast dams as they do not have backfill and they are able to persist in a wider range of stream slopes. They most commonly occur on steeper slopes, where flow is rapid and shallow. They may form under standing wave or antidune conditions, when grains in motion lodge against stationary grains. As these features are mobilised by flows that do not move surrounding materials, they must be generated at a

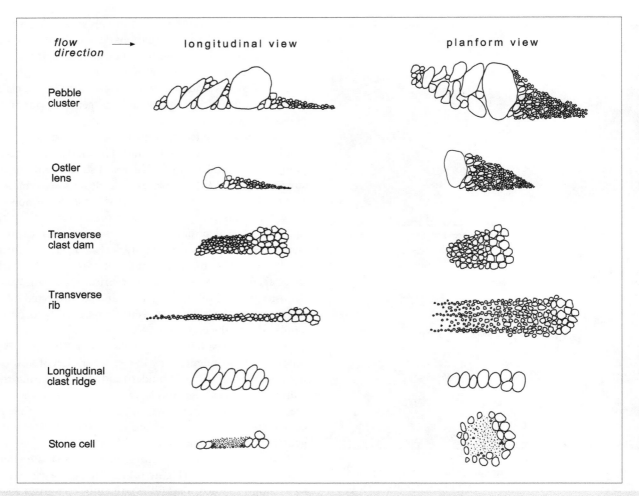

Figure 6.14 Bedform configurations in gravel-bed streams. These various forms of particle organisation, reflect differing flow–sediment interactions.

lower flow stage than imbrication. These features are commonly observed in braided outwash systems.

Longitudinal clast ridges are comprised of coarse, imbricated gravel (Figure 6.14). They run parallel to flow and are up to 7 m long and 1 m high (though most are much smaller). They are often very well sorted, with little particle size gradation along a ridge. They are commonly found on the steeper part of fans, but they may also be observed in the shadow of other obstacles, such as vegetation.

Stone cells are interconnected, irregular cells, with a characteristic diameter of around 1 m (Figure 6.14). The cell border consists of pebble/cobble-size material (occasionally boulders). The centre of the cell contains very poorly sorted material, finer than that of the border. These features are associated with declining sediment supply during flood events. They increase bed resistance to subsequent entrainment and transport of material.

Sediment transport in mixed-load river systems

The bed material of natural streams is characteristically non-uniform, with a mixture of sediments of various grain sizes in transport at any one time. Materials on the channel bed are subjected to size-selective entrainment and transport. Different size fractions become mobile at different shear stresses or flow stages. This theory of *partial mobility* suggests that as larger particles require a greater shear stress to become mobile due to their greater inertia, smaller particles are selected for transport. As a result, only some of the surface grains are mobilised at any given time. The competence of the flow determines the threshold for transport of the selected sizes. The grain-size distribution and arrangement of bed materials are key controls on the degree of size-selective transport. For example, the development of an armour layer promotes size-selective bed material transport, as finer grained materials are win-nowed. Partial mobility is one mechanism by which down-stream fining of materials occurs along river courses. Gravel transport is considered to be size selective with a propor-tional relationship between the particle size and the shear stress exerted by the flow.

The hypothesis of *equal mobility* proposes that threshold conditions for transport are independent of grain size. Although fine-grained materials are more readily trans-ported, coarse grains are also readily mobilised as they are more exposed to entrainment forces. Factors such as the protrusion of large grains into the flow are considered to be sufficient to compensate for their greater submerged weight. Under these conditions, all grain sizes are mobilised within a narrow range of shear stresses. Once large clasts are removed from the bed, the smaller grains around them are also entrained. Therefore, all grains become mobile at approximately the same shear stress and grains of differing sizes are entrained in the proportion that they are found on the bed. As transport rates increase, the transported material remains in that same proportion. In this case, the size distribution of transported material does not change with flow stage, resulting in equal sediment mobility.

Full bed mobility occurs when all surface grains are in motion. Full mobilisation of a size fraction typically occurs at around twice the bed shear stress needed to entrain that size fraction. Typically, equal and full mobility only occur in large-magnitude, low-frequency flow events.

The relationship between flow energy and the propor-tion of the bed with inactive, partially mobile and fully mobile surfaces must be viewed in terms of the increase in flow stage relative to the height of differing geomorphic surfaces. During exceptionally high floods, periods of equal mobility may occur. Full mobilisation of the bed surface occurs infrequently. Full sediment mobility occurs more frequently on lower morphological surfaces, such as the main channel, with lower frequency on low bar forms, while elevated units are largely inactive, experiencing partial mobility during even the highest magnitude flood events.

Sediment supply and transport-limited channels

The volume and calibre of sediment that is available to be transported by a channel, relative to the flow that is avail-able to transport that material, can be assessed to differen-tiate between supply- and transport-limited landscapes (Figure 6.15). *Supply-limited channels* are found where sediment supply/volume is significantly less than the capac-ity of the channel to carry the available sediment. This means that all sediment that is made available to the reach is transported. Excess energy may result in channel erosion along the reach. *Transport-limited channels* are found where sediment supply or sediment calibre is significantly greater than the capacity of the channel to transport that sediment. Transport-limited channels can be separated into two types. First, *capacity-limited channels* are those where the sediment supply (volume) is significantly greater than the capacity of the channel to carry that sediment. This means that the channel is 'overloaded' with sediment, deposition occurs and the channel bed aggrades. Second, *competence-limited channels* are those where the calibre of sediment is significantly greater than the ability of the channel to trans-port sediment of that size. The channel is unable to entrain its load and coarse calibre materials are stored on the channel bed.

The influence of material cohesiveness upon sediment movement in suspended-load river systems

Sediment transport relationships in suspended-load river systems differ markedly from the circumstances outlined for bedload or mixed-load rivers. The key issues here are the limited availability (or absence) of sediments coarser than sand-sized particles, and the cohesive properties of fine-grained materials that influence the way in which these sediments are mobilised and deposited in river systems. These effects are especially pronounced in low-relief accumulation zones, typified by lowland plains and fluvio-estuarine environments. Also, particular lithologies may induce such behaviour, because of their influence upon availability of materials of differing grain size (e.g. basalts and mudstones break down to generate fine-grained (mud or soft-bottomed) river systems).

In fine-grained systems, anomalous behavioural traits may be evident in some situations. Cohesive sediments have strong interparticle forces due to their surface ionic charges. As particle size decreases, surface area per unit volume (i.e. specific surface area) increases and interparti-cle forces, rather than gravitational forces, dominate the behaviour of the sediment. In these instances, settling velocity is no longer a function of only particle size (dis-cussed later). The weight of an individual fine-sediment particle is not sufficient to cause settling when the particle is suspended in water and small disturbances, such as tur-bulence fluctuation, are able to overcome the weight of the particle. Cohesive sediments tend to bind together (aggre-gate) to form large, low-density units called flocs (floccula-tions). This process is strongly dependent upon the type of sediment, the type and concentration of ions in the water, and the flow condition. Metallic or organic coatings on the particles may also influence the interparticle attraction of fine-grained sediments. Small particles may bind together to form larger flocs, which grow when they collide with other particles or other flocs. They may also be broken up by turbulent stress. As a result of this process, some mud-dominated rivers may demonstrate attributes of bedload

Sediment-transport-limited systems

capacity-limited channel

competence-limited channel

Sediment-transport-limited systems

supply-limited channel where incision
has occurred

supply-limited channel where channel
expansion has occurred

Figure 6.15   Sediment-supply- and transport-limited systems. Sediment transport-limited systems occur when sediment volume or size exceeds the capacity of the flow to transport it, resulting in sediment deposition and storage. Transport-limited systems can be competence (grain size is too large) or capacity (volume is too large) limited. Sediment-supply-limited systems occur when the flow is able to carry more sediment than is made available to it, resulting in erosion of sediment. Photographs: (a) Waiapu River, New Zealand (K. Fryirs), (b) Sangainotaki River, Japan (K. Fryirs), (c) Latrobe River, Victoria (G. Brierley), (d) Greendale Creek, NSW (K. Fryirs).

channels, despite the absence of sand or coarser particles to act as bedload materials. In these instances, colloidal properties of silt–clay sediments promote the development of aggregates that move as sand-sized bedload materials within highly sediment charged systems.

Deposition in river systems

Interpretation of the conditions under which sediments are deposited and preserved along valley floors is a significant skill in efforts to read the landscape. Typically, the channel

bed is composed of recently deposited sediments, while the distribution of sediments on floodplains is often a mix of contemporary and past depositional events. Depositional processes effectively remove sediment from the conveyor belt. These materials are not re-contributed until entrainment and transport subsequently occur. Deposition occurs when the flow no longer has the competence to maintain a grain in motion, such that the particle moves from the transport to the deposition phase of the Hjulström diagram (Figure 6.2).

The velocity at which a particle settles to a channel bed is known as the *fall velocity* V_0 or settling velocity. This is a function of the particle's grain size and density and the transporting fluid's viscosity and density. Falling under gravity, a particle reaches its fall velocity when the drag on the particle equals the submerged weight of the particle. In most cases the ratio of sediment density to water density is around 2.65. This is called the specific gravity. This means that, in still water, a particle in suspension has a downward vertical motion, since it is denser than water. When expressed in terms of grain diameter, submerged weight and viscosity, the fall velocity measures the rate at which sediments fall out of the sediment column based on their size. This is determined by:

$$V_0 = \frac{1}{18}D^2\frac{\rho_s - \rho}{\mu} \quad \text{for silts and clays} <0.0063\,\text{mm}$$
$$\text{(Stokes' law)}$$
$$V_0 = \sqrt{\frac{2}{3}Dg\frac{\rho_s - \rho}{\rho}} \quad \text{for gravels} >2\,\text{mm}$$

where V_0 (m s^{-1}) is the fall velocity, D (mm) is the characteristic grain size, ρ_s is the sediment density (assumed to be constant at 2650 kg m^{-3}), ρ is the water density (1000 kg m^{-3}), g is the acceleration due to gravity (9.81 m s^{-2}) and μ (N s m^{-2}) is the dynamic viscosity (which is affected by water temperature). For sand (0.063–2 mm) a composite law can be derived depending on particle size (see Table 6.3).

Deposition is a size-selective process. As flow velocity decreases, coarser sediment is deposited first, while finer grained particles remain in motion. Particle shape also affects the fall velocity of materials within fluid flow (see Figure 6.16). The greater the departure from a spherical shape, the lower the fall velocity. Very platy or flat clasts settle more slowly than spherical clasts of the same weight and density. Hence, particle shape should be viewed along with size in determinations of sediment movement in coarse-grained (gravel-bed and coarser) rivers.

In sand- and gravel-bed rivers, deposition occurs once the competence limit of any particular grain is reached. Depending on the grain size, deposition can occur with only a small drop in velocity (for bedload sediments) or after a more significant drop in flow velocity (for suspended-load sediments). In other words, deposition occurs when the bed shear stress is less than the critical shear stress. Only aggregates with sufficient shear strength to withstand the highly disruptive shear stresses in the near-bed region will be deposited. Deposition also occurs when the capacity limit of a flow is exceeded (as noted by the Exner equation and Lane's balance in Chapter 5). Differential settling of material in transit gives rise to sediment sorting and grading (see below) as well as paving and armouring.

In fine-grained systems, the floc structure (size, density and shape) determines the settling velocity. The effective settling rate is determined by the fall velocity multiplied by a hindrance factor (this represents the velocity reduction due to other particles). In simple terms, the settling velocity increases with sediment concentration at low concentration, then attains a maximum value and thereafter decreases due to hindered settling at intermediate concentrations and structural flocculation at high concentrations. Therefore, the depositional behaviour of cohesive sediment is not only controlled by bed shear stress and settling velocity, but also by turbulence processes in the zone near the bed, type of sediment, depth of flow, suspension concentration and ionic constitution of the suspending fluid. Partial deposition may occur if bed shear stress is greater than the critical shear stress for full deposition but smaller than the critical shear stress for partial deposition. At this range of bed shear stress, relatively strong flocs are deposited and relatively weak flocs remain in suspension.

Consolidation changes the thickness of active and inactive layers at the bed surface through changes in porosity.

Table 6.3 Fall velocity for sand-sized sediment in still water at 20 °C[a]

Grain size D (mm)	Wentworth scale	Fall velocity V_0 (mm s^{-1})
0.089	Very fine sand	0.005
0.147	Fine sand	0.013
0.25	Medium sand	0.028
0.42	Medium sand	0.050
0.76	Coarse sand	0.10
1.8	Very coarse sand	0.17

[a]Modified from Robert (2003).

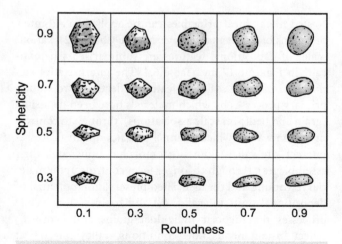

Figure 6.16 Classification scheme for grain shape (sphericity) and rounding. These schemes are used in sedimentological analyses of geomorphic surfaces, bank exposures and pit analysis.

It also changes the size fraction distribution within the bed. Primary consolidation is caused by the self-weight of sediment, as well as the deposition of additional materials. It begins when the self-weight of the sediment exceeds the seepage force induced by the upward flow of pore water from the underlying sediment. During this stage, the self-weight of the particles forces them closer together. The seepage force lessens as the bed continues to undergo self-weight consolidation. Primary consolidation ends when the seepage force has completely dissipated. Secondary consolidation is caused by the plastic deformation of the bed under a constant overburden. These are critical considerations in determination of potential subsidence within fine-grained depositional basins.

Interpreting sediment sequences as a tool to read the landscape

Analysis of river sediments (sedimentology) is key to the interpretation of depositional environments, thereby assessing the processes and conditions under which materials have been deposited, stored and preserved on valley floors. As noted in Chapter 2, these relationships vary markedly over differing spatial and temporal scales. Prior to outlining key principles in the development of a practical approach to river sediment analysis, scales of depositional features in river systems and contrasting sedimentary sequences in bedload, mixed-load and suspended-load depositional environments are briefly summarised, and key considerations in efforts to interpret sediment sequences when reading the landscape are outlined.

Scales of river sedimentary features

Nested hierarchical principles can be applied to aid interpretations of sedimentary sequences (see Chapter 2). Various scales of depositional features are summarised schematically in Figure 6.17 and Table 6.4. The interpretative significance of these scales of feature reflects controls upon primary processes by which materials have been deposited (grain and bedform scale associations), through to controls upon the way in which channel and floodplain processes interact on valley floors (geomorphic units and their assemblages), through to longer term controls upon material reworking and the associated preservation potential of depositional units (at valley fill and basin scales). While processes that deposit individual grains and bedforms reflect instantaneous flow conditions, valley fills reflect long-term system evolution dictated by tectonic history, climate change, base-level change and responses to disturbance events. It is difficult to ascribe specific dimensions to units in the hierarchy outlined in Figure 6.17 and Table 6.4.

Indeed, overlap in size for differing scales is inevitable, reflecting the dimensions of the river system under investigation. For example, an individual bar on a major river may extend over several kilometres – much longer than reaches of a smaller river system, in which bars may be just a few metres long. Each of these scalar issues can be interpreted to provide practical guidance in efforts to read the landscape.

The size and mineralogy of *individual grains* are indicative of the regional lithology and the energy of flow that transported and deposited materials. When viewed relative to other depositional sequences within the same river system, patterns of particle size and shape can be interpreted to provide insight into how far materials have been transported (i.e. distance from source) or the duration and intensity of any given flow event. Sorting of bed sediment is a result of mobility differences related to flow–particle interactions (see Figure 6.18). Sorting provides insight into the ranges of sizes of material present in a given setting, the frequency of reworking and the duration of the transport phase of the flow. Well-sorted sediments have been subjected to long periods of time in transport at velocities above the entrainment threshold. This allows sediments to be transported, organised and sorted before deposition. This tends to occur during the gentle waning stages of flows. Poorly sorted sediments tend to be deposited after short-duration flows with velocities above the entrainment threshold. Sediments are entrained and quickly deposited. Little time is available for the sediments to be organised or to settle from flow. Steep waning stages of flow events may be responsible for this condition. Similarly, particle shape (angularity/roundness) is indicative of the frequency with which a grain has been reworked and the distance it has travelled (Figure 6.16). Fluctuations in sediment availability and flow energy result in different grading patterns within an individual depositional (bedded) unit (Figure 6.19). Interpretation of these textural patterns provides guidance into formative flow conditions (specifically flow–sediment interactions). For example, this may be used to make inferences about flow depth at the time of deposition.

Sediment availability and flow energy determine what sizes of sediment are mobilised and the resulting type of *bedform* that is deposited. These depositional features are referred to as *facies*. Subsequent events determine the reworking and/or preservation of these deposits, and the stacking of additional bedforms. Facies are sedimentary structures that result from specific processes under a given range of flow and sediment transport conditions (e.g. see Figures 6.12 and 6.14). By characterising the facies assemblage of any landform, an interpretation of the environment of deposition can be made (Table 6.5). Silt–clay (fine-grained) facies occur in sediments <62 μm (Table

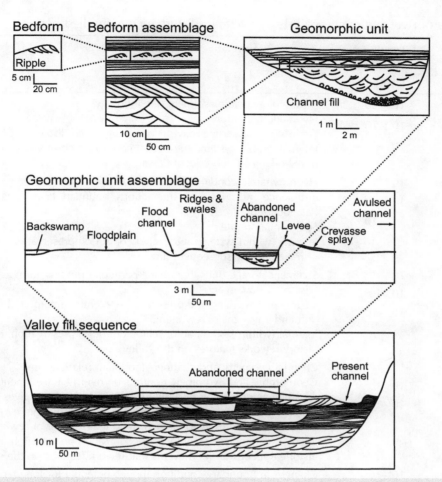

Figure 6.17   Scales of sediment analysis in river systems. In the constructivist approach outlined here, bedforms and their assemblages comprise geomorphic units. The geomorphic unit assemblage characterises reaches. These are inset within a larger valley fill sequence that preserves the history of river adjustment and change. Modified from Brierley (1996). © John Wiley and Sons, Ltd. Reproduced with permission.

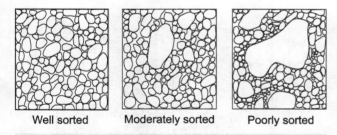

Figure 6.18   Classification scheme for grain sorting. These schemes are used in sedimentological analyses of geomorphic surfaces, bank exposures and pit analysis.

6.1). Field texturing is required to determine the relative proportions of silt and clay in the sediments (Table 6.2).

Fine-grained facies are low-energy, suspended-load deposits, typically found in floodplains (e.g. backswamps or abandoned channels). Bioturbation processes may destroy primary bedforms such that these deposits have a massive (relatively featureless) structure (i.e. Facies Fm and Fsm), although small laminations may form in some cases (Facies Fl).

Sand facies occur in sediments >62 μm but <2 mm in grain size (Table 6.1). Grain-size cards or sieving in the laboratory are required to determine the texture of the sediments. Bedform structures range from ripple structures (Facies Sr), to dune structures that form trough- and planar-cross beds (Facies St and Sp) to horizontally bedded plane beds (Facies Sh).

Gravel facies occur in sediments that contain clasts with *b* axes ≥2 mm (Table 6.1). Gravel facies can be formed by bedforms (Facies Gh or Gp) or channel fills (Facies Gt), but can also represent debris flow deposits that are either clast supported (Facies Gcm) or matrix supported (Facies Gmm). Debris flows occur when masses of poorly sorted materials surge down hillslopes onto valley floors.

Table 6.4   Scales of river sediment analysis[a] (no human disturbance is implied here)

Depositional unit	Length (m) Width (m) Depth (m)	Primary interpretative significance
Individual grain	10^{-6}–10^{0} 10^{-6}–10^{0} 10^{-6}–10^{0}	Nature of available sediment (i.e. lithology) Hydraulic conditions Grain size and sorting may provide insight into distance from source Internal (within system, autocyclic) control on preservation, reflecting the ease of reworking by subsequent flows
Bedform-scale facies (ripples, dunes, etc)	10^{-3}–10^{1} 10^{-1}–10^{2} 10^{-3}–10^{1}	Hydrodynamic conditions at time of deposition (e.g. instantaneous discharge, turbulence and fluid–grain interactions, boundary-layer dynamics, flow depth) Palaeocurrent direction Sedimentation rate Internal (within system, autocyclic) control on preservation, reflecting the ease of reworking by subsequent flows
Geomorphic unit (channel and floodplain units such as point bar, backswamp)	10^{1}–10^{6} 10^{1}–10^{5} 10^{-1}–10^{2}	Sediment storage units that reflect particular process–form associations History of flow events Degree of reworking and preservation potential (interpreted from geometry of unit and any signs of reworking) Internal (within system, autocyclic) control on preservation, reflecting how flood history reworks features on the valley floor
Geomorphic unit assemblage (reach-scale associations)	10^{2}–10^{6} 10^{2}–10^{5} 10^{1}–10^{2}	Channel geometry and channel–floodplain relationships Geomorphic and hydrologic controls on river type (sediment and flow regimes) Assemblage of features and their stacking arrangement gives insight into depositional environment, flood history and geomorphic evolution based on assemblages of preserved elements and their degree of reworking Internal (within system, autocyclic) control on preservation, reflecting the historical record of flood events that rework features on the valley floor
Valley fill sequence (basin scale)	10^{4}–10^{7} 10^{3}–10^{6} 10^{2}–10^{4}	Changes in boundary condition induce potential for long-term preservation and sediment storage Regional controls on river type: topography, tectonics, climate, vegetation Subsidence External (allocyclic) control on preservation, reflecting how tectonic setting and climatic factors affect the nature and rate of sedimentation

[a]Modified from Brierley (1996) © John Wiley and Sons, Ltd.

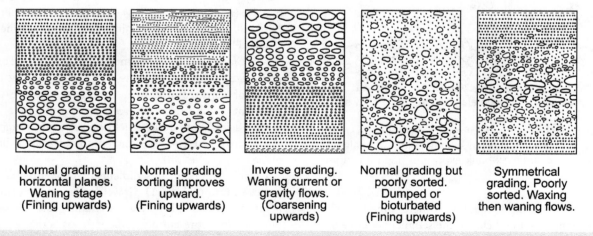

Normal grading in horizontal planes. Waning stage (Fining upwards)

Normal grading sorting improves upward. (Fining upwards)

Inverse grading. Waning current or gravity flows. (Coarsening upwards)

Normal grading but poorly sorted. Dumped or bioturbated (Fining upwards)

Symmetrical grading. Poorly sorted. Waxing then waning flows.

Figure 6.19   Classification scheme for grain grading. These schemes are used in sedimentological analyses of bank exposures and pit analysis.

Table 6.5　Facies coding scheme[a]

Facies code	Facies	Sedimentary structures	Interpretation	Appearance
Gmm	Massive, matrix-supported	Weak grading	Plastic debris flow (high strength, viscous)	
Gh	Clast-supported, crudely bedded gravel	Horizontal bedding, imbrication	Longitudinal bedforms, lag deposits, sieve deposits	
Gcm	Clast-supported, massive gravel	None (mixed)	Pseudoplastic debris flow (inertial bedload, turbulent flow)	
Gt	Gravel, stratified	Trough cross-beds	Minor channel fills	
Gp	Gravel, stratified	Planar-cross-beds	Transverse bedforms, deltaic growths from older bar remnants	
St	Sand, fine to very coarse, may be pebbly	Solitary or grouped trough cross-beds	Sinuous-crested and linguoid (3-D) dunes	
Sp	Sand, fine to very coarse, may be pebbly	Solitary or grouped trough cross-beds	Transverse and linguoid bedforms (2-D) dunes	
Sr	Sand, very fine to coarse	Ripple cross-lamination	Ripples (low-flow regime)	
Sh	Sand, fine to very coarse, may be pebbly	Horizontal lamination parting or streaming lineation	Plane-bed flow (critical flow)	
Sl	Sand, fine to very coarse, may be pebbly	Low angle (<15°) cross-beds	Scour-fills, humpback or washout dunes, antidunes	
Se	Sand, fine to very coarse, may be pebbly	Crude cross-bedding	Erosional scours with interclasts (e.g. around vegetation)	
Ss	Sand, fine to very coarse, may be pebbly	Broad, shallow scours	Scour fill	
Sm	Sand, fine to coarse	Massive or faint lamination	Sediment–gravity flow deposits (quick deposition)	
Fl	Sand, silt, mud	Fine lamination, very small ripples	Overbank, abandoned channel or waning flood deposits	
Fsm	Silt, mud	Massive	Backswamp or abandoned channel deposits	
Fm	Mud, silt	Massive, desiccation cracks	Overbank, abandoned channel or drape deposits	
Fr	Mud, silt	Massive, roots, bioturbation	Root bed, incipient soil	
C	Coal, carbonaceous mud	Plant, mud films	Vegetated swamp deposits	
P	Palaeosol carbonate (calcite)	Pedogenic features: nodules, filaments	Soil with chemical precipitation	

[a]Modified from Miall (1985).

Analysis of depositional units at the landform scale is the key tool for interpretation of process–form relationships that reflect the type of river under investigation. These landforms are three-dimensional building blocks that make up a river reach and are referred to as *geomorphic units*. Sedimentological terms for this scale of depositional feature include elements, architectural elements, morpho-stratigraphic units or morphogenetic units. Depending on their composition and position on the valley floor, deposits within geomorphic units can reside in the landscape for short periods of time (days) to thousands of years. Analysis of the facies assemblages that make up individual geomorphic units can be used to assess the history of depositional events that created that feature. In many cases, geomorphic units are comprised of specific assemblages of sedimentary structures. For example, backswamps largely comprise Fsm or Fm facies, while point bars comprise sequences of Sr and Sh facies capped by finer grained facies (Fl or Fm). Critically, however, geometry, spatial pattern and stacking arrangement of geomorphic units can be used to interpret the history of events that generated sediment sequences.

Much can be gained by analysing and interpreting the three-dimensional geometry and bounding surfaces that define geomorphic units within sediment sequences. Ero-sional boundaries represent a disjunct in the history of sedimentation and/or reworking. This may reflect phases within an individual flood event. Alternatively, it may record an erosive event that removed/erased a significant proportion of the depositional record. The record of these disruptive events is partially recorded as unconformities in the depositional record. In some cases, the sediment archive provides a very selective record, as thousands of years of near-continuous sedimentation may be removed by a single flood event. Hence, analysis of disjuncts in depositional sequences provides critical guidance in the analysis of river history. Obviously, sediments removed from one part of a landscape may become incorporated within depositional sequences downstream.

Geomorphic unit assemblages provide the most reliable basis with which to interpret the type of river under which sediment sequences have been deposited and preserved (see Chapter 10). Preservation of bar and channel fill features, along with floodplain deposits, is conditioned by the position of these deposits within a reach, the history of subsequent flood events and the aggradational nature of the environment within which sediments are accumulating (Chapters 8 and 9). The reach-scale pattern of these features in contemporary environments (i.e. their juxtaposition) provides guidance into the stacking arrangement with which the deposits from these features are likely to be observed within sedimentary sequences. This accords with Walther's law of the correlation of facies, whereby depositional units which are juxtaposed in contemporary field situations are likely to rest atop each other in vertical sediment sequences. Interpretation of the range, pattern and association of channel and floodplain geomorphic units provides the most reliable indicator as to whether the depositional environment reflects a bedload, mixed-load or a suspended-load river. Bedload-dominated rivers generate sediment sequences that are comprised primarily of sheet-like channel fills. Typically, coarse sediments are deposited near the base of the channel, atop which floodplains are deposited. Deposits of suspended-load rivers are character-ised primarily by vertically accreted floodplain deposits, with occasional stringers of sediment reflecting well-defined channels that were abandoned by avulsion proc-esses (see Chapter 9). Fine-grained sediment (fine sand, silt and clay) settles out in backwater areas in the channel and low-velocity flows over the floodplain. Composite bank deposits are a characteristic attribute of mixed-load rivers.

Valley fill sequences are three-dimensional assemblages of genetically and spatially related depositional units. Geo-morphic factors control the geometry, preservation and stacking arrangement of features at the basin scale. Various geologic and climatic controls determine whether element-scale depositional units ultimately become part of the basin fill. Indeed, it is only within the lowland accumulation zone (or offshore) that the depositional sequence is more complete, presenting a spatially and temporally aggregated record of sediment movement across the catchment as a whole. Analysis of depositional records in these locations provides insight into the averaged or pulsed rates of sediment accumulation over differing timeframes.

Sediment sequences in differing landscape settings: deposits of bedload, mixed-load and suspended-load rivers

Sedimentological investigations must be carefully framed in their spatial and temporal contexts. In geographic terms this reflects landscape setting, position in a landscape (catchment) and relationships to other features in that setting (i.e. landscape connectivity and determination of whether adjacent features are genetically linked or not). History matters, in the sense that what is seen today is a product of past and present processes, specifically the dep-ositional record itself, the variable preservation potential of features and the sequence of events that determine what is left today, alongside considerations such as tectonic history or subsidence that may affect long-term preservation (see Table 6.4).

Inevitably, analysis of depositional records is most sub-stantively and effectively performed in accumulation zones (Chapter 3). By definition, the depositional record in steep, confined valley settings is virtually non-existent. Partly confined valleys have a spatially patchy record of depositional features preserved within floodplain pockets. These materials are subject to reworking because the relatively narrow valleys in these settings are subjected to frequent, high-magnitude and high-energy events that can flush materials on valley floors. Hence, there is selective preservation of floodplain deposits at valley margins. These deposits are typically found in low-energy zones associated with downstream changes in valley alignment or fans, or at tributary confluences. Even within these settings, the deposits only provide a selective record of floodplain-forming processes, with limited insight into the nature of the channel itself. In contrast, a full suite of channel and floodplain features is found in laterally unconfined valley settings. Analyses of depositional sequences for alluvial river systems should first differentiate among deposits of bedload versus mixed-load versus suspended-load rivers. These rivers are quite different in terms of both their formative processes and the ease/frequency with which materials are likely to be reworked.

In bedload-dominated rivers, channels have high width/depth ratio because of their non-cohesive banks. Flows recurrently rework gravel and sand bedforms, such that their preservation potential is limited unless the channel shifts position to another part of the valley floor. These rivers are sensitive to adjustment, and disturbance events are frequent. Commonly, a series of topographic surfaces is inundated and reworked at differing flow stages. Floodplains are typically small features in sheltered sections of the valley floors. Long-term preservation is only likely in aggradational settings. Channel fill deposits make up a large proportion of the valley (basin) fill (see Figure 6.20a).

In mixed-load rivers, channels rework materials cross the valley floor in a more predictable and systematic manner, relative to bedload rivers. Composite banks are more stable than their non-cohesive counterparts. As a result, channels have a lower width/depth ratio and commonly adopt an asymmetrical geometry. This reflects accretion on the shallow convex slope and erosion of the concave bank, with a sinuous channel outline (see Chapter 7). In contrast to bedload rivers, bank-attached geomorphic units are more common than mid-channel geomorphic units (see Chapter 8). The floodplains of mixed-load rivers are produced by a mix of lateral (within-channel) and vertical (overbank) accretion processes, with an array of geomorphic units (see Chapter 9). Basins subjected to long-term preservation are characterised primarily by a mix of channel and floodplain depositional units, the nature of which can be used to infer the type of river that created the depositional sequences (see Figure 6.20b; Chapter 10).

As noted for bedload and mixed-load rivers, mutual interactions between channel and floodplain processes also fashion the character and behaviour of suspended-load rivers. These rivers are found in low-relief, low-energy settings, often with very extensive, slowly accreting floodplains in subsiding basins. The lack of bedload-calibre materials significantly restricts the range of channel and floodplain geomorphic units found in these settings. Fine-grained, cohesive banks induce relatively stable channels with low width/depth ratios. Channel stability, in turn, limits the capacity for channel adjustment on the valley floor, resulting in limited reworking of floodplain materials in these vertically accreting, aggradational environments. Floodplain flows may occur frequently, but energy is dissipated across wide valley floors, so erosion is limited and deposition is dominant. These floodplain-forming processes, in turn, deposit the materials that subsequently become channel boundaries, thereby perpetuating this type of river behaviour. It is difficult to rework these cohesive, fine-grained sediments. As a consequence, basin fills are dominated by fine-grained floodplain deposits, with channel fills observed as threads called stringers (Figure 6.20c).

Primary considerations in the use of sediment analyses to assist efforts to read the landscape are outlined in the following section.

Guiding principles in efforts to read the landscape by interpreting sediment sequences

Analysis and interpretation of river sediments are key skills in efforts to read the landscape. By definition, such practices provide an incomplete guide to the suite of processes that create landscapes, as they focus on depositional products rather than erosional processes (and sources of sediment) or on the transport mechanisms by which these materials were brought to any given depositional site. However, it is sometimes possible to make inferences about sediment sources, based on interpretations of mineralogical and textural attributes that reflect particular lithologies. Alternatively, geochemical traits can sometimes be used to trace the erosional origins of materials (see Chapter 14). Similarly, inferences about the energy of transport/depositional processes, and associated environments of deposition, can be made from facies-scale analyses of textural and structural (bedform-scale) attributes of depositional units.

Detailed analysis of sediment sequences, framed in relation to their position in a landscape and their juxtaposition (relationship) to adjacent/other depositional units, provides significant insight into formative processes and controls upon reworking and/or the preservation of differing features on the valley floor. Such investigations provide the primary evidence with which to interpret the behavioural regime for that type of river (Chapters 10 and 11), as they

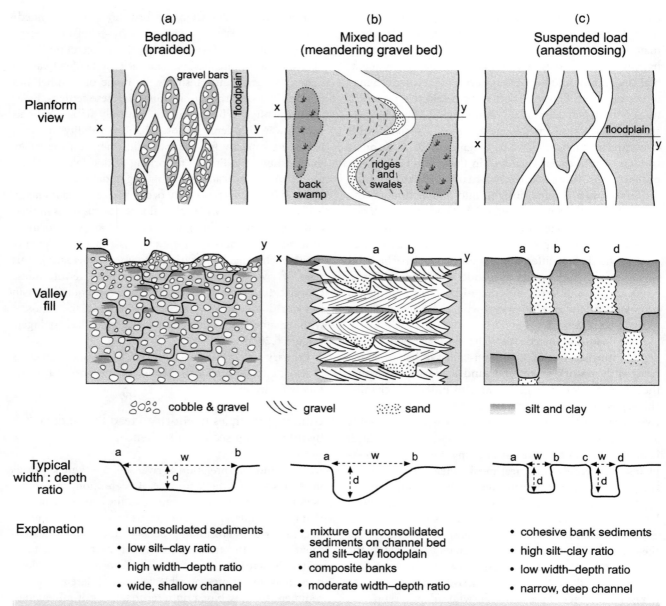

Figure 6.20   Channel width/depth ratio and its relationship to sediment mix. Bedload, mixed-load and suspended-load rivers have channels of varying size and shape. (a) The channel width/depth ratio is highest in unconsolidated loose sediments such as those found in braided rivers. (c) Cohesive, fine-grained sediments produce low width/depth ratio channels such as those in anastomosing rivers. (b) Intermediate shapes and sizes are formed in mixed-load rivers with composite banks that have sand or gravels along the bed, such as some meandering rivers.

record deposits from the range of site-specific formative and reworking events. As such, they provide critical guidance into how the channel and/or floodplain respond to flows of differing magnitude (and sequence), enabling interpretations of river behaviour at low flow, bankfull and overbank flow stages. This entails analysis and interpretation of depositional attributes of landforms (geomorphic units or elements) that make up the channel and floodplain compartments along the valley floor (outlined in detail in Chapters 8 and 9). Relationships between the process domains that fashion patterns and assemblages of deposits in channel and floodplain compartments, in turn, are influenced by channel geometry (e.g. the role of bank strength is a key consideration in the differentiation of bedload, mixed-load and suspended-load rivers; see Chapter 7).

Just as important as these interpretations, however, are determinations of whether any given reach has behaved as a different type of river at some stage in the past, and whether this change in river type occurred as a product of natural variability (e.g. tectonic activity, climate change) or as a response to human disturbance (Chapters 12 and 13 respectively). Analysis of the range, character and pattern of geomorphic units on the valley floor, and their material (sedimentological) properties, is a key step in assessment of river evolution, thereby providing important guidance in efforts to read the landscape. For example, terraces along valley margins represent old river (channel and floodplain) deposits (see Figure 6.21). The materials that make up the terrace may be poorly sorted gravels, indicative of a bedload-dominated river that had non-cohesive channel boundaries. The contemporary channel and floodplain may have attributes of a mixed-load system, with non-cohesive gravels and sand on the channel bed, but fine-grained (silt and clay) cohesive sediments that reflect deposition from suspension overlying these materials on the floodplain (and hence reflected in the channel bank). This simple scenario can be interpreted as three stages of activity:

1. A bedload-dominated river was aggrading on the valley floor.
2. Incision resulted in the channel cutting into its bed (degrading), as the river adjusted to changes in the flow–sediment balance as expressed on the Lane balance diagram. The terrace was formed.
3. The river has subsequently adopted the contemporary character and behaviour of a mixed-load system, with a differing assemblage of channel and floodplain fea-

tures and deposits relative to the former river system that created the terrace.

Detailed field investigations and applications of dating techniques would be required to develop a timeframe (chronology) for these evolutionary adjustments, providing a basis with which to explain river changes in relation to altered boundary conditions (see Chapter 12).

The scenario outlined above may be an accurate representation of reality. However, it may be a gross simplification of what actually happened, as it is only possible to interpret sequences (or phases) of river history/evolution from the depositional record that remains. Many other events may have affected the river to a significant degree, reworking a large proportion of stored sediments on the valley floor. The only remaining evidence for such activity may be the erosional contact between depositional units that indicates that 'something' happened in the intervening period. Therefore, the characteristics of the boundaries between units, and the geometry of these 'bounding surfaces', are a key indicator used to guide palaeoenvironmental reconstructions. Important inferences can be derived from the nature of erosional or depositional contacts between features. For example, sharp contacts between units reflect an intervening erosion episode between depositional events, while onlapping deposits reflect accordant deposition with no intervening episode of erosion. In some instances it may not be possible to infer whether erosion was the product of one event or multiple events, and whether deposits built up recurrently at that site only to be subsequently reworked.

Erosion at one site may influence the depositional record at downstream sites. Inevitably, this record is far more

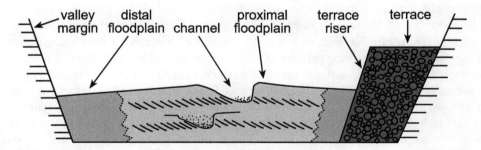

Figure 6.21 The relationship between channel, proximal and distal floodplain, and terrace features. A floodplain is formed under the contemporary flow regime, whereas a terrace is a palaeo-floodplain formed under a previous flow regime. An erosional boundary and terrace riser separates these two landforms. The proximal floodplain is the area adjacent to the channel bank where channel adjustment is most likely (e.g. channel migration and abandonment). Distal areas away from the channel tend to be depositional zones where backswamps form. Initial inferences for reading the landscape can be made by examining what types of landforms are present and active, what the shape and size of the channel are, and what sediments make up the landforms at different positions on the valley floor.

complete in lowland and offshore environments that are subjected to ongoing (continual) deposition. The vast majority of this record lies buried within deep subsiding basins. Thus, it is important to frame basinwide-scale (valley fill) analyses of depositional records in relation to their catchment context, interpreting the original sources of sediments, how they have been transported and transformed along their pathway to their depositional site (diminution in grain size, angularity/roundness, etc.) and how these materials have been altered once they arrived at their final destination. This requires interpretation of whether materials have been packed, consolidated, cemented or deformed in the period post-deposition. It is also important to consider the prospective preservation potential of the materials that are being analysed. Clearly, there is a marked difference in the likelihood of preservation of floodplain deposits relative to channel materials that are more frequently reworked (drapes of sand-sized materials, for example, are recurrently reworked). Preservation is ultimately a product of the aggradational nature of the river, where sediments build up on valley floors over time, relative to degradational environments, where incision may remove a significant part of the valley fill. The manner and ease with which a channel adjusts its position on the valley floor is another key determinant of the nature of depositional sequences and their preservation potential (i.e. likelihood for reworking).

In light of these considerations, significant care and attention must be given to the design and implementation of sediment analyses in the field. A practical approach to analysis of river sediments is outlined in the following section.

A practical approach to river sediment analysis

Targeted approaches to sediment analysis are applied within a carefully considered plan or strategy. Obviously, this is constrained by what is available. Site selection should strive to be as representative as possible, with a clear and explicit rationale as to why analyses are being performed at any given locality. In contemporary depositional environments, targeted sampling strategies select representative cross-sections for any given reach and then the sedimentological attributes for each landform (geomorphic unit) within that cross-section are analysed.

It is important to conduct these sediment analyses and evolutionary interpretations across a range of sites, in order to assess the continuity of depositional contacts and the associated geometry of depositional units whenever possible, producing a two-dimensional picture of the depositional record. This is often very difficult, as the only available exposure sites may be isolated remnants, reflect-

ing a patchy record of preserved sediments. Subsurface scanning technologies, such as seismic surveys and ground-penetrating radar, can be used to analyse boundaries between depositional features and construct dimensional geometries of basin fills across varying scales of enquiry. Such data sets provide a sound platform for targeted and representative sediment analyses using drilling techniques, auger hole analysis or simply by digging a pit and analysing the sediments. Whenever possible, recourse is taken to analysing exposures along road cuts or if pipelines are dug across floodplains. More typically, sediment sequences are analysed and interpreted in bank exposures at channel margins. In many instances, this provides critical guidance into the processes that formed the floodplain at some period in the past. Ironically, however, these deposits have limited preservation potential. The very fact that they are exposed in a bank indicates that erosion is underway, and future flood events are likely to continue to rework these materials.

Obviously the resolution of information that can be gleaned through analyses is markedly different for one-dimensional data from pits or auger/drill holes relative to bank exposures or trench analyses, as insight into lateral or longitudinal continuity of units is constrained. Indeed, great care must be taken in making inferences about the continuity of any given bed, as it must always be remembered that the record of deposits from any given event may vary with position on the valley floor. For example, a flood may deposit coarse sands on a bar at the channel margin, silty sands on levees atop the floodplain (proximal floodplain) and silty clays in backswamps at the valley margin (distal floodplain). This is referred to as event stratigraphy.

For one-dimensional data, representative pit and/or drill/auger hole locations are selected in each geomorphic unit under investigation. Multiple pits and/or drill/auger holes are needed to trace boundaries and trends in sediment characteristics. The resolution of analyses is much greater from a pit relative to a drill or auger hole. The nature of contacts between bedded units can be assessed, as can the scale of bedform features such as grading. Bedding can be assessed from drill hole (core) data, alongside some small-scale bedform attributes and sorting/grading trends, but these features cannot be examined from auger data, where all structures and grading are destroyed. In this case, analyses are limited to broad-scale textural considerations.

When selecting and using bank exposures to interpret depositional sequences, it is beneficial to seek continuity of exposure such that bounding surfaces and the geometry of features can be determined. Where bank exposures are available, these are cleaned to reveal the stratigraphy. Working from a bank exposure or a pit/trench face, procedures used to analyse sediment sequences are summarised in Table 6.6. Simplified versions of these procedures can be used to analyse drill hole or auger hole data. Data are typically processed as sediment columns, bank exposures and fence or block diagrams.

One of the key issues to be addressed in sediment analysis is the depth to which analyses will be undertaken. Seismic techniques are able to extend to a much greater depth than hands-on applications, where enormous constraints are imposed by technical limitations of drilling and auger equipment, the stability of sediment exposures in pits (especially in sand-sized materials) and depth to the water table. For practical reasons, many applications stop when gravel bedload materials or the water table is reached, and material properties of bedded units are analysed above this basal contact.

Interpretations of depositional units are best performed at the landform (geomorphic unit/element) scale. The architecture of a depositional body is evaluated in terms of its external shape and bounding surfaces, and its internal composition (see Chapter 8). Facies that make up each element provide insight into formative flows (Table 6.5). Working within the channel, bed sediments may be indicative of a range of flow events, from formative flows that shape channel geometry through to fine-grained depositional drapes from the waning stage of the last flow event. Inevitably, what is seen at any given time is influenced by the recent history and sequence of flows. Key behavioural differences are evident between boulder, cobble, gravel, sand and fine-grained channels. Particular note should be made of packing arrangements and surface–subsurface (armouring) textural attributes.

Table 6.6 Procedures used to analyse sediment sequences in a pit or bank exposure

1. Select representative sites and determine which techniques will be used to analyse subsurface sediments (pits, auger holes, cores, bank exposures, etc.).
2. Identify bedded units, demarcated by changes in facies type, grain size, bedding characteristics and sorting.
3. Trace the boundaries of bedded units. Boundaries between units may be distinct or gradual. Erosion surfaces may be identified.
4. Differentiate between bedded units and element (geomorphic unit) boundaries. Determine and measure the geometry of the latter.
5. Record the depth of geomorphic unit boundaries within a representative vertical column.
6. Document the geometry of the unit. Identify which attributes have been used to determine the geomorphic unit type (geometry, position, bounding surface, etc).
7. Record the depth/thickness of individual bedded units that make up each geomorphic unit.
8. Record pertinent attributes for each bedded unit, including:
   (a) *Grain size*. This may involve using field texturing (Table 6.2), sand grain-size cards and/or gravel *b*-axis measurements (Table 6.1). Alternatively, representative samples can be taken for laboratory analysis. The range of grain sizes within the bedded unit is recorded. For well-sorted materials, the modal grain size is recorded. For poorly sorted sediments the coarsest fraction is noted, as well as the modal grain size (most commonly a matrix). Grain size is an indicator of the magnitude or the flow required for transport of the sediment.
   (b) *Determine sorting, sphericity and roundness* (see Figures 6.16 and 6.18). Roundness can be used as an indicator of distance from source, while sorting is an indicator of the type and duration of the flow that has deposited the bedded unit.
   (c) *Sediment mix*. Is there any evidence of grading (Figure 6.19)? Is the sediment loose or cohesive? Sediment mixes provide indicators as to the dynamics of the flow responsible for the deposition of the unit.
   (d) *Record facies type* (see Table 6.5). Dimensions of sedimentary structures and any palaeocurrent indicators should be recorded. Inferences about the environment of deposition and the flow characteristics can be made from interpretation of these structures.
   (e) *Shape (bedding) of sedimentary unit*. Record whether the shape of the unit is horizontal, lens-like, irregular, etc. These can reflect uniform deposition, channel fills or erosion scours that have subsequently infilled, etc.
   (f) *Boundaries between sedimentary units*. Determine through visual assessment the definition and shape of boundaries between each bedded unit. Distinct is <1 cm in thickness; gradual is 1–2 cm in thickness; diffuse is barely perceptible. The shape of the boundary (and the unit) defines erosion and depositional surfaces and processes at the start and end of a flow event.
   (g) *Presence/absence of organics*. Preserved roots or organic matter layers can indicate that a palaeosurface was exposed for some time. Evidence of bioturbation indicates homogenisation of bedded units over some period of time.
9. Pull the information together as a sediment column, bank exposure, block diagram or fence diagram.

Analysis of channel bank sediments provides the key guide to the differentiation of bedload, mixed-load and suspended-load rivers (see Figure 6.20). Stacked sediment sequences may be observed in the banks of incised channels, potentially providing insight into former depositional conditions and river type in that valley setting. However, such conditions cannot be detected from the contemporary banks of an aggrading river, where evidence of former river activity has been buried.

The geomorphic behaviour of alluvial rivers is often best interpreted from analysis of floodplain features, as floodplain geomorphic units and their sedimentological attributes reflect longer term adjustments (typically over decades, centuries or millennia). Insights into formative processes, reworking events and changes in environmental conditions can be gleaned from the assemblage and pattern of units, their stacking arrangement on the valley floor and boundaries between these features. Changes in depositional sequences may reflect altered formative processes, perhaps indicating transitions in the way floodplains are deposited and/or reworked, or altered channel–floodplain relationships. This may be indicated by textural and facies attributes of geomorphic units, their geometry and the boundaries between units. As noted in Figure 6.21, such transitions may be especially marked in analyses of terrace deposits, as these are indicators of former river conditions and palaeoenvironments.

Essentially, geomorphic interpretation of river deposits entails two key issues. First, what were the primary transport mechanisms and depositional processes responsible for a bedded unit? Second, how and why did this feature become preserved as part of the depositional record (i.e. what processes and sequence of events brought about the longevity of that deposit)? In combination, these insights enable interpretations of river behaviour and evolution, including assessment of responses to flood history, channel movement and floodplain evolution. In these analyses, it is critical to separate appraisal of the contemporary behavioural regime of the river (deposits from the contemporary type of river; see Chapter 11) from attributes that reflect and record evolutionary adjustments from a different type of river (i.e. river change; see Chapter 12). For example, transitions from a jumbled, chaotic sequence of sediments to a well-ordered sediment sequence may provide indicative evidence of transition from a bedload to a mixed-load river. The importance of channel geometry within these considerations is highlighted in the following chapter.

Conclusion

Relationships among sediment entrainment, transport and deposition determine the relative mix of erosion, transport

and deposition processes in different parts of a catchment, and within a given reach. Rivers can only transport the sediment made available to them. As noted from the Lane balance, transport- and supply-limited conditions exert a key control upon the degradational or aggradational tendency of a reach. Bed and bank material properties can be used to differentiate among bedload rivers (non-cohesive banks, loose coarse fraction on the bed), suspended-load rivers (low width/depth, fine-grained cohesive banks) and mixed-load rivers (bedload channels with composite banks covered by vertically accreted fine-grained sediments). In many instances, bed surface materials protect the underlying mobile (active) fraction. Entrainment, as such, is a threshold-driven phenomenon.

River landforms are shaped by erosion, movement (transport) and deposition of sediment. The Hjulström diagram summarises phases of sediment entrainment, transport and deposition in river systems in relation to flow velocity and bed material size. Rivers transport most of their sediment as dissolved or suspended load, but the bedload fraction is the primary determinant of river morphology.

Caution should be used when applying bedload transport equations to predict rates of sediment transport. There is immense scatter in the amount of sediment transported for a given discharge. In general, a far greater volume of sediment is transported on the rising limb of a flood than at the falling stage. Sediment rating curves are used to predict the amount of sediment transported at a given flow stage.

Bedform-scale facies have limited preservation potential (i.e. they are readily reworked) and their interpretation can only provide insight into flow conditions at the time of deposition. As such, their analysis needs to be placed in a geomorphic context, related specifically to the landform-scale features that they are a part of. Interpretation of sedimentary sequences at the geomorphic unit and valley fill scales is a key tool for reading the landscape.

Key messages from this chapter

- River channels act as conveyor belts that transfer sediments from headwaters to the sea.
- Alluvial landforms are shaped by erosion, transport and deposition of sediment. The Hjulström diagram summarises phases of sediment entrainment, transport and deposition in relation to flow velocity and bed material size.
- Entrainment is the process by which grains are picked up or plucked from the channel bed. The Shields number quantifies sediment mobility and, therefore,

the entrainment threshold at which grains will be moved. Bed shear stress is a key control on bed material entrainment.

- Sediment transport can occur as dissolved load, suspended load or bedload. Rivers transport most of their sediment as suspended load or dissolved load, but the bedload fraction is the primary determinant of river morphology. Bedload rivers carry more than 11% of their load as bedload materials. Channels have non-cohesive banks. Mixed-load rivers carry 3–11% of their load as bedload and also carry a large volume of suspended load. Channels have composite (upward fining) banks. Suspended-load rivers carry less than 3% of their load as bedload. Channels have cohesive fine-grained banks.
- There is immense scatter in the amount of sediment transported for a given discharge. Factors such as sediment size and availability exert a major influence upon sediment transport. In general, a far greater volume of sediment is transported on the rising limb of a flood than at the falling stage. Sediment rating curves are used to predict the amount of sediment transported at a given flow stage.
- Caution should be used when using bedload transport equations to predict rates of sediment transport. A wide range of field, analytical and modelling techniques can be used to estimate rates of bedload and suspended-load movement in rivers.
- Bed material organisation (packing arrangements, surface–subsurface relationships, vegetation cover, etc.) inhibits entrainment. Once entrainment is initiated, the entire bed may be mobilised.
- Selective entrainment and abrasion result in progressive downstream gradation in bed material size along the longitudinal profile of river systems. Sediment inputs at tributary junctions or from hillslope failures may disrupt this pattern.
- The range of bedforms formed under differing flow energy conditions can be related primarily to bed material size, flow depth and flow velocity. Sand-bed rivers produce bedforms such as ripples, dunes, plane bed and antidunes. Sediment slugs and waves may be generated if the channel is overloaded with sediment. In gravel-bed rivers, pebble clusters, cluster dams and transverse ribs bedforms may be formed.
- In mixed-load rivers with a mixture of sediments of various size in transport at any one time, partial mobility or equal mobility can occur depending on the magnitude of the flow event.
- Some river systems are sediment supply limited (transport capacity greater than sediment supply) and others are transport limited (either capacity or competence limited).
- In suspended-load rivers, flocs of fine sediment can be formed, resulting in aggregates acting more like bedload.
- Deposition occurs when flow no longer has the competence to maintain a grain in motion.
- Analysis of river sediments (sedimentology) is used to interpret depositional environments and is a critical skill for reading the landscape. Sediment analyses occur at the scale of the individual grain, geomorphic unit or valley fill.
- The types of sediments and the landforms produced vary for bedload, mixed-load and suspended-load rivers.
- A wide range of sediment features (facies) can be differentiated. They provide some guidance into flow conditions at the time of deposition, and can be used to interpret environments of deposition. Recourse is best taken to analysis of architectural elements (i.e. preserved deposits at the landform (geomorphic unit) scale)), with emphasis on geometry, location, bounding surfaces, juxtaposition to other features, etc.
- Analysis of sediment sequences can be undertaken using a range of methods, including pit, auger and bank exposure analysis as well as seismic surveys and ground-penetrating radar. A combination of techniques is typically employed within a systematic sampling strategy. However, reworking and selective preservations constrains the record that can be assessed.

CHAPTER SEVEN

Channel geometry

Introduction

Channel shape and size are among the most obvious attributes of river systems. Casual observations of valley floors note the presence or absence of a channel, the number of channels, the variability of their form and whether they 'fit' for that given landscape setting. Channel types can be differentiated into forced (e.g. bedrock river) and self-adjusting (i.e. alluvial) morphologies.

Channel geometry refers to the three-dimensional form of a channel. In general terms, channel size and shape are fashioned by flow and sediment conditions. The components of channels that can adjust include the width (including bed width and water surface width at bankfull stage), depth (including mean and maximum depth), slope (channel gradient, determined by sinuosity) and cross-sectional area (width multiplied by depth; see Chapter 4). As these components vary in space and time, so too do indices such as the width/depth ratio, the wetted perimeter and the hydraulic radius of channels (see Chapters 4 and 5).

Channel size and shape are fashioned by the distribution and effectiveness of erosional and depositional processes along the bed and bank. These processes influence, and in turn are influenced by, processes that form and rework floodplains. Collectively, channel and floodplain adjustments affect the resistance to flow. Channel geometry, sinuosity and the number of channels affect boundary and form resistance. These relationships vary with flow stage and flow alignment (thalweg position) within the channel or across the valley floor.

Ultimately, processes acting on the channel bed influence what happens on the banks. As indicated by the Lane balance diagram, incision and aggradation dictate bed stability, thereby affecting processes acting on the banks. As noted in Chapter 5, these relationships vary at differing positions along longitudinal profiles, with marked differences in source, transfer and accumulation zones. As a consequence, there is pronounced variability in channel geometry at the catchment scale. Atop this, however, reach-specific factors may induce particular sets of process–form relationships that influence channel geometry. Considerable within-reach or site-specific variability may be evident.

Features that make up channels can be represented as a nested hierarchy of forms, in which the broader (bankfull-stage) macrochannel contains a mix of erosional and depositional features (instream geomorphic units). Mid-channel and bank-attached features, in turn, are comprised of differing textures and associated sedimentary structures (see Chapter 6).

Bank erosion processes are differentiated into entrainment and mass movement mechanisms. Sediment composition, especially material size and cohesion, influences the relative role of these processes, and resulting bank morphology. Channels with non-cohesive banks tend to have high width/depth ratios (i.e. bedload dominated rivers), whereas rivers with cohesive banks tend to have low width/depth ratio channels (i.e. suspended load rivers).

Discharge, flow hydraulic relationships and sediment transport mechanisms that influence channel geometry were summarised in Chapters 4, 5 and 6 respectively. This chapter examines how interactions between these factors affect channel size and shape. Channel morphology can be differentiated into symmetrical, asymmetrical, compound or irregular forms. Similar forms can be generated by different sets of processes (the principle of equifinality or convergence). *Hydraulic geometry* is the study of how width, depth and velocity components of channels change with flow stage (at-a-station) and as you move downstream. Principles of at-a-station and downstream hydraulic geometry assume that discharge Q is the dominant independent variable that affects the dependent variables that are the component parts of discharge (width, depth and velocity; see Chapter 4). In this framework, changes in channel size/shape are related to variability in discharge over time at a given cross-section or moving downstream along a river course. Prior to discussing principles of hydraulic geometry, the bed and bank processes that influence channel size and shape are outlined.

Geomorphic Analysis of River Systems: An Approach to Reading the Landscape, First Edition. Kirstie A. Fryirs and Gary J. Brierley.
© 2013 Kirstie A. Fryirs and Gary J. Brierley. Published 2013 by Blackwell Publishing Ltd.

Bed and bank processes that influence channel shape

Variability in bed and bank materials along a river influences the capacity for channel adjustment and resulting channel morphology. The inherent strength and stability of the bed and banks determine the ease with which a channel is able to adjust, whether vertically (the depth dimension) or laterally (i.e. channel width). Hence, channel size and shape are essentially functions of bed and bank processes. These processes are affected by, and in turn determine, the composition of the channel boundary itself. As indicated on the Lane balance diagram, processes on the bed dictate channel stability.

Bed processes that influence channel shape: incision and aggradation

The channel bed can adjust in two primary ways. Vertical adjustments reflect either bed degradation (i.e. channel incision/lowering of the channel bed) or channel aggradation (i.e. accumulation of sediment on the channel bed) (Figure 7.1). Channel incision occurs where the bed is destabilised, or scoured, resulting in bed lowering and

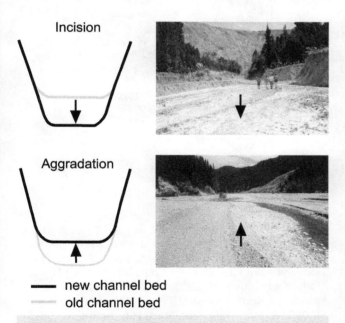

— new channel bed
--- old channel bed

Figure 7.1 Vertical adjustments in channels. Incision and aggradation induce bed-level instability, lowering the bed via erosion and raising the bed through deposition respectively. The photographs are from the Tarndale gully site, Waipaoa Catchment, New Zealand, where recurrent cutting and filling of tributary networks (top photograph) supplies sediment to the trunk stream that aggrades (lower photograph). Photographs (K. Fryirs).

channel deepening (see Chapter 4). Elsewhere, degradation entails accentuated erosion of sculpted geomorphic units such as potholes (see Chapter 8). Incision is commonly initiated by a reduction in the availability of bed material. As noted by the Lane balance and Exner equation (Chapter 5), a delicate balance is required here, as bed materials provide the erosional tools that promote incision. If there are too many materials, the bed is protected. In contrast, if there are too few materials, rates of erosion are inhibited. Degradation may occur through downstream- or upstream-progressing mechanisms. Downstream-progressing degradation is typically associated with a decrease in bed material load or increase in water discharge. Alternatively, upstream-progressing degradation typically reflects a fall in base level. Upstream-progressing degradation generally proceeds at a much faster pace than its downstream counterpart, because the former mechanism increases slope while the latter mechanism decreases slope. In contrast, channel aggradation results in shallower channels with an array of depositional geomorphic units on the channel bed (Figure 7.1; see Chapter 8).

Bank processes that influence channel shape: bank erosion and morphology

Bank erosion processes

Bank erosion entails two phases, namely detachment of grains from the bank and subsequent entrainment (removal) of that material. Lift and drag forces may be entirely responsible for detachment and entrainment, but in most instances, especially in cohesive materials, aggregates and soil micropeds are loosened and partially or completely detached by weakening or preconditioning processes prior to entrainment. The three most important weakening mechanisms are pre-wetting, desiccation and freeze–thaw activity. Hence, subaerial processes are important determinants of bank erosion. *Pre-wetting* influences rates of bank erosion as cohesive materials become more erodible when wet. When very wet, seepage can cause sapping of localised areas of the bank face. Piping may also be evident (see Chapter 4). The role of *desiccation* reflects the nature of clay fabrics that make up cohesive, fine-grained banks. Desiccation can encourage higher bank retreat rates. Materials derived from direct spalling of drier upper bank surfaces collect at the foot of the bank and become available for entrainment at higher flow stage. Cracking up and incipient exfoliation of bank surfaces allow flood water to seep around and behind unstable soil structures (referred to as peds). Slaking refers to 'bursting' of bank peds during saturation. This reflects build-up of air pressures created by the influx of water into the soil pores during rapid immersion. *Freeze–thaw* processes may be an important preconditioning agent. This mechanism is especially prominent for later fluid entrainment of cohesive bank materials in climate settings that oscillate around 0 °C. Given that water expands by approximately 9 % when frozen, ice lenses may reduce cohesion by wedging peds apart. Cantilevers of ice attached to the bank and ice rafts cause serious damage to some rivers during spring thaw.

Hydraulic action (also referred to as fluvial entrainment or corrasion) and mass failures are the primary mechanisms of bank erosion (Figure 7.2). *Hydraulic action* refers to grain-by-grain detachment and entrainment. It is typically associated with non-cohesive bank materials. Grain removal by hydraulic action is closely related to near-bank flow energy conditions, especially the velocity gradient close to the bank and local turbulence conditions, as these determine the magnitude of bank shear stress (see Chapter 5). *Fluvial entrainment* occurs when individual grains are dislodged or shallow slips occur along almost planar surfaces. In some instances a distinct notch may be left in the bank following a flood event, indicating the peak stage achieved. *Undercutting* occurs when velocity and boundary shear stress maxima occur in the lower bank region. High rates of bank retreat at the apex of a bend are explained by steep velocity gradients and high shear stresses generated within large-scale eddies against the concave bank of a bend. During the process of undercutting, flow not only entrains material directly from the bank face, it also scours the base of the bank. This leads to oversteepening, and eventually gravitationally induced mass failure.

The effectiveness of hydraulic action is determined by the balance between motivating forces, which include the downslope component of submerged weight and fluid forces of lift and drag, and resisting forces, which include the interparticle forces of friction and interlocking (see Chapter 6). These effects are especially pronounced along composite banks, where preferential erosion of more erodible basal layers generates overhangs that promote collapse of overlying blocks of cohesive material. Fluid entrainment of basal material following collapse is vital to the effectiveness of this mechanism. Hence, the stability of the lower bank is crucial to the stability of composite banks.

Cohesive, fine-grained bank materials are usually eroded by entrainment of sediment aggregates which are bound tightly together by electrochemical forces, rather than as individual particles. These behavioural traits are heavily dependent upon the physical properties of the materials, such as their mineralogy, dispersivity, moisture content and particle size distribution, and on properties of the pore and eroding fluid, such as temperature, pH and electrical conductivity. Entrainment occurs when the impelling (motivating) forces overcome the resisting forces of friction and cohesion. Hard, dry banks are very resistant. However, wet banks are relatively easy to erode, especially if loosened by repeated wetting and drying or frost action.

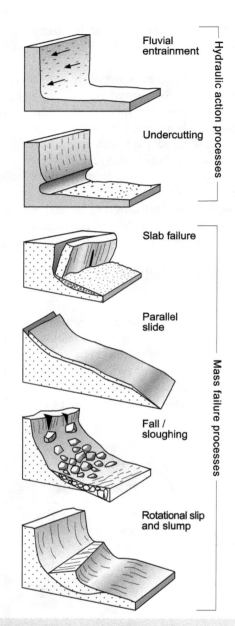

Figure 7.2 Bank erosion processes. Hydraulic action involves the shearing of sediment from a channel bank as water flows across it (entrainment and undercutting). Slab failure, parallel slide, fall/sloughing and rotational slumps are forms of mass failure whereby sections of bank fall into the channel. Modified from Thorne (1999) in Brierley and Fryirs (2005). © John Wiley and Sons, Ltd. Reproduced with permission.

The susceptibility of banks to *mass failure* depends on their geometry, structure and material properties. Deep-seated failures are rare in non-cohesive banks, where basal scour, oversteepening and collapse mechanisms are favoured (Figure 7.2). Along more cohesive banks, weakening and weathering processes reduce the strength of bank material, thereby decreasing bank stability. The effectiveness of these processes is related to soil moisture conditions. Cycles of wetting and drying cause swelling and shrinkage of fine-grained materials, leading to the development of interpedal fissures and tension cracks which encourage failure. Seepage forces can reduce the cohesiveness of bank material by removing clay particles. This may promote the development of pipes in the lower bank. The downslope force of the weight of a potential failure block is the primary motivating force. An increase in this force occurs when fluvial erosion leads to an increase in the bank height or bank angle. Catastrophic failure occurs when the critical value of height or angle is reached. Block mass is greatly influenced by moisture content. The switch from submerged to saturated conditions following flood events can double the bulk unit weight of the sediment, prospectively triggering drawdown failures even without the generation of excess porewater pressures. If rapid drawdown does generate positive porewater pressures, friction and effective cohesion are reduced. In extreme circum-

stances, this can lead to liquefaction, a threshold-induced response that may induce complete loss of strength and flow-type failures.

Shallow parallel slips occur in cohesionless banks, while deep-seated *rotational slip* and *slab failures* are the dominant mechanisms in banks of high and low cohesiveness respectively. Rotational slips occur along a curved failure plane that induces rotational movement via slipping of material down the bank face. Retreat of near-vertical banks via slab failure occurs when blocks of sediment are released from the face of the bank and topple into the channel. Other mass failure processes include *fall* or *sloughing*, where small quantities of material dislodged from the top of the bank accumulate at the base, and *parallel slide*, where slices of material parallel to the bank slip down the bank face (Figure 7.2).

The effectiveness of these various bank erosion processes is greatly enhanced by basal scour or bed incision, which effectively increases bank angle and height (Figure 7.3). In

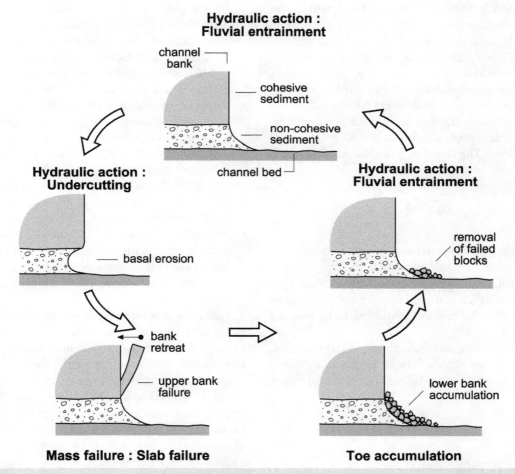

Figure 7.3 The cycle of bank retreat processes. The influence of hydraulic action and mass movement processes varies at different stages in the bank erosion cycle. Following mass failure, bank retreat cannot recommence until materials at the toe of the bank have been removed. Note how bank morphology changes at each phase of adjustment. From Brierley and Fryirs (2005). © John Wiley & Sons, Ltd. Reproduced with permission.

composite banks, where cohesive materials overlie non-cohesive sands or gravels, undercutting of the lower bank by hydraulic action generates an overhang or cantilever in the upper layer. These overhanging materials are prone to fail and blocks of bank material slide or fall towards the toe of the bank during bank failure and collapse. They remain at the toe of the bank until they are broken down or entrained by flow. Once a failed block is deposited on the channel bed, it locally modifies the morphology and flow. Failed blocks, in turn, may temporarily protect the toe of the bank from erosion. This pseudo-cyclic process plays an important role in controlling bank form, stability and rate of retreat (Figure 7.3).

In some cases, channel width and depth do not adjust independently. Adjustments to channel depth via incision or aggradation may trigger secondary adjustments via bank erosion. Positive feedback mechanisms may accentuate these patterns of adjustment following exceedance of critical threshold conditions. For example, banks become higher and are increasingly oversteepened if bed incision exceeds *critical bank height* (Figure 7.4). Indeed, in some instances, channels become so deep that hydraulic action only impacts on lower parts of the bank, and mass failure mechanisms become the dominant form of bank erosion. Undercutting is not necessarily a precursor to this process.

A range of techniques can be used to measure bank stability. Channels with cohesive banks can be characterised as either bank shear or bank height constrained. Banks of shear-constrained channels are eroded principally by fluvial entrainment processes and failure occurs at critical bank shear stress. The factor of safety for a critical bank shear can be measured as:

$$FS_\tau = \frac{\tau_{crit}}{\tau_{bank}}$$

where FS_τ is the factor of safety with respect to bank shear, τ_{crit} (kPa) is the mean bank shear stress and τ_{bank} (kPa) is the critical bank shear stress for bank sediment. Banks are considered to have limited stability with respect to mass failure when $FS_\tau = 1.0$ and are considered to be unstable when $FS_\tau < 1.0$.

Bank-height-constrained channels are subject to erosion by mass failure once critical bank height is exceeded. Critical bank height can be defined as a factor of safety, whereby:

$$FS_h = \frac{H_{crit}}{H} = \frac{N_s C_u}{\gamma_s}$$

where FS_h is the factor of safety with respect to bank height, H_{crit} (m) is the critical bank height, H (m) is the vertical bank height, N_s is a dimensionless stability number

$(3.83 + 0.052(90 - \theta) - 0.0001(90 - \theta)^2$, where θ is the slope angle), γ_s (kN m^{-3}) is the saturated unit weight of soil and C_u (kPa) is the undrained cohesion. Banks have limited stability with respect to mass failure when $FS_h = 1.0$ and are considered unstable when $FS_h < 1.0$.

Bank morphology

Bank morphology may reflect a range of differing conditions or processes. In some instances, similar forms may be produced by differing processes (the principle of equifinality). Hence, significant caution may be required in making interpretations of why banks have adopted a particular morphology. Appraisals of underlying causes must consider a range of possible scenarios, giving due regard to:

- The aggradational/degradational balance of the reach, i.e. bed processes that are operating.
- Bank position within the reach and its relation to flow alignment at differing flow stages.
- The balance of erosional and depositional processes operating on or adjacent to banks.
- The mix of materials that make up the bank and their sedimentology (including the presence of bedrock). This may include sediments that were deposited under a different flow regime and now represent inherited forms that are preserved in the banks.
- The stage of evolution of the bank, including the rate of delivery of materials to the toe of the bank and subsequent rates of removal.
- The primary origin of depositional features that may line (and protect) the bank (whether derived from further up the bank or from upstream sources).
- The combination of fluvial erosion and mass failure mechanisms that erode the bank.
- Vegetation associations and faunal interactions may influence the nature and effectiveness of bank-forming processes. This includes biogeochemical alteration of bank materials, biofilms, etc.

Various bank morphologies are portrayed in Figure 7.5. Banks with a homogeneous sediment mix commonly have a uniform morphology. This may take a near-vertical form, where fluvial entrainment processes are effective (Figure 7.5a), or various forms of inclined bank (Figures 5b–d). In general terms, the sediment mix dictates bank angle. Cohesive sediment forms steeper banks, while sandy banks have a gentler angle of repose. Uniform bank morphologies may reflect hydraulic action processes such as fluvial entrainment or mass failure processes such as parallel sliding or rotational slips. Along convex-upwards banks, gradual mass movement mechanisms may be inferred (Figure 7.5b). Alternatively, overflow mechanisms from the floodplain to the channel during the waning stages of floods may modify patterns of sedimentation on the bank, especially around vegetation. Graded banks represent a planar condition that may reflect grain-by-grain movement downslope at the angle of repose or parallel sliding down a slip face (Figure 7.5c). A concave-upwards bank profile may arise following removal of rotational slip and slump materials from the base of the bank (Figure 7.5d).

Banks with coarsening upwards profiles, such as sand deposits overlying fine-grained, cohesive sediments, tend to have zones of sediment accumulation along the bank toe (Figure 7.5e). Mass failure processes, such as falling and slipping/slumping, deliver sediment to the toe of the bank, where it accumulates until it is removed by fluvial entrainment. In contrast to banks with bank-attached geomorphic units, the toes of these banks are comprised of materials derived directly from the adjacent bank. As such, sediment

Figure 7.4 The influence of bed degradation upon bank-forming processes. If incision exceeds critical bank height, mass failure occurs and the channel widens. Incision of Jones Creek, Victoria, Australia, has triggered significant mass failure of banks. Photographs courtesy of Tim Cohen.

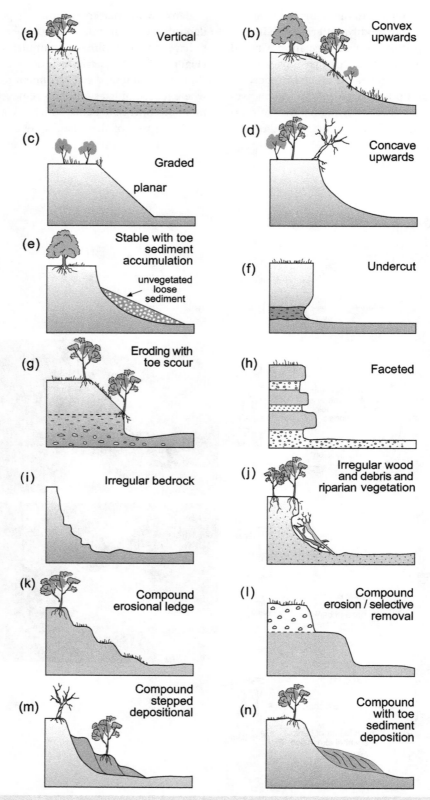

Figure 7.5   Bank morphology. Bank morphology reflects factors such as bank material properties (often inherited from former phases of river activity) and the nature/balance of erosion and deposition processes occurring along the bank. Similar forms may result from differing processes. Modified from Thorne (1999) in Brierley and Fryirs (2005). © John Wiley & Sons, Ltd. Reproduced with permission.

composition is similar to that in the bank, whereas materials derived from upstream may have a quite different sedimentary structure, texture and morphology; see Figure 7.5m and n). In the case of slips and slumps, where entire sections of the bank are removed, the bank structure may be maintained during displacement.

Undercut banks are commonly associated with upward-fining sediment sequences. Commonly, basal gravel lags are overlain by sand or mud deposits. Less cohesive units at the base of the bank are often undercut or eroded by hydraulic action (see Figure 7.5f). Toe scour leaves an overhanging bank of cohesive, finer grained materials. Mass failure by slab failure or fall results in sediment accumulation at the base of the bank. Differing stages of adjustment in this process have quite different bank morphologies. The resultant bank morphology following undercutting reflects the initial condition of the bank prior to commencement of toe scour. For example, Figure 7.5g represents a planar bank that is subsequently subjected to toe scour.

Banks that comprise multiple layers of sediment of varying texture tend to have a complex morphology, as coarse materials are selectively reworked. This produces a faceted bank with multiple overhangs (Figure 7.5h). Hydraulic action via fluvial entrainment and undercutting are common bank erosion processes that act as precursors to mass failure by slab failure or fall. In contrast, channels with bedrock margins tend to have an imposed, irregular morphology (Figure 7.5i). Irregular bank morphologies also occur along channels where resisting/forcing elements, such as wood and riparian vegetation, induce local variability in patterns of scour and deposition (Figure 7.5j).

Finally, compound bank morphologies with stepped profiles may result from differing sets of erosional and depositional processes. They are commonly found in reaches subjected to bed instability (whether degradation or aggradation). For example, ledge features may be observed along incised channels (see Chapter 8; Figure 7.5k). These erosional forms may reflect a slot channel inset within the broader trough. They are typically associated with phases of channel expansion. Alternatively, compound bank morphologies may reflect selective removal of less cohesive materials from the upper part of a bank profile. For example, reaches that are subjected to valley floor aggradation in response to upstream disturbance events commonly have coarser materials atop the banks. Subsequent channel expansion may selectively erode these less cohesive materials, producing a stepped bank morphology (Figure 7.5l). A compound bank morphology may also be derived from depositional mechanisms. For example, a stepped morphology may reflect multiple phases of deposition adjacent to the bank, in the form of inset features called benches (see Chapter 8; Figure 7.5m). These features are typically associated with phases of channel contraction.

These various scenarios that consider a range of bank morphologies and the circumstances that may be responsible for their formation tend to overemphasise the complexity of bank forms. In many instances, banks reflect relatively simple sets of depositional processes that generate bank-attached bars. These accretionary forms range from lateral and point bars to compound bar features (see Chapter 8). Depositional features along banks tend to be flat-topped or gently graded forms that effectively reduce bank angle. These features protect the banks from subsequent erosional activity. Resulting bank morphologies have a compound shape (Figure 7.5n). This commonly creates a low-slope, stepped morphology, often with a convex-upwards profile. Vegetation colonisation of these various surfaces can induce considerable complexity to the resulting form.

Controls on bank erosion processes and rates

The amount, periodicity and distribution of bank erosion are highly variable as they are influenced by a multitude of factors. In general terms, bank erosion is accentuated under higher *discharge conditions* (bankfull stage), but the effectiveness of these flows is determined by bank condition at the time of any given event (*preconditioning and wettedness*). Seasonal variability in bank erosion processes may be evident, wherein large summer floods induce little bank erosion on dry banks and smaller winter flows cause considerable bank retreat as they act on thoroughly wetted banks. The effectiveness of weakening, fluvial erosion and mass failure processes induce considerable variability in rates of bank erosion, instability and/or retreat. Rates of lateral channel adjustment range from 0.02 m yr^{-1} to in excess of 1000 m yr^{-1}. For disturbed rivers with highly modified vegetation cover, mean channel migration rate is around 3% (range 0.07–25%) of channel width per year, with a median of 1.6% of channel width per year. However, the wide range of variability is probably more important than these mean or median rates (i.e. these are best viewed as reach-specific situations).

Material properties exert a key control on bank erosion processes and rates. Fine-grained banks tend to be cohesive because of the electrochemical properties of these materials. Vertically stratified (composite) banks are typically characterised by a coarser grained basal layer overlain by fine-grained alluvium. In these instances, the strength of less cohesive basal materials controls the angle of repose and the tensile strength of the bank (and therefore channel width). Differential physical properties of cohesive and non-cohesive materials generate marked differences in erosion processes, critical bank height and failure modes and resulting erosion rates. Although fine-grained materials are resistant to fluid shear, they tend to have low shear strength and are susceptible to mass failure. Unlike cohesionless sediment, the erodibility of cohesive fine-grained bank material may vary because of its susceptibility to weakening.

Change in one part of a system may alter the pattern and/or rate of bank erosion in adjacent reaches. Hence, *position in catchment* is important. Maximum erosion rates tend to occur in middle (transfer) reaches, related to the peak in stream power and bank textural attributes (see Chapter 3). In general terms, bank materials become finer grained and more uniform downstream, especially where well-developed floodplains are found. Stream power also declines in a downstream direction, such that the dominant bank erosion process is transitional from hydraulic action to mass failure.

Vegetation cover, especially the role of root networks, may reinforce bank materials, thereby increasing resistance to erosion. Vegetated banks tend to have a more open fabric and are better drained. The sediment is strong in compression but weak in tension. Roots are the reverse, so they reinforce the tensile strength of sediments by up to an order of magnitude relative to root-free samples. These impacts may be offset, in part, by the additional loading applied to banks by vegetation cover. In large-scale cohesive banks, critical failure surfaces may extend well below the root zone. Vegetation structure also influences patterns and rates of flow dynamics adjacent to the banks. Stems and trunks of bank vegetation alter the distribution of near-bank velocity and boundary shear stress. The spacing (density) and pattern of trees and the distribution of wood

may exert a significant influence on the distribution of form drag, influencing the potential for detachment and entrainment. If discharge, slope, bend curvature, bank texture and bank heights are constant, then a river migrating through cleared or cultivated floodplain may erode at almost twice the rate of channels reworking forested floodplain. Non-vegetated banks may be five times more likely to undergo notable erosion compared with vegetated banks.

Other forms of human impact that affect rates of channel bank erosion include channelisation (typically concrete banks), cattle trampling, wave action and rehabilitation or bank protection works (see Chapter 13). The key consideration when reading the landscape is to recognise that bank erosion is a natural process along rivers, but when it occurs in unexpected places or at a rate that is above natural or expected levels, it suggests that anomalous processes are occurring along the reach.

Channel shape: putting the bed and banks together

Typical channel characteristics for channels with different boundary textures are shown in Table 7.1 (also see Figure 6.20). Sand or gravel bedload rivers with noncohesive bank materials tend to have wide, shallow channels with high width/depth ratios. In contrast, suspended-load rivers with cohesive fine-grained (silt–clay) bank materials have narrow, deep channels and low width/depth ratios. Com-

Table 7.1 Classification of channel boundary conditions[a]

Primary type	Characteristics
Bedrock channel	Imposed (forced) morphology (bed and banks). Typically have an irregular cross-section. May have a partial cover of consolidated material. Common in steep headwater reaches.
Non-cohesive boulder bed	Irregular channel morphology as flow moves around coarse boundary materials that are only mobilised during major floods. A surface armour is common. As bed materials may be relatively 'locked in place', many attributes of boulder streams are essential 'forced'.
Non-cohesive gravel bed	Mixed-load systems characterised by intermediate width/depth ratios. Differing bedload fractions are transported at differing flow stages. A surface armour may protect underlying materials. Banks commonly have cohesive finer grained materials atop the bedload material base (i.e. these are composite banks). Channel shape is largely determined by channel alignment and the recent history of formative events. In meandering reaches, for example, the channel is asymmetrical at bends and symmetrical at points of inflection.
Non-cohesive sand bed	'Live-bed' channels are active over a wide range of discharges. Bedform attributes reflect flow stage (depth and velocity) and grain-size relationships. Channels have a high width/depth ratio.
Cohesive silt-clay channel	Suspended-load systems with a limited capacity to rework their boundaries and adjust their form. Channels have a low width/depth ratio, and commonly have a symmetrical form. Once entrained, materials are maintained in transport even if flow energy decreases significantly.

[a]Modified from Knighton (1998).

posite banks along mixed-load rivers have a mix of bedload and suspended-load materials, producing channels with intermediate width/depth ratios. In general, rivers on steeper slopes, or systems that transport large volumes of coarse bedload with divided or braided channels (i.e. capacity- or competence-limited rivers; see Chapter 6), tend to develop wide, shallow channels with higher width/depth ratios. Conversely, supply-limited rivers are able to move all materials made available to them, and channels tend to be narrow and deep with bank-attached depositional forms (e.g. meandering rivers; see Chapter 10).

As inherited controls dictate channel shape and size in confined or partly confined valley settings, the width/depth ratio is not very meaningful measure of river character, and along discontinuous watercourses it is meaningless (Figure 7.6). Similarly, it may be difficult to define channel width and depth for irregularly shaped channels or for compound and multi-channel networks (Figure 7.6).

In reality, channel shape is a function of the energy available to erode or deposit materials of different calibre along the bed and banks at different flow stages. Bed and bank processes are not necessarily in phase with each other and may comprise a combination of features that reflect both contemporary and inherited processes. For example, as incised channels cut through alluvium they expose older bed and bank deposits which may have been deposited under a former depositional regime. These materials have been inaccessible to the channel for extended periods of time. Alternatively, the prevailing flow regime may no longer be able to mobilise coarse boulders activated under extreme events in the past. Also, ancient fine-grained, cemented materials that record suspended load deposition on floodplains and/or terraces under a former climatic regime may limit the capacity for adjustment of the contemporary channel. These factors can constrain contemporary channel size and shape (i.e. the river retains a memory of the past). In general terms, however, transient bed materials are younger than materials exposed along the banks. It is not unusual for the bed and banks to comprise different material mixes. Indeed, many channels have cohesive banks and a non-cohesive bed. Unlike the bed that is typically reworked on a semi-regular basis, bank materials may be artefacts of history. This hiatus between bed and bank materials varies markedly from reach to reach and from system to system, reflecting the aggradational/degradational balance of the channel and the history of formative and reworking events.

Although bed and bank material texture is a primary consideration in analysis of channel shape, many other factors must also be considered. For example, fluctuations in discharge and sediment load fashion the balance between bed aggradation and degradation, and the resulting patterns and rates of bank erosion and deposition in any given

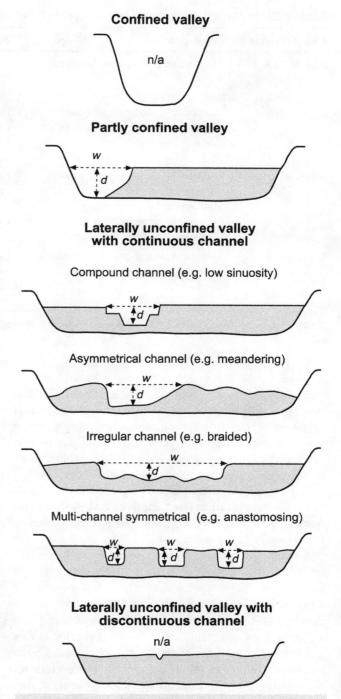

Figure 7.6 Schematic representations of channel shape and width/depth ratio for a range of river types. Width/depth ratio in confined rivers and discontinuous watercourses is a less meaningful measure than it is in more freely adjusting rivers with floodplains. In partly confined rivers, the depth component is often constrained by bedrock, whereas a range of channel sizes and shapes can result in laterally unconfined rivers where both the bed and the banks are comprised of deformable alluvial sediment.

Table 7.2 Putting the beds and banks together to assess channel shape[a]

Channel shape	Bank process	Bed process	Energy distribution
Symmetrical	Erosion (e.g. fluvial entrainment).	Erosion (e.g. headcut formation) or deposition (e.g. sand sheets).	Evenly distributed across channel.
Asymmetrical	Deposition along convex bank (point bar); erosion along concave bank (often undercut).	Deposition (point bar formation along convex bank), and erosion (pool scour along concave bank).	Thalweg along concave bank. Secondary circulation cells are common.
Irregular	Imposed condition, such as a bedrock or coarse boulder stream. Influenced by erosion (e.g. slumping) or deposition (e.g. bank-attached bar formation).	Erosion (e.g. sculpted pools or cascades) or deposition (e.g. mid-channel bars and island formation).	Unevenly distributed around alluvial materials or bedrock outcrops. Rough surfaces generate turbulent flows.
Compound (macrochannel)	Deposition (e.g. bench formation) or erosion (e.g. ledge formation).	Deposition (e.g. sand sheet formation) or erosion (e.g. headcut formation that creates an incised (slot) channel).	Similar to symmetrical channels, but varies with flow stage as differing surfaces are inundated.

[a]From Brierley and Fryirs (2005). © John Wiley and Sons, Ltd. Reproduced with permission.

cross-section or along any reach. Flow alignment and the distribution of flow energy at differing flow stages can result in a wide range of bed and bank morphologies that adjust over time. Ultimately, however, bed morphology is determined by patterns of sculpted/erosional geomorphic units and mid-channel depositional geomorphic units, while bank morphology reflects the balance of bank erosion processes and the assemblage of bank-attached depositional geomorphic units (see Chapters 8 and 9). Combinations of these factors at any given site determine channel shape (see Table 7.2 and Figure 7.7).

A channel with a given width and depth can be characterised by a wide range of possible shapes depending on the array of mid-channel and bank-attached geomorphic units. This reflects the distribution of energy within the channel (which is a function of slope, channel size and flow alignment), the sediment flux of the reach (i.e. the calibre and volume of available materials, and whether the reach is transport limited or supply limited) and process interactions with instream vegetation and forcing elements. Any adjustment to the sediment flux or energy distribution that alters bed material calibre and organisation or flow characteristics may modify the geomorphic structure of the reach and, hence, channel shape.

Differing combinations of bed and bank processes are key determinants of channel shape (Table 7.2). *Symmetrical* channels tend to be characterised by banks with uniform or upward-fining sediments and a near-homogeneous bed. The channel has a high width/depth ratio if it is bedload dominated (Figure 7.7a) and a low width/depth ratio if it is suspended-load dominated (Figure 7.7b). Channels tend

to be relatively free of depositional features other than relatively uniform sheetlike deposits, as flow energy is spread evenly across the bed. Symmetrical channels commonly occur at the inflection points of bends, along low-sinuosity channels, along fine-grained suspended-load rivers with cohesive banks or in incised channel situations.

Flow energy in *asymmetrical* channels is concentrated along the concave bank in bends, such that erosion occurs along one side of the bed while deposition occurs on the other (Figure 7.7c). Bank erosion via fluvial entrainment or undercutting on the concave bank and deposition/reworking of bank-attached geomorphic units along the convex bank (commonly point bars) promote lateral migration of the channel. Asymmetrical channels are also commonly observed in partly confined valley settings, where discontinuous pockets of floodplain line one bank (with point bars or point benches common), while the concave bank abuts against bedrock (Figure 7.7d). This situation is associated with downstream translation of bends (Chapter 11).

In *compound channel* situations, a smaller channel is inset within a broader macrochannel. This is commonly observed in cut-and-fill landscapes or along rivers that are subjected to significant (seasonal) variability in discharge. The range of geomorphic units that makes up the stepped cross-sectional morphology reflects phases of channel expansion and/or contraction, or they record river activity at different flow stages. Formation of one or more inset levels (i.e. benches) at channel margins may reflect depositional phases associated with channel contraction (Figure 7.7e). Alternatively, channel expansion may be recorded by

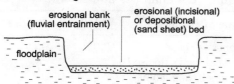

(a) Symmetrical: coarse-grained

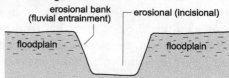

(b) Symmetrical: fine-grained

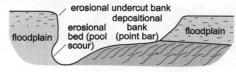

(c) Asymmetrical: meander bend

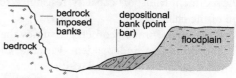

(d) Asymmetrical: partly confined valley

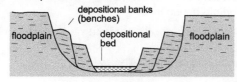

(e) Compound: depositional benches

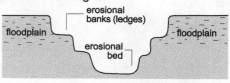

(f) Compound: erosional ledges

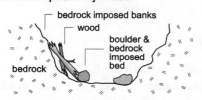

(g) Irregular: bedrock imposed by wood

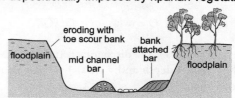

(h) Irregular: depositionally imposed by riparian vegetation

Figure 7.7 Channel shape. Symmetrical, asymmetrical, compound and irregular channels reflect differing combinations of bed and bank components, and associated mixes of erosional and depositional processes. Some morphologies are free-forming (alluvial); others are imposed by bedrock and/or ancient boundary materials and/or other forcing elements such as riparian vegetation and wood. Modified from Brierley and Fryirs (2005). © John Wiley & Sons, Ltd. Reproduced with permission.

the formation and/or reworking of ledges (Figure 7.7f). Elsewhere, compound channels may reflect long-term river evolution recorded by terraces that may have formed under different environmental conditions to those experienced today. Terraces perched above the contemporary river system may reflect changes to the flow/sediment regime, tectonic uplift or isostatic rebound, among many considerations.

Irregular channels do not have a clearly defined shape that has been moulded by a particular set of flow–sediment interactions. Rather, channel shape is locally variable, reflecting site-specific characteristics. In confined valleys, imposed controls on bed/bank morphology induce an irregular channel shape (Figure 7.7g). These are *forced* river morphologies. Flow energy is distributed unevenly around bedrock or coarse substrate, generating sculpted or erosional geomorphic units. In more *alluvial* rivers, mid-channel geomorphic units and either erosional or depositional banks can induce an irregular channel shape (Figure 7.7h). In

some instances, an irregular channel shape is inherited from the past and is out of phase with contemporary processes. Alternatively, significant heterogeneity is often evident along forested streams, where wood and riparian vegetation induce irregularities in channel shape (Figure 7.7h). Elsewhere, an irregular channel shape may indicate that the river has yet to become fully adjusted to the prevailing flow and sediment conditions, such that a chaotic pattern of depositional forms is found. For example, the irregularly shaped macrochannel of braided rivers reflects formation and dissection of mid-channel bars.

Hydraulic geometry and adjustments to channel morphology

The nature and rate of adjustments to channel geometry vary markedly through space (from reach to reach) and over time. Hydraulic geometry principles build upon *regime theory*, which was originally devised to appraise flow efficiency in canals. These empirical equations are based on the theory that channels strive to attain an equilibrium form by adjusting their cross-section (width, depth and velocity) and slope to transfer the discharge and sediment supplied from a catchment. Mutual interactions among width, depth and velocity mean that alterations to discharge or sediment load result in adjustments to channel form. These changes may also be mediated by alterations to the sediment transport capacity, resistance/roughness or slope of the channel.

Analysis of hydraulic geometry assumes that discharge Q is the dominant independent variable and that the dependent variables of channel width w, mean depth d and mean flow velocity v are related to it in the form of simple power functions:

$$w = aQ^b$$
$$d = cQ^f$$
$$v = kQ^m$$

From the continuity/discharge equation (see Chapter 4):

$$Q = wdv = aQ^b cQ^f kQ^m$$

where Q (m^3 s^{-1}) is the discharge, w (m) is the width, d (m) is the channel depth and v (m s^{-1}) is the mean velocity. It follows that:

$$ack = 1 \quad \text{and} \quad b + f + m = 1$$

At-a-station hydraulic geometry considers channel responses to temporal variations in flow at a cross-section. The exponents b, f and m indicate the proportional increase in that component with increasing discharge. These relationships vary markedly with bed and bank material attributes (see Table 7.3). Increases in depth are the primary response to increasing flow stage in channels with cohesive banks. Reduced frictional resistance with increasing flow stage, in turn, promotes an increase in velocity. However, width scarcely changes. Quite different relationships are found in channels with non-cohesive banks, as width increases markedly, with much lower rates of increase in flow depth and velocity. Many other factors influence these relationships, such as vegetation cover. Hence, rates of change in at-a-station exponents are highly variable and are subject to section-specific controls.

Hydraulic geometry principles can be used to assess adjustments to channel shape and size over time. By definition, these adjustments affect the periodicity of floodplain inundation, formation and reworking. These appraisals relate channel adjustments to alterations in flow and sediment conditions that affect the Lane balance. Empirical

Table 7.3 Width, depth and velocity components of different channels at-a-station[a]

Type of channel	Width b	Depth f	Velocity m
Cohesive, near-vertical banks (incised channel), fine-grained bed	0.05	0.45	0.50
Cohesive, near-vertical banks; sand and gravelly beds	0.01	0.52	0.47
Cohesive but non-vertical banks; sand and gravelly beds	0.08	0.50	0.42
Non-cohesive, readily erodible boundaries	0.54	0.26	0.20
One cohesive, one non-cohesive bank; sand and gravelly beds	0.40	0.31	0.29

[a]Modified from Knighton (1974, 1998).

relations that outline adjustments for channel width w and depth d, channel slope s and median bed material texture D_{50} in fully alluvial channels are summarised in Table 7.4. In general, an increase or decrease in discharge changes the dimensions of the channel and its gradient. However, an increase or decrease in bed material load at constant mean annual discharge changes both channel dimensions and the channel shape (width/depth ratio). An increase in discharge is accommodated by an increase in channel width and depth, a coarsening of material transported on the bed and an accompanying decrease in channel slope. This may occur as a result of water releases from reservoirs and dams. The converse situation applies if discharge is decreased, such that channel width and depth decrease, bed material texture becomes finer and slope increases. This may occur as a dam is closed and water and sediment discharge are trapped in the reservoir. If sediment load increases but discharge remains constant, then channel width may increase, depth decreases, bed slope increases and bed material texture becomes finer. This situation may arise in response to increased erosion in the catchment or the formation of a sediment slug. Once more, the converse applies if sediment load decreases and may arise in response to reduced erosion or storage of sediment in the catchment. In reality, discharge or sediment load seldom change in isolation. For example, altered ground cover not only modifies various components of the hydrological cycle and

Table 7.4 Morphological responses to changes in discharge and sediment load[a]

Relationship	Equation[b]	Types of triggers
Decrease in discharge alone	Q_s^0, $Q_w^- \approx s^+$, D_{50}^-, d^-, w^-	Abstraction of water from channel resulting in narrower channel.
Increase in discharge alone	Q_s^0, $Q_w^+ \approx s^-$, D_{50}^+, d^+, w^+	Water release from a reservoir.
Increase in bed material load	Q_s^+, $Q_w^0 \approx s^+$, D_{50}^-, d^-, w^*	Increased sediment supply from catchment.
Decrease in bed material load	Q_s^-, $Q_w^0 \approx s^+$, D_{50}^-, d^-, w^-	Reduced erosion in the catchment. Riparian revegetation.
Significant decrease in bed material load	Q_s^{--}, $Q_w^0 \approx s^-$, D_{50}^+, d^+, w^*	Sediment starvation.
Decrease in bed material load and decrease in water discharge	Q_s^-, $Q_w^+ \approx s^-$, D_{50}^+, d^+, w^-	Intensification of vegetation cover. Shift in climate from semi-arid to temperate conditions. Diversion of water into a river channel, e.g. from a dam.
Increase in bed material load and increase in water discharge	Q_s^+, $Q_w^+ \approx s^*$, D_{50}^*, d^*, w^{+-}	Land use change from agricultural to semi-urban.
Increase in bed material load and significant increase in water discharge	Q_s^+, $Q_w^{++} \approx s^-$, D_{50}^+, d^+, w^+	Increasing frequency and magnitude of water and sediment discharge (e.g. urbanisation, sequence of flood events).
Significant increase in sediment supply and no change in discharge.	Q_s^{++}, $Q_w^0 \approx s^+$, D_{50}^-, d^-, w^*	Sediment slug formation and movement.
Significant increase in bed material load and increase in water discharge	Q_s^{++}, $Q_w^+ \approx s^+$, D_{50}^-, d^-, w^+	Land use change from forested to agricultural. Sediment discharge increases more rapidly than water discharge. Bed changes from gravel to sand, wider shallower channel.
Decrease in bed material load and decrease in water discharge	Q_s^-, $Q_w^- \approx s$, D_{50}^*, d^*, w^-	Dam construction.
Decrease in bed material load and significant decrease in water discharge	Q_s^-, $Q_w^- \approx s^+$, D_{50}^-, d^-, w^-	Dam closure.

[a]Schumm (1969).

[b]Q_s: sediment load; Q_w: water discharge; s: slope, D_{50}: median grain size; d: flow depth; w: flow width; +/–: increase or decrease; 0: no change; ++/——: change of considerable magnitude; *: unpredictable.

associated runoff/discharge conditions, it also modifies the distribution, extent and rate of erosion. Four primary combinations (and a number of sub-combinations) of changing discharge and sediment load can be considered. For example, if both discharge and bed material load increase, channel width is likely to increase, whereas channel depth, bed slope and bed material texture may either increase or decrease. When both discharge and sediment load decrease, the opposite relations are noted. In many instances, alterations in discharge and sediment load do not occur in the same direction. For example, a shift in climate from semi-arid to temperate conditions is likely to increase discharge but decrease sediment load because of enhanced ground cover. As a result, channel width decreases and channel depth increases. Alternatively, a climate shift in the opposite

direction (i.e. from temperate to semi-arid conditions) reverses this relationship. It is critical to remember that relationships outlined in Table 7.4 are gross simplifications of reality, and complex (changing) interactions among discharge, sediment flux, vegetative interactions and human disturbance are the norm. Adjustments to channel form may involve many variables, the interdependence of which is not always clear. As such, it may be difficult to isolate the specific role of a single variable. This highlights the importance of framing site- or reach-specific analyses of river behaviour and change in their catchment context (see Chapters 11 and 12).

Unlike at-a-station hydraulic geometry, which deals with temporal variations in channel properties at a given site, *downstream hydraulic geometry* examines how channel size varies at different positions along a river course. In simple terms, as discharge increases, the mean channel depth and width increase, i.e. channel cross-sectional area increases downstream. Channel dimensions adjust to the quantity of water moving through the cross-section through erosional and depositional processes. In general, the downstream proportional increase in width is greater than the down-stream proportional increase in depth and velocity with exponents of b, f and m of 0.5, 0.4 and 0.1 respectively. This suggests that width is the most readily adjustable component of channel geometry in a downstream direction, gen-erating channels with a higher width/depth ratio. The rate of increase in flow velocity in a downstream direction is negligible. The slight increase reflects more laminar (less turbulent) flow in larger channels downstream. While these relationships provide a general sense of how channel size adjusts in response to increasing discharge, in reality the exponent values for the width, depth and velocity compo-nents vary significantly, reflecting the multivariate nature of controls on channel form.

Different relationships for suspended-load systems, gravel-bed systems and sand-bed systems reflect the impact of bed and bank texture on channel size. As discharge increases, channels flowing through gravel substrates tend to become relatively wider than deeper relative to channels with a high proportion of silt–clay materials in their banks. Hence, gravel-bed channels have a lower depth exponent f than suspended-load rivers. Complications arise when the other controlling factors are considered. For example, where tributaries join the trunk stream, significant increases in discharge may (or may not) occur, resulting in disrup-tion of the systematic downstream increase in channel size. As another example, as rivers flow downstream, bed mate-rial texture tends to become finer and the slope of the channel bed decreases. This results in a situation where the banks become more cohesive downstream and width increases more slowly than depth, giving rise to a lower width/depth ratio channel than expected under the influ-ence of discharge alone. As a consequence, channel capacity of alluvial rivers can actually decrease in a downstream direction and the recurrence of overbank flooding increases. In many cases, this situation is historically induced (i.e. materials deposited along the river course reflect anteced-ent controls). In the case shown in Figure 7.8, estuarine fills and materials deposited under past flow regimes exert a key control on this relationship.

Discharge and sediment load are the dominant controls of channel form adjustment. These independent variables which integrate the effects of climate, vegetation, soils, geology and basin physiography in any particular region. Hence, hydraulic geometry relationships are best described at a regional level, for particular landscape settings with similar hydrologic and sediment (lithologic) conditions

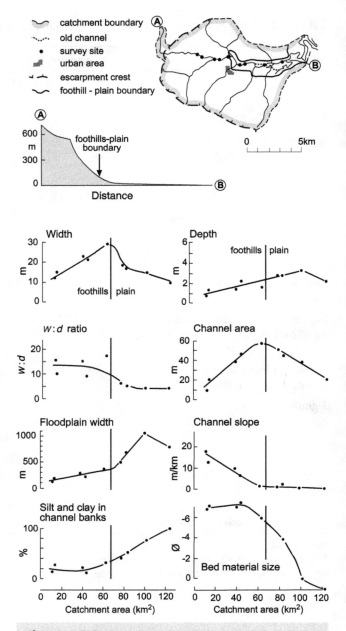

Figure 7.8 Downstream changes in hydraulic geometry along the Minamurra River in the Illawarra, NSW, Australia. This system demonstrates a down-stream decrease in channel size once the river exits the foothills and enters the lowland plain. The key controls driving this relationship are the bank material texture which is fine grained and cohesive, and the extensive floodplain width of the lowland plain over which flows are regularly dispersed. This means that channels do not have the erosive energy to form large channels in downstream reaches. Modified from Journal of Hydrology, 52 (3–4), Nanson, G.C. and Young, R.W., Downstream reduction of rural channel size with contrasting urban effects in small coastal streams of southeastern Australia, 239–255, © 1981, with permission from Elsevier.

and equivalent riparian vegetation associations. These principles only apply to fully alluvial rivers where the channel bed and banks are readily deformable. Channel geometry is forced or imposed in many other settings. In addition, discontinuous watercourses, local variations in bed and/or bank texture, disturbance (e.g. removal of vegetation or wood), amongst other things, produce irregularities that disrupt downstream hydraulic geometry relationships. As such, each river must be viewed in its landscape context, considering notions of downstream connectivity in water and sediment regimes, antecedent controls and local factors that may shape channel morphology and size.

Conclusion

Analysis of bed conditions is the key to interpretation of channel geometry. Bank processes vary markedly in degradational and aggradational environments. As a consequence, downstream patterns of channel shape and size reflect differing flow/sediment balances, and associated mixes of erosional and depositional processes, in source, transfer and accumulation zones. The imposed (forcing) role of bedrock tends to diminish with distance downstream, with more alluvial (self-adjusting) rivers evident in lower slope, wider valley settings. Site-specific and reach-scale variability must be assessed in relation to catchment-scale controls. Analysis of bank materials and the role of forcing elements such as riparian vegetation and wood aid assessment of local (site-specific), within-reach and downstream patterns. Human disturbance may modify these relationships, either directly (e.g. channelisation) or indirectly (e.g. changes to flow/sediment load due to deforestation; see Chapter 13).

Timescales of adjustment must also be considered. Changes with flow stage may reflect differing flow energy, alignment (especially the position of the thalweg) and inundation/reworking of differing materials that make up the bank. Implicit to this understanding is an appreciation that the make-up of sediments along the bank is a product of past depositional conditions and environments. Changes to flow–sediment conditions, or other controls, will induce alterations to channel geometry. Hence, meaningful analysis of channel geometry is a system-specific exercise that must be framed in light of catchment (spatial) and evolutionary (temporal) considerations (e.g. timing of analysis in relation to flood events). Caution should be heeded in the prescriptive use of hydraulic geometry relationships, though they may be useful for comparative purposes.

Key messages from this chapter

- Channel geometry refers to the three-dimensional form of a channel. The balance of impelling and resisting forces along a reach determines channel geometry. Bedrock reaches have an imposed morphology. Alluvial reaches are self-formed and self-adjusted.
- Erosional and depositional processes along the bed and bank produce channel size and shape. As indicated by the Lane balance incision and aggradation processes determine the stability of the channel bed. Bank erosion processes are differentiated into entrainment and mass movement mechanisms. Critical bank height determines a factor of safety. Deposition is prominent adjacent to many banks.
- Material size and cohesion are key influences on the nature and effectiveness of bed and bank processes. Channels with non-cohesive banks tend to have high width/depth ratios, whereas rivers with cohesive banks tend to have low width/depth ratios.
- Vegetation cover/composition and the loading of wood exert a primary influence on channel geometry.
- Channel morphology can be differentiated into symmetrical, asymmetrical, compound or irregular forms. Similar forms can be generated by different sets of processes (the principle of equifinality or convergence).
- Hydraulic geometry is the study of how width, depth and velocity components of channels change with flow stage (at-a-station) and with distance downstream.
- Hydraulic geometry relationships should be viewed in relation to catchment-specific (spatial) and evolutionary (temporal) considerations. Marked differences in channel geometry relationships are induced by inherited (forcing) factors, along with local-scale considerations such as bed/bank material size, vegetation cover/composition and the loading of wood.

CHAPTER EIGHT

Instream geomorphic units

Introduction

Patterns of erosional and depositional processes in river systems reflect the catchment-scale distribution of impelling and resisting forces, and resulting textural segregation, along longitudinal profiles (Chapter 3). Downstream transitions from imposed bedrock conditions in source zones to self-adjusting alluvial channels in accumulation zones reflect the shift from degradational to aggradational processes. This downstream trend is typically mirrored by progressive adjustment in valley confinement from confined to laterally unconfined valley settings. While incisional processes are dominant in steep confined settings of headwater streams, the wider valleys of low-relief lowland settings promotes dissipation of flow energy and differentiation of channel and floodplain processes and forms. This represents a shift in process domain from forced to fully alluvial rivers. Downstream trends from bedload through mixed-load to suspended-load rivers, and associated floodplain features are common (Chapters 6 and 9). These transitional relationships result in a range of landforms that are produced and reworked along valley floors. In this book, these erosional and depositional features are referred to as geomorphic units.

Geomorphic units are the building blocks of river systems. They are defined by their morphology (shape and geometry), sedimentary composition, bounding surfaces and position on the valley floor. Differentiation is made between instream (channel) and floodplain forms (Chapters 8 and 9 respectively). Erosional or depositional processes, or a range thereof, produce these features. Analysis of process–form relationships that generate and rework landforms can be used to interpret river behaviour. Processes affect the form, while the form affects the effectiveness of process relationships. These mutual interactions are referred to as morphodynamics. Correctly identifying a geomorphic unit and examining its form and sedimentology can be used to interpret the processes that form and rework that feature.

The availability of sediment and the potential for it to be reworked in any given reach determines the distribution of geomorphic units and resulting river structure. Some rivers comprise forms that are sculpted or eroded, reflecting a dominance of reworking processes. Other rivers comprise sets of geomorphic units that are the result of short- or long-term sediment accumulation. Characteristic forms are found at characteristic locations within river systems, at both catchment and reach scales. Geomorphic units are created by distinct sets of processes at different positions along a river course (i.e. erosional processes dominate in source zones, while depositional forms are dominant in accumulation zones). Within any given reach, distinct features are found at particular locations on the valley floor (e.g. point bars are found on the inside of bends, levees are found at channel margins).

Instream geomorphic units are found along a slope-induced energy and textural gradient. A continuum of features extends from high-energy erosional (sculpted) forms in bedrock and boulder settings to mid-channel depositional units and bank-attached forms. Forced units are produced when flow patterns and available energy are disrupted by obstructions such as bedrock, vegetation or wood that induce irregularities along the valley floor, thereby creating resistance. Sculpted forms may also be formed in low-energy, fine-grained settings. Unit features are products of single depositional events, whereas compound features reflect a range of flow conditions and reworking events. Inevitably, the pattern of features is intimately tied to channel geometry (shape and size; Chapter 7).

Characteristic assemblages of features analysed at the reach scale determine the character and behaviour of differing types of rivers (Chapters 10 and 11 respectively). Different features are formed and reworked at differing flow stages, dependent upon their position and elevation on the valley floor and how this affects the distribution and use of flow energy. Energy use varies with slope and confinement (and associated total/unit stream power; Chapter 5),

Geomorphic Analysis of River Systems: An Approach to Reading the Landscape, First Edition. Kirstie A. Fryirs and Gary J. Brierley.
© 2013 Kirstie A. Fryirs and Gary J. Brierley. Published 2013 by Blackwell Publishing Ltd.

sediment availability and calibre (Chapter 6) and hydrological considerations (Chapter 4). Many geomorphic units are *genetically linked*, whereby the processes that form or rework one feature affect the formation or reworking of adjacent units. For example, steps and pools tend to occur together along steep, bedrock-controlled rivers, whereas pools and riffles tend to occur together along meandering rivers. Differentiation of spatially and genetically connected or disconnected geomorphic units is a key attribute in analysis of river adjustment. Analysis of reach-scale patterns of geomorphic units enables interpretation of landscape evolution. For example, terrace features reflect a different phase of river activity relative to the contemporary floodplain or channel zone (see Chapter 12).

This chapter focuses on the process–form associations that generate and rework instream geomorphic units. Although individual geomorphic units may be observed along reaches in a range of river types (e.g. pools are common along many variants), specific ranges and assemblages of geomorphic units tend to occur along different types of rivers. Analysis of the full set of processes responsible for the formation of instream and floodplain geomorphic units facilitates interpretations of *river behaviour* at different flow stages. Appraisal of river character and behaviour based on assemblages of geomorphic units can be viewed as a building block, or a *constructivist approach*, to analysis of river systems.

Categories of geomorphic units and measures used to identify them in the field

Geomorphic units are initially categorised according to their *position* along a river course. *Instream* geomorphic units are found within the channel zone itself (considered at bankfull stage). They comprise *sculpted/erosional* (*bedrock* or *fine-grained*) forms, *mid-channel depositional* forms, and *bank-attached depositional* forms. *Floodplain* geomorphic units, described in Chapter 9, occur outside the channel zone and are formed and reworked at flow stages beyond bankfull when the floodplain is inundated by overbank flows.

The type of geomorphic unit is identified on the basis of its *morphology* (*shape* and *relative size*) and *position*. For example, an elongate unit located in the middle of the channel that scales roughly to the width of the channel is likely to be a longitudinal bar. An arcuate landform attached to the convex channel bank and scaled to the curvature of the bend is likely to be a point bar. A ridge-shaped unit elevated above the surrounding floodplain and positioned on top of the channel bank is likely to be a levee.

Sediment analysis is used to determine the internal composition of the geomorphic unit and interpret the processes

that formed the feature (Chapter 6). In most cases, geomorphic units can be identified using visual estimates of texture, differentiating the relative mix of bedrock, boulders, gravels, sands or silt/clay. More detailed assessments of river process and behaviour examine the architecture of a depositional body, interpreting the facies assemblage that makes up the geomorphic unit (i.e. its *morphostratigraphy*). Each geomorphic unit is comprised of characteristic facies assemblages. Analysis of depositional sequences can be used to interpret the flow conditions under which a geomorphic unit was formed and reworked. In many cases, geomorphic units are comprised of specific assemblages of sedimentary structures. For example, longitudinal bars are typically comprised of coarser material at the bar head (typically gravel), grading down-bar in terms of bed material texture and sedimentary structures (Sr and Sh facies) reflecting downstream growth of the bar. Gravel/sand point bars in actively migrating channels have cross-bedded sedimentary structures that dip towards the channel (facies Sp, St and Sh). Levee deposits are typically comprised of stacked sequences of sand and silt that thin away from the channel. This reflects the progressive vertical accretion of the feature during overbank flow (Sh and Fm facies). Levees typically grade to floodplain and backswamp environments that are typically comprised of Fsm or Fm facies.

The depositional or erosional origin of geomorphic units is reflected in their *bounding surfaces* which define their geometry. Analysis of sediment exposures can be used to appraise the formation and reworking of preserved features. For example, erosional basal contacts are demarcated by distinct unconformities in sediments. The position and relative juxtaposition (spatial association and stacking arrangement) of surrounding geomorphic units aids interpretation of the history of primary deposition and subsequent reworking. In general terms, longitudinal bars are aggradational features that lie atop channel bed deposits. These deposits may be reworked and dissected by chute channels which leave erosional contacts within the bar deposits. Burial of these deposits by further longitudinal bars is likely demarcated by a vertical transition to coarser sediments. Point bars lie conformably atop basal lag materials from the channel, reflecting progressive lateral migration of the channel. Ridge (accretionary features, depositional contact) and swale features (erosional contact) often indent the surfaces of point bar deposits. Levee deposits lie conformably above floodplain deposits at channel margins. These accretionary (depositional) surfaces are typically longitudinally continuous, with elongate boundaries. Basal contacts are often relatively flat, while upper surfaces are gradually inclined, tilting away from the main channel as the feature thins laterally. Crevasse channel and splay features occasionally dissect levee surfaces with erosional contacts.

Analyses of geometry, bounding surfaces, spatial associations (location, genetic linkages and interactions with adjacent units) and sedimentary attributes provide fundamental insight into the processes that form and rework geomorphic units.

Process–form associations of instream geomorphic units

Four categories of instream geomorphic units are differentiated here:

- sculpted, erosional bedrock and boulder units;
- mid-channel, depositional units;
- bank-attached, depositional units;
- sculpted, erosional fine-grained units.

Sculpted, erosional bedrock and boulder instream geomorphic units

Process-form attributes of sculpted, erosional geomorphic units are summarised in Table 8.1. Bedrock and boulder geomorphic units are often non-deformable channel features around which flow and sediment accumulations locally adjust. These features are shaped by antecedent controls such as structural and/or lithological considerations and the impacts of major flood events. Forced morphologies tend to form in reaches with steeper gradients (high

Table 8.1 Sculpted, erosional geomorphic units[a]

Unit	Form	Process interpretation
Bedrock step (waterfall) 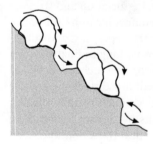	Locally resistant bedrock that forms channel-wide drops. Transverse waterfalls >1 m high separate a backwater pool from a plunge pool downstream.	Erosional features formed and maintained as highly turbulent flow falls near-vertically over the lip of the step. Steps are major elements of energy dissipation. These locally resistant areas may represent headward-migrating knickpoints. Equivalent features may be forced by wood.
Step–pool	Formed on steep slopes (0.03–0.10 m m⁻¹) channel-spanning stair-like features comprise bedrock, or boulder and cobble clasts or wood separated by areas of quieter flow in a *plunge pool* downstream. The risers of individual steps are generally made up of several large boulders, or keystones.	A cyclic pattern of acceleration and deceleration characterises the flow regime as water flows over or through the boulders forming each step before plunging into the pool below. Such tumbling flow is supercritical over the step and subcritical in the pool. Turbulent mixing results in considerable energy dissipation. Step development is strongly influenced by local sediment supply (availability of keystones) and transport conditions.
Cascade 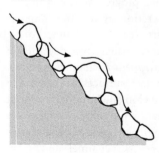	Very stable, coarse-grained or bedrock features observed in steep, bedrock-confined settings. Comprise longitudinally and laterally disorganised bed material, typically cobbles and boulders. Flow cascades over large boulders in a series of short steps about one clast diameter high, separated by areas of more tranquil flow of less than one channel width in extent.	More than 50% of the stream area is characterised by supercritical flow. Typically associated with downstream convergence of flow. Near-continuous tumbling/turbulent and jet-and-wake flow over and around large clasts contributes to energy dissipation. Finer gravels can be stored behind larger materials or wood. During moderate flow events, finer bedload materials are transported over the more stable clasts that remain immobile. Local reworking may occur in high-magnitude, low-frequency events.

Table 8.1 (*Continued*)

Unit	Form	Process interpretation
Rapid	Very stable, steep, ridge-like sequences formed by arrangements of boulders in irregular transverse ribs that partially or fully span the channel in bedrock-confined settings. Rapids in bedrock channels may be analogous with riffles in alluvial systems. Individual particles break the water surface at low flow stage.	Boulders are structurally realigned during high-energy events to form stable transverse ribs that are not associated with either divergent or convergent flow. Typically, 15–50% of the stream demonstrates supercritical flow.
Run (glide, plane-bed)	Stretches of uniform and relatively featureless bed, comprising bedrock or coarse clasts (cobble or gravel). These smooth flow zones are either free-flowing or imposed shallow channel-like features that connect pools. They occur in both alluvial and bedrock-imposed situations. Individual boulders may protrude through otherwise uniform flow.	Plane-bed conditions promote relatively smooth conveyance of water and sediment. Slopes are intermediate between pools and riffles.
Forced riffle	Longitudinally undulating gravel or boulder accumulations that act as local steps. Irregular spacing is dictated by the distribution of bedrock outcrops, wood or hillslope sediment inputs along the river. They tend to occur at wider sections of valley in bedrock-confined systems (e.g. at tributary confluences).	Flow is characterised by high-energy turbulence over lobate accumulations of coarse bedload materials, wood and bedrock outcrops. At the lower end of the energy spectrum, riffle–pool spacing in bedrock-confined settings may reflect purely rhythmic hydraulic processes of sediment transport.
Forced pool	These deeper areas along longitudinal profiles are scour features associated with irregularly spaced bedrock outcrops, wood and forced riffles. A *backwater pool* may form immediately upstream of a bedrock step.	These areas of tranquil flow within high-energy settings may accumulate finer grained materials at low–moderate flow stage, but they are flushed and possibly scoured during extreme events. At the lower end of the energy spectrum, riffle–pool spacing in bedrock-confined settings may reflect purely rhythmic hydraulic processes of sediment transport.
Plunge pool	Deep, circular, scour feature formed at the base of a bedrock step.	As flow plunges over a step, its energy is concentrated and scour occurs by corrosion, cavitation and corrasion processes. Erosion may be aided by preweakening by weathering.
Pot hole	Deep, circular scour features occur in bedrock.	Pot holes are sculpted from bedrock by corrasion (i.e. hydraulic and abrasive action of water). The effectiveness of this process is determined by the volume and hardness of particles that are trapped in the pothole. Abrasion is induced by these particles, which deepen and widen the pothole.

[a]Modified from Brierley and Fryirs (2005). © John Wiley & Sons, Ltd. Reproduced with permission.

transport capacity) and/or lower sediment supply relative to their free-forming counterparts. Sculpted or erosional forms typically reflect the operation of high-energy processes. Erosion of bedrock occurs via the chemical action of water (corrosion), the mechanical (hydraulic and abrasive) action of water armed with particles (corrasion) and the effects of shock waves generated through the collapse of vapour pockets in a flow with marked pressure changes (cavitation). The largest clasts are customarily exposed above the water surface. These features contribute to considerable energy loss during flood events (Chapter 5).

Downstream transitions in channel slope, bed material size and stream power conditions typically induce a continuum of variants of sculpted, erosional instream geomorphic units, including waterfalls, step–pools, cascades, rapids, runs, forced riffles and pools (see Table 8.1; Figure 8.1). There is considerable overlap in the range of conditions/settings in which individual features form. This reflects local combinations of factors such as slope, discharge characteristics (or history), range of sediment availability and bed material calibre, or forcing elements such as imposed bedrock steps or constrictions, changes in valley alignment, or loading of wood. Hence, interpretations of controls on form–process associations must relate general principles to site-specific considerations.

Waterfalls or *bedrock steps* are characterised by falling flow over bedrock or boulder steps that have a near-vertical drop greater than 1 m. *Plunge pools* are circular scour features that form when flow becomes concentrated at the base of waterfalls, steps or obstacles. The force of the flow induces corrasion and cavitation. *Potholes* are deep, spherical features sculpted into bedrock. Once initiated, bedload particles trapped within the pothole induce scour by corrasive erosion during turbulent flow, widening and deepening the feature.

Step–pool sequences are a commonly observed form of genetic association of geomorphic units in steep headwater rivers with gradients between 0.03 and 0.10 mm^{-1}. These channel-spanning stair-like features comprise bedrock, or boulder and cobble clasts or wood, separated by areas of quieter flow in a *backwater pool* upstream from a *plunge pool* downstream. The risers of individual steps are generally made up of several large boulders, or keystones. Each step is like a boulder jam. When $D/d \sim 1.0$ and the width of the channel is less than an order of magnitude greater than the diameter of the largest stones within it, keystones form stone lines that define steps. These stonelines act as a framework against which smaller boulders and cobbles are imbricated. The tightly interlocking structure of these features results in considerable stability, such that steps are only likely to be disturbed during extreme floods. Given the need for one or more keystones, step development is strongly influenced by local sediment supply (availability

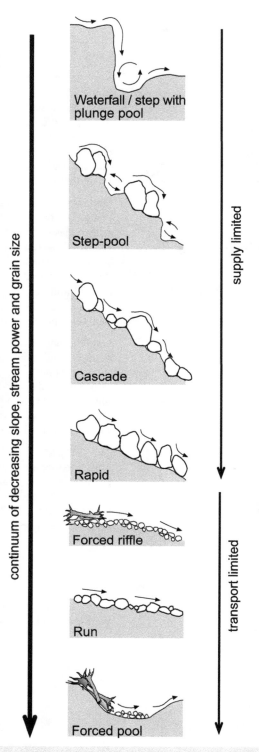

Figure 8.1 The continuum of sculpted, erosional instream geomorphic units. Downstream decreasing slope, stream power and grain size induce a transition from supply-limited features such as waterfalls, step–pool sequences, cascades and rapids through to transport-limited forms such as forced riffle–pool sequences and runs.

of keystones) and transport conditions. Small pools between steps store finer grained bedload material, creating a contrast in sediment size which is much sharper than that between riffles and pools. Step–pool elements are spaced about two to three channel widths apart. A cyclic pattern of acceleration and deceleration characterises the flow regime as water flows over or through the boulders forming each step before plunging into the pool below. Such tumbling flow is supercritical over the step and subcritical in the pool. Turbulent mixing results in considerable energy dissipation. Further energy is expended by form drag exerted by the large particles that make up the steps. Thus, step–pool sequences have an important resistance role.

Cascades occur on steep slopes (>0.1 m m^{-1}) and are characterised by longitudinally and laterally interlocked bed material, typically cobbles and boulders. Near-continuous tumbling/turbulent and jet-and-wake flow occurs over and around individual clasts in a series of short steps about one clast diameter high separated by areas of more tranquil flow of less than one channel width in extent. More than 50% of the stream area is characterised by supercritical flow. A stair-like morphology may develop in settings where the materials are better organised. These features induce significant energy dissipation. Finer gravels can be stored behind larger materials or wood. During moderate flow events, finer bedload materials are transported over the more stable clasts that remain immobile during these flows. Localised reworking may occur in high-magnitude, low-frequency events.

Rapids are ridge-like arrangements of boulders on steep slopes. Individual particles are numerous enough or large enough to break the water surface at mean annual discharge. Rapids form by transverse movement of boulders at high flow stage (recurring perhaps once every few years). Series of ridges of coarse clasts are spaced proportional to the size of the largest clast. Rapids in bedrock channels may be analogous with riffles in alluvial systems. They can be differentiated from riffles by their increased steepness, their greater areal proportion of supercritical flow and the arrangement of boulders into transverse ribs that span the channel.

Runs (or *glides*) are generally uniform, near-featureless forms with trapezoidal cross-sections. Individual boulders may protrude through otherwise uniform flow along long stretches of bedrock and coarse clasts. Runs are typically generated under plane-bed conditions on moderate slopes of 0.01–0.03 m m^{-1}. In these areas, the volume of coarse sediment inputs exceeds the transport capacity of the channel, such that aggradation induces a relatively homogeneous bed profile. These features can form in either bedrock-dominated or fully alluvial settings.

The transition from runs to *riffles* and *pools* tends to be accompanied by increased sediment supply and/or decreased transport capacity. Pools and riffles in confined valleys are typically forced features (see below). These sculpted, longitudinally undulating features typically form on slopes <0.01 m m^{-1}. Unlike their free-forming counterparts, these features are generally irregularly spaced. Quiet flow through deeper areas (pools) is often separated by turbulence over lobate accumulations of coarse bedload materials in intervening shallow riffles. The formation of forced pools and riffles may be induced by wood accumulations or downstream changes in bedrock resistance, which controls variations in bed topography, valley width or alignment. Alternatively, sediment input from tributaries or mass movement inputs from hillslopes may fashion the pattern of riffles and pools. At the lower end of the energy spectrum, riffle–pool spacing in bedrock-confined settings may reflect purely rhythmic hydraulic processes of sediment transport. In these cases, the primary riffles may remain anchored in place or may migrate slowly along the system dependent upon the relative mobility of the material forming the channel bed and the valley configuration. Abrupt changes in valley alignment or confinement may anchor otherwise migratory sediment accumulations.

Pool shape may vary markedly, especially in bedrock-controlled reaches, or any area where forcing elements such as wood or large boulders promote scour. Pools have low flow velocities and low water-surface gradients. *Bluff pools* are characterised by poorly sorted sand- to large boulder-sized bed material, v-shaped cross-sections and bedrock or coarse talus banks. *Lateral pools* have gravel- to cobble-sized bed material, asymmetrical cross-sections and banks that comprise alluvial materials. In bedrock-controlled reaches, pool morphology is largely imposed by the effectiveness of erosion and scour processes. This reflects factors such as lithologic variability (i.e. hardness) and changes in valley alignment. Pronounced variability may be evident in pool depth. These features are often the last remaining waterholes when flow diminishes in ephemeral systems. In many settings, shallow elongate pools at low flow stage act as runs (or glides) at moderate flow stage.

Mid-channel, depositional instream geomorphic units

Mid-channel geomorphic units tend to scale relative to the dimensions of the channel in which they form. These features have strong relationships with other morphological attributes of rivers, notably channel shape and channel planform. Systematic downstream changes in bed configuration reflect the tendency for bed material and slope to decrease, and discharge to increase downstream. A range of mid-channel depositional forms is presented in Table 8.2 and Figure 8.2.

Table 8.2   Mid-channel geomorphic units[a]

Unit	Form	Process interpretation
Riffle and pool 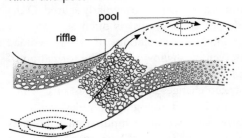	*Riffle* Topographic highs along an undulating reach-scale longitudinal profile. They occur at characteristic locations, typically between bends (the inflection point) in sinuous alluvial channels. Clusters of gravel (up to boulder size) are organised into ribs, typically with a rippled water surface at low stage. Alluvial riffles are alternating shallow step-like forms that span the channel bed. These sediment storage zones tend to comprise tightly imbricated bed materials, suggesting the action of local sorting mechanisms. They induce local steepening of the bed.	*Riffle* Riffles are zones of temporary sediment accumulation which increase roughness during high flow stage, inducing deposition. Concentration of coarser fractions at high discharges (bankfull and above) produces incipient riffles, while lower flows (up to bankfull) may be sufficiently competent to amplify and maintain the initial undulations once they have reached a critical height. In subsequent high discharges, deposition occurs as the resistance of these features induces a reduction in velocity over the riffle surface. At high flow stage the water surface is smooth, as bed irregularities are smoothed out. Riffles are commonly dissected during the falling stage of floods, when the water surface is shallow and steep, and the stepped long profile is maintained. Although very stable, with 5–10% of the stream area in supercritical flow and some small hydraulic jumps over obstructions, riffles may be mobile at and above bankfull stage. Indeed, they may be removed and replaced during extreme floods, as they reform at lower flow stages (velocity reversal hypothesis).
	Pool Pools may span the channel, hosting tranquil or standing flow at low flow stage. Alluvial pools are alternating deep areas of channel along an undulating reach-scale longitudinal bed profile. Pools tend to be narrower than riffles and act as sediment storage zones. They form at characteristic locations, typically along the concave bank of bends in sinuous alluvial channels.	*Pool* At high flow stage, when flow converges through pools, decreased roughness and greater bed shear stresses induce scour and flushing of sediment stored on the bed. Subcritical flow occurs during divergent flow at low flow stage. Pool infilling subsequently occurs, as pools act as areas of deep, low flow velocity and near-standing water conditions. Pools and riffles are genetically linked in alluvial rivers. Velocity reversal at high flow stage maintains these features.
Longitudinal bar (medial bar) 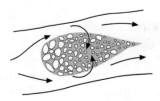	Mid-channel, elongate, tear-drop-shaped unit bar, aligned with flow direction in gravel- and mixed-bed channels. Bar deposits typically decrease in size downstream, away from a coarser bar head. May contain distinctly imbricated materials.	As flow diverges around the coarse bedload fraction it is no longer competent to transport sediment and materials are deposited in mid channel. Finer materials are trapped in the wake. Alternatively, there is too much sediment for the channel to transport (i.e. exceedence of a capacity limit under highly sediment-charged conditions) and material is deposited.

Table 8.2 (*Continued*)

Unit	Form	Process interpretation
Transverse bar (linguoid bar)	Mid-channel unit bar, oriented perpendicular to flow, generally found at points of abrupt channel and flow expansion points in sand-bed channels. They have a lobate or sinuous front with an avalanche face. The upstream section of the bar is characterised by a ramp which may be concave in the centre and form an arcuate shape.	Formed via flow divergence in highly sediment-charged sandy conditions. Flow moves over the centre of the bar, diverges and is pushed up the ramp face. Sediment is pushed over the avalanche face and deposited on the lee side. As a result, the bar builds and moves downstream as a rib.
Diagonal bar (diamond bar)	Mid-channel unit bar, oriented diagonally to banks in gravel- and mixed-bed channels. These bars commonly have an elongate, oval or rhomboid planform. Particle size typically fines down-bar. Commonly associated with a dissected riffle.	Formed where flow is oriented obliquely to the longitudinal axis of the bar. May indicate highly sediment-charged conditions or reworking of riffles.
Expansion bar	Coarse-grained (up to boulder size) mid-channel bar with a fan-shaped planform. A streamlined ridge forms trail behind obstructions. Foreset beds commonly dip downstream with a very rapid proximal–distal grain-size gradation. Often occur downstream of a bedrock constriction that hosts a forced pool. May be colonised and stabilised by vegetation.	As flow expands abruptly at high flood stage in high-energy depositional environments, it loses competence and induces deposition. Dissection is common at falling stage. These bars remain fairly inactive between large floods, constraining processes at lower flow stages.
Island	Vegetated mid-channel bar. Can be emergent at bankfull stage. Generally compound forms, comprising an array of smaller scale geomorphic units. They are commonly elongate in form, aligned with flow direction scaling to one or more channel widths in length.	Generally form around a bar core that has been stabilised by vegetation. This induces further sedimentation on the island. Islands are differentiated from bar forms by their greater size and persistence, reflecting their relative stability and capacity to store instream sediments. The pattern of smaller scale geomorphic units reflects the history of flood events and processes which form and rework the island.
Boulder mound	Linguoid-shaped boulder feature with a convex surface cross-section. Comprise a cluster of boulders without matrix, fining in a downstream direction.	Deposited under high-velocity conditions. When the competence limit of the flow drops, the coarsest boulders are deposited, forming obstructions to flow. Secondary lee circulation occurs in the wake of the coarse clasts. Finer boulders and pebbles are subsequently deposited downstream of the core clasts, resulting in distinct downstream fining.

(*Continued*)

Table 8.2   *(Continued)*

Unit	Form	Process interpretation
Bedrock core bar	Elongate bedrock ridge over which sediments have been draped and colonised by vegetation. Sediments become finer grained downstream, and the age structure of the vegetation gets younger.	During the waning stages of large flood events, sediments are deposited on top of an instream bedrock ridge. When colonised by vegetation, additional sediment is trapped and accumulates on top of the bedrock core. Over time the bar builds vertically and longitudinally as sediments are trapped in the wake of vegetation.
Sand sheet	Relatively homogeneous, uniform, tabular sand deposits which cover the entire bed. May consist of an array of bedforms, reflecting riffle, dune or plane-bed sedimentation.	Formed when transport capacity is exceeded or competence is decreased and bedload deposition occurs across the bed. Generally reflect transport- and capacity-limited conditions due to an oversupply of sediment. Bedforms are regularly reworked and replaced as the sand sheet moves downstream as a pulse.
Gravel sheet (basal or channel lag)	Relatively homogeneous, thin/tabular bedload sheets that are deposited across the bed. Often coarse grained and poorly sorted. May consist of an array of gravel bedforms such as pebble clusters and ribs.	Deposited under uniform energy conditions in highly sediment-charged rivers. Generally indicates transport- and capacity-limited or competence-limited conditions due to oversupply of sediment. Surficial gravel bedforms are frequently reworked as the sheet moves downstream as a pulse. May represent residual deposits that form a basal lag or a diffuse gravel sheet, reflecting rapid deposition and/or prolonged winnowing. May be armoured.
Forced mid-channel bar (pendant bar, wake bar, lee bar)	A mid-channel bar form that is induced by a flow obstruction (e.g. bedrock outcrop, boulders, large wood, vegetation). The resultant bar form often has a downstream-dipping slip face as the bar extends downstream.	Perturbations in flow and subsequent deposition are induced by obstructions. The resultant bar morphology is shaped by the flow obstruction, which forces flow around the obstruction and deposition in its wake in secondary flow structures. Depending on flow stage, these secondary flow structures may locally scour the bed. These bars build in a downstream direction and may become vegetated.
Compound mid-channel bar chute channel ridge scour hole ramp contained within a chute channel	A mid-channel bar that comprises an array of smaller scale geomorphic units. Variable morphology reflects material texture, flow energy and the history of flood events that induce formation and subsequent reworking, producing chute channels, ramps or dissection features. Further deposition may form ridges and lobes. If vegetation colonises parts of the bar, additional depositional features result, producing an island.	The assemblage of geomorphic units is dependent largely on channel alignment (and associated distribution of flow energy over the bar surface at different flow stages) and patterns of reworking by flood events. Formed initially from the lag deposition of coarser sediments (a unit bar). At high flow stage the bar may be reworked or material may be deposited around obstructions. At low flow stage the bar may have finer depositional features deposited on top of the bar platform. The range of bedforms reflects sediment transport across the surface.

[a]Modified from Brierley and Fryirs (2005). © John Wiley & Sons, Ltd. Reproduced with permission.

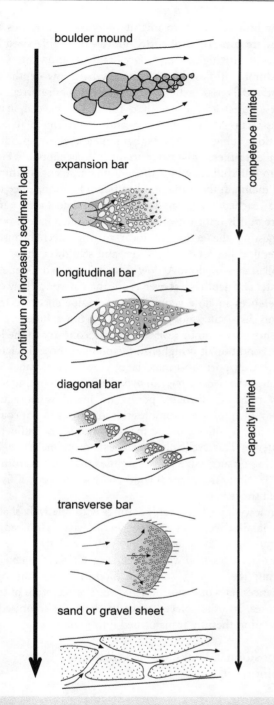

boulder mound

expansion bar

longitudinal bar

diagonal bar

transverse bar

sand or gravel sheet

continuum of increasing sediment load

competence limited

capacity limited

Figure 8.2 The continuum of mid-channel geomorphic units. The continuum occurs along a gradient of increasing sediment load, extending from coarse-grained boulder mounds in competence-limited conditions, to longitudinal and transverse bars and sand sheets under capacity-limited conditions.

The most common mid-channel geomorphic units are accumulations of deposits referred to as *bars*. These free-forming depositional features are areas of net sedimentation of comparable size to the channels in which they occur.

They tend to have lengths of the same order as the channel width or greater and have heights comparable to the mean depth of the generating flow. Bar form and configuration provide key indicators into formative processes, reflecting the ability of a channel to transport sediment of variable calibre. In most cases, depositional sequences in bars reflect lateral or downstream accretion of sediment. Bars interact with, and influence, the patterns of flow through a reach. Flow divergence produces a zone of low tractive force and high bed resistance, which accentuates sediment deposition.

Coarse materials often make up the basal platform of bars. Shifts in channel position rework these bedload materials. Mid-channel forms are more likely to be reworked than bank-attached features. Long-term preservation of bar deposits is conditioned by the aggradational regime and the manner of channel movement. Bar forms are more a reflection of sediment supply conditions and channel-scale processes than local fluid hydraulics.

The formation of mid-channel bars reflects circumstances in which the coarse bedload fraction can no longer be transported by the flow (exceedence of a competence limit) or there is too much material for the flow to transport and instream deposition occurs (exceedence of a capacity limit). These conditions tend to be associated with gravel- (or coarser) and sand-bed channels respectively. Most mid-channel bars are characterised by downstream fining of deposits, where the coarsest fraction is deposited at the head of the bar and finer materials are deposited in the lee by secondary flow currents. Mid-channel bars tend to accrete in a downstream direction.

Bed material character and the competence of flow to transport it determine the formation of *longitudinal* bars. These features form as flow divides around a tear-drop-shaped structure, depositing materials in the lee of the coarser bar head. Subsequent deposition leads to downstream extension and vertical accretion, resulting in an elongate, oval or rhomboid planform. When flow is oriented obliquely to the long axis of the bar, a *diagonal* feature is produced. The upstream limb of these bars may be anchored to the concave bank, reflecting a dissected riffle. Bed material sorting tends to be rudimentary or absent in these bars.

In highly sediment-charged sand-bed conditions, flow divergence forms *transverse* or *linguoid* bars which extend across rather than down the channel. These features have a broad, lobate or sinuous front with an avalanche face. A concavity in the central part of the upstream ramp forms when flow moves over the centre of the bar, diverges and is pushed up the ramp face. Sediment falls over the avalanche face, depositing material on the lee side, and the bar moves downstream. *Expansion bars* are coarse-grained, fan-shaped bars that form in areas of abrupt flow expansion downstream of a forced pool. Alternatively, the entire channel bed may comprise a homogeneous *sand* or *gravel sheet*, where a continuous veneer of sediment moves along the channel as a very low amplitude bedform.

Areas of channel widening or local slope decrease along confined valleys may induce the development of low-relief, elongate or linguoid-shaped *boulder mounds*. These features form under high-velocity conditions by the same mechanism as longitudinal bars, but with much coarser sediments. Over time, a preferred single channel tends to become established. Following abandonment of a side channel, boulder mounds may evolve into flat, gravel–boulder sheets that are attached to the bank.

Alluvial *riffles* and *pools* are oscillatory bed features, in which patterns of scour and deposition produce a more or less regular spacing between consecutive elements (Figure 8.3). These bar-like undulations in bed elevation and grain size are for the most part expressed in the longitudinal rather than the lateral direction. Pool–riffle patterns usually

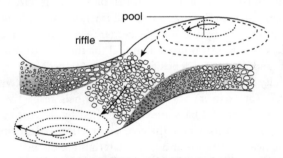

Figure 8.3 Riffle–pool formation in meandering alluvial rivers. Accentuation of erosion on the concave bank of meander bends produces pools. Riffles form at the inflection points between bends, where sediment is deposited in shallower parts of the channel. Point bars occur on the insides of the bend as secondary flow moves sediment onto the bar surface. Down-bar (around the bend) gradation in grain size is commonly observed on these bar surfaces. Oparau River, New Zealand. Photograph: K. Fryirs.

show little tendency to migrate downstream. Riffles and races are comprised of sorted gravel- or cobble-sized bed material with high water surface gradients and white-water conditions. Riffles characteristically have low depths and trapezoidal cross-sections. Races (sometimes referred to as coarse runs) are deeper than riffles and have u-shaped cross-sections. Concentration and deposition of coarser fractions at high discharges (bankfull and above) produces incipient riffles, while scour occurs in adjacent pools. Flows up to bankfull may be sufficiently competent to amplify and maintain the initial pool–riffle undulations once they have reached a critical size. Riffles tend to have coarser, more tightly imbricated bed materials than adjacent pools, suggesting the action of local sorting mechanisms. In general, riffles tend to be wider and shallower than pools at all stages of flow. At low flow, velocity and slope are greater and depth is less over a riffle than in a pool. However, differences in flow geometry and competence are more evenly distributed along a reach at high flows. Indeed, competence may even be reversed so that, contrary to the low-flow condition, it is higher in the pools at those discharges which transport most material in gravel-bed streams. This is called the *velocity reversal hypothesis*. In combining high-flow transport through pools and low-flow storage on riffles, such reversal promotes the concentration of coarser material in riffles and the maintenance of the riffle–pool sequence. This cyclic character has a more or less regular spacing of successive pools or riffles that ranges from 1.5 to 23.3 times the channel width, with an average of five to seven times.

In low-slope, low-energy settings with relatively shallow alluvial fills, accretionary forms may develop atop bedrock (Figure 8.4). These *bedrock core bars* are characterised by bedrock ridges atop which alluvial materials are deposited during the waning stages of floods. Vegetation cover enhances rates of deposition and vertical accretion of these features. In other places, wherever bars are colonised by vegetation they are transformed into *islands*.

Depositional, bank-attached instream geomorphic units

The geometry of channel margins reflects a combination of bank erosion processes, as channels rework floodplain deposits or inset features, and depositional processes that generate a range of bank-attached geomorphic units (see Table 8.3).

Lateral bars are elongate features attached to banks along relatively straight channels (Figure 8.5). They commonly alternate on opposite sides of the channel along a reach. Several platform levels may be evident, separated by steep slipfaces, reflecting lateral accretion and/or downstream migration. In some instances, lower platforms are draped

Figure 8.4 Other types of mid-channel geomorphic units. (a) Islands are vegetated bars, (b) bedrock core bars/islands occur where vegetation grows on finer substrates deposited on bedrock outcrops and (c) forced bars occur where obstructions such as wood or vegetation induce deposition in the middle of the channel bed. (a) Tanana River, Alaska. Photograph: Roger W. Ruess, Institute of Arctic Biology, University of Alaska; www.iab.uaf.edu/research/research_project_by_id.php?project_id=171. (b) Kruger National Park, South Africa. Photograph: G. Brierley, (c) Faulkenhagen Creek, western NSW. Photograph: K. Fryirs.

by progressively finer grained materials during intermittent, successive stages of flood recession.

Point bars typically have an arcuate shape that reflects the radius of curvature of the bend within which they form

(Figure 8.5). An array of forms may be determined, reflecting bend curvature and bed/bank material texture. Point bars are attached to the inner (convex) bank and are inclined towards the centre of the channel, reflecting the asymmetrical channel geometry at the bend apex. They form when helicoidal flow is pushed over the bar surface as the thalweg shifts to the outside of the bend at high flow stage (see Chapter 5). Traction processes move sand or gravel bedload towards the convex slopes of bends, building the bar with lateral accretion deposits. The around-the-bend set of sedimentary structures and grain-size trends reflect the tightness of the bend (i.e. its radius of curvature) and the associated distribution of secondary circulation cells, as determined by the position of the thalweg. In general terms, the coarsest materials are deposited at the bar head, where the thalweg is aligned closer to the convex bank. Further around the bend, the thalweg moves towards the concave bank and finer grained sediments (a bedload and suspended-load mix) are deposited. The most recently accumulated deposits are laid down as a bar platform.

In some instances, *scroll* bars are deposited in the shear zone between the helicoidal flow cell in the thalweg zone and the separation zone adjacent to the convex bank of a bend. As these features build vertically and the channel shifts laterally, scroll bars become incorporated into the floodplain as lateral accretion deposits. These accretionary *ridges* and intervening *swales* form on the inside of bends. Swale fills are narrow, arcuate sediment bodies with a prismatic cross-section. Their fine-grained fill thickens downstream, with a concave-upward basal surface. Series of ridges and swales record former positions of the channel and the pathway of migration.

Concave bank-benches may form in the upstream limb of obstructed or tight bends in laterally constrained situations. In these settings, suspended-load slackwater sediments are deposited in a separation zone at high flow stage. Just like point bars, these features may become incorporated into the floodplain as the bend assemblage translates downstream. Alternatively, *point dunes* may form on the top of point bars during high-energy flow stages in some laterally constrained bends, presenting an alternative to around-the-bend depositional patterns.

In low- to moderate-sinuosity sand-bed channels, oblique-accretion *benches* may form as sand or mud deposits are lapped onto relatively steep convex banks (Figure 8.5). During the rising stage of flood events, bedload materials are deposited atop these step-like features. Suspended load materials cap these deposits at waning stage, forming flood couplets. Similar low-energy, falling-stage mud drapes form *point benches* along the convex banks of slowly migrating fine-grained channels with high suspended-load concentrations. Sediments that form bench and point bench features are typically quite distinct relative to the

Table 8.3 Bank-attached geomorphic units[a]

Unit	Form	Process interpretation
Lateral bar (alternate or side bar) 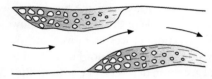	Bank-attached unit bar developed along low-sinuosity reaches of gravel- and mixed-bed channels. Bar surface is generally inclined gently towards the channel. These bars occur on alternating sides of the channel. They are generally longitudinally asymmetrical and may or may not have an avalanche face on the downstream side.	Flow along a straight reach of river adopts a sinuous path. Bar length and width are proportional to these flows. Bar height is dictated by flow depth. Bars form by lateral or oblique accretion processes, with some suspended-load materials atop (i.e. typically upward fining depositional sequence). They generally migrate in a downstream direction.
Scroll bar 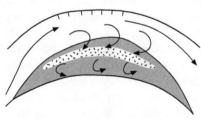	Elongate ridge form developed along the convex bank of a bend. Commonly develop on point bars with an arcuate morphology.	Formed by two-dimensional flow paths on the inside of a bend. Adjacent to the thalweg, sand or gravel bedload material is moved by traction towards the inner sides of channel bends via helicoidal flow. This is accompanied by a separation zone adjacent to the bank formed at near-bankfull stage, which shifts flow alignment adjacent to the bank. Convergence of these flow paths leads to the deposition of a ridge-like feature on the point bar surface. Associated with laterally migrating channels, scroll bars reflect the former position of the convex bank. With progressive channel shift and stabilisation by vegetation, scrolls develop into ridge and swale topography.
Point bar 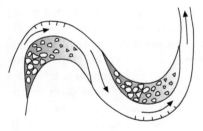	Bank-attached arcuate-shaped bar developed along the convex banks of meander bends. Bar forms follow the alignment of the bend, with differing radii of curvature. The bar surface is typically inclined towards the channel, as are the sedimentary structures. Grain size typically fines down-bar (around the bend) and laterally (away from the channel). Typically these unit bar forms are unvegetated.	Result from lateral shift in channel position associated with deposition on the convex bank and erosion on the concave bank. Sand or gravel bedload material is moved by traction towards the inner sides of channel bends via helicoidal flow. Differing patterns of sedimentation are imposed by the radius of curvature (bend tightness) as well as the flow regime and sediment load. The coarsest material is deposited from bedload at the bar head, where the thalweg is aligned adjacent to the convex bank (at the entrance to the bend). As the thalweg moves away from the convex bank down-bar, lower energy suspended-load materials are deposited in secondary flow circulation cells, as the propensity for deposition is increased. Secondary flow also forces material up onto the face of the bar, building it laterally.

Table 8.3 *(Continued)*

Unit	Form	Process interpretation
Tributary confluence bar (channel junction bar, eddy bar) 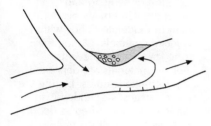	Formed at, and immediately downstream of, the mouth of tributaries. These delta-like features have an avalanche face. They generally comprise poorly sorted gravels and sands with complex and variable internal sedimentary structures. They represent a form of slackwater deposit (interbedded sands and mud) that is not elevated above the channel and is prone to reworking.	Typically form at high flood stage in reaches where a comparatively minor tributary enters the trunk channel. Flow separation and generation of secondary currents in the backwater zones promote sedimentation in sheltered areas under low flow velocity conditions.
Ridge and chute channels (cross-bar channels) 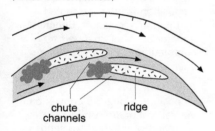	*Ridge* Linear, elongate deposit formed atop a bar platform on a mid-channel or bank-attached bar. May be curved or relatively straight. Deposits tend to fine down-ridge. May be formed downstream of vegetation or other obstructions on the bar surface.	*Ridge* Ridge morphology and alignment atop bar surfaces reflect the character of channel adjustment over the bar at high flow stages. Vegetation promotes ridge development with sediment being deposited in the wake.
	Chute channel Elongate, relatively straight channel that dissects a bar surface. Usually initiated at the head of the bar. Commonly found on bank-attached, mid-channel bars and islands, where they form compound features.	*Chute channel* Develop as water flows over a formerly emergent bar surface. Scour occurs during the rising stage of over-bar flows. If the bar is short-circuited, flow energy is concentrated, inducing scour that reworks the bars and forms a chute channel.
Ramp (chute channel fill) and point dune 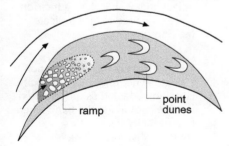	*Ramp* Coarse-grained, ramp-like feature that partially infills a chute channel. Formed at the upstream ends of bends and rise up from the channel to the bar surface.	*Ramp* Under high flow conditions flow alignment over the bar short-circuits the main channel. A relatively straight channel is scoured. Sediment is subsequently ramped up the chute channel, partially infilling this feature with high-energy deposits, such as gravel sheets or migrating dunefields.
	Point dune Dune bedforms that accrete along convex banks, generally atop compound point bars. These features have a down-valley alignment, rather than reflecting around-the-bend trends.	*Point dune* Produced when high-magnitude flow is aligned down-valley rather than around the bend, typically in sand-bed streams. Formed at high flood stage, when the thalweg shifts to the inside of the bend (over the point bar). Preserved in the falling stages when the thalweg switches back along the concave bank of the channel bend.

(Continued)

Table 8.3   (*Continued*)

Unit	Form	Process interpretation
Bench and point bench (oblique-accretion bench) 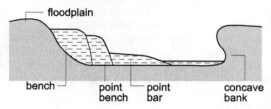	*Bench* A distinctly stepped, elongate, straight to gently curved feature that is inset along a bank. These significant in-channel sediment storage units are often situated atop bar deposits. They may comprise obliquely attached mud-rich drapes with a convex geometry in suspended-load systems, or obliquely and vertically accreted sand deposits in bedload systems. Sedimentary structures tend to be quite distinct from the floodplain.	*Bench* Formed by oblique- and vertical-accretion of bedload and suspended-load materials during small to moderate floods within widened channels. During the rising stage of flood events, bedload materials are deposited atop step-like features. During the waning stages, suspended-load materials are deposited as flood couplets atop bedload materials. Oblique accretion benches represent low-energy falling-stage suspended-load deposition in sand-bed and mud-rich streams. Deposition is often promoted by riparian vegetation. Benches are a major agent of channel contraction in overwidened channels.
	Point bench Distinctly stepped, bank-attached unit developed along the convex bank of a sinuous channel. Typically has an arcuate planform with a planar surface that is elevated above the point bar.	*Point bench* Deposition along the convex bank via vertical and/or oblique accretion of interbedded sands and mud indicates slow lateral migration or lateral accretion within an overwidened bend.
Ledge 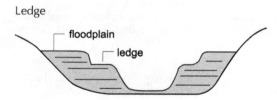	Distinctly stepped, flat-topped, elongate, bank-attached unit. Has a straight to gently curved planform, flanking the banks. Composed of the same materials as the basal floodplain (i.e. sediments are laterally continuous from the ledge to the floodplain).	Formed by channel expansion processes where flows selectively erode the upper units of the floodplain as the channel incises and expands. Unpaired ledges reflect lateral shift during incision, whereas paired ledges indicate incision only.
Boulder berm (boulder bench) 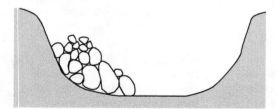	Elongate, bank-attached stepped feature. Can have a convex cross-section. Comprised of coarse, boulder bedload materials with limited finer grained matrix.	Formed from bedload deposition in a single event under high-velocity conditions. Materials are accreted (or dumped) along the bank where flow velocity decreases substantially. Reworking is restricted to subsequent high-velocity events that have the competence to mobilise the boulders.

Table 8.3 (*Continued*)

Unit	Form	Process interpretation
Concave bank bench (convex bar) 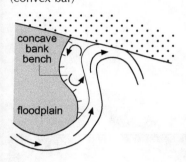	Bank-attached unit, often with a low ridge across the central portion parallel to the primary channel. Located along the upstream limb (i.e. along the concave bank) of relatively tight bends that abut bedrock valley margins or a flow obstruction. Often inset against the floodplain. Comprise slackwater sediments (interbedded sands and mud) and organic materials.	Associated with flow separation and generation of secondary currents at high flood stage. Sedimentation occurs in sheltered backwater zones of relatively low flow velocity. Form from flow separation when the primary flow filament continues around a bend. At flood stage, flows separate from the primary filament, circulating back around the bend. This is often channelled by a shallow ridge. During the rising stages of flood events, this process may accentuate scour on the surface of the bench. Deposition of suspended-load materials subsequently occurs during waning stages.
Compound bank-attached bar 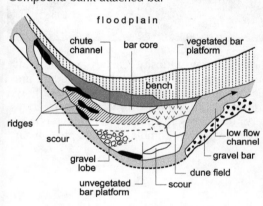	Bank-attached bar that is comprised of an array of smaller scale geomorphic units. Generally composed of laterally accreted sand or gravel, but may include silt or boulders. Variable morphology reflects material texture, flow energy and the history of flood events that form and rework the bar. If a bar is reworked by chute channels, ramps or dissection features may form. Deposition may create ridges and lobes. If vegetation colonises parts of the bar, additional depositional features result. Forms of bank-attached compound bar include compound point bars and compound lateral bars.	Development of compound lateral or point bar forms is dependent on channel alignment (and associated implications for the distribution of flow energy over the bar surface at different flow stages) and associated patterns of reworking by flood events. Formed initially from the lag deposition of coarser sediments (a unit bar). At high flow stage the bar may be reworked or material deposited around obstructions. At low flow stage the bar may have finer depositional features deposited on top of the bar platform, or a range of bedforms preserved, reflecting sediment transport across the bar surface.
Forced bank-attached bar 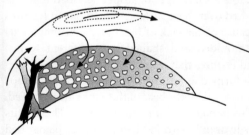	Bank-attached bar that is induced by a flow obstruction (e.g. bedrock outcrop, boulders, large wood, vegetation). The resultant bar form often has a downstream fining sediment sequence.	Perturbations in flow and subsequent deposition are induced by obstructions. The resultant bar morphology is shaped by the flow obstruction, which promotes turbulence and deposition in its wake via secondary flow structures. Depending on flow stage, secondary flow structures may induce local scour around the obstruction.

[a]Modified from Brierley and Fryirs (2005). © John Wiley & Sons, Ltd. Reproduced with permission.

Figure 8.5 Bank-attached geomorphic units. (a) Lateral bars are formed in low-sinuosity reaches and (b) point bars on meander bends. (c) Benches are depositional features reflecting channel contraction, while (d) ledges are erosional features reflecting channel expansion. (e) Concave bank benches form in separation zones where flow impinges against bedrock on a concave bank. (a) Lerida Creek, NSW. Photograph: K. Fryirs. (b) Washington, USA. Photograph: Dave Montgomery from www.uvm.edu/~geomorph/gallery. (c) Tuross River, NSW. Photograph: K. Fryirs. (d) Faulkenhagen Creek, western NSW. Photograph: K. Fryirs. (e) Luangwa River, Zambia. Photograph: Gilvear *et al.* (2000) © John Wiley & Sons, Ltd. Reproduced with permission.

adjacent floodplain against which they have formed. In contrast, *ledges* are erosional forms that produce a distinct step along channel margins during phases of channel incision and expansion (Figure 8.5). Ledge sediments are continuous with the adjacent floodplain from which the feature has been sculpted.

Additional types of bank-attached depositional features may form in local settings (Figure 8.6). For example, open-framework *boulder berms* are step-like features with concave cross-sections that are attached to the bank of high-energy, boulder-bed streams. They may form in the zone of large velocity gradient at bank crests at peak flood stage and are often deposited in one event. *Channel junction* bars commonly develop as delta-like features, as backwater effects induce slackwater deposition downstream of tributary confluences.

Analysis of these bank-attached depositional forms alongside assessment of bank erosion processes (Chapter 7) provides a coherent basis with which to appraise variability in channel geometry.

Sculpted, erosional fine-grained instream geomorphic units

Sculpted, erosional fine-grained instream geomorphic units are scoured from surrounding fine-grained materials (Figure 8.7; Table 8.4). Drapes of suspended load material may be deposited over these surfaces. The range of features is limited because of the lack of bedload-calibre materials that can be moulded and shaped into different forms. Bank-attached features that resemble *lateral* and *point bars* and *ledges* are formed as sediment is eroded from the tops of banks and materials at the base of the banks are moulded. Mid-channel *scour pools* and *runs* are common. These units are shaped by hydraulics around bends or over planar bed surfaces.

A pool is not a pool is not a pool!

The principle of equifinality indicates that similar looking features may be created by different sets of processes. For

Figure 8.6   Other types of bank-attached geomorphic units. (a) Bars may occur at tributary confluences, or as (b) berms and (c) point dunes. (d) Forced bars occur where obstructions such as wood or vegetation lead to deposition against the channel bank. (a) Tongariro River, New Zealand. Photograph: P. Chappell. (b) Faulkenhagen Creek, western NSW. Photograph: K. Fryirs. (c) Kangaroo River, NSW. Photograph: K. Fryirs. (d) Washington, USA. Photograph: Dave Montgomery from www.uvm.edu/~geomorph/gallery.

Figure 8.7   Fine-grained sculpted geomorphic units. Erosion and scour processes create forms that are similar to depositional counterparts such as bars, ledges, pools and riffles. Photographs: Cooper Creek, Central Australia. (K. Fryirs).

Table 8.4 Fine-grained sculpted geomorphic units

Unit	Form	Process interpretation
Sculpted lateral bar	Bank-attached feature that resembles a depositional lateral bar. Forms along the low-sinuosity reaches of fine-grained channels. Bar surface is inclined towards the channel but at low angles. Localised form that is not as continuous as a ledge.	Flow along a straight reach of river adopts a sinuous path. In fine-grained systems, these features are sculpted from the surrounding sediment of the adjacent channel bank. Sediment may subsequently be draped over the toe of the bank during waning-stage flows and smoothed by flows that inundate it.
Sculpted point bar	Bank-attached arcuate-shaped bar that resembles a point bar along the convex banks of bends. The bar surface is typically inclined towards the channel but at a low angle.	These fine-grained features are sculpted from the surrounding sediment of the adjacent channel bank. Sediment is subsequently draped over the low-lying toe of the convex graded bank. May form by oblique accretion of suspended-load sediments during the waning stage of flows. Scour of adjacent pools accentuates the morphology of these features.
Ledge	Distinctly stepped, elongate, flat-topped, bank-attached unit. Has a straight to gently curved planform, flanking one or both banks. Composed of the same materials as the basal floodplain (i.e. sediments are laterally continuous from the ledge to the floodplain). These erosional units reflect channel expansion.	Formed by bankfull flows stripping the surface layers of the bank, leaving a step adjacent to the bank. Oblique and vertical accretion may deposit small amounts of sediment atop these steps during the waning stages of flow.
Sculpted run	Stretches of uniform and relatively featureless bed, comprising fine-grained sediment. These smooth flow zones are free-flowing areas that connect pools.	Plane-bed conditions promote relatively smooth conveyance of water and sediment in these linking features. Slopes are intermediate between pools and riffles.
Scour pool	Pools may span the channel, hosting tranquil or standing flow at low stage. Alluvial pools are alternating deep areas of channel along an undulating longitudinal bed profile. Pools tend to act as sediment storage zones that occur at characteristic locations, such as the concave bank of bends in sinuous alluvial channels. In fine-grained systems they also occur in expansion zones where flows exit from more confined, laterally stable sections or where flow converges at anabranch confluences.	At high flow stage, when flow converges through pools, decreased roughness and greater bed shear stresses induce scour and flushing of sediment stored on the bed. Subcritical flow occurs at low flow stage, when divergent flow occurs. Pool infilling subsequently occurs, as pools act as areas of deep, low flow velocity and near-standing water conditions.

example, pools in bedrock, gravel, sand or fine substrates may vary markedly in terms of their spacing pattern and the dominant scour action (i.e. vertical or laterally oscillatory). In bedrock settings where cascades and steps dominate, pools are formed by vertical scour and are bedrock floored. In these reaches the pools are randomly spaced, although cascade-dominated reaches tend to contain small pools between boulders spaced less than one channel width apart. In step (waterfall)-dominated reaches, pool development tends to occur every one to four channel widths. Further down the longitudinal profile, pools formed alongside glides, runs and riffles in gravel-bed rivers are created by lateral scour. In straight reaches dominated by glides and runs, pools tend to be large and elongate with spacing controlled by local controls such as valley expansion zones or wood. These pools may be bedrock- and gravel-based features. In alluvial settings with self-adjusting channels and distinct riffles, pools are formed by lateral scour and spacing is typically five to seven channel widths apart. Point bars, pools and riffles are well connected via mutual sediment transfer processes in these reaches. These pools may comprise a range of gravel, sand or fine-grained sediments.

Unit and compound instream geomorphic units

Geomorphic units are not typically simple unit features formed in one particular erosional or depositional event. In most instances compound forms reflect multiple phases of deposition and reworking under a range of flow conditions. For example, compound bars may be comprised of a range of geomorphic units (e.g. bar platform, ridge, chute channel), indicating phases of primary and secondary deposition and reworking. In simple terms, *unit point bars* comprise a relatively homogeneous bar platform deposited in a single flow event. In contrast, *compound point bars* comprise a range of geomorphic units that result from depositional and/or erosional processes in which surficial (supraplatform) deposits are reworked.

Compound point bars are often comprised of an assemblage of *depositional* features, with characteristic upstream–downstream, around-the-bend and lateral trends (Figure 8.8). The upstream portion of the bar (bar head) is usually topographically higher than the tail (downstream) portion. Bar platform deposits at the bar head tend to form at high flow stage and are the coarsest part of the bar. In many settings these materials form an extension to riffle features. Unit bars and structures added to this feature may retain their identity or become incorporated into the bar structure. Platform deposits form the basal core of the bar. *Supraplatform* deposits, such as *diffuse gravel sheets*, ride across over this basal surface and are shaped into various geomor-

phic units. The bar tail records changes in flow pattern on the lee face of the bar. A range of finer grained depositional structures may form in this area during the recessional stage of flood events.

Reworking of compound point bars may generate an irregular array of geomorphic units. This may result in multiple platforms with a range of erosional and depositional forms. These differing features commonly have variable textures, reflecting activity at differing flow stages. Multiple phases of bar reworking, bar expansion, lateral migration and downstream translation may be evident. The resulting array of erosional and accretionary patterns reflects the direction and rate of bend adjustment. *Chute channels* may short-circuit the bend, cutting a relatively straight channel from the head of the bar at high flow stage. Coarser grained *ramp* deposits may be input into this feature. Enlargement of a chute channel and plugging of the old channel may generate a *chute cut-off*. Because of the small angular different between the old channel and the chute channel, flow continues through the old channel for some time, depositing bedload sediment at the upstream and downstream ends and on the sides until terminal closure of the cut-off is complete, leaving a remnant feature on the floodplain. Chute channel fills are notably straighter in outline than either meander cut-offs or swales. They commonly comprise high-energy ramp deposits comprised of coarse gravels with a steep upstream-facing surface. *Ridges* are elongate features parallel to flow direction that form where deposition occurs around obstacles such as vegetation. Alternatively, scour may accentuate erosion within a chute channel, leaving a perched ridge-shaped feature on the bar surface. These features tend to record the alignment of high stage flows over the bar.

Many of the features described above also occur on *compound mid-channel bars*. Dissecting channels may produce a chaotic assemblage of features. Resulting sediment sequences exhibit little trend in grain size or facies, whether down-bar, laterally or vertically. *Islands* typically comprise an array of units that are scoured or deposited around vegetation. Other bank-attached geomorphic units, such as benches and ledges, commonly have a compound structure with chute channels and/or ridges evident. In some instances, formerly erosional ledges can be transformed into depositional benches due to alterations in sediment regime.

Forced instream geomorphic units

Reading the landscape strives to interpret formative processes and explain why they occur where they do. Key distinction is made between free-forming and forced morphologies. Forced geomorphic units form when an

(a) oblique photograph of a compound point bar

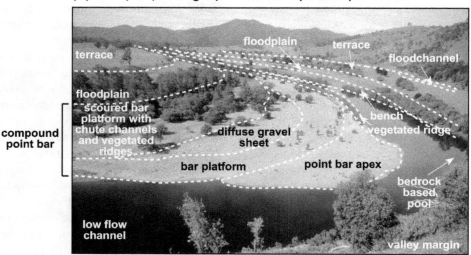

(b) planform view of a compound point bar

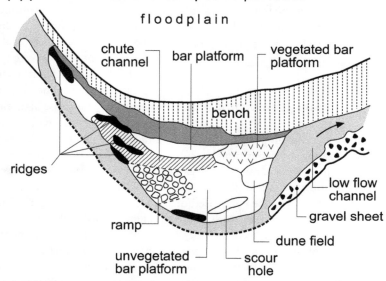

Figure 8.8   The structure of compound point bars. (a) An oblique view of a compound gravel point bar on the Manning River, NSW, Australia. Geomorphic units that make up this compound bar are depositional (e.g. ridges) and erosional (e.g. chute channels). Photograph: G. Brierley. (b) A planform view of a typical compound point bar showing the juxtaposition of a range of geomorphic units.

obstruction or irregularity in the channel induces local erosion and/or deposition. Common obstructions that disrupt flow dynamics and induce erosion include wood, bedrock or coarse clasts and vegetation. *Forced pools* may reflect scour adjacent to shifts in bedrock valley alignment or accumulations of wood in a log-jam. Elsewhere, obstructions may induce local deposition. For example, *forced mid-channel bars* may reflect deposition adjacent to an obstruction such as a large clast or vegetation-induced secondary flow circulation. Sediment accumulation in the lee

of an obstruction may create a *forced longitudinal bar*. Deposition in the lee of vegetation or wood on/adjacent to a channel bank may form *forced bank-attached bars*.

The continuum of instream geomorphic units and transformations in type

The range of instream geomorphic units is represented along a continuum of energy (as determined by flow and slope) and available sediment (primarily the calibre and volume of material) in Figure 8.9. This continuum of instream geomorphic units extends from sculpted, erosional bedrock units through mid-channel depositional units, to bank-attached depositional units, and sculpted, erosional fine-grained units.

Sculpted, erosional bedrock instream geomorphic units occur under steep, high-energy conditions in which erosion dominates over deposition. These units are typically scoured from bedrock or are comprised of boulder material. The continuum from waterfalls to rapids, cascades, riffles, runs and pools reflects the transition from supply- to transport-limited conditions, with some gravel forms towards the lower energy end of this category (Figure 8.10).

A transition to depositional instream geomorphic units occurs further down the continuum of energy and sediment availability conditions. Within this category, mid-channel geomorphic units form under higher energy conditions than bank-attached variants. Depositional, mid-channel variants occur where coarse sediments cannot be moved and the channel is competence limited or where there is high sediment load and the channel is capacity limited (Figures 8.9 and 8.10). These units tend to be formed of boulders, gravel and sand. Different types of units tend to develop under particular sets of flow energy and bed material texture conditions, resulting in a typical down-valley transition in mid-channel forms from boulder mounds to longitudinal, diagonal and linguoid bars. Depositional, bank-attached instream geomorphic units reflect a decline in energy and/or sediment load. Lateral and point bars are less frequently reworked than mid-channel forms. Landforms that are sculpted from fine-grained sediments in low-slope, low-energy settings are found at the lowest end of the continuum of instream geomorphic units.

The continuum presented in Figure 8.9 provides a conceptual synthesis of the diversity of instream geomorphic units. Overlap in the environmental domains within which these units are found ensures that this is a simplification of reality. Local conditions dictate the type and formation of individual geomorphic units along any reach. Repeated or disrupted patterns of features may be evident. Characteristic patterns of features are associated with different types of river (Chapter 10). Identification and interpretation of reach-specific patterns of features in relation to their formative processes generates an understanding of the range of geomorphic behaviour of the reach (Chapter 11).

Conclusion

Geomorphic units are the building blocks of river systems. Their morphodynamic attributes provide a critical basis to interpret river character and behaviour. As such, analysis of geomorphic units is a key interpretative tool in efforts to

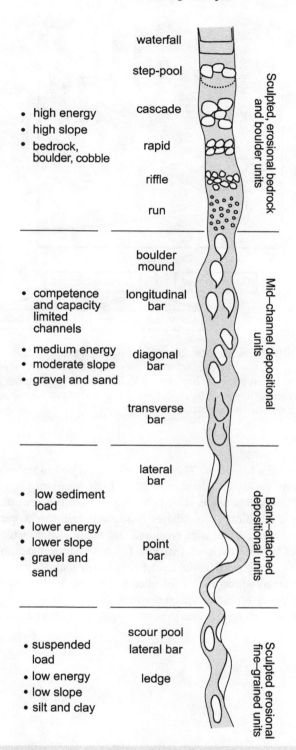

Figure 8.9 The longitudinal continuum of instream geomorphic units. Transition from sculpted, erosion bedrock and boulder forms in high-slope, high-energy conditions, to mid-channel, bank-attached and sculpted fine-grained units in lower slope, lower energy conditions is shown. This continuum does not occur along all rivers. Rather, it reflects local slope, grain size, sediment supply conditions and the presence of forcing features.

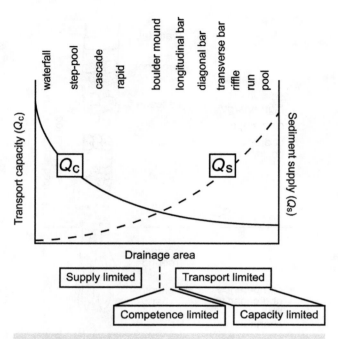

Figure 8.10 Sediment supply and transport capacity relationships for sculpted, erosional and mid-channel instream geomorphic units. Modified from Montgomery and Buffington (1997). © Geological Society of America. Reproduced with permission.

read the landscape. These features are defined in terms of their geometry (shape and size), their location/position within a landscape, their bounding surface (i.e. erosional or depositional boundaries) and their sedimentological attributes. A range of erosional and depositional features can be differentiated. Instream (within-channel) geomorphic units range from forced (erosional, sculpted) forms in imposed, bedrock-controlled settings through an array of freely formed, alluvial features. Alluvial forms can be viewed along an energy continuum from mid-channel

to bank-attached forms through to fine-grained features. Instream geomorphic units provide considerable insight into the range of channel-forming processes along a reach. Assessment of genetic linkages between adjacent features provides insight into the range of behaviour and reworking along a reach. In general terms, however, instream geomorphic units have limited preservation potential as they are readily reworked. As such, although they provide guidance into river behaviour, analysis of longer term sediment stores that make up geomorphic units on floodplains is required to provide insight into longer term river evolution.

Key messages from this chapter

- Geomorphic units are the building blocks of rivers. They are defined by their position, morphology (shape and geometry), sediment composition and bounding surfaces.
- Each geomorphic unit has a distinct form–process association. Assemblages of geomorphic units can be used to interpret river behaviour.
- Instream geomorphic units are found along a slope-induced energy and textural gradient. The continuum of instream geomorphic units extends from high-energy erosional (sculpted) forms found in bedrock and boulder settings, to depositional units that are located in mid-channel locations, to depositional units that are attached to the bank. Sculpted forms are found in low-energy, fine-grained settings.
- Unit features are products of single depositional events, whereas compound features reflect a range of flow conditions and reworking events.
- Forced units form as a result of flow disruption around an obstruction such as bedrock, vegetation or wood.
- Genetic associations of geomorphic units reflect circumstances where the processes that form one unit affect the formation of adjacent units.

CHAPTER NINE
Floodplain forms and processes

Introduction

Floodplains are areas of sediment accumulation made up of alluvial materials between the channel banks and the valley margin. Floodplains accumulate sediment when the sediment supply during overbank flow events exceeds the transport capacity of the flow and sediment is deposited (Chapter 6). Areas adjacent to the channel banks are referred to as the *proximal floodplain*, while areas furthest from the channel, adjacent to the valley margins, are termed the *distal floodplain* (see Figure 6.21). Floodplains are often poorly drained, acting as stilling basins in which fine-grained suspended-load sediments settle out from overbank flows. They typically comprise tabular, prismatic bodies of horizontally bedded materials that are roughly rectangular in cross-section and are elongated parallel to the channel. Although the basal or distal parts of the floodplain may contain elements from prior flow regimes, these materials are typically formed or reworked by contemporary processes. Older, elevated floodplain deposits along valley margins are referred to as terraces. These palaeo-floodplain surfaces are not actively formed or reworked under the current flow regime. In many instances floodplains and/or terraces preserve a sedimentary archive that extends back many thousands of years. As such, analysis of their sedimentology can provide key insights into river evolution and palaeoenvironmental conditions.

The presence of floodplains along a longitudinal profile marks a transition in the process domain along an energy gradient (Figure 9.1). This transition reflects a shift from source-zone activity, characterised by within-channel processes of erosion and transport of materials with occasional short-term stores of coarse sediment, to transfer-zone activity, in which out-of-channel processes create floodplain sediment stores. Floodplain sedimentation occurs because reduced slopes and greater accommodation space promote dissipation of energy, enabling suspended-load materials to be stored along channel margins. These deposits often lie atop former within-channel deposits. Segrega-

tion in the transfer and storage of the river's load creates sediment stores outside the channel (i.e. on the floodplain). This typically occurs within partly confined valleys when slope decreases below $0.008\,\mathrm{m\,m^{-1}}$. Floodplains typically occur as isolated pockets in the source or transfer zone in the middle–upper catchment and as discontinuous, alternating pockets in the transfer zone. Initially, floodplain pockets may be bedrock controlled, wherein sediments are trapped behind bedrock spurs or at sites of local valley widening (e.g. at tributary confluences). Eventually, these pockets alternate, as the river switches from one side of the valley to the other, creating planform-controlled floodplain pockets within partly confined valley settings (Figure 9.1). Significant pocket-to-pocket variability in floodplain forms may be evident. This reflects a range of localised controls such as changes in the nature/degree of valley confinement or differing flow alignments over floodplain surfaces in response to changes in valley or channel alignment. As slope decreases further downstream, and the valley widens further, floodplain pockets become more frequent, eventually becoming continuous along both banks in laterally unconfined valleys of the accumulation zone. In these settings the river flows within its own deposits (i.e. these are fully alluvial rivers).

Some floodplains accrete primarily via lateral (within-channel) deposits (bedload or mixed-load systems). Elsewhere, solely vertical accretion deposits characterise floodplains adjacent to laterally stable channels (i.e. suspended-load rivers). Channel marginal elements separate channel and floodplain processes, providing insights into the lateral connectivity of the system (i.e. frequency and nature of inundation). Lateral gradation in sediment size and material properties on floodplains reflects the operation of different sets of processes on levees at the channel bank (the proximal zone) relative to backswamps adjacent to the valley margin (the distal zone). Bedload materials may be launched onto floodplains adjacent to the channel, while distal areas are inundated by lower energy flows which deposit suspended-load materials at the waning

Geomorphic Analysis of River Systems: An Approach to Reading the Landscape, First Edition. Kirstie A. Fryirs and Gary J. Brierley.
© 2013 Kirstie A. Fryirs and Gary J. Brierley. Published 2013 by Blackwell Publishing Ltd.

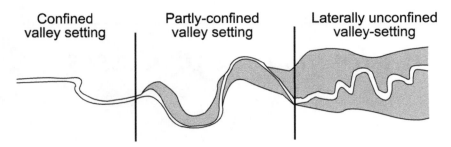

Figure 9.1 Floodplain distribution in confined, partly confined and laterally unconfined valley settings. In plan-form view, confined valleys have no floodplains, partly confined valleys have discontinuous floodplains that alternate along the valley floor and laterally unconfined valleys have continuous floodplains along both channel banks.

stage of floods. These process relationships are greatly influenced by what happens at channel margins. Some rivers have prominent levees, whereas others do not. Some channels are connected to their floodplains on a regular basis, whereas others are not. As floodplains may be subjected to a range of formative and reworking processes (floodchannels, stripping, etc.), their morphology is not always flat. Indeed, a mosaic of depositional and erosion forms can be produced.

Distinct packages of floodplain geomorphic units reflect a combination of energy conditions (function of slope and valley width relative to upstream catchment area), availability of sediment (the calibre and volume of sediment relative to the accommodation space along the valley), history and sequence of development (the range of floodplain-forming and reworking processes that have shaped the floodplain morphology and affect contemporary flow alignment over floodplains). A continuum of forms is evident, ranging from high-energy non-cohesive floodplains to low-energy cohesive floodplains. Differing assemblages of geomorphic units are evident for these different types of floodplains. Sediment sequences in floodplains and terraces can be studied to provide insights into river behaviour and evolution. Hence, analysis of the pattern and extent of floodplains along a river course provides an initial entry point in efforts to read the landscape.

This chapter starts by describing floodplain formation processes, differentiating lateral accretion (within-channel), vertical accretion (overbank) and other mechanisms. Processes that rework floodplain deposits are then outlined. Following documentation of the array of depositional and erosional forms on floodplains (i.e. process–form relationships of floodplain geomorphic units), floodplain types are framed along an energy spectrum including high-, medium- and low-energy variants (defined in terms of specific stream power). Finally, patterns of floodplains along longitudinal profiles are outlined, highlighting the

use of floodplain deposits as a basis to interpret river behaviour and evolution.

Floodplain formation processes

Sediments that comprise floodplains can take a number of forms. First, *bedload deposits* often make up the basal parts of floodplains. These materials may be launched onto floodplain surfaces during high-energy events, typically in sheetlike forms, or they may be deposited within floodchannels. Second, *intermittently suspended-load deposits* are sand-sized sediments that are moved by saltation and are, therefore, well sorted. They are usually deposited above bedload deposits. Finally, *suspended-load deposits* are the fine-grained sediment (silt, clay and fine sand) that comprise the bulk of floodplain deposits. These sediments settle out from suspension in backwater areas and low-velocity zones on the floodplain.

Lateral accretion and overbank vertical accretion are the primary processes that form floodplains. Related, less-prominent mechanisms include braid channel accretion, oblique accretion, counterpoint accretion and abandoned channel accretion. Each of these floodplain formation processes is described in turn.

Lateral accretion

Lateral accretion occurs when bedload and intermittently suspended-load deposits on the convex slope of bends are incorporated into the floodplain as the channel migrates across the valley floor or translates downstream. Lateral accretion deposits can comprise 60–90% of some floodplain sediments. These materials are exposed at the surface in proximal floodplain locations, but they are typically buried by vertical accretion (overbank) in distal parts of floodplains.

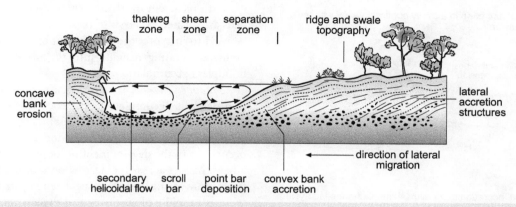

Figure 9.2 Lateral accretion. Cross-section view at the apex of a meander bend where scroll bar deposition on the inside of a bend is accompanied by erosion of the concave bank. The channel moves laterally across the valley floor for this actively meandering river. Ridges and swales on the floodplain represent past channel positions and migration pathways. The flow structures are shown for bankfull stage.

During bankfull conditions, the high-velocity filament of flow is located along the concave bank of a bend (Figure 9.2). Spiral or helicoidal flow develops along the concave banks of bends. The strength of helicoidal flow depends on water depth, bend radius and friction. This motion is anticlockwise when facing downstream for right-turning bends. Flow filaments are deflected from the thalweg as flow travels around the bend. This forms a shear zone along the convex bank of the channel, whereby sediment is transferred from the thalweg zone to the point bar such that a scroll bar accretes laterally and vertically within the shear zone. Eventually, the surface of the scroll bar approaches the elevation of the older part of the floodplain. As the channel shifts laterally, the scroll bar becomes incorporated into the floodplain. Over time, a number of ridges with intervening swales may form. The hummocky appearance of former channels is retained on the floodplain on the inside of the bend until overbank deposits smooth out the floodplain surface.

Oblique accretion is related to lateral accretion, in that intermittently suspended-load or suspended-load deposits are deposited as muddy drapes or sand deposits along the inner accretionary bank. As these deposits onlap the channel margin, and surfaces build, inset floodplains or benches are formed and are incorporated into the floodplain. Oblique or dipping sedimentary structures result from this form of accretion. This process is prevalent along non-migrating rivers or along channels that are actively contracting.

Vertical accretion

Vertical accretion results from overbank deposition of suspended-load materials during floods. As a channel over-

tops its banks, it loses power due to the greatly reduced depth and energy of the unconfined sheetlike flow on the floodplain surface. Deposition of suspended load materials builds the floodplain vertically over time. In general, these horizontally bedded, fine-grained, suspended-load materials dominate floodplain sequences beyond the active channel zone. In many instances, vertical accretion deposits overlie lateral accretion deposits.

Vertical accretion of floodplains is especially dominant for those rivers that have sufficient accommodation space (valley width) to allow floodplains to develop, yet the channel(s) are relatively stable such that the proportion of within-channel deposits on the valley floor is small. Channel stability may reflect the influence of bedrock or consolidated alluvial sediments in partly confined valley settings. Vertical accretion is especially dominant along fine-grained laterally unconfined rivers where cohesive material along channel banks inhibits lateral channel migration (i.e. suspended load rivers).

Two forms of sediment sorting generally characterise vertical accretion deposits: distal and vertical fining (Figure 9.3). When channelised flow breaches its banks, suspended-load sediments are deposited atop the bank. The coarsest sediment is deposited at channel margins. As the transport capacity of the flow decreases significantly with distance from the channel, progressively finer grained materials are deposited at floodplain margins. This proximal–distal gradation in grain size of vertically accreted floodplain sediments is referred to as *distal fining* (Figure 9.3a). Vertical accretion deposits also grade with depth (Figure 9.3b). This is typically expressed as repeated cycles of *upward-fining flood couplets*. Individual flood events tend to generate flood cycle deposits reflecting the rising and falling stages of flow over the floodplain. These units commonly fine

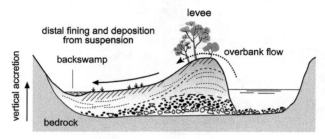

(a) vertical accretion in a partly confined valley with levee and backswamp

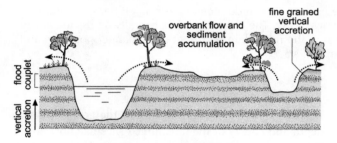

(b) vertical accretion in an alluvial multi-channelled system

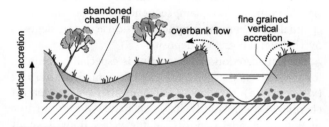

(c) vertical and abandoned channel accretion in an alluvual meandering system

Figure 9.3 Vertical accretion. Sediments deposited during overbank flow build the floodplain vertically. (a) In a partly confined valley the coarsest floodplain sediments are deposited on a levee and the finest in a backswamp, resulting in distal fining. In these settings, floodchannels may short-circuit the floodplain pocket. (b) In an alluvial multichannelled river, fine-grained sediments settle out of suspension during overbank sheet flow, promoting vertical accretion. (c) Vertical accretion also occurs in palaeochannels, progressively filling the abandoned channel over time. These sediments are typically finer grained than the original bed materials that were transported along that channel.

upwards from an erosive base, reflecting scour during the rising stage of a flood, followed by deposition during waning-stage velocities. Coarser grained suspended-load sediments are deposited first, followed by finer grained sediment as the flow wanes. Resulting flood couplets often grade internally from fine–medium sand to silt–clay depos-

its. With recurrent overbank events, cyclical flood couplet beds are formed. Analysis of these sediment sequences can be used to interpret the flow conditions during overbank events, generating insights into the history and rate of floodplain accumulation.

Abandoned channel accretion occurs when vertical accretion deposits accumulate in palaeochannels and cut-offs during overbank flow (Figure 9.3c). Neck cut-off is the primary mechanism of meander loop abandonment. This occurs late in the development of loops, either by gouging a new channel across the narrow neck of land or through the capture of one loop by the next bend upstream. Bedload sediment rapidly plugs the end of the abandoned channel to produce an oxbow lake (billabong). These features have well-defined morphologies. Overbank flows generate upward-fining particle size trends within abandoned channel fills. Gravel and fine sands grade into mud and/or swamp deposits as the palaeochannel or cut-off is filled with sediment. Over time, progressively larger flows are required to occupy the abandoned channel. The nature of the sediment that infills abandoned channels, especially clay plugs, may resist subsequent movement of the channel, affecting meander morphology and migratory pathways on the valley floor. In general, the rate of infilling of palaeochannels and cut-offs reflects their antiquity (i.e. the longer the period since their abandonment, the greater the degree of infilling). In some instances, however, the rate of infilling of palaeochannels and cut-offs reflects their alignment relative to the contemporary channel, and their geometry and position on the floodplain. Straighter channels are subjected to higher flood velocities and tend to be actively reworked, while backwaters flood more sinuous cut-offs. Sediments that infill abandoned channels are typically much finer than the palaeo-bedload materials that formed the original channel bed. In many instances, abandoned channel fills can be excavated to reconstruct palaeochannel dimensions in river evolution analyses.

Braid channel accretion

Braid channel accretion is observed along actively adjusting multichannelled rivers such as braided rivers (Figure 9.4). Deposition of bedload materials atop mid-channel bars during large flood events promotes the development of topographic surfaces that are beyond the reach of small–moderate flood events. Preferential flow orientation down one of the channels may lead to abandonment of the other channel. Subsequent infilling of this channel by overbank sediments may result in the incorporation of the former bar into the floodplain. Alternatively, shifting of primary braid channels to another position on the valley floor allows infilling of old braid channels with overbank sediments. Finally, incision of primary channels may perch

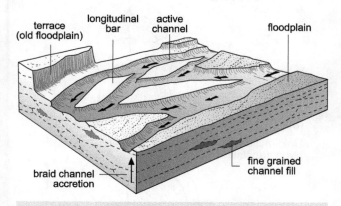

Figure 9.4 Braid channel accretion. Thalweg shift results in reoccupation and abandonment of channels across the valley floor. Channels in relatively sheltered areas may become incorporated into the floodplain. As a result, the valley floor is comprised of a series of topographic surfaces that are inundated and reworked at differing flow stages.

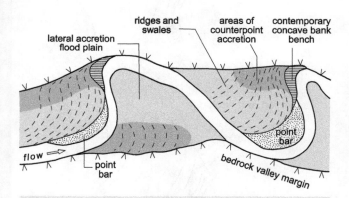

Figure 9.5 Counterpoint accretion. This occurs where the outer banks of meander bends are laterally constrained by bedrock or terrace features. A separation zone occurs as flows impinge against the obstruction. Suspended load deposition within this secondary circulation cell forms a concave bank bench. This feature is incorporated into the floodplain as the bend translates downstream. Modified from Page, K.J. and Nanson, G.C. (1982) Concave bank benches and associated floodplain formation. Earth Surface Processes and Landforms 7, 529–543, © 1982, with permission from Elsevier.

braid belts on the floodplain. Sediment patterns for braid accretion floodplains reflect the incorporation of bars into the floodplain. Sequences of channel bar deposits may be overlain by vertically accreted intermittently suspended-load sediments.

Counterpoint accretion

Counterpoint accretion occurs on the upstream limb of the concave bank of tightly curved, laterally constrained bends (Figure 9.5). Obstructions on the outsides of bends promote flow separation and the development of secondary flow circulation at bankfull stage. Deposition of vertically accreted suspended-load materials within these secondary circulation cells is referred to as counterpoint accretion. This process generates concave bank benches. These deposits are comprised of suspended-load or intermittently suspended-load deposits. These features are incorporated into the floodplain as the channel translates down-valley over time. Downstream translation of bends can result in counterpoint accretion deposits making up 20% of the floodplain material along laterally constrained meandering channels.

Floodplain reworking processes

Many floodplains are purely aggradational forms. This is especially the case for laterally unconfined settings dominated by vertical accretion processes. However, some floodplains show evidence of reworking, such that resulting floodplain morphology is as much a product of processes that mould the surface as it is a product of the constructive processes that created them. Mechanisms of floodplain reworking are determined by the pathway and rate of channel movement over the valley floor. The operation of these processes reflects valley setting and flow alignment over the floodplain surface.

Lateral migration

Floodplain reworking via lateral migration is a by-product of the lateral accretion process described above. Lateral migration occurs as a meander bend progressively moves across the valley floor. As bend curvature increases, the potential for thalweg scour increases, leading to greater bank erosion, more rapid rates of lateral migration and floodplain reworking, and wider spacing of ridges and swales (Figure 9.6). The rate of lateral bend migration is greatest when the ration of the radius of curvature to the channel width $r_c/w \approx 3.0$ (Figure 9.6).

Patterns of channel migration and floodplain reworking can be unidirectional or multidirectional. Series of accretionary ridges and intervening swales record the migration pathway, marking former positions of the channel (Figure 9.7). The direction of migration depends on where concave bank erosion is most accentuated. The focal point of bank erosion in any given bend reflects channel alignment (i.e. sinuosity) and flow stage. Commonly, maximum flow

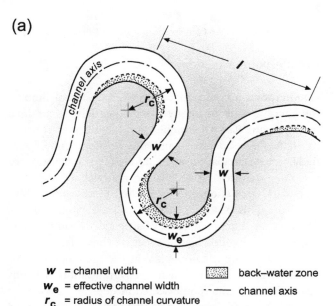

w = channel width
w_e = effective channel width
r_c = radius of channel curvature
l = meander wavelength

▓ back–water zone
— · — channel axis

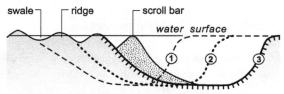

swale — ridge — scroll bar
water surface

① channel boundary (period 1)
② channel boundary (period 2)
③ present channel boundary

- · –①– channel boundary (period 1)
- ·②·· channel boundary (period 2)
- ⊤③⊤ present channel boundary

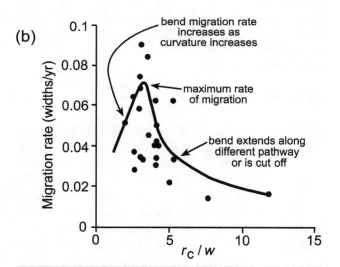

Figure 9.6 Controls on lateral migration. (a) Plan view and cross-sectional views indicating how to measure the radius of curvature to width ratio for meander bends. Ridge and swale topography can be used to determine rates of lateral migration in freely meandering rivers. (b) Meander migration rates for gravel-bed rivers are highest when the ratio of the radius of channel curvature to width is around three (modified from Hickin and Nanson (1975)). © Geological Society of America. Reproduced with permission.

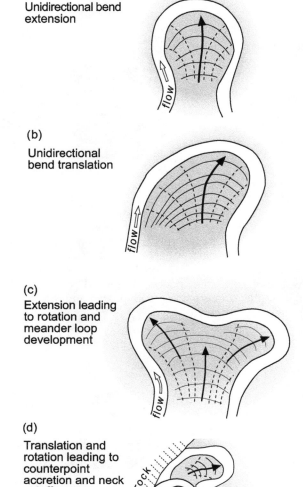

(a) Unidirectional bend extension

(b) Unidirectional bend translation

(c) Extension leading to rotation and meander loop development

(d) Translation and rotation leading to counterpoint accretion and neck cutoff

Figure 9.7 Lateral migration pathways. Bends can extend, translate, rotate or be subjected to combinations of these trajectories as they migrate laterally.

velocity impinges on the concave bank progressively further downstream, increasing the sinuosity and generating tortuous meanders. At low flow stage, high flow velocity occurs towards the upstream end of the bend apex, inducing unidirectional bend extension (Figure 9.7a). At high flow stage, concentration of erosion downstream of the bend apex promotes bend translation (Figure 9.7b). Varying phases and patterns of erosion at different flow stages promote multidirectional bend migration, with various combinations of extension and translation leading to bend rotation and the development of meander lobes (Figure 9.7d). Accentuation of bends increases channel sinuosity. The downstream translation or rotation of bends may be hindered if the migration path comes into contact with obstructions such as bedrock valley margins or cohesive sediments (e.g. clay plugs of abandoned channel infills) at the margin of a meander belt. This produces an irregular meandering pattern.

Cut-offs

Cut-offs form whenever a meandering stream shortens its course by cutting off a bend, leaving an abandoned channel on the floodplain. Shortening of the channel course results in local increase in slope, which increases channel instability. Neck cut-off is the primary mechanism of meander loop abandonment (Figure 9.8). Such cut-offs occur late in the development of the loops, as a result of tightening of a bend (i.e. accentuated sinuosity) via extension and translation. During trough migration, one bend 'catches up' with another such that flow is captured and a meander loop is abandoned. A new channel erodes the narrow neck of land between two loops. Cut-off formation is a form of lateral instability and floodplain reworking that starts as a progressive adjustment, but can be catastrophic at the time of the cut-off. In many instances bedload sediment rapidly plugs the ends of the abandoned channel to produce an oxbow lake (or billabong).

Chute cutoffs occur where part of a bend is short-circuited at high flow stage by erosive flows that are aligned down-valley over a point bar or through a floodchannel, rather than around the bend. This generates a relatively straight chute channel. Differing degrees of infilling of palaeochannels and meander cut-offs reflect varying rates of sedimentation from overbank flows.

Avulsion

Avulsion refers to a relatively sudden shift in the course of a river over a considerable length – say, several meander bend wavelengths. A range of avulsion mechanisms may be observed in different settings (Figure 9.9). These process relationships are strongly associated with aggrading floodplains. Avulsion is relatively common in multichannelled river networks, but the underlying mechanisms by which avulsion occurs may be quite different. Coarse bedload braided rivers with non-cohesive banks and high stream power are susceptible to relatively sudden shifts in channel position (Figure 9.9a). Creation of new channels or reoccupation of old channels may be accompanied by the abandonment of former channel threads. This type of river behaviour is referred to as thalweg shift (Chapter 11). In contrast, laterally stable anastomosing channel networks are prone to channel abandonment if one of the channel threads becomes blocked or infilled (Figure 9.9b). Secondary floodplain channels are reoccupied on a regular basis in these fine-grained vertically accreted multichannelled networks.

Avulsion may also be prominent in aggrading meandering rivers where the active channel and levee zone within the meander belt, also referred to as an alluvial ridge, becomes perched above the adjacent floodplain because of accentuated deposition in these proximal (channel-marginal) environments (Figure 9.9c). This creates a 'gradient advantage' between the channel and the adjacent floodplain, such that any event that cuts through the levee may induce avulsion, whereby the channel shifts to a lower position on the valley floor. Finally, log-jams may block the primary channel thread, promoting scour of the adjacent floodplain, initiation of a new channel course, and abandonment of the original channel (which subsequently infills with overbank deposits; Figure 9.9d).

The location and timing of avulsions are contingent on local conditions and histories and on the timing/sequence of flood events. Wholesale abandonment of channels and adoption of a new channel course via avulsion processes is recorded through preservation of palaeochannels. As the new channel develops and builds the adjacent floodplain, the palaeochannel is infilled via overbank flows. Analysis of channel size, alignment and bed material size within palaeochannels can be used to make inferences about the behavioural regime of the abandoned channel and associated palaeoenvironmental conditions (e.g. discharge, flow energy).

Floodplain stripping

Floodplain scour or stripping occurs when flow energy at high flow stage becomes sufficiently concentrated such that it is able to remove a layer of alluvium from floodplain pockets. This mechanism is particularly pronounced in partly confined valley settings. The resultant morphology

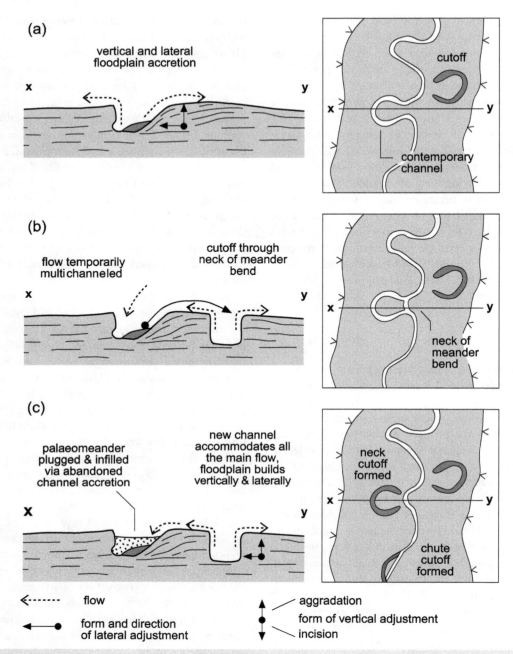

Figure 9.8   Neck and chute cut-off formation. (a, b) During high-magnitude events, narrow necks on bends may erode, ultimately leading to channel abandonment and the formation of neck cut-offs. (c) When flow short-circuits a meander bend over a bar or along a floodchannel, chute cut-offs may form. Modified from Brierley and Fryirs (2005). © John Wiley & Sons, Ltd. Reproduced with permission.

varies depending on the position of the floodplain pocket on the valley floor and the alignment of flow over the pocket. Two common morphologies result: stepped, flat-topped floodplains, which comprise terraces and inset floodplain geomorphic units, and levee–floodchannel floodplains. Vertically accreted mud tends to form a flat-topped

morphology, whereas vertically accreted sands are common in the levee–floodchannel variant. In the accretionary phase, the floodplain progressively builds vertically. The resultant morphology may be relatively flat or inclined towards the valley margin (Figure 9.10a). If the rate of floodplain accretion is greater than the rate of channel bed

(a) Thalweg shift

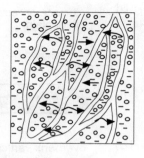

(b) Channel formation and reoccupation

channels preferentially shift position in overbank events

palaeochannels progressively infill in overbank events

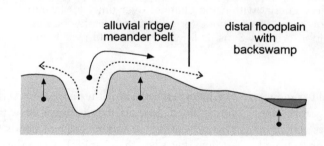

floodplains build vertically with fine grained material

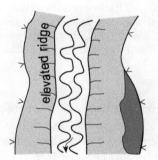

(c) Alluvial ridge-meander belt formation

alluvial ridge/ meander belt

distal floodplain with backswamp

(d) Blockage-induced avulsion

infilling of palaeochannel from overbank and down valley flows

log jam blockage

incision of new channel and vertical accretion of new proximal floodplain

abandonment and infilling of original channel

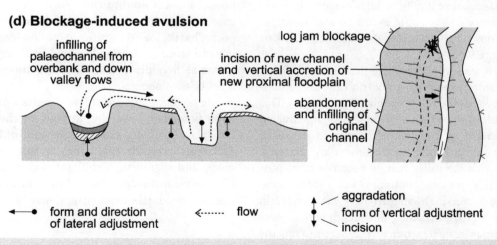

←——• form and direction of lateral adjustment

←------ flow

aggradation

form of vertical adjustment

incision

Figure 9.9 Channel avulsion. Wholesale shifts in channel position can occur in a range of different scenarios. (a) Thalweg shift brings about abandonment of some channels in braidplains. (b) Channels may be cut and reoccupied in multichannelled systems. (c) Development of an alluvial ridge accentuates the lateral gradient into the distal floodplain, eventually promoting the channel to shift to this lower elevation, abandoning the former channel. (d) Log-jams or other obstructions may instigate erosion of a floodchannel that subsequently becomes the dominant channel, abandoning the former course.

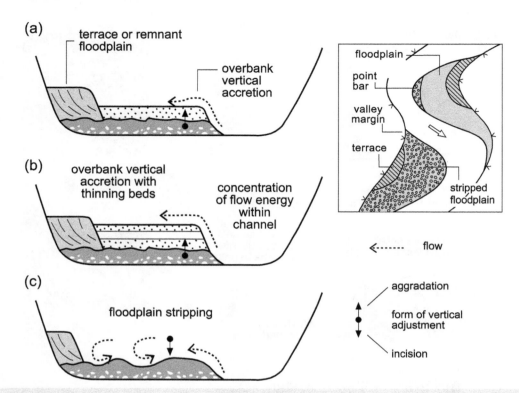

Figure 9.10   Catastrophic floodplain stripping in partly confined valley settings. (a, b) Progressive vertical accumulation of floodplain deposits increases flow energy within the channel over time. (c) Eventually, a significant proportion of the floodplain may be reworked during a major flood (or series of moderate flood events; after Nanson (1986)). From Brierley and Fryirs (2005). © John Wiley & Sons, Ltd. Reproduced with permission.

accretion, progressively larger floods are required to produce overbank deposits (Figure 9.10b). Hence, in general, the thickness of depositional units (beds) decreases vertically. Over time, flow energy becomes increasingly concentrated within the channel. Eventually, a high-magnitude flood event (or sequence of moderate events) reworks (strips) the floodplain down to basal materials, whether bedrock or gravel lag (Figure 9.10c).

Floodchannels tend to occur where there is a significant dip in floodplain morphology away from the contemporary channel towards the valley margin (Figure 9.11a). This can be the result of levee formation or simply accentuated deposition adjacent to the channel relative to the distal floodplain. As the proximal floodplain builds vertically, the incline towards the valley margin becomes more pronounced. As this process continues, subsequent flood events become increasingly aligned over the floodplain pocket (Figure 9.11b). Increasing flow depth within these floodchannels may instigate scour, further accentuating the elevation difference between the levee and the distal floodplain at the valley margin. Floodchannel depth tends to increase down-pocket. In many instances, the basal section of a floodchannel is elevated above the low flow channel

(i.e. it lies perched within the floodplain). Vertically accreted fine-grained deposits may accumulate within the floodchannel during the waning stage of floods (Figure 9.11c). At flow stages less than bankfull, when the entrance to the floodchannel is not breached, suspended-load deposition may occur via backfilling, as the downstream end of the pocket has the lowest entrance to the floodchannel. Reworking and scour in the floodchannel and accretion atop the proximal floodplain accentuate the pronounced lateral relief of the floodplain.

The presence/absence of levees exerts a primary control upon floodplain reworking through mechanisms such as floodplain stripping and the behaviour of floodchannels. This reflects considerations such as valley width (confinement) and alignment. Significant pocket-to-pocket variability in the make-up of floodplain features, and associated channel–floodplain relationships, may be evident.

Floodplain geomorphic units

The type and mix of floodplain formation and reworking processes determines the range and pattern of geomorphic

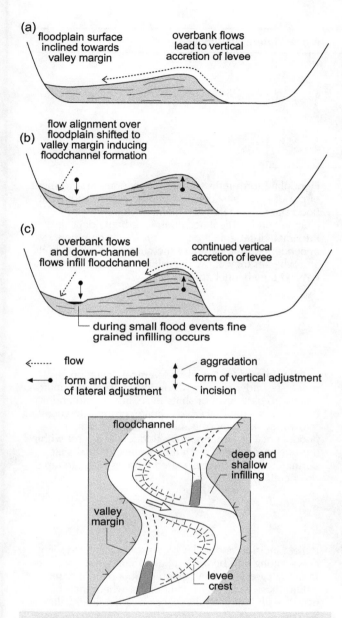

Figure 9.11 Levee–floodchannel formation in partly confined valley settings. (a) Proximal–distal topographic variability is accentuated by levee–floodchannel complexes. (b, c) Over time, floodchannels may scour or be partially infilled. From Brierley and Fryirs (2005). © John Wiley & Sons, Ltd. Reproduced with permission.

units found along any given floodplain. The range of floodplain geomorphic units reflects the type of river under consideration and the history of formative and reworking events. Floodplain geomorphic units are differentiated primarily on the basis of their shape, position, bounding surfaces and depositional sequences. Pronounced differences are evident between floodplains comprised largely of non-

cohesive alluvium (gravel and sand) and those comprised of cohesive alluvium (fine sand, silt and clay).

Floodplains in bedload-dominated rivers tend to occur in isolated pockets in sheltered locations at valley margins (e.g. adjacent to tributary confluences, downstream of fans or shifts in valley alignment). Given the coarse, non-cohesive nature of materials that line the valley floor, floodplain pockets are prone to reworking and are only likely to be preserved in rapidly aggrading environments. By definition, the basal component of floodplains for mixed-load rivers is made up of within-channel (bedload) deposits (also known as bottom stratum deposits). Finer grained overbank (suspended-load) deposits (also known as top stratum deposits) cap these materials. Together, these deposits make up composite banks. In contrast, floodplains made up solely of vertically accreted suspended-load materials are fine grained and cohesive. These deposits impose major constraints upon lateral channel adjustment. Significant pocket-to-pocket variability in floodplain forms and processes may be evident along a river course. Process–form associations for floodplain geomorphic units are summarised in Table 9.1.

Ridge and swale topography records lateral accretion pathways of the channel. Their form is related to the radius of curvature of the bend and associated channel sinuosity (see Table 9.1). Ridge and swale topography is genetically related to *point* and *scroll bars* that form on the insides of bends (Figures 9.2 and 9.6).

Benches are the result of oblique accretion processes that occur within the channel. They form inset floodplains within the channel zone that can eventually be incorporated into the floodplain. Concave bank benches are produced from counterpoint accretion in laterally constrained bends where the bend translates down-valley over time.

Floodplains that are dominated by vertically accreted fine-grained overbank deposits tend to be relatively flat and featureless. However, floodplains with a range in sediment size may have significant topographic variation. For example, *levees* are raised, elongate, prismatic landforms with an asymmetrical cross-section that form at channel margins (Figure 9.11). These wedge-shaped ridges exert a critical control on the pathway and velocity of overbank flows, thereby shaping patterns and rates of sedimentation and reworking of deposits in proximal–distal zones. In general, levee dimensions scale relative to the size of the adjacent channel. They are best developed at the concave bank of bends, where they commonly form steep high banks. Levees are typically comprised of a stacked sequence of upward-fining vertical accretion deposits with rhythmites of sand and silt–clay deposits deposited from decelerating floods. Deposition rate of these flood-couplet deposits is greatest close to the channel and declines down-levee, resulting in the slope into the floodbasin.

Table 9.1 Floodplain geomorphic units[a]

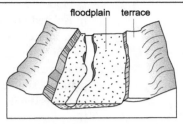

Unit	Form	Process interpretation
Floodplain (alluvial flat)	Lies adjacent to or between active or abandoned channels and the valley margin. Typically tabular and elongated parallel to active channels, but can be highly variable, ranging from featureless, flat-topped forms to inclined forms (typically tilted away from the channel) to irregularly reworked (scoured) forms. Floodplains are the principal sediment storage unit along most rivers. May be coarse grained, fine grained or intercalated. Can be separated into *proximal* (channel-marginal) and *distal* (valley margin) zones.	Floodplain form reflects the contemporary arrangement of out-of-channel sediment build-up and reworking at flood stage. Formed from lateral accretion (within-channel) and vertical accretion (overbank) deposits. Proximal–distal gradation in grain size is common, dependent on the nature of the channel-marginal units and whether they allow deposition of coarse sediments beyond the channel zone.
Alluvial terrace (fill terrace)	Typically a relatively flat (planar), valley marginal feature that is perched above the contemporary channel and/or floodplain. These abandoned floodplains are no longer active. They are separated from the contemporary floodplain by a steep slope called a terrace riser. Flights of terraces are common. They may be paired or unpaired. Terraces may be of great age (e.g. Tertiary terraces are not uncommon). Terraces often confine the contemporary channel, in a manner that is analogous to bedrock valley margins.	Initially formed by lateral and vertical accretion processes under prior flow conditions. Tectonic uplift, a change to base level or shifts in sediment-load and discharge regime (linked to climate) prompt downcutting into valley floor deposits, abandoning the former floodplain. The contemporary floodplain is inset within these terraces. Unpaired terraces reflect lateral shift during incision, whereas paired terraces indicate rapid downcutting only.
Strath terrace	Typically a relatively flat, valley marginal feature that is perched above the contemporary channel or floodplain. These erosional surfaces have a bedrock core, often with a thin alluvial overburden. Strath terraces often confine the channel, analogous to valley margins.	Reflect incision and valley expansion associated with downcutting into bedrock, abandoning terrace surfaces. In many cases, the contemporary floodplain is inset within these terraces. In other cases, where incision occurs with little lateral expansion, a confined valley is formed.

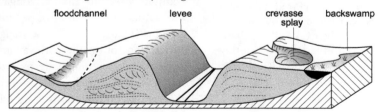

Table 9.1 (*Continued*)

Unit	Form	Process interpretation
Levee	Raised elongate asymmetrical ridge that borders the channel (i.e. along the proximal floodplain). The channel margin is steeper than the floodplain margin. Levees scale in proportion to the adjacent channel. Levee crests may stand several metres above the floodplain surface or be relatively shallow, laterally extensive features. Composed almost entirely of suspended-load sediments (dominantly silt, often sandy).	Levee form is influenced by, and in turn influences, the channel–floodplain linkage of biophysical processes, influencing the lateral transfer of water, sediment, organic matter, etc. Levees are produced primarily from overbank suspended-load deposition at high flood stage. During overbank events, flow energy dissipates when flows spread out over the floodplain. Under these conditions, the flow has insufficient energy to carry its load. The marked reduction in velocity results in deposition of fine-grained bedload and suspended load materials on proximal floodplains. Interbedded flood-cycle deposits, termed flood couplets, reflect rising- and falling-stage sedimentation. Finer materials are carried into the distal parts of the floodplain. Highly developed levees along extensive fine-grained floodplains infer a laterally fixed channel zone and well-defined segregation of water and sediment transfer between the channel and floodbasin. As the levee grows, the deposition rate of coarser sediment near the crest is reduced, leading to a generally fining-upward sequence of deposits within the levee profile.
Crevasse splay (crevasse channel-fill)	A sediment tongue fed by a crevasse channel that breaches the levee. Crevasse splays have a lobate or fan-shaped planform with distal thinning away from the levee. The surface of the crevasse splay may have multiple distributary channels, producing hummocky topography. Composed of bedload material, predominantly sand, sometimes gravel. The crevasse channel fill has a symmetrical, lenticular geometry and low width/depth ratio. Upward-coarsening gradation of grain sizes is common, as is proximal–distal gradation away from the channel.	Crevasse channels breach and erode the levee, taking bedload materials from the primary channel and conveying them onto the floodplain at high flood stage. Deposition reflects the rapid loss of competence beyond the channel zone. Flow velocity is able to carry relatively coarse material, which is spread outward onto a fan-shaped area of floodplain which thins away from the levee. The angle of trajectory increases with high levee backslopes and/or decreases with higher flow velocity. Crevasse channel fills represent bedload plugging of old crevasse channels, indicating an aggradational environment. Their formation may be linked with the formation of an alluvial ridge.
Floodchannel (back channel)	Low-sinuosity subsidiary channel with a defined bed and banks. Entrance height approximates bankfull stage. Commonly observed at valley margins. Floodchannel depth tends to increase down-pocket with the basal section of the floodchannel elevated above the low flow channel (i.e. it lies perched within the floodplain).	Flow alignment along the valley floor short-circuits the channel during high discharge events, steepening the down-valley flow trajectory and inducing scour. At lower flood magnitudes, when the entrance to the floodchannel is not breached, suspended-load deposition may occur via backfilling. Flood channels do not necessarily lead to meander cut-offs, but may situate future (or past) avulsion channels.
Flood runner	Relatively straight depression that occasionally conveys floodwaters. Tends to have a relatively uniform morphology.	Acts like a chute within a depressed tract that short-circuits flow atop floodplain surfaces.
Backswamp (distal floodplain, floodplain wetland, floodpond, floodplain lake)	The distal floodplain, at valley margins, is typically the lowest area of the valley floor. They are major storage units of fine-grained, vertically accreted, suspended-load sediments. Morphology is typically fairly flat (or has low relief), with depressions. Ponds, wetlands and swamps commonly form where lower order tributaries drain directly onto the floodplain.	Forms when the reduction in energy gradient from the proximal to distal floodplain only allows suspended-load materials to be transferred to the backswamp. This results in slow rates of fine-grained vertical accretion in these settings. A distinct gradation in energy with distance from the channel may result in pronounced textural segregation across the floodplain. Backswamps, wetlands, lakes and pond features are common in these poorly drained (unchannelled), low-energy, vertically accreting environments. Dense aquatic/swamp vegetation traps fine-grained suspended-load sediments, promoting accumulation of cohesive, organic-rich mud. Materials are typically highly bioturbated.

(*Continued*)

Table 9.1   *(Continued)*

Unit	Form	Process interpretation

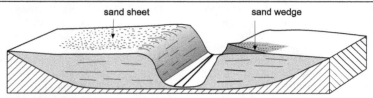

Sand wedge	Sandy deposits with wedge-shaped cross-section at channel margins in non-levee settings. Typically have a scoured basal contact. Cross-beds are transitional to finer grained flood cycle interbeds.	Sand wedges reflect bedload deposition, differentiating them from levees. They form atop the proximal floodplain in moderate–high-energy environments. As flows go overbank, velocity is sufficient to carry relatively coarse material. Energy is spread outward onto a wedge-shaped area of the floodplain, depositing sand.
Floodplain sand sheet	Flat, tabular laterally extensive sheets in non-levee settings with massive, often poorly sorted facies. Show little lateral variation in thickness, mean grain size or internal structure. Surface expression generally conforms to the underlying floodplain. Differentiated from splays by their shape, extensive area, and lack of distal thinning.	Associated with rapid sediment-charged bedload deposition on the floodplain during extreme flood events. Requires competent overbank flows for bedload materials to be deposited on the floodplain in sheetlike forms that cover the entire surface. Deposited as planar, homogeneous sequences. Common in sandy ephemeral streams. Often formed downstream of transitions from confined to unconfined flows and associated with a break in slope (as on alluvial fans). Sand sheets build the floodplain vertically.
Palaeochannel (prior channel, abandoned, ancestral channel)	An old, inactive channel on the floodplain. May be partially or entirely filled. Extends over more than one meander wavelength (thereby differentiating it from a meander cut-off). Can have a wide range of planforms, from elongate and relatively straight to irregular or sinuous, reflecting the morphology of a former primary channel. Low-sinuosity palaeochannels may be overprinted with flood channels. Upward-fining fills typically comprise a channel lag of coarser material with finer, suspended-load materials atop.	Caused by a sudden shift in channel position (avulsion), generally to a zone of lower elevation, abandoning a channel on the floodplain. The palaeochannel may subsequently fill with suspended-load sediments derived from overbank flooding. They record palaeoplanform and geometry of the avulsed channel. If this is markedly different from the contemporary channel, it may indicate a shift in sediment load, discharge or distribution of flood power within the system.
Ridge and swale topography	Ridges are scroll bars that have been incorporated into the floodplain. Swales are the intervening low-flow channels. These arcuate forms have differing radii of curvature, reflecting the pathway of lateral accretion. Ridge and swale topography may indicate phases of palaeo-migration paths, palaeo-curvature and palaeo-widths of channel bends.	During bankfull conditions the high-velocity filament of flow is located along the concave bank of a bend. Helicoidal flow erodes the concave bank of the bend and transfers sediments to the point bar. Eddy flow cells occur in a separation zone along the convex bank. A shear zone between these secondary flow circulation patterns pushes sediments up the point bar face to form a ridge (or scroll bar). At bankfull stage this scroll bar accretes laterally and vertically. As the channel shifts laterally, the scroll bar becomes incorporated into the floodplain, forming ridge and swale topography. Subsequent overbank deposits smooth out the floodplain surface and the former channel position is retained on the inside of the bend.

Table 9.1 (*Continued*)

Unit	Form	Process interpretation
Valley fill (swamp, swampy meadow)	Relatively flat unincised surface. May have ponds and discontinuous channels or drainage lines. Composed of vertically accreted mud. Organic-rich deposits may develop around swampy vegetation. Sand sheets may form downstream of discontinuous gullies.	These sediment storage features are typically formed by flows which lose their velocity and competence as they spread over an intact valley floor, and deposit their sediment load. Vertically accreted swamp deposits are derived by trapping of fine-grained suspended-load sediments around vegetation. Mud beds may alternate with laterally shifting floodout and sand sheet deposits.
Floodout	Lobate/fan-shaped sand body that radiates downstream from an intersection point of a discontinuous channel (i.e. where the channel bed rises to the level of the valley fill). Tend to have a convex cross-profile, and fine in a downstream direction. Comprise sand materials immediately downstream of the intersection point, but may terminate in swamps or marshes as fine-grained sediment accumulates downstream.	Formed when a discontinuous channel supplies sediment to an unincised valley fill surface. Sands are deposited and stored as bedload lobes which radiate from the intersection point of the discontinuous channel. At this point there is a significant loss of flow velocity. Beyond the floodout margin, fine-grained materials are deposited in seepage zones. Deposition associated with breakdown of channelised flow may reflect transmission loss and low channel gradient. Floodout lobes shift over the floor of the valley fill, preferentially infilling lower areas with each sediment pulse.
Meander cutoff (neck cutoff, oxbow, billabong)	A meander bend that has been cut through the neck, leaving an abandoned meander loop on the floodplain. The bends have an arcuate or sinuous planform (generally one meander loop). Horseshoe or semicircular forms are common, reflecting the morphology of the former channel bend. May host standing water (i.e. oxbow lake or billabong) or be infilled with fine-grained materials.	Associated with channel adjustment in meandering streams. Formed by the channel breaching the meander bend (possibly linked to flow obstruction upstream) or through the development of a neck cut-off during high flow conditions. Reductions in sinuosity shorten the stream length, steepening the water slope at flood stage. As the palaeo-meander loop becomes plugged with instream materials the abandoned meander becomes isolated from the main channel. The loop may infill with fine-grained, suspended-load materials and develop into a billabong. These features record the palaeo-planform and geometry of the channel.
Chute cut-off	Straight/gently curved channel that dissects the convex bend of the primary channel, short-circuiting the bend. This may occur through a point bar. This chute subsequently becomes the primary channel. Chute cut-offs have a straighter planform than meander cut-offs.	Short-circuiting of the primary channel reduces sinuosity and stream length, steepening the water slope at flood stage. Concentrated flow with high stream power cut across the bend. Chute cut-off enlargement may result in abandonment of the bend with the chute becoming the primary channel. The old channel bend is filled mostly with bedload deposits. Chute cut-offs generally occur in higher energy settings than meander cut-offs.
Anabranch (secondary channel)	Pattern of coexistent channels that repeatedly bifurcate and rejoin. Given the fine-grained depositional environment, channels have low width/depth ratio. These open channels remain connected to the trunk stream(s).	Formed during high flow conditions events when flow reoccupies and reactivates former channels. Avulsion may also occur. A multichannelled network is retained. These channels are dominated by low-energy, suspended-load deposits.

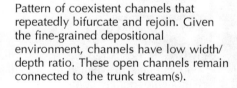

Upward-coarsening sequences reflect increase in bed shear stress in some events. As levees are only occasionally inundated, they commonly have vegetated surfaces. Succession of vegetation from pioneer to riparian forest is common, resulting in increased hydraulic resistance, sediment trapping and added stability. As a consequence, levee deposits may have a significant proportion of roots and organic material. Bioturbation destroys small-scale cross-laminations, resulting in massive (i.e. homogenous) deposits. Levees may induce clear textural segregation between channel and backswamp deposits, whereby coarser materials are deposited on the levee crest and fine-grained suspended-load materials accrete slowly via vertical accretion in the lowest elevation areas of floodplains in *lakes*, *floodponds* or *backswamps* (Figure 9.12). Backswamps generally have a distinctive wetland vegetation association. In some instances peat may accumulate.

As levees build above the floodplain, crevasse splays and floodchannels may form (Figures 9.11 and 9.12a). *Crevasse-splays* are narrow to broad, localised tongues of sediment which are sinuous to lobate in plan. A crevasse channel that cuts the channel-marginal levee feeds the crevasse splay. Once a crevasse channel is initiated, flood waters may deepen the new course and develop a system of distributive channels on the upper slope of the levee. Splays generally extend well beyond the levee toe onto floodbasin deposits. These units become thinner with distance from the channel. Coarser sands may override the fine-grained deposits of a previous crevasse unit, with little evidence of basal scour. Small-scale cross-bedding is dominant, with some small-scale cut-and-fill structures.

Floodchannels and *floodrunners* short-circuit a floodplain pocket by scouring a channel or depression into the floodplain surface. These relatively straight, depressed tracts occasionally convey floodwaters. The entrance height of a floodchannel tends to approximate bankfull stage, while depth tends to increase down-pocket (Figure 9.11). The basal section of floodchannels lies above the low flow channel. Following flood events, backfilling may partially infill these features via deposition of suspended-load materials from ponded water. Scour is better accentuated in floodchannels (i.e. floodrunners can be viewed as potential (incipient) floodchannels).

Levee construction and restriction of a stream to a meander belt may elevate an *alluvial ridge* above the floodplain surface (Figures 9.9 and 9.12b). Perching of the channel above its floodplain enhances the prospect that crevassing and avulsion will occur. In river systems without levees, especially those with shallow channels, there is considerable potential for bedload-calibre deposits to be launched onto the adjacent proximal floodplain in the form of *sheets* or *wedges*.

Vertical accretion deposits also accumulate in *palaeo-channels* and *cut-offs* (Figures 9.8 and 9.9). Typically,

(a) Levee–crevasse splay scenario

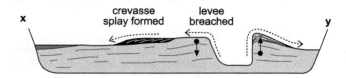

(b) Meander belt scenario

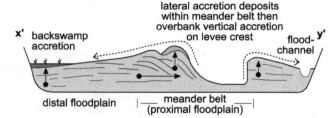

(c) Laterally–stable scenario

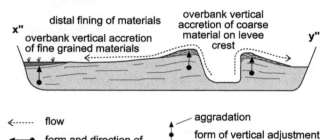

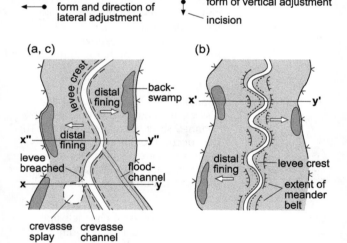

Figure 9.12 Crevasse splays, alluvial ridges and floodchannels as examples of differing forms of channel–floodplain connectivity. (a) Crevasse splays form as levees are breached, launching bedload materials onto the floodplain surface. (b) Accretion of alluvial ridges via meander belt formation may perch the proximal floodplain above the distal floodplain. (c) Backswamps may develop in distal areas of floodplains in relatively wide, laterally unconfined settings. These features may be texturally segregated from the channel, comprising an array of fine-grained suspended-load deposits or even peat. From Brierley and Fryirs (2005). © John Wiley & Sons, Ltd. Reproduced with permission.

upward-fining sequences of gravel and fine sands grade into mud and/or swamp deposits via abandoned channel accretion. Palaeochannels extend over more than one meander wavelength and can have a wide range of planforms, from elongate and relatively straight to irregular or sinuous. Cut-offs are abandoned meander loops. They are generally horseshoe or semicircular in planview. Both these forms result from floodplain reworking processes in the form of either progressive lateral migration or downstream translation of trains of meanders, to cut-offs of varying form (*meander* and *chute cut-offs*).

A distinct set of geomorphic units occurs along unincised river courses where *valley fill* deposits accrete vertically over time. *Ponds* tend to be relatively elongate, scour features formed along preferential drainage lines. *Floodouts* are lobate/fan-shaped depositional features composed largely of bedload-calibre materials that radiate downstream from an intersection point of a discontinuous channel (i.e. where the channel bed rises to the level of the valley fill). These deposits are associated with the breakdown of channelised flow, reflecting transmission loss and low channel gradient. They tend to have a convex cross-profile and fine in a downstream direction. Beyond the floodout margin, fine-grained materials are deposited in seepage zones. Over time, floodout lobes shift over the valley floor, preferentially infilling lower areas with each sediment pulse.

Distinct assemblages of floodplain geomorphic units characterise process–form relationships for different types of channel adjustments (see Chapters 10 and 11). As noted for many other attributes of river systems, the array of floodplain types can be represented along an energy gradient.

The energy spectrum of floodplain types

An array of formation and reworking processes produces different floodplain types in different valley settings. The patterns and rates of floodplain formation and reworking are determined by the stage of floodplain development, the frequency of inundation and the calibre of sediments that make up the floodplain. A spectrum of stream power/energy conditions can be used to differentiate among high-energy non-cohesive floodplains, medium-energy non-cohesive floodplains and low-energy cohesive floodplains.

High-energy non-cohesive floodplains

High-energy non-cohesive floodplains form in headwater confined and partly confined valley settings (see Chapter 10) where there is limited capacity for the channel to shift on the valley floor and unit stream power exceeds $300\,W\,m^{-2}$. Although bedload deposits of sand, gravel and some boul-

ders are the dominant sediments on the valley floor, vertical accretion is the primary floodplain building process. Limited space inhibits the capacity for lateral accretion processes. Floodplains occur as occasional or discontinuous pockets that alternate along the valley floor. Given their high-energy setting within partly confined valleys with steep slopes, floodplain reworking via floodplain stripping, floodchannels and local channel widening is prominent. This often results in stepped, stair-like floodplains. Geomorphic unit assemblages may include levees and floodchannels, with occasional backswamps at distal floodplain locations or sheltered behind bedrock spurs. Gravel and sand sheets launched onto floodplains in non-levee situations are comprised of horizontally or trough/planar cross-bedded deposits that lie atop the proximal floodplain with a scoured basal contact. These sequences grade vertically into finer grained flood cycle deposits consisting of rippled and laminated interbeds.

Medium-energy non-cohesive floodplains

Medium-energy non-cohesive floodplains form in high-energy, moderate to high-slope alluvial settings. Typical planforms include braided, wandering and meandering rivers (see Chapter 10). In these settings there is significant capacity for the channel to adjust on the valley floor. Unit stream power ranges from 10 to $300\,W\,m^{-2}$. Floodplains build from an array of lateral and vertical accretion mechanisms. They characterised by a mix of bedload and intermittently suspended-load deposits, with sediment sequences dominated by sand and gravel with some silt and organic materials. Floodplain reworking is common. Braid channel accretion is the dominant floodplain formation process along braided rivers. Reworking occurs via avulsion, with vertical accretion deposits infilling abandoned (palaeo)channels. Lateral accretion and abandoned channel accretion are the dominant forms of floodplain building along gravel- and sand-bed meandering rivers, while lateral migration and cut-offs are the dominant forms of floodplain reworking. Geomorphic units such as ridges and swales and cut-offs are common. Counterpoint accretion may be evident if lateral migration is constrained. Floodplain geomorphic units include levees, floodchannels, cut-offs, concave bank benches and backswamps. Distal fining of grain size is common. Vertical accretion and abandoned channel accretion are the dominant floodplain building processes along wandering gravel-bed rivers. Significant floodplain reworking occurs via avulsion (creating palaeochannels) and floodchannel processes.

Low-energy cohesive floodplains

Low-energy cohesive floodplains tend to form in laterally unconfined valley settings with low-slopes. Unit stream power in these broad open plains is less than $10\,W\,m^{-2}$. Suspended-load deposits are comprised of fine sand, silt and clay. Given the mud-dominated nature of materials transported in these rivers, thick, uniform sequences of vertically accreted deposits form flat-topped floodplains. Typical geomorphic units include low-lying levees, palaeochannels, swamp wetlands and ponds. In some cases floodplain materials may be organic rich. While there is accommodation space for the channel to shift on the valley floor, the channel is relatively stable, as it flows within cohesive floodplain sediments that have been deposited over long periods of time. As such, these floodplains are produced almost entirely by vertical accretion processes. Given their low-energy condition, floodplain reworking is restricted to occasional avulsion. These floodplains are commonly observed along anastomosing, anabranching and passive meandering rivers, as well as along discontinuous watercourses (see Chapter 10).

Patterns of floodplains along longitudinal profiles

The energy and sediment calibre spectrum of floodplain types provide a convenient way to categorise floodplain forms and processes. However, it must always be recognised that catchment-specific patterns of longitudinal profile (slope) and valley morphology (accommodation space) determine the downstream sequence of floodplain types along any given river. Progressive downstream sequences from high-energy, through medium-energy to low-energy floodplain types may not be evident within any particular river. Also, given the localised occurrence of many floodplain formation and reworking processes, significant *pocket-to-pocket variability* in floodplain forms may be evident. This reflects factors such as variability in valley confinement and local flow alignment at differing flood stages. As such, adjacent floodplain pockets may have quite different proportions of lateral and vertical accretion deposits and be subjected to a wide array of floodplain reworking processes.

Given the importance of site-specific considerations, analysis of floodplain formation and reworking is best achieved through interpretation of the assemblage and connectivity of geomorphic units. This approach to reading the landscape incorporates assessment of process–form relationships that fashion the formation and reworking of individual features, whilst enabling assessment of the range of processes that have fashioned the array of floodplain geomorphic units over time. In this way, contemporary process–form relationships are related to the historical imprint upon riverscapes. Analysis of stacked depositional sequences can provide insight into phases of formation and reworking. The deposits themselves provide a record

of formative processes. Bounding surfaces between geomorphic units provide insight into linkages between these features. Erosional and depositional contacts indicate whether reworking has occurred, or progressive sediment accumulation is evident. Terraces at valley margins, or buried within valley fills, provide a record of former river activity. Most importantly, however, the constructivist, building-block approach to analysis of river systems relates contemporary process relationships in the channel zone to process activity on floodplains, thereby providing insight into river character, behaviour and evolution (see Chapters 10–12).

Conclusion

Efforts to read the landscape build upon an appreciation of process differentiation in channel and floodplain compartments, and understanding of transition zones that affect the relationships between these compartments in river systems. Hence, interpretation of the pattern and extent of floodplains provides an initial point of entry into analysis of riverscapes. Transitions from sections of river course in which floodplains are absent, occasional, discontinuous (alternating/recurrent) or continuous provide critical insight into the process domain of a river. Key factors that fashion these transitions are downstream changes in slope and lateral constraints on the river (i.e. valley confinement). Collectively, these factors fashion the available energy of the river and the way in which that energy is used along any given reach.

The assemblage of geomorphic units can be interpreted to provide guidance into the formative and reworking processes that generated any given floodplain. Floodplains formed by lateral and vertical accretion mechanisms have distinct morphologies and depositional sequences. Patterns of geomorphic units provide insight into mechanisms of channel adjustment on the valley floor, and associated reworking of floodplain features. The presence or absence of levee features acts as a key control on channel–floodplain connectivity. Thalweg shift and channel abandonment are dominant floodplain formation mechanisms in high-energy settings. Along actively migrating meandering rivers, progressive lateral channel migration is evident in some situations, while downstream translation of bends is dominant elsewhere. Passive meandering or anastomosing rivers are characterised by stable channels and vertically accreting fine-grained floodplains in low-energy settings. Immense diversity in river character and behaviour can be discerned based on the interactive role of channel and floodplain compartments. These mutual interactions are fashioned by both contemporary and historical considerations.

Key messages from this chapter

- Floodplains are defined as areas of sediment accumulation made up of river deposits (alluvial materials) between the channel bank and the valley margin.
- Floodplains are formed by a different set of processes to those occurring in the channel. Textural segregation is evident, differentiating sediments transported along the channel (bedload-dominated) from suspended-load deposits that make up the floodplain.
- Channel marginal elements separate channel and floodplain processes, providing insights into the lateral connectivity of the system (i.e. frequency and nature of inundation; i.e. whether depositional or erosive events). Marked differences may be evident between proximal and distal floodplain forms.
- Floodplains can occur as isolated pockets (typically in the source or transfer zone in the middle–upper catchment), discontinuous pockets (typically in the transfer zone) or as continuous features along both channel banks (typically in accumulation zones).
- Significant pocket-to-pocket variability in floodplain forms may be evident, reflecting a range of localised controls such as changes in the nature/degree of valley confinement or differing flow alignments over floodplain surfaces.
- Lateral and vertical accretion processes are the primary forms of floodplain formation. In certain circumstances and along certain types of rivers, other formation processes such as braid channel accretion, braid channel accretion, oblique accretion, abandoned channel accretion and counterpoint accretion may be prominent.
- Floodplain reworking processes include lateral migration and cut-off formation, avulsion, floodplain stripping and floodchannel scour.
- A mix of floodplain formation and reworking processes dictates the shape and structure of floodplains. As a result, floodplain morphology is not always flat. A mosaic of depositional and erosional forms can occur. In many instances a lateral gradient is evident from levees at the channel bank (the proximal floodplain) to backswamps adjacent to the valley margin (the distal floodplain).
- Floodplain forms and processes occur along an energy continuum from high-energy non-cohesive floodplains to low-energy cohesive floodplains. Distinct assemblages of geomorphic units occur along differing types of floodplains.
- Analysis of the assemblage of geomorphic units on any given floodplain is a key component in efforts to read the landscape, as these features provide insight into both the contemporary process regime of the river and the imprint of historical (evolutionary) influences.

CHAPTER TEN

River diversity

Introduction

Although rivers are made up of relatively simple components, as flow moves over and around sediment and/or bedrock and interacts with vegetation, process–form interactions generate a bewildering array of river types. This remarkable diversity can be attributed to the wide range of environmental settings, and associated combinations of boundary conditions fashioned by geologic, topographic, climatic, biogeographic and anthropogenic factors, among many considerations. Effective description of river character is a fundamental prerequisite in efforts to meaningfully explain why a river looks and behaves in the way that it does. Reaches are made of assemblages of geomorphic units. Analysis of process–form relationships at the geomorphic unit scale, alongside interpretations of interactions among geomorphic units at the reach scale, can be used to appraise the range of processes and events that determine the range of behaviour for any given reach (Chapter 11). Each reach should be viewed in its catchment context. This chapter pulls together the various components of river systems outlined in previous chapters to provide a meaningful basis to interpret river character. This is a fundamental component of efforts to read the landscape.

Figure 10.1 shows examples of a range of rivers. These rivers fall along a continuum from steep upland settings to relatively flat lowland plains, from confined valley settings to open valley floors. While some of these river types may be considered representative or common (e.g. gorges and braided rivers), others may be considered uncommon or 'unusual' (bedrock-anastomosing, meandering-anabranching or chain-of-ponds). The latter have characteristics that span several 'types' along the continuum of river diversity. The key to reading the landscape is to characterise rivers within an open-ended approach to analysis. This chapter provides a framework with which to analyse river diversity in any field situation. This is achieved through a set of questions to be asked when analysing a river, whether in the field, from aerial photographs or satellite images, or on a map. For example, does the river have a single channel, multiple channels or no channel? Are the channels relatively straight, sinuous, irregular or discontinuous? Is there a floodplain, along one or both banks of the channel? Is the floodplain relatively flat and featureless, or does it provide a record of past adjustments (e.g. abandoned channels or a different size, shape and sinuosity)? The process of interpretation outlined here moves beyond prescriptive categorisation of rivers based upon notionally representative 'types' towards a set of principles that enables any riverscape to be 'pieced together' and interpreted using concepts described earlier in this book. This allows for unique river types to be identified alongside more common variants.

The continuum of river types occur from bedrock to fully alluvial variants occurs along a slope, bed material calibre and energy gradient. Confined rivers in steep upland settings reflect an imposed condition with a slope-induced gradient of sculpted instream geomorphic units, while floodplains are absent (or are only found as isolated pockets). Partly confined rivers contain discontinuous floodplain pockets that are either bedrock- or planform-controlled. Rivers in accumulation zones freely adjust within their own sediments in laterally unconfined valley settings. These alluvial rivers are assessed in terms of their channel planform, defined as the configuration of a river in plan view. Channel planform is measured in terms of (a) number of channels, (b) sinuosity and (c) lateral stability. Six primary planform types are characterised here: boulder-bed, braided, wandering gravel-bed, meandering, anastomosing and discontinuous watercourses. However, the key premise in this chapter is that there is no magic number of river types. The approach to analysis of river character outlined here entails interpretation of the assemblage of

Confined valley

Partly confined valley

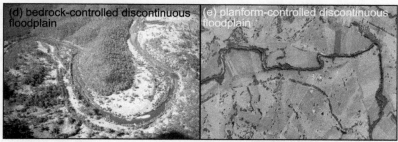

Laterally unconfined valley with continuous channel

Laterally unconfined valley with discontinuous channel

Figure 10.1 The diversity of river types (See Colour Plate 2). A range of rivers is found in different valley settings, ranging from fully confined rivers to partly confined rivers with pockets of floodplain, to freely adjusting alluvial (or laterally unconfined) rivers. Discontinuous watercourses also occur in some settings. (a) Steep headwater in New Zealand; photograph: D. White. (b) Gorge in Grand Canyon, USA; photograph: P. Chappell. (c) Gorge in south-west Tasmania, Franklin River, Australia; photograph: K. Fryirs. (d) Bedrock-controlled discontinuous floodplain, Clarence River, NSW; photograph: R. Ferguson. (e) Bedrock-controlled discontinuous floodplain, Williams River, NSW; photograph © Google Earth 2011. (f) Braided river, New Zealand; photograph: D. White. (g) Anastomosing river, Cooper Creek, central Australia; photograph: G. Nanson. (h) Bedrock-based anastomosing river, Sabie River, Kruger National Park, South Africa; photograph: G. Brierley. (i) Meandering-anabranching river, tributary of Upper Yellow River, China; photograph: G. Brierley. (j) Chain-of-ponds, Macquarie Marshes, NSW; photograph: K. Fryirs. (k) Upland swamp, Budderoo National Park, NSW; photograph: K. Fryirs.

geomorphic units within any given reach. As geomorphic units are products of erosional and/or depositional processes, analysis of these features provides a consistent approach to differentiation of river types across the spectrum of environmental settings.

This chapter is structured as follows:

1. The cross-scalar approach to analysis of rivers in this book is revisited. Insights from previous chapters on bed material size, bedforms and channel and floodplain geomorphic units are brought together in a framework that can be used to analyse any given river.
2. The diversity of rivers is framed along a continuum that extends from bedrock-dominated erosional landscapes through to alluvial rivers.
3. The character of bedrock and partly confined rivers is outlined.
4. The range of alluvial rivers is documented in relation to measures of channel planform.
5. The constructivist approach to river analysis is outlined, viewing reaches as assemblages of channel and floodplain geomorphic units.
6. Discriminating functions used to differentiate among river types are summarised.
7. A brief description of a generic approach to analyis of river systems, the River Styles framework, is provided.

Framing rivers as assemblages of cross-scalar features

As noted in Chapter 2, river systems can be differentiated into a nested hierarchy of features across various scales. Features observed and analysed at differing scales offer different insights into how a river system looks and works. This section draws together several of these threads to develop a coherent set of guidelines with which to interpret river diversity.

Rivers operate along an energy continuum from their headwaters to the mouth. However, the gradation in process types and dominance is not always smooth and unidirectional. Patterns of features may be disrupted or repeated. Irregular or anomalous features may be observed at a particular location, whether as a product of a unique set of circumstances or perhaps as a product of the river's history. There is inordinate complexity in the way that different scales of features can combine and interact at different positions within a catchment.

Geomorphic features at differing scales can be brought together to characterise river reaches. As outlined in Chapters 5 and 6, a range of bed material sizes and bedform attributes reflect the interaction of flow of a particular energy acting upon available materials on the channel bed.

The type and range of features observed are indicative of either the flow conditions at that time or they may be inherited from previous higher flow conditions. Geomorphic units are made up of differing assemblages of these small-scale features, wherein flow energy and alignment within the channel influence the nature and distribution of bed material sizes and bedform features. Morphodynamic relationships fashion the mutual interactions of process–form linkages that determine the character, behaviour and pattern of geomorphic units. In general terms, the energy gradient of erosional and depositional forms is marked by differing combinations of channel and floodplain components along a river course. The gradient from bedrock channels in source zones through to laterally unconfined alluvial settings in accumulation zones results in quite differing assemblages of features. However, this is not always a deterministically prescribed set of downstream relationships, and anomalies are common.

As differing scales of features provide differing insight into river character and behaviour, their interpretation is useful for different purposes (Chapter 6). Although bed material size and bedform assemblages are fundamental considerations in analysis of sediment transport, their configuration along the bed at any given time is largely a function of recent flow events. As such, analysis of these features does not provide fundamental guidance into controls upon river diversity and patterns of river types in any given catchment. Reach-scale analysis of river diversity, framed in terms of the assemblage of channel and floodplain geomorphic units along a reach, provides the most insightful basis to interpret river character and behaviour. As all rivers are comprised of differing combinations of erosional and depositional geomorphic units, these features provide a unifying element with which to analyse all types of river.

Defining reach boundaries

A reach is defined as a length of river that operates under relatively consistent and characteristic boundary conditions such that a relatively uniform morphology results. This may extend over several hundreds of metres or over tens or hundreds of kilometres of river length. There is no specific length of river that defines a reach. While it is relatively easy to define the boundaries of some reaches, the designation of boundaries elsewhere may prompt significant controversy, conjecture and uncertainty. The transition from a gorge to an alluvial river may be relatively abrupt, but differentiation of reaches based on the pattern of sediment stores within partly confined valleys may be contentious. Inevitably, rivers adjust to disturbance events over differing timeframes. In some instances the position of reach boundaries may change frequently. Examples include the operation of headcuts or the downstream movement of sediment slugs. In general terms, however, reaches are defined by the assemblage and pattern of geomorphic units. Although the position of these features may vary over time, the assemblage of features remains consistent unless there is a change in river type (see Chapters 11 and 12). If a reach boundary is to be important then it should delimit adjustments not only in terms of river character, but also in terms of its behaviour. Hence, the transition from a reach without a floodplain to a reach with a floodplain typically demonstrates a shift in the way flow energy is utilised by a river. In all instances, practitioners must be able to articulate specifically what is different upstream and downstream of a boundary, and why this is important.

Figure 10.2 demonstrates different types of reach boundaries. The abrupt boundary shown in Figure 10.2a represents a fundamental shift in river character and behaviour brought about by a dramatic change in valley width. Where there is a gradual, progressive adjustment in river character and behaviour, such that river character and behaviour are quite different either side of the boundary, it is not always easy to define a specific point at which the transition occurs. In these instances the boundary is gradual or diffuse, and should be conveyed as a transition zone. A different situation is shown in Figure 10.2b. In this instance there is a repeated pattern of features along relatively small-scale lengths of river, with notable boundaries between these various sections of river. These sub-reaches

can be meaningfully designated as a segment (i.e. repeated reaches). The example shown would be defined as alternating reaches of gorge and partly confined valleys with bedrock-controlled discontinuous floodplain.

As noted previously, the key to effective description of a river system lies in meaningfully capturing differences in river character and behaviour as one moves downstream, not only in terms of the types of river noted, but also in terms of the type of transition (whether abrupt or gradual). Analysis of the location of reach boundaries helps interpret controls upon downstream patterns of rivers. For example, transitions in river type could reflect changes in geology and, hence, slope and valley confinement. Alternatively, reach boundaries may reflect altered balances of flow and sediment downstream of tributary confluences. As noted recurrently, transition in river character and behaviour typically indicate adjustments to the balance of erosional and depositional processes along a river. These transitions can be appraised in broad terms in relation to position within a catchment. In essence, identification of reach boundaries is an artefact of the practical desire to break the continuum of river forms and processes into meaningful units.

The continuum of river form

The spectrum of river diversity can be considered as a continuum extending from the colluvial interface at which sediments begin to accumulate and flow becomes

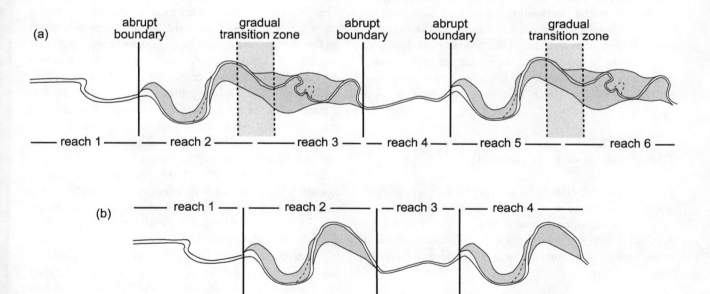

Figure 10.2 Defining reaches and segments. (a) A reach is defined as a length of river with a near-uniform character and behaviour. (b) A segment is an alternating (repeating) sequence of reaches.

channelised on valley floors through to coastal margins or inland lakes where rivers terminate. Several key transitions in river type typically occur along this continuum:

1. The point at which colluvial (hillslope) processes and deposits can be differentiated from river processes and reworked sediments.
2. The point at which a definable slot channel is cut within bedrock.
3. The point where sediment accumulation occurs on the channel bed.
4. The point where distinct floodplain pockets occur along valley margins, delineating the creation of sediment stores outside the channel such that bed and bank deposits are evident.
5. The point when floodplains occur along both banks of a channel, such that bedrock no longer exerts a discernible influence upon contemporary channel processes. However, channel behaviour may be affected by older sediments stored on the valley floor, whether in the form of floodplain deposits laid down under a different set of conditions to those experienced today or older terraces that may induce a particular river configuration.
6. The point at which the unidirectional (downslope) of rivers interacts with tidal flows at coastal margins (i.e. the tidal limit).

Essentially, these transitions reflect downstream changes in the balance of erosional and depositional processes along the course of a river. As noted on the Lane balance (see Chapter 5), a fundamental transition is demarcated by the shift from the degradational part of a catchment in which erosional processes are dominant to the aggradational section of a river in which depositional processes predominate. In broad terms, the former setting is characterised by bedrock rivers, while the latter setting comprises alluvial rivers (i.e. rivers that flow within their own deposits).

The spectrum of river diversity

This section briefly describes an array of river types found along the continuum of stream power/slope, sediment calibre and valley confinement that shapes patterns of deposition and the potential for reworking of sediment stores along river courses. There is considerable overlap among variants.

Valley settings

Valley confinement is a primary control on the differentiation of sediment source, transfer and accumulation zones along rivers (see Chapter 3). Topographic controls such as slope and valley confinement influence the capacity for sediment storage or reworking along a reach. Bedrock rivers tend to occur in the incisional, degrading parts of landscapes, typically characterised by long-term sediment source or transfer zones. Structural and lithologic factors induce lateral and vertical control on river character and behaviour. Local-scale forcing elements impose major constraints on river processes and resulting forms. Moving downstream, valley confinement fashions the formation and reworking of floodplain types. Different types of rivers are found in confined (no floodplain), partly confined (discontinuous floodplain) and laterally unconfined (continuous floodplain) valley settings (Chapter 9).

Bedrock or terraces are observed along both channel banks over more than 90% of reach length in *confined valley settings*. Bedrock also imposes a vertical control on the capacity for channel adjustment, as bedrock lines the channel bed. The river course has either no floodplain or floodplains are restricted to isolated pockets (<10% of reach length). Channel planform is imposed by valley configuration. For example, if long-term landscape evolution has resulted in a deeply incised and sinuous bedrock valley, the channel must conform to this configuration, producing a gorge. Elsewhere, gorges may be straight, as they follow the geologic structure of a region (e.g. along fault lines). In other instances, the channel can be fully contained within terraces or ancient, cemented alluvial deposits that line the valley margin.

Given their steep slopes, confined rivers tend to have high stream power and high sediment transport capacities. Channels are strongly coupled to adjacent hillslopes, which act as major sources of sediment. The exposure of bedrock on the channel bed reflects high transport capacity relative to sediment supply. Deep, narrow cross-sections encourage macroturbulent flow and cavitation during floods. Mobile bedload or suspended-load materials are readily flushed through these reaches. Large particles that line the bed exert significant resistance, impacting upon river character and behaviour. In some instances these materials are only mobilised during catastrophic events.

Between 10 and 90% of the channel abuts directly against bedrock or ancient, cohesive materials in *partly confined valley settings*. Bedrock also imposes significant base-level control, with outcrops common along the bed. Discrete floodplain pockets commonly occur on alternating sides of the channel. The position of the channel relative to the valley margin determines the differentiation of river types in partly confined valley settings. This influences how often and over what length of river course the channel impinges on the valley margin. The distribution of floodplain pockets may be bedrock-controlled or planform-controlled. Valley morphology is a primary influence on the distribution of specific (unit) stream power at differing flow stages, thereby shaping patterns of sediment deposition and reworking along valley floors. Remarkable pocket-to-pocket variability in floodplain sediments may be evident. The assemblage of geomorphic units on any given floodplain pocket is largely determined by downstream changes in valley configuration and flow alignment.

Less than 10% of the channel margin abuts against bedrock or terrace features in *laterally unconfined valley settings*. Alluvial channels are laterally unconstrained, flowing atop and within their own deposits with continuous floodplains along both channel banks. Banks are deformable, such that the channel is able to mould and rework its boundaries. In many instances channels have significant capacity to adjust on the valley floor. Rivers in laterally unconfined valley settings are differentiated in terms of planform attributes (number of channels, sinuosity and capacity for adjustment on the valley floor). Reaches with a continuous channel are differentiated from those where the channel is discontinuous or absent.

Rivers in confined valley settings

Steep headwater rivers

Steep headwater areas are the sediment source zone of river systems. Many mountainous areas are shaped by hillslope and glacial activity, delivering large volumes of sediment to the valley floor in highly dissected and coupled landscapes (Figure 10.3). Given the confined valley setting, most sediments are flushed downstream efficiently. Channel processes accentuate incision into the bedrock valley floor, enhancing the imposed river condition. Forests often contribute high loadings of wood into narrow streams, acting as an additional forcing element upon river morphology. Steep slopes induce a gradation of downstream geomorphic units extending from waterfalls, bedrock steps, rapids, cascades and bedrock pools into run/glide and imposed pool–riffle sequences (see Chapter 7). The site-specific assemblage of these features is determined by local variability in channel slope and the distribution of forcing elements.

Gorges

Gorges are found in some of the world's most spectacular landscapes. Most gorges are associated with one of two scenarios. In some settings, rivers 'cut back' into bedrock settings via headward retreat mechanisms. Knickpoints are demarcated by significant waterfalls (Figure 10.4). This typically occurs when tectonic plate margins pull apart, and streams cut into terrain as they respond to base level changes. Elsewhere, geologically uplifted blocks create superimposed drainage networks in which streams cut down through uplifting terrain, often in plateau settings.

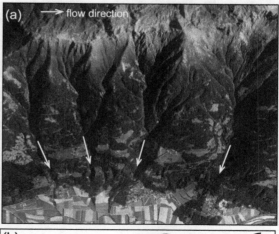

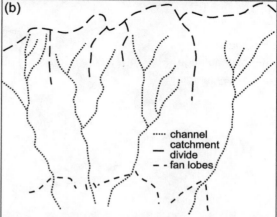

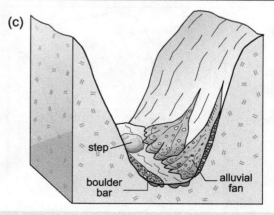

Figure 10.3 Steep headwater. (a, b) Planform view of the Tyrol Mountains in Austria from © Google Earth 2011. Note limited sediment stores on the bedrock valley floor, with occasional inputs of materials from coupled hillslopes. (c) Block diagram of a steep headwater river.

In both instances a confined, slot-like channel results. As with steep headwater streams, concentration of flow energy promotes bed incision over valley expansion. However, gorges are not necessarily located in highly

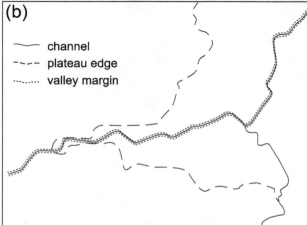

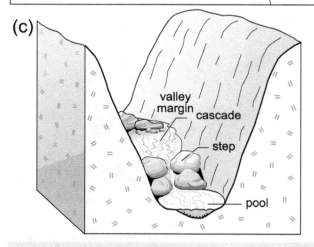

Figure 10.4   Gorge. (a, b) Planform view of the Kangaroo River at Carrington Falls, NSW, Australia. Note the imposed geomorphic structure within this confined valley setting. Discontinuous sediment stores line the valley floor. (c) Block diagram of a gorge river.

connected landscapes, as hillslopes are often set back from the gorge by a plateau. As such, the channel is primarily 'fed' by sediments from upstream. Hence, volumes of sediment delivered to the channel are less per unit catchment area relative to headwater settings. Despite the lower slopes that may prevail in these areas, the large catchment area of many gorges generates sufficient discharge and associated specific stream power to flush most sediments through these reaches. Typical geomorphic units include waterfalls, cascades, rapids and elongate runs and plunge pools. Bedrock forcing exerts a key control upon the morphology, position and patterns of these features. Plunge pools form as scour features downstream of steps, waterfalls or localised inputs of coarse sediment from hillslopes or small tributaries. Rapids comprising coarse boulders form from local influx of materials from tributaries or mass movement processes on adjacent hillslopes. These features exert a clear impact on the longitudinal profile. Bedrock steps may represent secondary knickpoints that act as local base-level controls. If subjected to significant sediment supply, beds can temporarily store materials in bars or behind forcing elements and flow obstructions such as wood.

Confined valley with occasional floodplain pockets rivers

Discrete pockets of floodplain may be evident in some confined valley settings (Figure 10.5). These are typically associated with tributary confluence zones or abrupt shifts in valley alignment. Channel zones may comprise bedrock steps, scour pools and runs. Shallow floodplain pockets may have coarse-grained levees, sand and gravel splays, chute channels, scour holes and abandoned channels covered by thin overbank deposits of fine-grained alluvium. These non-cohesive, high-energy floodplains are formed by a combination of lateral, vertical and abandoned channel accretion processes (see Chapter 9). Given the confined valley setting, deposits are prone to stripping or reworking by chute cutting or channel avulsion. Abandoned channels may infill with coarse sediments.

Rivers in partly confined valley settings

As valleys widen, space is provided for sediments to be deposited along valley floors, separating the channel from floodplain pockets. This results in textural segregation, with the channel comprising the coarser bedload fraction, while the high-energy floodplain is made up of both bedload and suspended load materials (see Chapter 9). Critically however, bedrock continues to exert a prominent influence upon river forms and processes in partly confined valleys. This influence tends to diminish downstream,

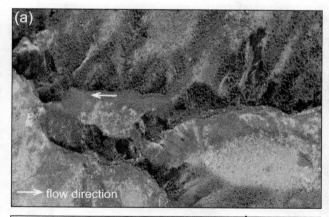

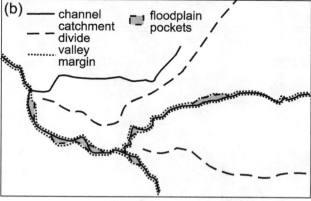

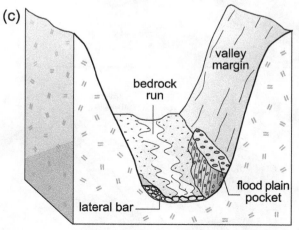

Figure 10.5 Confined valley with occasional floodplain pockets. (a, b) Planform view from near Machu Picchu, Peru (from © Google Earth 2011). Floodplain pockets typically form along areas of local valley widening, or immediately downstream of changes in valley alignment. These local areas are prone to river adjustment, while the geomorphic structure of confined areas are bedrock controlled (i.e. they have a forced morphology). (c) Block diagram of a confined valley with occasional floodplain pockets river.

marking a transition from bedrock- to planform-controlled floodplain pockets.

Rivers in partly confined valley settings with bedrock-controlled discontinuous floodplains

Rivers with bedrock-controlled discontinuous floodplains tend to be found in sinuous valleys where bedrock spurs on the insides of bends protrude onto the valley floor (Figure 10.6a). The channel is forced to flow along the outsides of bends, with the channel running adjacent to the valley margin along 50–90 % of its length. Floodplain pockets tend to form in sheltered areas of low flow velocity on the insides and downstream sections of bends. These pockets often alternate along the valley floor.

Given the prominent bedrock influence, channels in partly confined valleys with bedrock-controlled discontinuous floodplains tend to have an irregular shape with steps, boulder bars and bedrock-controlled pools commonly observed. Deep bedrock-controlled pools may be induced by enhanced scour at flood stage. Compound point bars with ramps, chute channels, ridges and sheets are formed and recurrently reworked on the insides of bends. Floodplain pockets often emerge from boulder bars deposited on the downstream end of bedrock-controlled bends. Vertically accreted medium- to fine-grained gravel and mud sediments typically lie atop boulder bar deposits. This relatively thin cap of deposits flattens off the floodplain surface. Although within-channel deposits make up the majority of the floodplain, overbank deposits are an important additional feature. Floodplain reworking is common in these partly confined valleys. Depending upon the stage of floodplain development, the surface may be irregular (the boulder bar itself) or relatively flat (if sufficient finer grained overbank sediments have been deposited). Vegetation cover may exert a significant influence upon these relationships.

Rivers in partly confined valleys with planform-controlled discontinuous floodplains

Inevitably, as valleys widen, the floodplain becomes an increasingly dominant proportion of the valley floor and the ratio of channel width to valley width decreases. This marks a progressive reduction in the influence of bedrock control on channel and floodplain processes. In partly confined valleys with planform-controlled discontinuous floodplains there is a mix of imposed bedrock influences with a self-adjusting channel and floodplain. These valleys are more open than their bedrock-controlled counterparts. The channel impinges against a valley margin along between 10–50 % of its length. The nature and distribution of discontinuous floodplain pockets is determined by how the channel has shifted on the valley floor (i.e. its

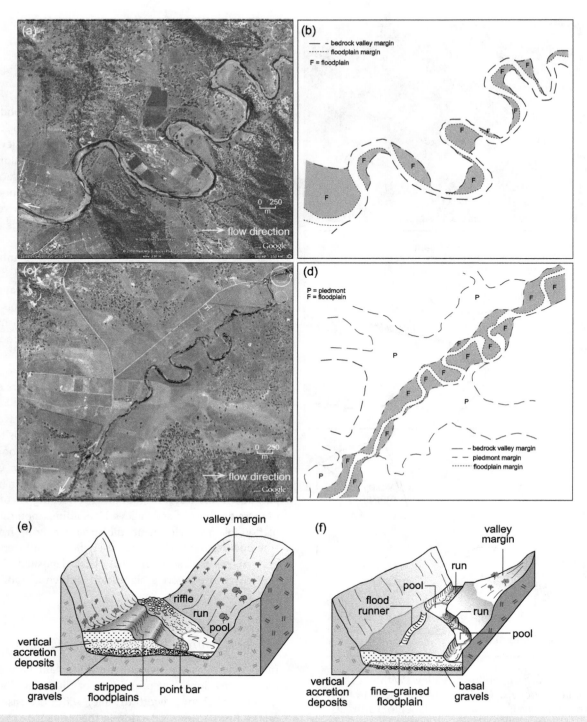

Figure 10.6 Rivers in partly confined valley settings. (a, b) Bedrock-controlled discontinuous floodplain pockets typically form at sheltered locations in sinuous valleys. Example from Pages River, NSW, Australia (from © Google Earth 2011). (c, d) The distribution of planform-controlled discontinuous floodplain pockets is determined by the position of the channel on the valley floor and its planform. This meandering planform-controlled variant is from Dart Brook, NSW, Australia (from © Google Earth 2011). (e, f) Block diagrams of bedrock-controlled and planform-controlled variants respectively.

planform). Two planform variants are commonly observed: meandering and low sinuosity (see Figure 10.6c and f). Partly confined valleys with meandering planform-controlled floodplains are characterised by a sinuous channel with bends that touch the valley margin in a systematic (alternating) or irregular manner. Although the channel has a meandering outline, lateral migration is not a major component of river behaviour. Rather, lateral confinement imposes downstream translation of bends, whereby obstructions at concave margins result in the formation of concave bank benches at the outside of bends and point bar deposits on the inside of bends. Channels tend to be asymmetrical at bends (shallow on the convex slope, deeper adjacent to the concave bank) and uniform (symmetrical) at points of inflection. Floodplain pockets are comprised of both within-channel and vertical accretion deposits. Reworking tends to be restricted to cut-offs and floodchannels in bends.

Some channels in partly confined valleys have a low-sinuosity planform, where the channel hugs one valley margin for a while and then abruptly shifts to the opposite valley margin. This results in an alternating sequence of floodplain pockets. The channel is typically stepped (i.e. compound) with benches and/or ledges prominent. The channel bed is often relatively smooth, with elongate shallow pools, numerous glides/runs, poorly defined (often alternating) bars and occasional riffles. Sandy substrate conditions promote relatively flat-topped floodplain pockets with sand sheets prominent.

Alluvial rivers

Alluvial rivers are self-adjusting channels that flow within river-borne sediments in laterally unconfined valley settings. The active channel zone can be clearly separated from floodplains along *both* channel margins. Distinction can be made between systems in which channels are continuous (whether this is a single- or multiple-channelled system) and those in which channels are either absent or are discontinuous. The latter category of river is referred to as a cut-and-fill river system. The 'fill phase' represents a period of sediment accumulation in which channels are absent or discontinuous and fine-grained materials accrete vertically on the valley floor. This phase is often characterised by a wetland or swamp condition. Eventually, an incised channel cuts through these deposits (the 'cut phase', sometimes referred to as a channelised fill), but this channel commonly refills and the valley floor is transformed back to an intact surface once more.

Alluvial rivers with continuous channels are differentiated on the basis of their planform, defined as the configuration of a river in plan view. Three attributes are used to measure channel planform: number of channels, channel sinuosity and lateral stability of the channel(s).

Number of channels

Although assessment of the number of channels sounds intuitively very straightforward, several issues must be considered. What length of river should be analysed? How regularly (and at what interval) should measurements be made? As shown in Figure 10.7, analysis of the number of channels is best completed at the reach scale. Within any given reach, systematic measurements may be taken at a particular interval (say 10 measurements regularly spaced along the reach) and the average taken. Alternatively, the number of channels can be characterised in descriptive terms, such as mainly single channelled, with isolated instances where flow divides into two or three channels. Typically, differentiation of the number of channels is made between dominantly single-channelled, up to three channels and more than three channels.

Sinuosity

Channel sinuosity P is measured as the ratio of channel length λ to valley length Z for a representative reach of river (see Figure 10.8a). Within-reach variation in sinuosity should be recorded, as this aids interpretation of the stage of channel adjustment or the local-scale capacity for adjustment of the channel. Four classes of sinuosity are used (1–1.05 is straight, 1.06–1.30 is low sinuosity, 1.31–1.80 is sinuous and >1.80 is tortuous; Figure 10.8b). Tortuous meanders cut back upon themselves along the valley floor. Care should be taken when measuring sinuosity in partly confined or confined rivers (Figure 10.8a). Although the channel looks sinuous, it is in fact low sinuosity or straight because channel length is only marginally greater than valley length. It is important to remember that confined meanders are not alluvial rivers. Rather, the meandering channel outline in these settings is imposed by the shape of the bedrock valley or older alluvial sediments (e.g. terraces).

Various terms are used to describe different meandering patterns (Figure 10.8c). Smooth, systematic meanders indicate active and progressive lateral or downstream adjustment of the channel on the valley floor and are referred to as regular meanders. Channel adjustments in these mixed-load rivers include ridge and swale topography and chute and neck cut-offs (see Chapter 9). Elsewhere, the channel may be relatively inactive or passive, with little evidence of channel adjustment (i.e. features such as cut-offs or ridge and swale topography are absent). The channel may have relatively low sinuosity, but it could also demonstrate irregular or tortuous meandering patterns. The floodplain is typically relatively flat and featureless. Occasional

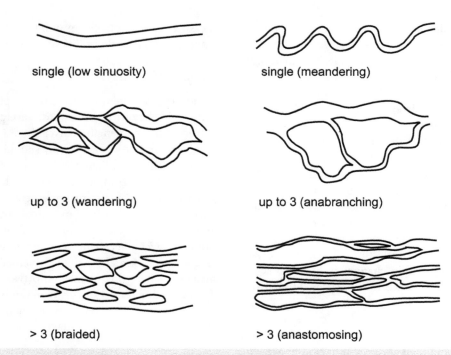

single (low sinuosity)   single (meandering)

up to 3 (wandering)   up to 3 (anabranching)

> 3 (braided)   > 3 (anastomosing)

Figure 10.7   Classification of the number of channels used in analysis of channel planform.

floodchannels may short-circuit bends. Near-vertical banks in these suspended-load rivers are comprised of cohesive, fine-grained overbank deposits. These are passive meandering rivers.

Some difficulties may be faced in measuring and analysing sinuosity for multichannelled rivers. In these instances, channel sinuosity is assessed for the dominant channel thread (i.e. the thalweg). This is not always easy, as numerous channels may have relatively equivalent flow. Once more, effective description is the key to reliable interpretation. If most channel threads have low sinuosity, but occasional threads are quite sinuous, the situation should be described as such. In these instances, an averaged-out statistical descriptor is meaningless.

Lateral stability

The third measure of channel planform, and the one that is most difficult to assess, is lateral stability (Figure 10.9). This measure defines how a channel adjusts on the valley floor. As such, it is a key indicator of channel behaviour (Chapter 11). Five forms of lateral stability are differentiated for alluvial rivers: stable channels, meander growth and shift, thalweg shift, avulsion, and channel expansion and contraction.

Laterally stable channels may be found in either single-channelled or multichannelled situations. The key consideration here is the low-energy vertically accreting settings in which these cohesive, suspended-load rivers are found. Examples include low sinuosity, passive meandering or multichannelled anastomosing rivers. Slot-like symmetrical channels typically have low width/depth ratios with vertical banks comprised of cohesive fine-grained materials. While vertical accretion is prominent, lateral adjustment is not.

Meander growth and shift is indicated by the degree and type of sinuosity observed along a reach. Enormous diversity in patterns and rates of meander growth and shift may be evident on bends (Figures 10.9 and 9.7). Meander extension refers to lateral migration that increases the amplitude of bends. Translation refers to progressive downstream movement of bends, typically associated with an obstruction at the outside of the bend. Bend rotation refers to deflection of the pathway of bend migration (i.e. neither across nor down the valley). Multiple phases of extension, translation and/or rotation may produce an array of ridges and swales across the floodplain. Short-circuiting of bends generates chute and neck cut-offs. Very sharp bends with strong banks scarcely migrate in passive meandering channels.

Thalweg shift occurs as low-flow channels shift position within a multichannelled network. It is commonly measured by the degree and type of braiding. The braided index B_i is a measure of the degree to which bars and/or islands separate multiple flow paths (Figure 10.9). It is measured as:

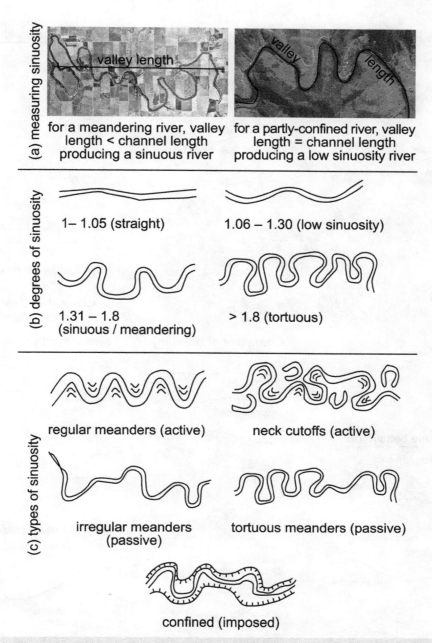

Figure 10.8   Classification of the degree of sinuosity used in analysis of channel planform.

$$B_i = \frac{2(\text{total bar length})}{\text{reach length}}$$

A value <1 occurs for single channels, whereas a value >5 reflects intensely braided rivers. The greater the degree of braiding (i.e. the higher the braided index), the greater the amount of in-channel sediment storage, such that the channel divides recurrently around bars and/or islands (see Figure 10.9c). Obviously, these measures are dependent upon flow stage (flood flows that submerge all bars/islands have a braid index of zero).

While the degree of braiding indicates the relative proportion of open water to sediment within a multi-channelled system, the character of braiding is indicative of the relative stability of sediment stores within a reach and the propensity for thalweg shift (Figure 10.9). Bar deposits reflect temporary storage of the bedload fraction. These readily mobilised sediments are likely to be altered during any moderate flow event. In contrast, islands are vegetated bars. These sediment accumulations may well have bedload materials at their core, but vegetation stabilises the bar surface and promotes deposition of

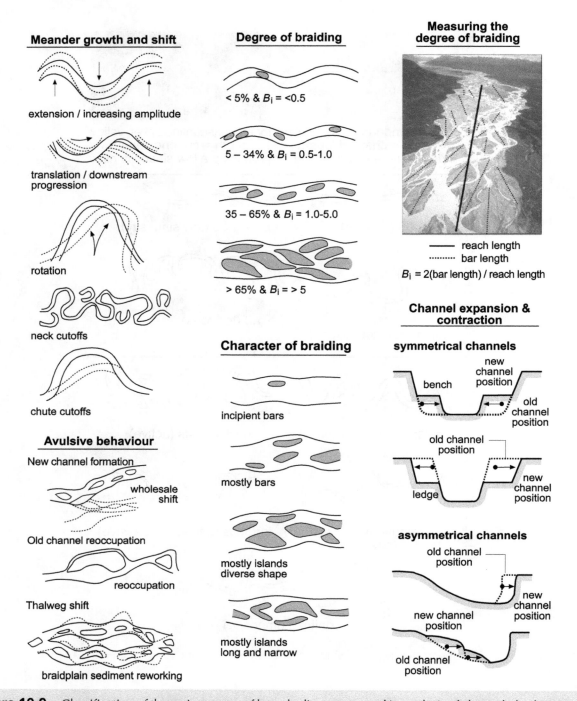

Figure 10.9 Classification of the various types of lateral adjustments used in analysis of channel planform. Degree and character of braiding are modified from Brice (1960). © Geological Society of America. Reproduced with permission.

fine-grained, suspended-load deposits. As such, the greater the proportion of islands relative to bars, the greater the lateral stability of a multichannelled reach.

The avulsive behaviour of a river describes the ability of a channel to jump from one thread to another or to create a new course on the valley floor (Figure 10.9). Thalweg shift

is by far the most common and recurrent form of avulsion, whereby flow cuts and divides bars and reoccupies former channel positions within a multichannelled braid plain. In less multichannelled but more sinuous situations, flow may preferentially reoccupy former channels in response to either changes in flow stage or altered patterns of sedimentation within channels. The most dramatic form of avulsion occurs when wholesale channel shift occurs along a significant length of river, resulting in an entirely new channel formed on the valley floor. This process typically occurs when flow breaches a levee and establishes a new course along a lower section of the floodplain. In other cases, the alluvial belt may aggrade and become elevated above the surrounding valley floor such that the channel shifts to occupy a position on the lower floodplain. In many instances former channel positions are reoccupied.

Channel expansion and/or contraction may be the only form of lateral adjustment experienced by a channel. Channel expansion refers to widening of a channel, and contraction refers to narrowing of a channel (Figure 10.9). Expansion of asymmetrical channels is marked by erosion of the concave bank without concomitant deposition on the convex slope of the bend. Alternatively, contraction is noted by deposition of a point bench or expansion of the point bar on the inside of the bend, without movement of the concave bank. Equivalent situations for symmetrical channels are evident whenever stepped banks with erosional planforms (i.e. ledges) are indicative of channel expansion, whereas deposition of benches along one or both banks is indicative of channel contraction (see Chapters 7 and 8). Sediment analysis is often needed to differentiate among these forms of adjustment, both of which generate compound channels.

Primary variants of alluvial rivers

Planform measures can be used to differentiate among variants of alluvial rivers. Primary examples of these different types of rivers are presented here along a hypothetical energy gradient from high- to low-energy conditions.

Boulder-bed streams

Boulder-bed rivers typically occur in areas of local valley widening immediately downstream of confined reaches. Instream geomorphic units in these steep-slope, high-energy settings include boulder mounds, boulder berms, cascades, rapids and islands. Floodplains tend to have a convex cross-profile with a fan-like morphology that thins downstream. Single-channel systems tend to have a low sinuosity, but they may avulse, leaving abandoned channels on the floodplain (Figure 10.10).

Braided rivers

Braided rivers have multiple channels (greater than three), low sinuosity (<1.3) and are laterally unstable (Figure 10.11). These bedload-dominated rivers are found in steep-slope, high-energy settings. Gravel-based systems are dominated by longitudinal bars, while sand-bed systems comprise transverse bars and dissected sand sheets. Lateral adjustment occurs via thalweg shift, as flow volume and orientation shift among channel threads at different flow stages. Flow diverges and rejoins around gravel and/or sand bars that scale approximately to the width of the channel. Channels are wide and shallow, reflecting the non-cohesive nature of bank materials and high sediment load. Bed/bank materials lack cohesive mud and/or binding vegetation. Non-cohesive, medium-energy floodplains (see Chapter 9) occur in isolated pockets in sheltered locations at valley margins (e.g. downstream of alluvial fans at tributary confluences or associated with shifts in valley alignment). In general, braided rivers have multiple topographic levels that reflect differing degrees of permanence and reworking.

Wandering gravel-bed rivers

Wandering gravel-bed rivers are transitional between braided and meandering river types (Figure 10.12). These rivers have up to three channels with variable sinuosity (some channels may have low sinuosity, while others may have moderate–high sinuosity). The lateral stability of these moderate-energy rivers is highly variable. Some sinuous channels are prone to lateral migration, while multichannelled areas with extensive bar areas are subjected to thalweg shift. Wandering gravel-bed rivers are prone to avulsion, whereby some channel threads are abandoned as flow preferentially shifts to a different position on the valley floor. This typically occurs via reoccupation of a former channel. A wide range of bar types may be observed, including both longitudinal and point bar forms. Some bars migrate downstream, others laterally. Compound bars are common. As mid-channel bars become vegetated they act as islands, around which flow divides and rejoins. Floodplains are far more prominent and continuous than along braided rivers. The mixed-load character of these rivers is noted by vertical accretion of fine-grained sediments (sands and silts) around vegetation on islands and on floodplains. Composite banks are more cohesive and, hence, more stable than the non-cohesive banks of braided rivers.

Meandering rivers

Meandering rivers are primarily single-channelled systems, though flow may locally divide where cut-offs occur. The key defining feature of these moderate-slope, medium-energy rivers is the sinuous nature of the channel. Ratios of channel length to valley length are ≥1.3. Lateral stability of meandering channels is determined primarily by the nature of valley floor sediments and vegetation associations. Mixed-load meandering channels which have a prominent gravel and/or sand bedload fraction adjust

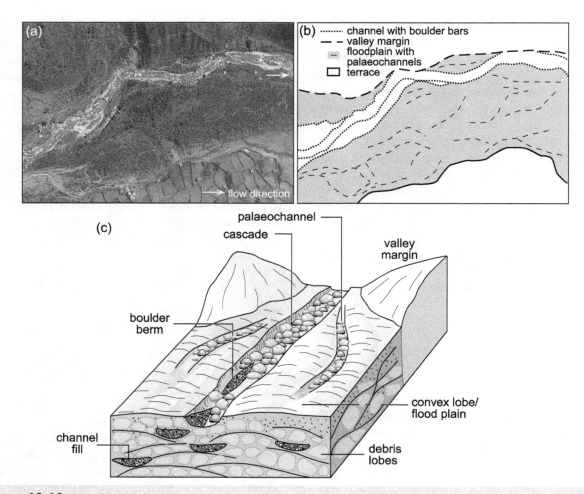

Figure 10.10 Boulder-bed rivers in a laterally unconfined valley setting. (a, b) Planform view of a stream draining the Himalayas from © Google Earth 2011. (c) The block diagram shows the avulsive nature of these streams as the valley widens in fan-like settings. Coarse-grained materials are reworked on an irregular, infrequent basis. Channels typically adopt a low sinuosity form.

laterally via bend migration processes (Figure 10.13a). Erosion of the outer (concave) bank is approximately compensated by deposition of point bars on the inner (convex) bank of a bend. These actively migrating channels have an asymmetrical shape. Meandering rivers tend to maximise their rate of alluvial reworking (lateral shifting) and sediment transport by optimising their bend curvatures. Floodplains are dominated by lateral accretion deposits capped by vertical accretion (overbank) deposits, with a wide range of geomorphic units such as ridges and swales, abandoned channels, billabongs (chute or neck cut-off channels), levees and backswamps.

In contrast, suspended-load meandering rivers are passive meandering systems, wherein channel position is relatively stable despite the sinuous or tortuous outline of the channel (Figure 10.13b). The cohesive nature of fine-grained banks, often accompanied by swampy conditions on floodplains, inhibits lateral erosion of channels. Hence, channels typically have a more uniform cross-sectional shape rather than an asymmetrical channel form. In these cases, there is little evidence for active erosion of banks. The lack of bedload materials limits the development of point bars. However, sculpted fine-grained geomorphic units line the channel. The nature and pattern of these features may be forced by instream and channel marginal vegetation and wood. Vertical accretion tends to create relatively flat-topped floodplains.

Anastomosing rivers

Anastomosing rivers are laterally stable multichannel systems, with sinuosity varying from channel to channel (Figure 10.14). They are found in low-energy, low-slope conditions, often in very wide alluvial plains (see Chapter

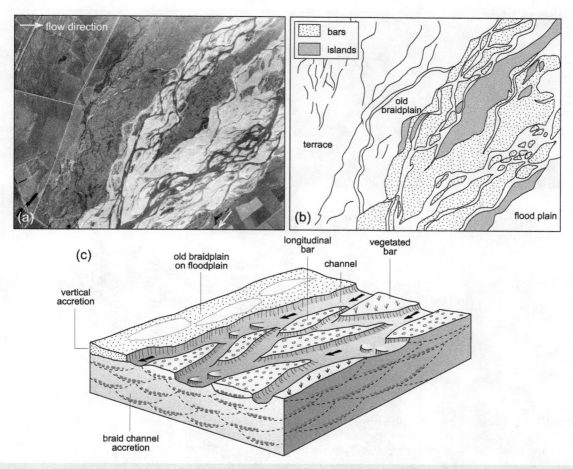

Figure 10.11 Braided rivers in a laterally unconfined valley setting. (a, b) Planform view of the Waimakariri River, New Zealand. (c) Channels are prone to thalweg shift. Floodplain pockets at valley margins are often associated with channel abandonment.

9). These suspended-load systems have a negligible bedload fraction. Hence, the array of instream geomorphic units is negligible and tends to be sculpted from the surrounding fine-grained sediment (see Chapter 8). As a result, channels tend to be uniform and slot-like, flowing within near-vertical banks comprised of vertically accreted fine-grained sediments. In many instances channels divide, run for several kilometres and rejoin around island and/or flood-plain segments. Differing channels across the valley floor may be occupied at differing flow stages. Vegetation may play a prominent role in promoting channel stability. In this context, anastomosing rivers are a type of anabranching river.

Cut-and-fill rivers – rivers with discontinuous channels

Some laterally unconfined, alluvial valley settings are characterised by valley floors upon which channels are discontinuous or absent. The key attribute of discontinuous watercourses is their capacity to switch from 'fill' phases, in which suspended-load sediments accumulate on the valley floor via vertical accretion, to 'cut' phases, in which an incised channel forms. Incision into the valley fill is accompanied by headcut activity. A subsequent phase of channel expansion occurs, until the channel becomes overwidened, flow energy dissipates, and deposition occurs and the next fill phase begins. A wide range of discontinuous water courses has been documented, including chains-of-ponds, floodouts and intact valley fills.

Cut-and-fill river systems are found in a wide range of environmental settings, ranging from tropical through temperate to arid conditions. They occur at various positions in a catchment, including upland and tableland settings and low-lying alluvial plains. Common aspects that influence their presence include relatively low slopes and a relatively wide valley (for the given catchment area), such that flow energy is dissipated across the valley floor. These suspended-load environments are dominated by

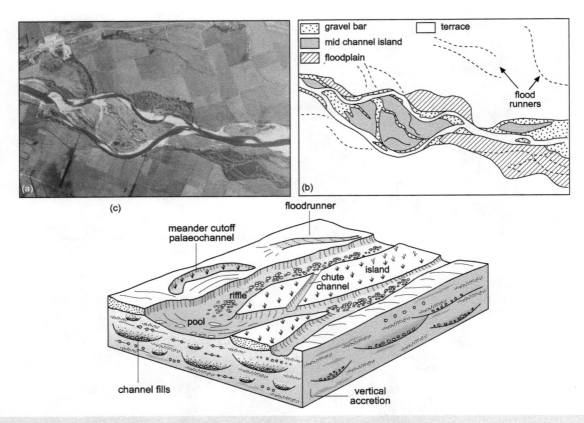

Figure 10.12   Wandering gravel-bed rivers in a laterally unconfined valley setting. (a, b) Planform view of the Waiau River, New Zealand. (c) Some bars adjust via downstream migration; others adjust laterally. Abandoned channels and bars may become incorporated within the floodplain.

silt and clay deposits, often with prominent accumulations of organic material. However, sand variants are also common. Essentially, during the fill phase, flow is unable to incise into the valley floor deposits, such that the vast proportion of flow is subsurface. Even under circumstances when the valley floor is saturated, flow has insufficient energy to become channelised, though preferential flow lines may be evident. Alternatively, areas with discontinuous channels simply cannot retain sufficient energy to transport available sediment. When drainage breakdown occurs, deposits are spread over the valley floor in a fan-like shape at the termination of the discontinuous channel; this is termed a floodout. Vegetation cover may play a prominent role in the development and maintenance of intact valley fills.

Cut-and-fill rivers in upland plateau-like settings are often described as 'upland swamps'. A wide range of forms may be evident, characterised primarily by the nature of valley floor sediments and vegetation associations (see Figure 10.15). Fine-grained (mud-rich) valley floor deposits often contain a significant proportion of organic, peat-like materials. Discontinuous water courses in sandier substrates may be referred to as dells. Fine-grained valley floors with irregular pools that act as groundwater windows are termed chains-of-ponds.

Discontinuous water courses are especially prominent along ephemeral streams that are unable to maintain their flow–sediment balance throughout a year such that a significant phase of sedimentation occurs. In some instances an abrupt termination in energy may be experienced, whereby the stream is unable to support its sediment load. Deposition atop unchannelised valley floors creates floodouts or terminal fans. These process relationships are especially common in arid settings. Elsewhere, large fan-delta system in low-energy, suspended-load settings may be characterised by amazing threads of multichannelled (anastomosing) and discontinuous channelled marshlands.

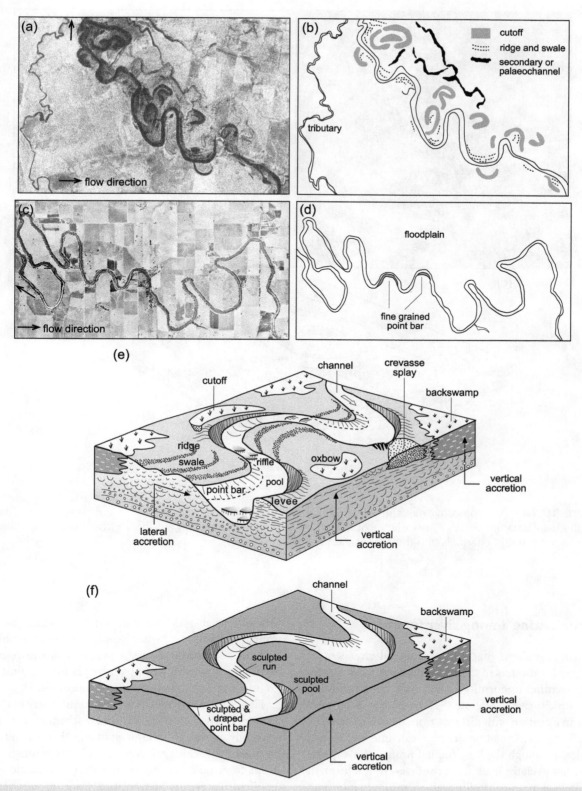

Figure 10.13 Meandering rivers in a laterally unconfined valley setting. (a, b) An active meandering river, Murray River, Australia. Ridge and swales are indicative of former channel positions. (e) In some instances, meander migration has generated cut-offs. (c, d) A passive meandering river, Goulburn River, Victoria, Australia. (f) There is no evidence of lateral adjustment in these fine-grained channels with cohesive bank sediments. Instream geomorphic units are often sculpted rather than deposited.

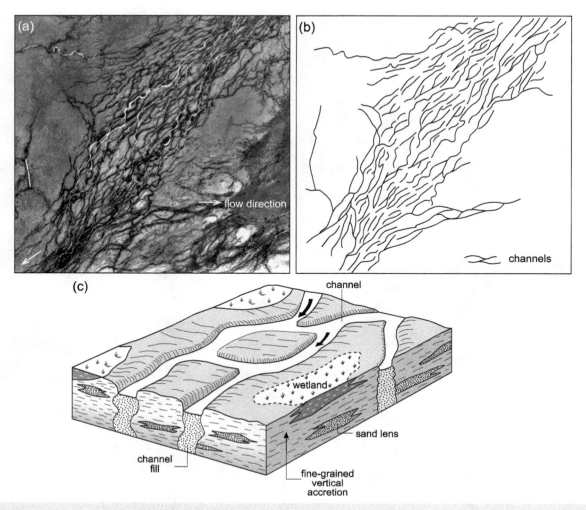

Figure 10.14   Anastomosing rivers in a laterally unconfined valley setting. (a, b) Planform view of the Channel Country near Innaminka, SW Queensland, Australia. (c) The block diagram shows the vertically accreting nature of these laterally stable channels and their associated floodplains.

Discriminating among river types

The energy gradient that characterises all river systems from their headwaters to their mouths conveys an impression of seeming uniformity in the underlying conditions under which rivers operate. However, the progressive decline in elevation with distance from source is one of the few unifying elements that describe catchment-scale relationships. Although the longitudinal profile of many river systems has a relatively uniform concave-upward form, the wide range of other controlling factors results in inordinate variability in the character, behaviour and pattern of river reaches. Critically, rivers are not only products of site-and/or reach-specific controls. Rather, processes and forms also reflect reach responses to prevailing (and past) fluxes, whereby reach-scale attributes in one section of a river course may influence what happens elsewhere within that system. As a consequence, generalised downstream patterns merely provide a comparative platform against which to interpret catchment-specific relationships.

Figure 10.16 provides a conceptual summary of the continuum of river diversity. Plan-view schematics are shown vertically along an energy gradient that extends from imposed high-energy bedrock conditions through a spectrum from high- to low-energy alluvial conditions. This gradient represents the transition from forced river morphologies through bedload-dominated (braided), mixed-load rivers (wandering gravel-bed and active meandering) and suspended-load conditions (passive meandering, anastomosing and discontinuous watercourses). The continuum

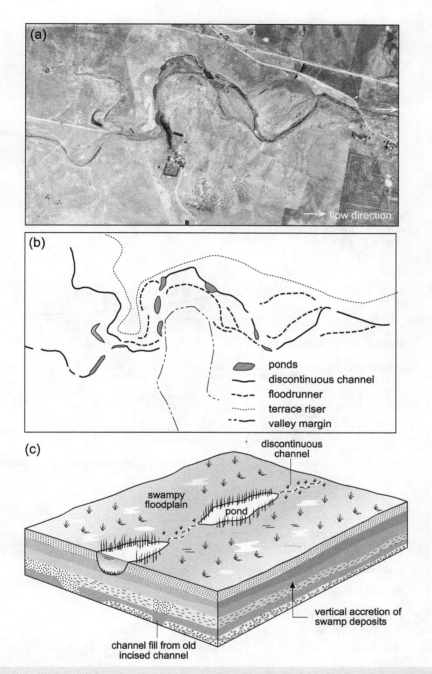

Figure 10.15 Cut-and-fill rivers in a laterally unconfined valley setting. (a, b) Planform view of a chain-of-ponds type along Mulwaree Ponds, NSW, Australia. (c) Vertical accretion of fine-grained sediments is the dominant contemporary process.

also reflects changes in valley width from confined through partly confined to laterally unconfined (fully alluvial) settings.

The trend in river planform schematics shown in Figure 10.16 also reflects a transition in bed material texture from coarse-grained to finer grained systems. Floodplains are absent in steep, confined valleys that generate high-energy, competence-limited bedrock rivers. The bed is largely immobile. Occasional sculpting of bedrock or organisation of boulder and cobble materials creates forced geomorphic units. Moving downstream there is a transition from irregular (isolated) floodplain pockets to discontinuous

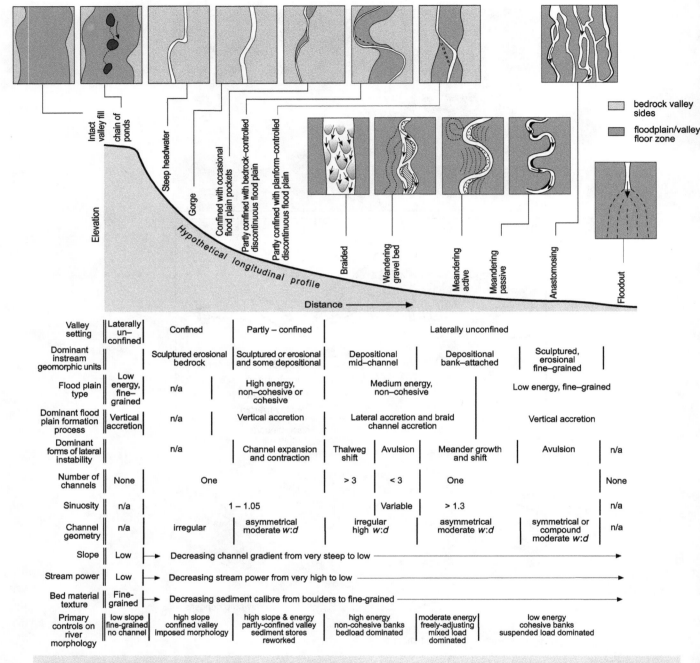

Figure 10.16 The continuum of river diversity along a hypothetical longitudinal profile, with associated controls on their formation.

pockets in partly confined settings. In these transfer zone settings the rates of sediment supply and throughput roughly equal sediment output. Channel and floodplain morphology in these coarse-grained, bedload-dominated systems is controlled directly by the shape of the bedrock valleys in which they are found. Bedload-dominated rivers in laterally unconfined valley settings are high-energy,

sediment-charged systems that entrain, transport and deposit coarser materials. The mobile bed is subjected to phases of degradation and aggradation, dependent upon the prevailing flow–sediment balance. Depositional, mid-channel geomorphic units dominate. Given the non-cohesive nature of the banks, channels have high width/depth ratios. Floodplains comprise a range of geomorphic units. In some set-

Step 1: Identify individual instream and floodplain geomorphic units and determine their process-form relationships.

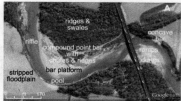

Example of interpretation
ridges & swales formed by helicoidal flow forming scroll bars, concave bank erosion and channel migration
compound point bar formed by deposition of gravel on the inside of a meander bend, flow realignment over the bar at bankfull stage and scour of chute channels. Sediment is deposited around vegetation to form ridges.

Step 2: Analyse and interpret the assemblage of geomorphic units at the reach scale and how they adjust over time.

Example of interpretation
low flow stage - flow aligned around bars and over riffles. Undercutting of concave bank occurs.
bankfull stage - point bars are short-circuited, pools are scoured, riffles are deposited, concave bank erosion and deposition on convex bank leads to channel migration.
overbank stage - flow aligned over the neck of meander bends, forming cutoffs

Past

Future

Present

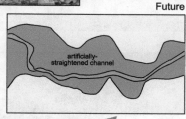

Step 3: Explain controls on the assemblage of geomorphic units, and 'natural' and human-induced impacts.

Example of interpretation
flux controls - transport-limited river with high sediment supply. Flashy flow regime.
imposed controls - tectonic activity resets base level.
human impacts - devegetation, cutoff formation and artificial straightening of channel.

Step 4: Integrate understandings of geomorphic relationships at the catchment scale.

Longitudinal profile and river patterns

Buffers, barriers and connectivity

Imposed controls (e.g. valley confinement)

Example of interpretation
*pattern of river diversity controlled by slope and valley confinement
*sediment supply is high due to landslide connectivity in headwaters
*role of disconnected tributaries varies over time (e.g. dams)
*channelised reaches convey sediment more efficiently

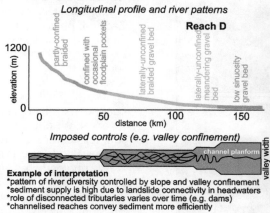

Plate 1 (Figure 1.4) An example of how to apply the reading the landscape approach to a real river system. The example used here is the Tagliamento River in Italy. The approach begins by interpreting process-form relationships for individual, then assemblages of geomorphic units along different reaches of river type. River behaviour is interpreted for a range of flow stages. The role of natural and human induced disturbance on river adjustments over time is considered when analysing river evolution. Finally, each reach is placed in its catchment context, analysing flux and imposed controls on river diversity along longitudinal profiles and how reaches to fit together in a catchment. Interpreting the efficiency of sediment flux at the catchment scale determines the (dis)connectivity of the catchment and associated off-site responses to disturbance. Maps constructed using Google Earth® 2012 images. Based on information in Bertoldi et al. 2009, Gurnell et al. 2000, Surian et al. 2009 and Tockner et al. 2003. The interpretation of river types, river evolution and connectivity are our own.

Geomorphic Analysis of River Systems: An Approach to Reading the Landscape, First Edition. Kirstie A. Fryirs and Gary J. Brierley.
© 2013 Kirstie A. Fryirs and Gary J. Brierley. Published 2013 by Blackwell Publishing Ltd.

Confined valley

(a) steep headwater

(b) gorge

(c) gorge

Partly confined valley

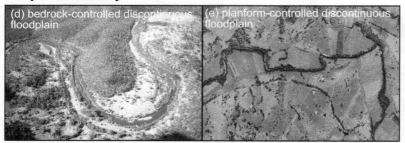

(d) bedrock-controlled discontinuous floodplain

(e) planform-controlled discontinuous floodplain

Laterally unconfined valley with continuous channel

(f) braided

(g) anastomosing

(h) bedrock-based anastomosing

(i) meandering-anabranching

Laterally unconfined valley with discontinuous channel

(j) chain of ponds

(k) upland swamp

Plate 2 (Figure 10.1) The diversity of river types. A range of rivers is found in different valley settings, ranging from fully confined rivers to partly confined rivers with pockets of floodplain, to freely adjusting alluvial (or laterally unconfined) rivers. Discontinuous watercourses also occur in some settings. (a) Steep headwater in New Zealand; photograph: D. White. (b) Gorge in Grand Canyon, USA; photograph: P. Chappell. (c) Gorge in south-west Tasmania, Franklin River, Australia; photograph: K. Fryirs. (d) Bedrock-controlled discontinuous floodplain, Clarence River, NSW; photograph: R. Ferguson. (e) Bedrock-controlled discontinuous floodplain, Williams River, NSW; photograph © Google Earth 2011. (f) Braided river, New Zealand; photograph: D. White. (g) Anastomosing river, Cooper Creek, central Australia; photograph: G. Nanson. (h) Bedrock-based anastomosing river, Sabie River, Kruger National Park, South Africa; photograph: G. Brierley. (i) Meandering-anabranching river, tributary of Upper Yellow River, China; photograph: G. Brierley. (j) Chain-of-ponds, Macquarie Marshes, NSW; photograph: K. Fryirs. (k) Upland swamp, Budderoo National Park, NSW; photograph: K. Fryirs.

tings, coarse-grained, non-cohesive features may be readily reworked, presenting significant potential for lateral and vertical adjustment. Further downstream, as energy decreases, mixed-load systems are characterised by a wide range of channel and floodplain geomorphic units. The bed and banks have contrasting sediment mixes. Within-channel sorting of materials formats bank-attached, coarse-grained geomorphic units, while the finer grained, suspended-load fraction is deposited on the floodplain. Composite banks have a coarse basal fraction overlain by interbedded coarse- and fine-grained fractions. Faceted bank morphologies are common. Selective reworking of coarser grained lenses presents significant potential for lateral adjustment. In the most downstream locations, suspended-load systems in low-energy, low-gradient environments are unable to maintain the transport of coarser materials, which tend to accumulate as bank-attached geomorphic units. Vertically accreted, fine-grained floodplains are prominent features of these aggradational environments. Cohesive banks inhibit bank erosion and lateral adjustment.

Various attempts have been made to quantify the range of conditions under which different river types are found. For example, floodplain pockets first occur along longitudinal profiles when slopes are below $0.008\,m\,m^{-1}$. Transition zones that characterise slope and catchment area (discharge) can be used to discriminate between river morphologies in steep headwater settings (Figure 10.17).

If slopes are greater than $0.2\,m\,m^{-1}$, then sediments are not retained on the valley floor and hillslope processes such as debris flows are dominant. These pulsed events result in irregular patterns of sediment stores, the majority of which are readily reworked, and a continuous cover of mobile materials is not found (i.e. bedrock is prominent on the bed). Channels with a slope less than $0.001\,m\,m^{-1}$ tend to be characterised by a continuous cover of finer sized (sand) materials along valley floors (i.e. these are alluvial rivers). For slopes between 0.2 and $0.001\,m\,m^{-1}$, the capacity for sediments to be stored on the channel bed is determined by upstream catchment area (and associated discharge). Potential for sediment storage decreases with increasing catchment area for a given slope, as the greater volume of flow is able to flush away bed materials. A transitional zone marks the transition from the bedrock-fluvial region, in which partly confined and confined rivers dominate, to coarse-grained alluvial rivers. These generalised relationships do not take account of local resistance factors on the valley floor, such as shifts in valley alignment or the role of forcing elements such as wood. Lithology also exerts an influence upon site-specific relationships, as it affects the erodibility of the valley walls which dictate valley morphology, and the size of sediments that are made available to the river.

Moving downstream and down-slope, discriminant analysis has been used to demonstrate the influence of

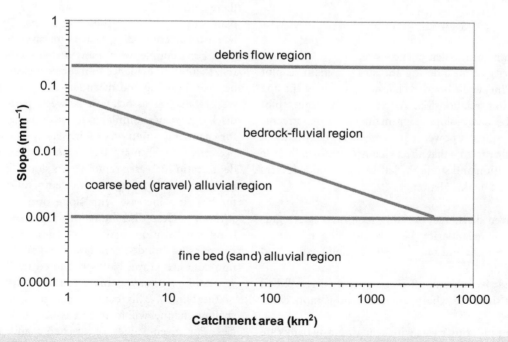

Figure 10.17 Slope–catchment area control on river type. The Sklar and Dietrich (1998) relationship discriminates among bedrock and alluvial river types. © American Geophysical Union (An edited version of this paper was published by AGU.). Reproduced with permission.

slope and catchment area (discharge) upon variants of alluvial rivers. As slope increases for a given discharge, channel sinuosity decreases and the number of channels increases, with associated increases in width/depth ratio, sediment load and sediment calibre. Hydraulic adjustments are marked by increases in flow velocity, tractive force and stream power. As a result, bedload transport increases and the lateral stability of the channel decreases. In general terms, this gradation of channel planform types reflects a declining energy gradient from braided through wandering gravel-bed, meandering and straight rivers.

In some ways, the distinct morphological characteristics of different alluvial rivers convey a sense that discrete river types can be identified, with characteristic thresholds or discontinuities between pattern states. This implies that a transition in channel pattern may occur if a river is close to a critical threshold slope for a given catchment area (i.e. discharge). An empirically derived discharge–slope relation for the threshold separating less-steep meandering and steep, braided streams is given by:

$$S = 0.0125Q^{-0.44}$$

where Q ($m^3 s^{-1}$) is the bankfull discharge and S is the threshold slope.

Because channel planform is also determined by the rate and mode of sediment transport, bed material size must also be incorporated within this equation, such that:

$$S = 0.52D_{50}^{1.14}Q^{-0.44}$$

where D_{50} (mm) is median particle size.

The majority of meandering and straight channels plot close to this line, while braided channels plot above the line defined by this relationship. Active gravel streams plot higher on a discharge–slope diagram than sand-bed streams because of the greater power requirements for bed-material transport. Other factors that affect channel planform include bank composition and strength. Self-formed braided rivers have very weak banks. Hence:

1. For a given discharge, braiding requires a larger gradient than meandering because braiding involves a greater amount and rate of channel modification and bank erosion.
2. Sand-bed rivers braid at lower slopes than gravel-bed rivers of similar discharge because sand is more easily entrained.
3. Active meandering rivers with sand or gravel banks are able to migrate and adjust much more readily than their fine-grained (passive) counterparts that have high bank strength.

Similar analyses have been conducted to differentiate between multichannelled rivers, in particular the braided and coarse-grained wandering gravel-bed rivers. One of the key distinctions between braided and wandering gravel-bed rivers is that while individual threads of a wandering gravel-bed channel are stable, the threads of a braided river are not. A channel system becomes fundamentally unstable (and hence braided) when the number of bifurcations/channels exceeds four. Therefore, the wandering gravel-bed–braided threshold reflects the transition between stable and unstable multiple thread channels, such that:

$$S_N = 0.40N^{0.43}\mu'1.41Q^{-0.43}$$

$$S_N = 0.72\mu'1.41Q^{-0.43} \quad \text{(when } N = 4\text{)}$$

where S_N is the threshold between stable and unstable multiple-thread channels, N is the number of channel divisions required for a channel with slope s to have stable anabranches and μ' is the dimensionless relative bank strength given by the ratio of the critical shear stress for entrainment of the channel banks to the critical shear stress for the channel bed. When the bed and banks are composed of similar material, like loose, noncohesive gravel with no fine-grained materials or vegetation effects, μ' can be set to a value of one.

Figure 10.18 synthesises these various interactions that fashion channel planform. While these relationships are useful for discriminating between gross river planforms, there remains significant overlap in the position of river planforms relative to these functions. Difficulties arise when other controlling factors on river type are considered. For example, within any one size class of bed material, there is no evidence to indicate a clear discrimination between braiding and meandering. Rather, there is only a weak statistical association between pattern and slope–discharge values. Similar relationships have been shown for channel geometry. For fairly steep sand-bed or gravel-bed streams within any bed-material size class, the width/depth ratio increases rapidly with slope. Since braiding is associated with a large width/depth ratio, opportunities for braiding increase with slope over a transition zone without a clear discrimination. Riparian vegetation and bank composition may affect these relationships. The meandering–braided threshold for gravel-bed rivers with non-cohesive gravel banks is altered by the influence of vegetation on bank stability (i.e. the erosional resistance of the banks). This controls the prospect that a wide–shallow channel will form and associated braiding tendencies. For given values of discharge and slope, if bank stability is decreased via the removal of bank and riparian vegetation, the meandering–braided threshold is effectively lowered and braiding is induced along rivers that

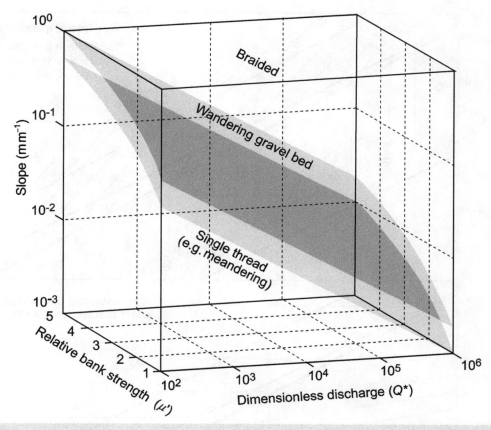

Figure 10.18   Slope, discharge and bank strength controls upon alluvial river types. Discriminating functions to differentiate between braided, wandering and single-thread (e.g. meandering) rivers. All axes are nondimensional and thus are scale free. Single-thread channels are limited to the region below the lower surface, braided channels are restricted to the region above the upper surface and wandering gravel-bed rivers are found between the two surfaces. Modified from Geomorphology, 120 (3–4), Eaton, B.C., Millar, R.G. and Davidson, S., Channel patterns: Braided, anabranching, and single-thread, 353–364, © 2010, with permission from Elsevier.

were once meandering. This relationship also works in reverse with increased forest cover or planting of riparian strips. However, this relationship seems to only hold true for braided and meandering rivers that are sensitive to change; that is, these reaches that sit close to the meandering–braided threshold. Rivers that sit away from the threshold tend to be relatively resilient to changes in bank and riparian vegetation density. No single threshold function can differentiate among planform variants; rather, a family of threshold curves reflects the sets of conditions within which different rivers operate.

These various relationships between slope, discharge and bed material size are used to differentiate among domains of channel planform and geometry for sand-bed rivers in Figure 10.19. Directions of morphological response to particular combinations of changes in discharge, slope and grain size can be identified. A regime channel in Region 1 is characterised by a flat slope, low

velocity, small bedload and flow resistance in the lower regime of ripples and dunes. The channel is relatively deep and narrow, with a width/depth ratio generally ranging from 4 to 20. All human-made stable canals are within this region, characterised by constant width and depth as well as long, straight reaches. Channel width is a function of both channel slope and bankfull discharge. However, the width is not sensitive to the slope and is essentially only a function of bankfull discharge. Braided rivers are found in Region 2. These are commonly found on steeper slopes where width/depth ratio is high. Rivers in Region 3 have slopes ranging from moderately steep to fairly steep, characterised by alternating riffles and pools at bankfull stage. Both channel width and depth are sensitive to slope, in sharp contrast to rivers in Region 1. Since an increase in slope is accompanied by rapid increase in width and decrease in depth, rivers in this region may be braided. The extent of braiding is in direct relation to the slope. Rivers

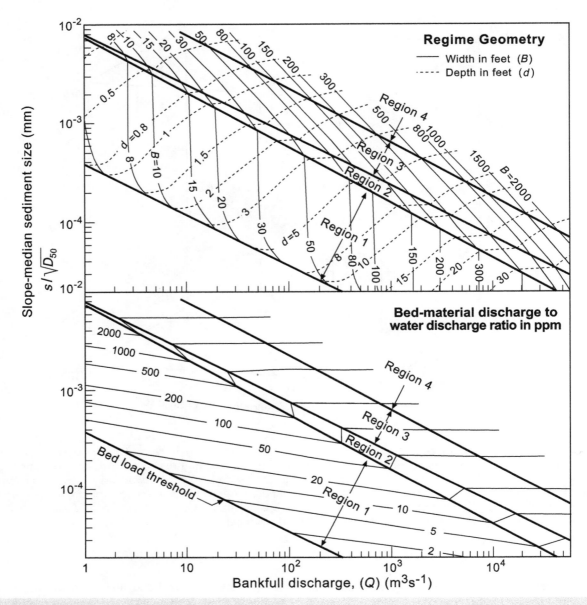

Figure 10.19 Slope–bed material size to discharge controls upon channel geometry and channel planform for alluvial rivers. Note logarithmic scale (see text for details). Modified from Chang (1988).

on less steep slopes are usually less braided but more sinuous; these are active meandering rivers. Those on steeper slopes are usually more braided and less sinuous; these are wandering gravel-bed rivers. The braided–meandering threshold described above falls within this region. Rivers in Region 4 have steep slopes. They are characterised by a large width/depth ratio and a highly braided channel pattern. The width/depth ratio is usually greater than 100. The braided channel pattern is usually straight, although some anabranches may be sinuous. Because of

the sensitivity of channel width to slope and discharge in Regions 3 and 4, a change in slope or discharge may be associated with a large change in channel width.

In some ways, the spectrum of rivers shown in Figure 10.16 provides a misleading sense of 'end member' situations. Remember, this is a continuum. In reality, distinctions between morphologic types are fuzzy and complex. There is significant range and overlap in the set of environmental conditions under which a particular type of river occurs. Indeed, unless this overlap occurred, river morphology would remain the same over time. River change reflects the ability for multiple river morphologies to exist under a given set of boundary conditions (Chapter 12). The primary factor that promotes change is adjustments to either the flow/sediment balance or to the nature/distribution of resisting elements along the valley floor (e.g. the type and distribution of vegetation). Braids can meander. Meanders locally braid. Many anastomosing channel networks have discontinuous channels. In other words, there is considerable overlap in the range of morphological attributes of these various river types. Discriminating functions should be viewed as indicators of the conditions under which a transition between planform type may occur. Distinct local conditions can result in unique morphological responses that do not conform to such analyses. For example, bedrock anastomosing rivers occur in low-slope, wide-valley settings with thin veneers of fine-grained alluvium atop bedrock. River history may play a vital role in determining contemporary morphology, as channels rework sediments laid down by former river networks. The role of riparian vegetation and wood as a control on channel geometry and planform varies markedly in differing environmental settings. Human activities may impose particular river morphologies.

Variants of channel planform are commonly not differentiated using consistent criteria. While meandering rivers are defined primarily on the basis of their sinuosity, braided channels are multichannelled (but unstable) and anastomosing river systems are differentiated by their laterally stable multichannelled configurations. Many reaches demonstrate different planform styles at different flow stages. Hence, individual channel planform types do not reflect specific geomorphic processes that occur under unique sets of circumstances. Rather, they reflect fluvial adjustment to combinations of interrelated variables, in which limiting factors may impose a particular morphologic response. The key to effective appreciation of river diversity is to meaningfully record what is observed in any given field situation, rather than indicate that the reach under investigation is closest to any particular type of river. A consistent but open-ended approach to analysis is required. The River Styles framework provides a generic set of procedures by which this can be achieved.

The River Styles framework

Meaningful approaches to classification are based on a consistent set of criteria. However, one of the difficulties faced in developing a coherent and comprehensive approach to classification of river systems is the fact that the primary attribute used to characterise different types of river varies across the spectrum of river diversity. Bedrock rivers are dominated by their imposed morphology, braided rivers by multiple laterally unstable channels, meandering rivers by their sinuosity and cut-and-fill rivers by discontinuous or absent channels. An alternative approach frames classification of river diversity in relation to the assemblage of channel and floodplain geomorphic units that make up a reach. These units comprise both erosional (forced/sculpted) and depositional features, thereby spanning the range from bedrock to fully alluvial conditions. Morphodynamic attributes of geomorphic units provide insight into the underlying processes that form and rework each feature.

A constructivist approach to river analysis assesses assemblages of erosional and depositional geomorphic units for each given field situation. Although recurrent patterns may be noted, this set of procedures enables each situation to speak for itself, such that new variants of river type can be added.

River Styles are identified at the reach scale. The key to differentiation of River Styles made on the basis of discernable differences in the range, extent or pattern of geomorphic units from one section of river to another. In this way, determination of reaches reflects differences in process relationships that reflect particular adjustments to local conditions along a river course. These principles are generic and can be applied to all river systems, irrespective of environmental setting or the nature/degree of human modification to rivers (e.g. forested or cleared catchments; regulated or non-regulated rivers; urban or rural streams).

As noted at the outset of the chapter, and throughout this book, river systems comprise a nested hierarchy of attributes. Geomorphic units are made up either of deposits of differing texture or they are forced features that are eroded/sculpted into bedrock or fine sediments. Differing assemblages of these features, in turn, make up the channels and floodplains of rivers. For example, a different subset of geomorphic units makes up the floodplain compartment of river systems in party confined and laterally unconfined valley settings. In those settings where the channel has significant potential to adjust its position on the valley floor, differing assemblages of geomorphic units are noted for rivers of differing planform type. Given these considerations, a nested hierarchical arrangement, building upon geomorphic units as a unifying principle, provides a logical basis for river classification. This thinking is applied in the River Styles framework. As the array of dominant features that characterise valley floors varies markedly in differing valley settings (e.g. presence/absence of floodplains; relevance of planform attributes), procedural trees used to interpret River Styles vary for each valley setting (see Figure 10.20).

In confined valley settings, three subsets of attributes are used to determine river types. First, the presence/absence

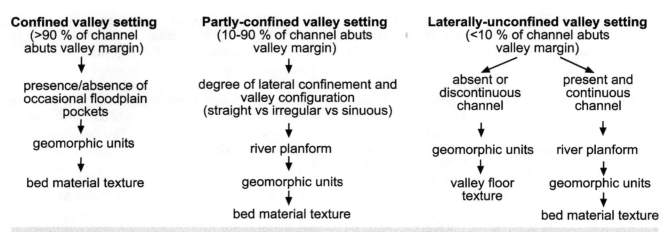

Figure 10.20   Procedures used to differentiate among river types as part of the River Styles framework. From Brierley and Fryirs (2005). © John Wiley & Sons, Ltd. Reproduced with permission.

of floodplain pockets is assessed. Second, the array of geomorphic units is measured. Erosional features are likely to dominate, with occasional depositional features. Third, bed material size is considered. In many instances bedrock dominates the valley floor in these settings. Elsewhere, these are primarily boulder- or cobble-bed streams.

In partly confined valley settings the first determination is whether the distribution of floodplain pockets is induced by the alignment of the valley itself (bedrock controlled), or whether the planform of the channel determines the pattern of floodplain pockets (planform controlled). In the latter instance, the planform type is noted. Beyond this, geomorphic unit assemblages and bed material size are recorded. In general terms, erosional geomorphic units (i.e. forced river morphologies) are more prominent in bedrock-controlled discontinuous floodplain river types than they are in planform-controlled discontinuous floodplain river types.

A much wider array of river types is likely to be observed in laterally unconfined valley settings. The first question to be addressed in these settings is whether the channel is continuous or discontinuous. Analysis of fully alluvial systems with continuous channels entails measurement of planform attributes, geomorphic unit assemblages in channel and floodplain compartments (primarily depositional forms) and bed material size. In laterally unconfined settings where the channel is absent or discontinuous there is no need to assess channel planform. Emphasis is placed upon analysis of valley fill (floodplain) geomorphic units and the discontinuous nature of the channel. Once more, the texture of valley floor sediments is a key determinant of river type.

Reach boundaries are differentiated on the basis of discernible changes in the assemblage of channel and/or floodplain geomorphic units. This differentiation is made

in the first instance in relation to landscape units and their associated valley settings. Maps, aerial photographs, satellite images or other forms of remotely sensed data are analysed to demarcate the patterns of floodplains, thereby separating confined, partly confined and laterally unconfined valley settings. This determines which branch of the River Styles tree to follow in assessing reach-scale attributes (Figure 10.20). Reach boundaries are marked by a discernible change in attributes (and associated processes). Boundaries may be distinct, gradual or intermittent. Once selected, discriminating attributes must be evident upstream and downstream of the boundary. Reach length may vary markedly from system to system. Alternating patterns of reaches are interpreted as segments.

Whenever practicable, reach boundaries are confirmed in the field. Representative reaches are selected to assess channel geometry (shape, size, variability) and appropriate attributes for characteristic geomorphic units that make up the reach (e.g. geometry, size, position, vegetation associations). Instream geomorphic units are differentiated into mid-channel and bank-attached forms. The pattern of erosional and depositional forms is appraised. Adjacent units are assessed to interpret whether they are genetically linked or not. Floodplains are analysed to determine whether they are flat and featureless, or if they contain a wide range of features such as billabongs, abandoned channels and backswamps. The presence/absence of a levee is used to appraise genetic linkages between channel and floodplain features, assessing proximal–distal relationships and linkages between primary depositional and reworked features.

The River Styles framework is an open-ended, learning approach to river analysis, striving to move beyond prescriptive or categorisational thinking. Rather than striving to present a categorical list of attributes for some notional sense of 'ALL' river types, the River Styles framework is designed as a flexible 'thinking tool' that can be applied to any catchment.

Tips for reading the landscape to interpret river diversity

Step 1. Identify individual landforms and their process–form associations

Reaches comprise an assemblage of geomorphic units (building blocks). For the river under consideration:

- What types of instream geomorphic units do you have? Where are they positioned in the channel?
- Are instream geomorphic units sculpted, erosional, mid-channel depositional, bank-attached depositional or fine-grained sculpted units?
- Is a floodplain present? If so, is the floodplain flat-topped or are other floodplain geomorphic units present?
- How are each of the instream and floodplain geomorphic units formed and reworked? What is their process–form association?
- Are individual landforms the result of erosional or depositional processes?

Step 2. Analyse river diversity at the reach scale

Given significant diversity in the range of river types that can be formed under a range of environmental conditions, there is no specific number of river types. Rather, a continuum of forms is generated along a gradient of slope, energy (stream power) and sediment calibre conditions.

A reach is defined as a length of river that operates under relatively uniform boundary conditions such that a consistent morphology and assemblage of geomorphic units occurs. Boundaries that separate reaches are determined by transitions in patterns of geomorphic units. Reaches upstream and downstream of a boundary have different mixes of geomorphic units. Boundaries may be distinct at changes in slope, valley confinement or tributary confluences, but they will be diffuse or gradual where changes to sediment and water flux relations occur over a considerable length of river course.

Analysis of river diversity can be made by assessing the following morphological attributes:

Valley setting (Chapter 9).
- Are floodplains absent, occasional, discontinuous or continuous?

- What is the pattern and extent of floodplain pockets along the river?
- What measures are used to assess river character in that valley setting?

Channel planform.
- If the river flows within a partly confined or laterally unconfined valley setting, how many channels are there, what is their sinuosity and how does the channel(s) adjust its position on the valley floor?
- What forms of lateral (in)stability occur?

Assemblage of channel and floodplain geomorphic units, and their relation to channel geometry (Chapters 7–9).
- What is the assemblage of landforms that makes up the floodplain?
- Is the floodplain flat and featureless, or does it contain a wide range of features such as billabongs, abandoned channels and backswamps (Chapter 9)?
- Does the river have a levee?
- Are proximal–distal relationships evident?
- Are instream geomorphic units erosional or depositional (Chapter 8)?
- If depositional, are they mid-channel or bank-attached forms?
- How does channel shape and size vary within this reach, relative to upstream and downstream reaches (Chapter 7)?
- Is this a discontinuous watercourse?
- Are process–form linkages for adjacent geomorphic units interlinked?
- How do these interactions and process linkages vary with flow stage.

Bed material size (Chapter 6).
- Is this a bedrock-dominated (forced) river, or is it a boulder, cobble, gravel, sand or fine-grained (silt–clay) system?

Importantly, this open-ended approach to analysis of river diversity does not force interpretations into differing classes or types. Rather new variants are characterised in relation to their assemblage of geomorphic units, framed within their valley setting.

Step 3. Explain controls on river diversity at the reach scale

Controls upon formation and reworking processes that generate packages of geomorphic units are analysed at the reach scale.

- Is river diversity imposed, or has the river adjusted to create that form?
- Is the reach dominated by erosional products (e.g. bedrock rivers) or depositional features (e.g. floodplains?

- Are instream depositional features primarily mid-channel or bank attached forms?

Emphasis here is placed upon determining why the reach has adjusted in this way. Downstream changes in channel geometry reflect the balance of imposed (forced) and self-adjusting (alluvial) conditions (Chapter 7). In the latter instance, it is important to determine whether the reach under investigation is a suspended-load, mixed-load or bedload-dominated river (Chapter 6).

Initial efforts to predict what type of river may be 'expected' in a particular setting may be appraised in relation to slope, discharge, bed material size and bank strength attributes using Figures 10.17–10.19. More thorough interpretation of river type relates river diversity to:

- *Valley setting* (Chapter 9). Variability in floodplain processes in differing settings is largely a product of flow energy.
- *Energy conditions under which the river is found* (Chapter 5). Slope, discharge and the influence of resisting elements on the valley flow dictate the energy available for geomorphic work in a reach. This is largely a product of landscape setting on the one hand, and position along the longitudinal profile (within a catchment) on the other hand. Climate setting fashions the variability in discharge. The impact of valley confinement upon the use of available energy is also critical.
- *Sediment regime.* Geologic and climatic controls upon the availability of materials and the energy of flow to transport these materials are key determinants of supply- and transport-limited landscapes, the prominence of bedrock relative to alluvial rivers, and the characteristics of bedload, mixed-load and suspended-load rivers (Chapter 6). These interactions determine the range and interactions among geomorphic units in the instream zone and on floodplains (Chapters 8 and 9) and associated channel geometry (Chapter 7).
- *Sediment analysis and bed/bank material properties* (Chapter 6). The calibre and mix of sediment that is carried by a river dictates, to some extent, the types of instream geomorphic units that will form and the manner/frequency with which they are likely to be reworked (Chapter 8). Bank strength may act as a key control upon channel geometry and planform.
- The balance of impelling and resisting forces (Chapter 5) determines the erosional and depositional processes operating on the bed and banks resulting in channels of different size and shape (Chapter 7). Alluvial channels adjust to create their own resistance. The sediment mix of the river influences channel size and shape. Bed and bank processes must be interpreted to assess controls upon channel geometry and planform.

Step 4. Explain how catchment-scale relationships affect river diversity

River diversity and patterns of river types along longitudinal profiles reflect the imposed and flux boundary conditions within which rivers operate at differing positions within a catchment (Chapter 3). Key transitions in river character and behaviour are demarcated by patterns of sediment storage along river courses, whether instream or in floodplains. This provides insight into the balance of erosional and depositional processes along a river, and associated patterns of imposed (forced) and self-adjusting (alluvial) morphologies. Primary influences to be assessed in framing reach scale variability in relation to catchment scale controls include appraisal of the landscape setting within which the reach is located, relationships to upstream and downstream controls (e.g. whether a large volume of coarse sediment is being input into the reach; whether base level has been affected by a landslide or construction of a dam downstream) and the behaviour of the reach as a source, transfer or accumulation zone.

Overarching, large-scale controls on patterns of rivers include geologic and climatic setting, catchment shape (configuration, morphometrics and tributary–trunk stream relationships), topography, relief, drainage density (landscape dissection), drainage pattern, hydrologic regime (Chapters 3 and 4) and evolutionary considerations (whether 'natural' or human-induced; Chapters 12 and 13). One way to interpret how these factors control the pattern of river diversity and to explain why certain rivers form where they do is to assess controls on the downstream pattern of reach boundaries for differing river types along a longitudinal profile. This provides a basis to assess reach relationships to slope, catchment area (discharge), stream power, valley confinement (valley setting), tributary inputs and sediment calibre key transitions to be explained include downstream changes in the nature, rate and effectiveness of erosional and depositional processes, the pattern of channel and floodplain geomorphic units, channel geometry/planform and associated river types. Site-specific relationships should be appraised in relation to reach variability and catchment-scale trends, differentiating local-scale impacts from broader scale boundary conditions. Particular attention should be given to assessment of the type and distribution of resistance factors in any given reach and how these attributes vary relative to upstream and downstream reaches. Often, reach boundaries relate to human-induced controls upon river character and behaviour, such as dams, channelised reaches, constructed levees and stopbanks, etc. Downstream patterns of river types can be compared with theoretical relationships such as the sequence shown on Figure 10.16 to determine if the system under investigation 'conforms' to classical notions. If possible, discriminating functions based on measures of channel slope, sediment calibre, discharge and bank strength are used to differentiate among river types.

Conclusion

Analysis of river diversity entails assessment of valley floor attributes across a nested hierarchy of scalar features, ranging from bed material sizes that fashion hydraulic units through process–form associations of channel and floodplain geomorphic units to characteristic assemblages of these features at the reach scale. A vast array of river morphologies exists, from imposed bedrock-controlled variants to fully self-adjusting alluvial forms that flow within river deposits. Critically, rivers are linked systems in which process relationships in one reach may affect river character and behaviour in other reaches. Position in the catchment is a key determinant of river type, as it determines reach location relative to the distribution of source, transfer and accumulation zones. Slope and valley setting are key determinants of the range of river type in any given landscape setting.

Valley setting is differentiated on the basis of the distribution of floodplain pockets along valley floors. Floodplains are either absent or restricted to isolated pockets in confined valley settings. These are bedrock-controlled rivers, such as steep headwaters and gorges. Rivers in partly confined valley settings are characterised by recurrent pockets of floodplain that are not continuous along both channel banks. Bedrock continues to exert a significant influence upon channel form. A slope-induced downstream transition from bedrock-controlled to planform-controlled floodplain pockets is commonly observed. Rivers in laterally unconfined valley settings flow within their own sediments. These alluvial rivers are differentiated into discontinuous watercourses (cut-and-fill rivers) and rivers with continuous channel(s). The latter category demonstrates the widest range of river types, as channels have the greatest range in degrees of freedom (i.e. capacity to adjust) because of deformable boundaries on both the bed and banks. Planform types are differentiated on the basis of the number of channels, their sinuosity and the manner and ease with which the channel is able to adjust its position on the valley floor. Braided rivers have multiple, relatively straight channels that adjust rapidly and recurrently via thalweg shift. These are bedload-dominated rivers. Meandering rivers are single-channelled and sinuous, but lateral stability ranges from active variants in mixed-load settings through to passive variants in suspended-load settings. Wandering gravel-bed rivers are intermediary between braided and active meandering rivers. Finally,

limited flow energy in low-relief suspended-load settings characterised by cohesive, fine-grained bank materials generates anastomosing rivers. These laterally stable multichannelled systems have variable sinuosity.

The interplay between discharge and sediment attributes in a valley setting with a given slope is the primary determinant of river type. A host of localised controls, such as vegetation cover and human impacts, may influence the reach-scale character of a river. Ultimately, river diversity is best viewed as a continuum, rather than as a spectrum of distinct types.

Reading the landscape is an open-ended 'learning' approach to the analysis of river systems. Analysis of erosional and depositional geomorphic units is framed in relation to valley setting, addressing different sets of questions with which to summarise river character depending upon the way in which the channel is able to adjust its position on the valley floor. Just as important as the descriptive overview of a river reach, however, is reliable assessment of how a river adjusts and works. Determination of the behavioural regime of different types of river and appraisal of the capacity for one type of river to change into another are considered in the following chapter.

Key messages from this chapter

- A reach is a length of river that operates under relatively uniform boundary conditions such that a relatively consistent morphology and assemblage of geomorphic units occurs. Reach boundaries can be distinct or gradual.

- Relationships along longitudinal profiles (linking slope, discharge, flow energy and bed material size to the aggradational/degradational balance of the river) are key determinants of the balance of erosional and depositional processes and resulting patterns of river types. Energy use along a river course reflects available energy (impelling forces) and the nature/distribution of resistance factors in any given reach.

- Rivers in erosive environments are forced, bedrock-controlled variants with imposed morphologies, whereas depositional environments in which channels flow and adjust within their own deposits are called alluvial rivers (i.e. the channel creates its own morphology).

- There is no specific number of river types. Rather, a continuum of forms is generated along a gradient of slope, energy and sediment calibre conditions. The river continuum extends from confined through partly confined to laterally unconfined (alluvial) variants. Confined valleys have no floodplain pockets, partly confined valleys have discrete, discontinuous floodplain pockets and laterally unconfined valleys have continuous flood-plains along both channel banks.

- Rivers in confined valley settings reflect an imposed condition and are typically high-energy, high-slope, bedrock and boulder systems. Types include steep headwater, gorge and occasional floodplain pocket rivers.
- Rivers in partly confined valleys are medium-energy, moderate-slope, gravel- and sand-bed rivers. Types include rivers with bedrock- and planform-controlled floodplain pockets.
- Alluvial rivers form in laterally unconfined valleys. Slopes of alluvial rivers are typically lower than in confined and partly confined valley settings. These rivers can deform their own boundaries. Substrate conditions extend from boulders through to gravels, sand and fine-grained (silt–clay) rivers. Some low-energy

alluvial rivers have discontinuous watercourses. These systems operate as cut-and-fill rivers.

- Channel planform, defined as the configuration of a river in plan view, is measured using three criteria: the number of channels, sinuosity and lateral stability. Examples of planform types include boulder-bed, braided, wandering gravel-bed, meandering and anastomosing rivers.
- Discriminating functions that differentiate among river planform types based on measures of channel slope, sediment calibre, discharge and bank strength provide a theoretical platform to interpret controls on river type.
- Reading the landscape is an open-ended approach to characterisation of river diversity based on analysis of assemblages of geomorphic units at the reach scale.

CHAPTER ELEVEN
River behaviour

Introduction

Analysis of river systems is not only concerned with what they look like, it also emphasises how they behave and why they adjust in the way that they do. Morphodynamic relationships fashion mutual interactions between river morphology and the processes that create and rework that form. This means that the processes shape the form, while the form influences the type and effectiveness of processes. Analysis of river behaviour entails consideration of how different types of rivers are able to adjust, the rate at which adjustment takes place and the permanence (or irreversibility) of those adjustments. These considerations determine the behavioural regime of a river reach.

Rivers are forever adjusting to disturbance events and prevailing flow and sediment fluxes. Vegetation conditions on valley floors greatly influence these interactions through their impact upon flow resistance. These relationships vary markedly in differing environmental settings. In addition, all river systems have a history of past disturbance events, whereby geologic, climatic or anthropogenic activities continue to influence contemporary river behaviour. Critically, adjustments in one part of a catchment affect reaches elsewhere within that system.

Any given reach is made up of a set of landforms (geomorphic units) that are created and reworked by a particular set of processes. Marked differences are evident for a confined river such as a gorge relative to a braided river or a swamp. For these rivers, different types of adjustments occur in different ways over different timeframes, in response to events of differing magnitude and frequency. The rate and extent of adjustment reflect the inherent sensitivity or resilience of a reach. These factors determine the 'natural capacity for adjustment' of a reach. This reflects the range of forms that reach may adopt in its given setting, and its associated behavioural regime.

Chapter 10 highlighted the remarkable range of river diversity in the natural world. The range of river behaviour and system adjustments to disturbance events is similarly broad. This reflects inherent variability in the capacity for adjustment of different types of rivers (i.e. their erodibility) on the one hand, and pronounced variability in the range of driving forces that promote river adjustment on the other hand (i.e. the erosivity of any given landscape). These factors determine how a river adjusts, how often it adjusts, what it adjusts towards and the likelihood that the river will be able to adjust back towards its previous state (i.e. whether these adjustments fall within the behavioural regime for that type of river or whether change has occurred).

Rivers adjust in vertical, lateral and wholesale dimensions. Vertical adjustment refers to the stability of the bed. Lateral adjustment refers to the ability of the channel to alter its banks. Wholesale adjustment refers to the manner and rate of alterations to channel position on the valley floor. Analysis of the package of vertical, lateral and wholesale adjustments defines the behavioural regime of a river. The capacity for river adjustment determines the ease with which a reach is able to adjust its form in these three dimensions. This is dictated by the relationship between available erosive flow energy and the nature/distribution of resistance elements along a reach. This is fashioned by, and in turn reflects, the texture and cohesiveness of channel boundary materials and the inherent roughness of the river (determined largely by channel geometry/planform, riparian vegetation and the loading of wood within the reach). The erosivity of a reach is determined largely by the nature of disturbance events, especially the magnitude, frequency and duration of flood events.

In general terms, different types of rivers are subjected to different forms of adjustment and have variable capacity to adjust. Rivers that can adjust in vertical, lateral and wholesale dimensions have greatest capacity to adjust. Hence, channels with non-cohesive bed and bank materials are particularly prone to adjustment. These rivers are considered to be sensitive to adjustment.

River behaviour varies markedly at differing flow stages. Analysis of the package of river adjustments at low flow, bankfull and overbank stages is used to define the behavioural regime of a river. Flow variability varies markedly in differing settings. In addition, magnitude–frequency relationships that fashion river form may vary for any given reach. As a result, there is pronounced variability in effective flow stage for differing river types. Furthermore, the geomorphic effectiveness of an event of a given magnitude varies not only with river type, it also varies with the state of the system at the time of the event. Factors such as sediment availability, vegetation cover and the time interval since the last flood event may induce variability in system responses to flood events. Some adjustments are progressive and predictable, while others reflect dramatic and unpredictable (threshold-breaching) circumstances. This notion of complex response highlights the importance of catchment-specific appraisals of river character and behaviour An open-minded approach to analysis is an important attribute in reading the landscape.

Adjustments in any given reach reflect the flux boundary conditions (flow and sediment delivery) in that system. Although generalised relationships can be established, this chapter develops a *thinking toolkit* with which to interpret how any given reach adjusts and behaves. This is performed through analysis of the assemblage of geomorphic units that make up the channel and the floodplain (see Chapters 8 and 9). Assemblages of these erosional and depositional forms in both channel and floodplain compartments are a manifestation of current and past geomorphic processes along a reach. Their analysis provides an interpretative tool with which to frame the contemporary behavioural regime of a river in its evolutionary context (Chapter 12). These relationships lie at the heart of efforts to read the landscape.

This chapter is structured as follows. First, the key concepts of river behaviour and change are defined. Second, vertical, lateral and wholesale forms of river adjustment are differentiated for differing types of rivers, drawing together insights into channel geometry, and channel and floodplain geomorphic units outlined in Chapters 7–9. Collectively, these differing forms of adjustment determine the natural capacity for adjustment for differing types of rivers. Some rivers respond sensitively to disturbance events, while others are quite resilient. Magnitude–frequency relationships that fashion these different forms of adjustment for different types of rivers are outlined. Third, assessment of the behavioural regime of differing types of rivers in differing environmental settings is summarised in terms of river adjustments at low flow stage, bankfull stage and overbank stage. Fourth, a conceptual framework for representing river behaviour is presented. This framework is called the river evolution diagram. Fifth, river responses to altered flow and sediment flux boundary conditions are briefly summarised.

River behaviour versus river change

Rivers continually adjust to a range of flow, sediment and vegetation interactions and the associated balance of impelling and resisting forces along valley floors. The nature and rate of river adjustment vary from system to system, and over differing timeframes. Differentiation can be made between river behaviour and river change.

Governing factors that dictate landscape evolution are considered to be sensibly constant over geomorphic timescales, constraining the forms of adjustment that can occur along any given river. These factors set the *imposed boundary conditions* within which rivers operate (see Chapter 2). River *behaviour* reflects ongoing geomorphic adjustments over timeframes in which flux boundary conditions (i.e. flow–sediment regimes and vegetation interactions) remain relatively uniform, such that a reach retains a characteristic set of process–form relationships. River behaviour is defined as adjustments to river morphology induced by a range of erosional and depositional processes by which water moulds, reworks and reshapes fluvial landforms, producing characteristic assemblages of landforms at the reach scale.

Alteration to the balance between impelling and resisting forces may induce a shift in the behavioural regime of a river whereby the reach evolves to a different type of river with a different characteristic form. This shift in process–form relationships along a reach is referred to as river *change*. River change may be reversible. However, irreversible evolution to a different type of river is to be expected over geologic timeframes. This may take the form of progressive evolutionary adjustments, or dramatic, near-instantaneous changes in response to catastrophic events. Alternatively, if the system lies close to a threshold condition, relatively small events can reconfigure the system to a different state. River change may be induced by natural or human disturbance (see Chapters 12 and 13).

Some systems are inherently more sensitive to physical disturbance than others. The *capacity for adjustment* of a river is a measure of the range and extent of geomorphic adjustments that can occur for that type of river (i.e. its natural range of variability). This can range from an imposed condition (natural or human induced) to a freely adjusting condition. For example, river forms and processes are geologically controlled within a bedrock-confined gorge, resulting in a naturally resilient configuration. Alternatively, regulated rivers downstream from dams or channelised rivers in urban settings are human-forced situations.

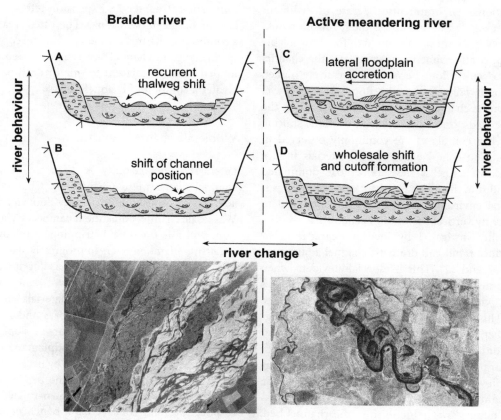

River behaviour - natural adjustments that occur for a particular river type
(e.g. lateral migration and cutoff formation along the meandering river;
thalweg shift for the braided river)

River change - when a wholesale change in river type occurs
(e.g. change from braided to active meandering)

Figure 11.1 Differentiating river behaviour from river change. Behaviour is expressed by thalweg shift and redistribution of bars in a braided river, and by progressive lateral migration of an active meandering river. Change is expressed by the transformation of a braided river to a meandering river. From Brierley and Fryirs (2008). © Island Press, Washington, DC. Reproduced with permission.

While a gorge or an urban channel is relatively insensitive to adjustment, a freely migrating meandering river, or the active channel zone of a braided river, may be very sensitive to adjustment. Rivers that have greatest capacity to adjust are especially prone to river change. A visual representation of the difference between river behaviour and river change is presented in Figure 11.1.

Dimensions of river adjustment

Different types of rivers are able to adjust in vertical, lateral and wholesale dimensions in different ways and to different degrees. The extent to which a river can adjust in any of

these dimensions provides a measure of its *capacity for adjustment*.

Vertical channel adjustment

Assessment of the stability of the channel bed and/or valley floor is a critical step in determination of the behavioural regime of a river. If the channel bed is unstable, then lateral and/or wholesale channel adjustments are also likely to occur. As noted in Chapter 7, the bed can build up over time (*aggrade*), be lowered over time (i.e. *incise* or *degrade*) or it can retain the same level (with or without bedload movement).

In aggradational environments the rate of sediment accumulation exceeds the rate of sediment reworking. As sediment supply rates are simply too high for the available flow to transport all sediments downstream, the channel bed and/or the valley floor become progressively higher. Relatively smooth valley floors with homogeneous gravel and/or sand sheets characterise these situations. Given the high sediment loads, braided rivers with multiple shifting channels and bar complexes are commonly observed in these settings. Non-cohesive boundary materials induce wide and shallow channels. These are bedload-dominated rivers. By definition, the fill phase of cut-and-fill settings is an aggradational environment in which sediments build up the valley floor by vertical accretion mechanisms.

Bed instability instigated by incision/degradation or uplift may induce rapid and dramatic channel adjustment (see Chapters 7 and 12). This is especially evident when headcuts eat into valley floors, consuming large volumes of sediment via headward retreat mechanisms (see Chapter 4). This threshold-driven phenomenon is triggered by either excess bed slope on the valley floor (intrinsic conditions) or excess flow energy associated with storm events or periods of sustained rainfall (extrinsic conditions). The flow has excess volume/energy relative to the amount of sediment that is readily available to be transported. This 'hungry' water incises into the channel bed, consuming energy by eroding and mobilising sediments. Bed instability often triggers bank instability, inducing bank erosion and channel expansion. Features such as headcuts and relatively uniform, slot-like channels with low width/depth ratios denote incision. Subsequent channel expansion may result in stepped (compound) channel geometries (as noted in Chapter 7, this morphology can result from a range of mechanisms).

Notionally 'stable' channel beds may reflect ongoing adjustments around a mean condition, whereby the system responds to variability in prevailing flow and sediment fluxes or an imposed condition forced by factors such as bedrock, coarse bedload materials that are infrequently reworked or riparian vegetation/wood associations. Short-term adjustments are imperceptible other than occasional shifting of materials on the bed. The assemblage of in-stream geomorphic units and the associated channel geometry reflect the way in which available energy moulds and reworks available materials, as noted in Chapter 8.

Lateral channel adjustments

In this book, lateral channel adjustment refers to alteration in channel shape and/or size, rather than alteration in channel position on the valley floor. The latter is referred to as wholesale channel adjustment. There are two primary forms of lateral channel adjustment: *channel expansion* and

contraction (Chapter 7). Expansion refers to widening via erosion of one or both banks. This often creates ledges with a compound channel geometry. Channel contraction refers to shrinkage in channel size caused by deposition of flat-topped benches adjacent to one or both banks. This also creates a compound channel form.

Wholesale channel adjustment

Lateral channel stability is discussed as a measure of channel planform in Chapter 10. Channels are able to adjust their position on the valley floor in a range of ways. Wholesale channel adjustment can occur via lateral migration, avulsion, floodplain stripping and thalweg shift.

Lateral migration refers to progressive lateral adjustment in the position of channel bends as floodplain deposits are reworked on the concave bank of bends and there is a concomitant accumulation of materials on point bars on the inside of the bend. Scroll bars and ridge and swale features are deposited on the convex slope of the asymmetrical channel. Sections of floodplain many be reworked, as channel cut-offs are generated by neck and chute cut-off mechanisms. If the outside of the bed is impeded (constrained), the bend translates down-valley creating a mix of point bar and concave bank bench deposits. Multiple phases of lateral migration may be preserved as abandoned suites of ridges and swales on the valley floor.

Channel *avulsion* refers to a wholesale shift in channel position whereby a portion of the flow, or the whole channel itself, occupies a separate course on the valley floor (see Chapter 9). As a result, the initial channel is abandoned, creating a palaeochannel. This feature may subsequently infill. Alternatively, it may be reoccupied at some stage in the future. The geometry and planform of palaeochannels provide insight into former environmental conditions.

Although it is a more localised occurrence than avulsion, *floodplain stripping* is probably the most dramatic form of wholesale channel adjustment (see Chapter 9). This process generally takes place along floodplain pockets that are located on the inside of bends in partly confined valley settings. Following a phase of long-term vertical accretion of floodplain pockets, the channel increasingly concentrates a greater proportion of total flow energy. Eventually, this energy becomes too great and the channel erodes the inside of the bend, effectively stripping the floodplain surface.

Finally, *thalweg shift* refers to wholesale adjustment as multiple bars and channels shift position on the valley floor, or where channels shift within a braid belt. Mid-channel bars and pockets of floodplain are readily reworked in these settings.

Natural capacity for adjustment of differing river types

The diversity of boundary conditions under which rivers operate, along with the continuum of flow, sediment calibre, slope and vegetation associations, ensures that there is considerable variability in which attributes of river morphology are able to adjust and how readily adjustments can occur for different types of rivers. In this book, this *natural capacity for adjustment* is defined as 'morphological adjustments of a river in response to alterations in flux boundary conditions such that the system maintains a characteristic state (i.e. morphology remains relatively uniform in a reach-averaged sense) and does not bring about a wholesale change in river type'.

Rivers in different valley settings are able to adjust their morphology in quite different ways (see Figure 11.2). By definition, bedrock rivers flow within confined valley

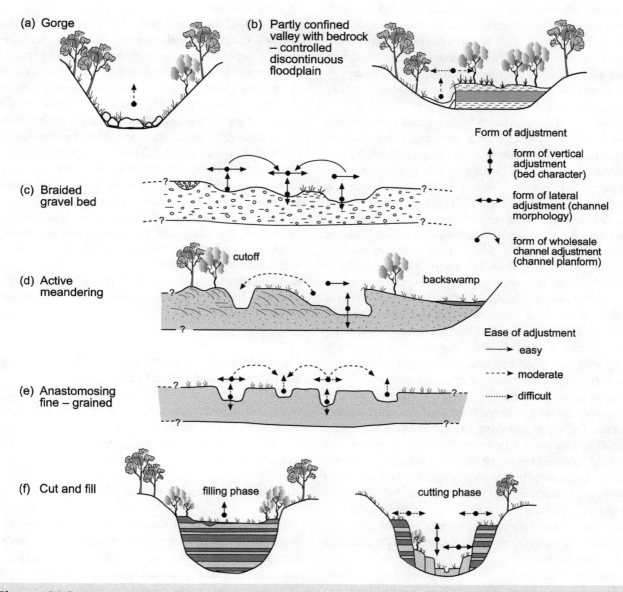

Figure 11.2 The capacity for vertical, lateral and wholesale adjustment of various river types. Arrows portray the forms of adjustment that can take place for different river types. These may be adjustments in bed character (the vertical dimension), channel morphology (the lateral dimension) and channel planform (the wholesale dimension). Some rivers have significant capacity to adjust in all three dimensions, while others adjust in one or two dimensions. The ease with which these adjustments take place is shown by the 'strength' of the arrows. Modified from Brierley and Fryirs (2005). © John Wiley & Sons, Ltd. Reproduced with permission.

walls and have an imposed morphology with very limited capacity for adjustment. Channel configuration in a gorge is ostensibly stable, with no potential for lateral or whole-sale adjustment. Vertical adjustment is restricted to local redistribution of materials around coarse substrate and the flushing of sediment downstream (Figure 11.2a). Extreme floods may sculpt erosional geomorphic units.

Rivers in partly confined valley settings flow over a mix of bedrock and alluvial materials. Bedrock limits the capacity for vertical channel adjustment (Figure 11.2b). Lateral adjustment is constrained by bedrock or terrace materials along one valley margin, but the floodplain pockets themselves are prone to adjustment. Valley margins also limit the capacity for wholesale lateral adjustment via channel migration, cut-off activity or avulsion. However, floodplain stripping may occur in these settings, as flow energy is concentrated at flood stage. This river has localised capacity for adjustment.

Truly alluvial rivers in laterally unconfined valley settings have the greatest capacity for adjustment. These channels have more degrees of freedom than rivers in confined and partly confined settings, as they are able to adjust in vertical, lateral and wholesale dimensions. However, there is significant variability in the capacity for adjustment within this class. In Figure 11.2c–e this is shown for high-, moderate- and low-energy settings (i.e. bedload, mixed-load and suspended-load streams), while Figure 11.2f conveys evolutionary stages for cut-and-fill rivers. This variability is not only fashioned by impelling forces. Conditions on the valley floor, and the way in which channels adjust their form to create flow resistance, also affect the type and ease of geomorphic adjustments. Hence, the character of sediments in channel and floodplain compartments, along with vegetation and wood (roughness) characteristics, exert a primary influence upon the ease with which channels are able to adjust.

Braided rivers are laterally unconfined, high-energy systems. These bedload-dominated rivers have significant natural capacity for adjustment in vertical, lateral and wholesale dimensions (Figure 11.2c). Each channel thread has significant potential to locally aggrade or degrade, to expand or contract, or to shift position on the valley floor via thalweg shift. Indeed, valley floors often comprise an assemblage of bars, islands, channels of varying geometry and differing topographic levels with various braidplains and palaeochannels.

Gravel-bed meandering rivers in laterally unconfined, medium-energy settings can adjust in both vertical and lateral dimensions and are prone to wholesale adjustment. Lateral migration, cut-off formation, abandonment of meander bends or channel avulsion can occur in active meandering rivers (Figure 11.2d). This type of mixed-load river has significant capacity to adjust.

Anastomosing rivers are found in laterally unconfined, low-energy settings. Although these suspended-load rivers are able to adjust in both vertical and lateral dimensions, and may be subjected to wholesale shifts in channel position on the valley floor, rates of adjustment are slow because of the cohesive nature of sediments that make up the bed and banks (Figure 11.2e). Adjustments occur as channel belts slowly accrete within wide plains and channels are subjected to incremental lateral expansion or contraction. Channel avulsion may occur via reoccupation of abandoned channels. These rivers have moderate to limited capacity to adjust.

Cut-and-fill rivers are found in laterally unconfined, low-energy settings. Channels may be continuous or discontinuous, dependent on the stage of adjustment. During fill phases the channel is either absent or discontinuous on the valley floor (Figure 11.2f). Over timeframes of hundreds or thousands of years valley floors are subjected to slow, pulsed yet progressive aggradation. There is limited capacity for adjustment at this stage. Should an erosional threshold condition be exceeded, this river has significant capacity to adjust both vertically via incision and laterally (initially channel expansion, but subsequently contraction). Eventually, the incised channel infills via aggradation.

These examples convey the range in both the forms of adjustment that can occur for different types of rivers in differing environmental settings, and the ease/recurrence with which these adjustments can take place. Prior to using these insights to guide analysis of the behavioural regime of rivers, the next section explores controls upon these different relationships.

Controls on the natural capacity for adjustment of different river types

Natural capacity for adjustment can be assessed in relation to the sensitivity/resilience of a system and magnitude–frequency relations of disturbance events that drive river adjustments.

River sensitivity and resilience

River behaviour is determined by the relationship between sediment supply and the relative energy that is available to transport or deposit that material. Channels can only move the sediments that are available to them. As noted on the Hjulstrom diagram, the most readily deformable channel boundaries are comprised of loose, non-cohesive medium–coarse sand with minimal vegetation cover (Chapter 5). Sensitivity is also influenced by the nature and distribution of resistance elements on the valley floor. Resistance is largely imposed by valley alignment and the roughness of the channel boundary along bedrock rivers, though instream wood and riparian vegetation may add further roughness. In contrast, alluvial rivers are masters of their own hydraulic efficiency, ranging from smooth, open channels to highly tortuous or divided (rough) channel boundaries. In these settings, resistance is a product of channel planform, channel geometry and bed material size/configuration, alongside vegetative influences.

Measures of river sensitivity reflect the ease with which adjustment can take place (i.e. the way in which the reach has adjusted its form to resist change) and the proximity to threshold conditions such that:

$$\text{sensitivity} = \text{capacity for adjustment}$$
$$+ \text{proximity to a threshold}$$

If an extrinsic threshold is breached, then a transition to a different type of river is likely to occur and river change results (Chapter 12). Breaching of an intrinsic threshold is more likely to result in a behavioural adjustment that forms part of the natural capacity for adjustment for that setting. Indeed, breaching of intrinsic thresholds is a natural part of the behavioural regime of many rivers. For example, gradual increases in valley floor slope associated with progressive aggradation in cut-and-fill settings may lower threshold conditions for incision over time, such that a relatively trivial event may incise the valley fill on the oversteepened valley floor. Similarly, the development of meander cut-offs may reflect threshold-exceeding conditions. Selective cut-offs may return a river to a more stable sinuosity in which there is a better balance between transporting ability and gradient. Finally, floodplain stripping in partly confined valleys represents a shift in state for the same type of river. In these settings, progressive build-up of sediments on the valley floor increasingly concentrates flow within the channel until a threshold condition is breached and stripping occurs.

Sensitivity analysis must also consider the ease with which a system or reach is able to adjust within its natural capacity for adjustment. Frequent perturbations typically result in minor adjustments to river character and behaviour without inducing a shift in state or change to a different type of river. At first glance, continual adjustment may be perceived as a form of instability, but this is not always the case. Rather, it simply means that the river is sensitive to adjustment and has capacity to adjust quite readily. For example, braided or wandering gravel-bed rivers continually adjust as bars and channels are recurrently reworked. In contrast, the behavioural regime of a more resilient river may be considered progressive. For example, gradual responses follow recurrent perturbations in passive (fine-grained) meandering or anastomosing rivers. Along bedrock rivers, even extreme flood events may bring about negligible geomorphic adjustment.

Sensitive rivers are readily able to adjust to perturbations as part of their natural capacity for adjustment, but are prone to dramatic change if significant thresholds are breached. Conversely, resilient rivers have an inbuilt capacity to respond to disturbance via mutual adjustments that operate as negative feedback mechanisms (see Chapter 2). In this scenario, long-term stability is retained because of the self-regulating nature of the system, which mediates external impacts. These rivers readily adjust to perturbations as part of their natural capacity for adjustment, without dramatic change in form–process associations.

Magnitude–frequency relationships

Flood events of a given magnitude, frequency and duration may have markedly differing consequences for different types of rivers, reflecting the capacity for adjustment and inherent sensitivity to change of a particular reach. The amount of geomorphic work performed by an event of a given magnitude and the geomorphic effectiveness of that event depend greatly upon the type of river under consideration and the recent history of responses to flow/sediment events at a particular site or reach. If a recent flow event has flushed away all readily available sediments, then there may be limited tools with which to perform erosional activity and depositional sequences from waning stages of flow events will be minimal. Alternatively, if recent flow deposited significant volumes of non-cohesive sediment at readily accessible locations within the reach, and vegetation cover (and associated roughness) is limited, considerable reworking, transport and redeposition of sediment are likely to occur during a major flood event.

Magnitude–frequency domains and associated flood-magnitude indices vary markedly for rivers in differing environmental settings (Chapter 4). Stark contrasts are evident between settings in which infrequent storms are the primary driver of river activity (e.g. arid/semi-arid regions), relative to Mediterranean climates that are subjected to marked seasonal variability (with geomorphic effectiveness most pronounced in late summer, when ground cover is minimised), to humid-temperate and/or tropical areas where perennial streams are active year-round, to polar or monsoonal regions where pronounced seasonality is marked by frozen ground and pronounced but predictable storms respectively.

The notion of a *dominant formative discharge* or *effective discharge* implies that regular, recurrent flow events are the primary agents of geomorphic work and effectiveness. Such uniformitarian thinking considers the contemporary flow regime to be the primary determinant of river character and behaviour. These relationships are well established for a subset of alluvial streams in humid-temperate settings. However, the situation is much more complex elsewhere. The dominant or effective discharge is not always the most geomorphically effective flow in terms of river adjustments either within channels or on floodplains. In many cases, the most effective flows occur irregularly but are greater than bankfull, while in other cases, regular and persistent flows less than bankfull result in significant adjustment within channels and the greatest volume of sediment transported. Bankfull discharge is not necessarily the most effective flow (see Chapter 4). River form is not the product of a single formative discharge but of a range of discharges. The sequence of flow events is also important, as past floods dictate the current channel and floodplain form and, therefore the effectiveness of subsequent events.

In some instances river behaviour is fashioned by the impact of major flood events, as these are the only conditions under which materials at channel boundaries can be reworked. This 'catastrophist' situation is especially common in areas subjected to pronounced variability in floods, with high flood-magnitude indices (e.g. Q_{100} : $Q_2 > 10$; see Chapter 4). For example, infrequent high-magnitude events are required to mobilise the bed of boulder streams. These are the only events that leave a persistent imprint upon the landscape. Elsewhere, for example in monsoonal areas or regions subjected to seasonal cyclonic activity, flood events are more regular and recurrent. These settings have low flood-magnitude indices (Q_{100} : $Q_2 < 10$), although rivers may demonstrate stark seasonal variability in their geomorphic behaviour (e.g. wet/dry tropics). Other rivers are fashioned primarily by progressive, incremental adjustments in response to recurrent events at and around bankfull stages (e.g. braided rivers, wandering gravel-bed rivers, active meandering rivers – though primary forms of adjustment vary for these differing alluvial rivers). Elsewhere, the stage of river adjustment may be a key determinant of river sensitivity to an event of a given magnitude. For example, some floodplain pockets along rivers in partly confined valleys are progressively accreted, prior to catastrophic stripping during a flood or sequence of flood events.

The length of time above a critical threshold condition is a key determinant of the geomorphic effectiveness of a particular event (see Chapter 2). Often, short-duration, highly peaked events that extend well beyond erosional thresholds are much less 'effective' in landscape-forming terms than are longer duration events that marginally exceed erosional threshold conditions. This reflects the amount of time that erosion and sediment transport take place.

The importance of roughness elements as a control on system sensitivity points to marked variability in river responses to disturbance events over time. If resistance is relatively low, the geomorphic effectiveness of an event is likely to be higher. Hence, human alterations to the ground cover, such as removal of wood or clearance of riparian vegetation, may greatly enhance the amount of erosion that takes place along a river (Chapter 13). Alternatively, systems may be especially sensitive to disturbance during a sequence of floods, as initial events knock out resistance elements such that subsequent events are more geomorphically effective. In this instance system sensitivity is fashioned by the recovery time between events (Chapter 2). Two recurrent lessons emerge here. First, there is marked variability in sensitivity dependent upon the type of river, determined especially by the amount of sediment that can be readily redistributed by that river. Second, each system has its own history of disturbance events, so catchment-specific investigations are required to generate meaningful insights into river character, behaviour and evolution.

Efforts to interpret magnitude–frequency relationships should consider the make-up of erosional and depositional features in any given reach, interpreting the range of processes that fashion and rework these landforms at differing flow stages. Magnitude–frequency associations that shape these interactions provide fundamental insight into river behaviour at the reach scale.

Interpreting the behavioural regime of different river types by reading the landscape

Differing types of rivers are subjected to differing forms of adjustment that operate with varying consequences over differing timescales. Rather than prescriptively categorising behavioural regimes for differing types of rivers, the approach adopted here emphasises analysis of river behaviour for any given reach, based largely on interpretation of the assemblage of erosional and depositional geomorphic units, and interpretation of the suite of processes that formed and reworked each feature, determining their position/pattern and interactions along the reach.

Differing processes dominate at differing flow stages in any given reach. Meaningful differentiation can be made between the forms and rates (sensitivity) of river adjustment that occur at *low flow* (i.e. adjustments to bed materials as flow covers the channel bed), *bankfull stage* (adjustments to channel geometry as the channel is filled to capacity) and *overbank* conditions (when the floodplain is inundated and activated; this is the stage that wholesale adjustments take place). Specific magnitude–frequency relations may vary markedly at these differing flow stages from system to system. However, analysis of the process relationships that operate at these differing flow stages provides a useful and consistent basis with which to interpret the behavioural regime of any given reach.

River behaviour at low flow stage

Analysis of *low flow stage* river behaviour appraises adjustments to bed material organisation and bedform-scale responses to variable flow energy over timescales of 10^{-1}–10^1 yr (Box 11.1). At low flow stage there is insufficient energy to adjust channel boundaries, but bed materials may be reorganised and deposition may occur. These vertical channel adjustments essentially reflect the behavioural adjustments of hydraulic units and bedforms (Chapters 5 and 6). Coarser grained fractions (bedload materials) are scarcely mobilised at this flow stage. Sand-sized bedforms may be remoulded, as these are the most readily mobilised grain-size fraction. However, deposition on the falling limb of flood events is the primary process that leaves a temporary imprint on the bed, as suspended-load deposits and organic materials are left as drapes of fine-grained sediment across parts of the channel bed, partially infilling pools.

Adjustment in grain size and/or distribution and the associated nature and pattern of bedforms, such as ripples, dunes, particle clusters, etc., provide useful insight into recent flow events. This may reflect various forms of sediment transport and deposition that dissect and rework sand/gravel forms, infill pools or develop an armour layer. Bed material size, organisation (including bedform type) and sorting are indicative of flow energy and the frequency of sediment transport and reworking (see Chapter 6). However, geomorphic activity is limited, and adjustments at low flow stage are seldom indicative of the formative processes that determine river morphology.

River behaviour at bankfull stage

River behaviour at bankfull stage reflects the conditions at which flow is contained within the channel without spilling onto the floodplain. Bankfull flows are the most hydraulically and geomorphically effective channel flow events, as specific stream power is at its highest (see Chapters 4 and 5). River behaviour at bankfull stage incorporates adjustments to the nature, pattern and rates of erosion and deposition on the channel bed and banks and associated modifications to the pattern of instream geomorphic units. Vertical and lateral adjustments modify channel shape and size via bed and bank processes (Chapter 7). Combinations of erosional and depositional processes result in differing assemblages of instream geomorphic units that characterise bankfull-stage river behaviour at the reach scale (Chapter 8). Analysis of river behaviour at this flow stage incorporates attributes outlined in Box 11.2.

Assessment of *bankfull-stage* behaviour comprises analysis of channel geometry (shape and size), as interpreted by vertical and lateral adjustments along the bed and banks respectively (see Chapters 6 and 7). Vertical adjustment of the bed via incision and aggradation has significant implications for erosional or depositional processes on the channel banks. Instream geomorphic unit type and assemblage can range from sculpted-erosional forms along bedrock and boulder rivers, to mid-channel and bank-attached features in gravel- and sand-bed rivers, to sculpted fine-grained units in suspended-load dominated rivers. Interpretation of bankfull-stage behaviour incorporates analysis of the *package* of processes recorded by these features, enabling insight into process responses on the rising and waning stages of flood events. Packages of instream geomorphic units provide critical insights into the balance of erosional and depositional forms and resulting channel behaviour over timescales of 10^0–10^2 yr.

River behaviour at overbank stage

Overbank flows may incorporate wholesale adjustments alongside vertical and lateral adjustments. Obviously, to go overbank there must be well-defined channel banks, so flows can only be defined in this way in settings with floodplains. At this flow stage the channel may be able to shift its position on the valley floor. Although some flow energy is dissipated through relatively shallow flows across potentially rough (vegetated) floodplain surfaces, total

---

Box 11.1 Analysing river behaviour at low flow stage

Bed material size and organisation
+
Surface flow characteristics around substrate on the channel bed
+
Interpretation of bedforms in sand- and gravel-bed systems

Box 11.2 Analysing river behaviour at bankfull stage

Assemblage of instream geomorphic units
+
Bed morphology and indicators of vertical adjustment (bed incision or aggradation)
+
Bank morphology and the type and extent of bank erosion that promotes channel migration or expansion/contraction (lateral adjustment)
=
Channel shape and size

stream power is highest at this flow stage (although specific stream power is not). Floodplain geomorphic units
are formed and/or reworked, prospectively adjusting
various attributes of channel planform. Analysis of floodplain geomorphic units highlights how valley confinement
induces differing capacity for adjustment and associated
diversity of floodplain types in confined, partly confined,
and laterally unconfined settings. These adjustments tend
to occur over timeframes of 10^1–10^3 yr. Analysis of river
behaviour at this flow stage incorporates attributes outlined in Box 11.3.

In general, different types of rivers in partly confined and
laterally unconfined valley settings have different floodplain types (see Chapter 8). Assemblages of geomorphic
units reflect the mosaic of depositional and erosional
forms. The primary component of overbank-stage analyses
entails assessment of how floodplains form, and the mix
and pattern of sediments derived from within-channel
and overbank mechanisms (see Chapter 9). Distinct geomorphic units may (or may not) be evident, such as levees,
crevasse splays, floodchannels and backswamps. Proximal–
distal relationships provide insight into how available
energy in overbank flows is utilised on floodplain surfaces.
Significant pocket-to-pocket variability in floodplain forms
may reflect localised controls such as flow alignment over
floodplain pockets and valley confinement. Floodplain
reworking may ensue. Adjustments to channel position on
the valley floor may be exemplified by alterations to channel
multiplicity, channel alignment (i.e. sinuosity, meander
pattern or wavelength, bend radius of curvature), lateral
stability of channel(s), or floodplain character (as measured by the assemblage of floodplain geomorphic units).
Floodplain features such as active cut-offs, floodchannels,
crevasse splays, sand sheets, avulsion channels, floodplain
stripping, etc provide field evidence for these process
interactions.

The behavioural regime as a package of flow-stage adjustments

Analysis of river behaviour entails assessment of the full
suite of processes occurring in the channel as flow stage

rises and spills over the channel banks, attains peak stage
over the floodplain and then wanes. Various geomorphic
units are formed and reworked at different flow stages.
These relationships are not only determined by process
interactions within a given reach, they also reflect the way
in which that reach is linked (connected) to other parts of
the river system (and associated fluxes). Analysis of reach-
scale river character and behaviour interprets links between
formative processes, associated forms of adjustment and
the range of features that make up the reach. This assessment must be viewed in a dynamic context, assessing morphological adjustments at differing flow stages. The pattern
of reaches in any given catchment and the effectiveness
with which reaches are connected to each other (i.e.
upstream–downstream and tributary–trunk stream relationships) determine the flux boundary conditions (i.e.
flow and sediment) to which each reach adjusts. Analysis
of reach-scale adjustments must be framed in relation to
changes in flux over time. Appraisal of these catchment-
specific relationships is meaningfully informed by understandings of generalised interpretations of behavioural
regimes for different types of river.

Examples of behavioural regimes for differing types of rivers

Any given reach has its own set of process–form relationships. Typical assemblages of interactions can be determined for particular types of rivers in differing landscape
and environmental settings. Behavioural regimes are fashioned by the range of geomorphic adjustments experienced
at differing flow stages. This section documents characteristic process–form relationships that summarise the behavioural regime of various river types in confined, partly
confined and laterally unconfined valley settings.

The behavioural regime of rivers in confined valley settings

Rivers in confined valley settings have limited capacity for
adjustment in vertical, lateral and wholesale dimensions
and are considered resilient (Figure 11.3).

Low flow stage

Grain size, sorting and hydraulic diversity are constrained
by bedrock, restricting adjustments to local reworking of
transient bedload flux. The distribution of erosional and
depositional geomorphic units is fashioned primarily by
local-scale variability in flow energy (determined primarily
by local slope and valley width), the volume and calibre
of bed material, and erosional resistance of the bedrock.

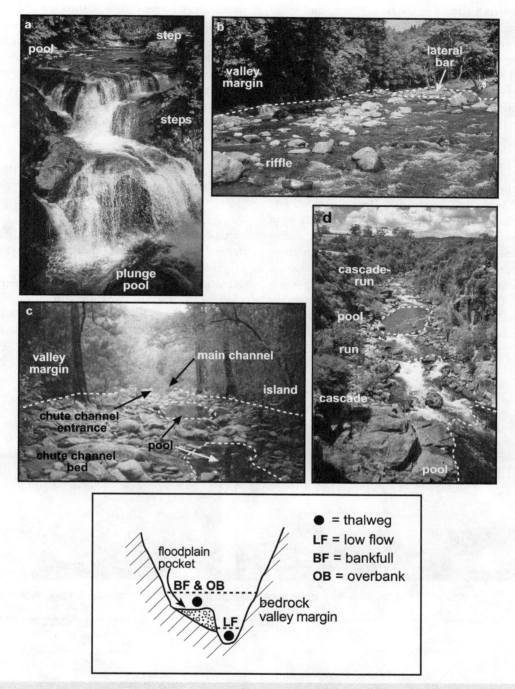

Figure 11.3 River behaviour in bedrock-confined rivers. Instream geomorphic units are only activated at high flow stage. A forced morphology prevails. Photographs: (a, b) Sangainotaki River (Three Steps of Waterfall), Japan; (c) upper Kangaroo River, NSW; (d) gorge near Launceston, Tasmania (K. Fryirs). Modified from Brierley and Fryirs (2005). © John Wiley & Sons, Ltd. Reproduced with permission.

Flow paths reflect patterns of coarse substrate and scour features in bedrock. Although these reaches are hydraulically diverse, little sediment reworking occurs at this flow stage, exerting negligible impact on the geomorphic structure of these rivers. Bedrock and boulder geomorphic units dissipate flow energy. Fine-grained materials that locally accumulate in pools and behind obstructions are flushed by subsequent higher magnitude flow vevents.

Bankfull and overbank stages

Rivers in confined valley settings do not have readily definable channel banks and floodplains. Hence, bankfull and overbank behaviour is interpreted for flows that span the valley and inundate all instream surfaces. Even during high-magnitude events, bedrock-confined rivers have limited ability to adjust. In a sense, this forced morphologic condition constrains river behaviour at anything other than extreme flow stages. Coarse basal materials, typically cobbles or boulders, are the only materials that are retained in these settings for any length of time. Bed materials may be locally redistributed, and finer materials are stored in transient forms or are flushed through these reaches. Reworking of irregular channel margins is negligible. Channel size and shape are imposed by bedrock and forcing elements such as wood. Bank erosion is negligible. Geomorphic units are largely imposed forms that are generated, locally redistributed and reshaped during high-magnitude–low-frequency flow events. Bedrock may be sculpted through corrasive action, forming potholes and plunge pools.

The behavioural regime of river in partly confined valley settings

Rivers in partly confined valley settings have localised capacity for adjustment. Bedrock may constrain lateral and vertical adjustment along significant parts of these reaches (Figure 11.4). The distribution of high-energy instream geomorphic units is influenced by local channel bed slope. Floodplain pockets typically comprise upward-fining depositional sequences in which coarse within-channel deposits are covered by vertically accreted suspended-load materials. These are moderately resilient rivers.

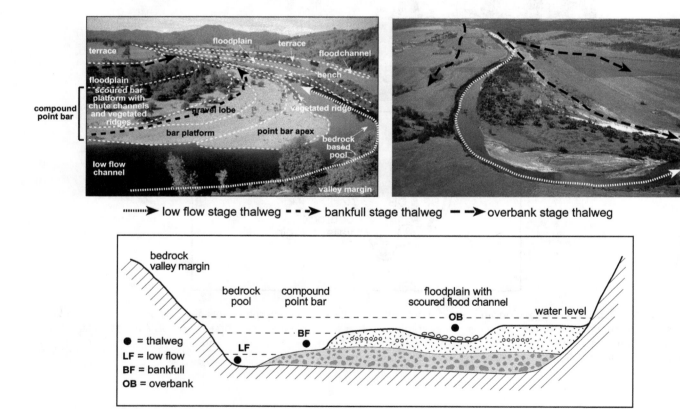

Figure 11.4 River behaviour in a partly confined valley. As flow alignment varies with flow stage over this bedrock-controlled floodplain pocket, differing phases of erosion and deposition generate and rework an array of channel and floodplain forms. Photographs: left, Macleay River, NSW; right, Clarence River, NSW (R. Ferguson).

Low-flow-stage behaviour

At low flow stage, flow is confined to forced pool–riffle–run sequences. Fine-grained materials accumulate in pools. Textural segregation of instream geomorphic units induces significant hydraulic diversity. Bed adjustments are dependent on material availability and the history of bedload transporting events.

Bankfull-stage behaviour

Lateral adjustment to channel width and shape is restricted to channel contraction (deposition) or expansion (erosion) adjacent to floodplain pockets. Flow alignment and the location of bedrock outcrops are key determinants of energy distribution. Channels tend to be relatively narrow and deep. Composite banks may facilitate channel expansion through undercutting and slumping at higher flow stages. Dissipation of energy at channel margins promotes the formation and reworking of bank-attached geomorphic units adjacent to floodplain pockets (Figure 11.4). These features can be erosional or depositional forms. Hence, the assemblage of instream geomorphic units typically comprises a mix of depositional forms – such as compound bank-attached bars (point and lateral) and benches – and/or erosional forms – such as ledges, forced pool–riffle sequences and bedrock steps.

Instream geomorphic units are formed and reworked at bankfull stage. Riffles are reworked and pools are scoured. In some instances, high-energy flows may sculpt bedrock, typically by corrasion. Lateral bars and benches may form at channel margins, producing a compound channel shape. These flat-topped, elongate surfaces create steps that reduce channel dimensions. Asymmetrical channels in bedrock-controlled bends induce scour in deep bedrock-based pools. Compound point bars and riffles may be reworked at bankfull stage. Chute channels and ramp features may form as flow short-circuits the bend. During waning stages, fine-grained sediment may accumulate around vegetation, often forming ridges on compound bars.

Overbank-stage behaviour

Instream and floodplain features are formed and reworked during overbank flows. Flow alignment over the floodplain pocket shifts with flow stage, becoming increasingly less influenced by the channel and more influenced by the alignment of the valley itself (Figure 11.4). Floodchannels may be scoured and/or infilled. Catastrophic erosion may strip floodplain pockets down to the basal gravel lag. In some settings the downstream translation of bends is recorded by the formation (and reworking) of concave bank benches. However, the capacity for wholesale channel adjustment is limited. Vertical accretion of floodplains occurs during the waning stages of floods.

The behavioural regime of laterally unconfined, high-energy rivers

Laterally unconfined, high-energy rivers are formed on high slopes and tend to be bedload dominated (e.g. braided rivers) (Figure 11.5). Given the readily mobilised materials on the bed and the non-cohesive nature of bank materials, these rivers are highly sensitive, with significant capacity to adjust in vertical, lateral and wholesale dimensions. Compound (macro) channels are wide and shallow with multiple topographic surfaces. Riffles and runs are commonly observed between bars, while pools form at points of flow convergence. Mid-channel bars build downstream and vertically as materials are deposited around a coarse bar head. Continued deposition in the lee of the bar head results in downstream fining. Individual bars may show a range of accretionary patterns, reflecting long-term aggradation, and downstream and lateral migratory tendencies. Recently deposited platforms of less-coarse material guides insight into patterns of accretion and the duration and intensity of flow. Once established, bar position and shape have a significant effect on flow alignment and the formation of adjacent geomorphic units during bankfull flows. Dramatic adjustments may occur during bankfull and/or overbank events.

Low-flow-stage behaviour

Sediment reorganisation of bed materials occurs recurrently under low flow conditions, as transient bedload flux induces significant local adjustments. Downstream gradations in grain size and increased sorting may occur along bars. Fine-grained materials are deposited in pools. Grain size, sorting and hydraulic diversity may be constrained by coarse sediments that armour the bed.

Bankfull-stage behaviour

Channels have significant capacity to adjust in vertical and lateral dimensions. Mobile channels are prone to vertical adjustment, exemplified by net aggradation of braidbelts. Mid-channel bars and channel margins are recurrently reworked, resulting in a wide range of instream and bank-attached geomorphic units.

Bankfull flows readily adjust channel size and shape. Bars are transient and unvegetated features that are regularly reworked as flow aligns to an increasingly down-valley orientation during flood events. Bar dissection and modification produces compound features with a range of platforms, chute channels and ridges. The nature and pattern of these features provide insight into geomorphic

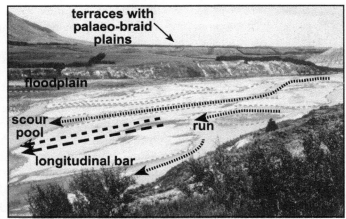

▸ low flow stage thalweg   - - ▸ bankfull stage thalweg
- - ▸ overbank stage thalweg

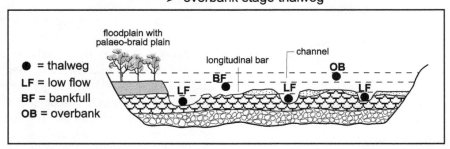

Figure 11.5   River behaviour in laterally unconfined, high-energy rivers – braided river type. Inundation of bar surfaces instigates erosion, reworking and redeposition, via both bar extension and thalweg shift. Photograph: Rakaia River, New Zealand (G. Brierley).

adjustments at differing flow stages up to bankfull and the history of flow events. Islands are formed if vegetation becomes established on these surfaces.

Overbank-stage behaviour

Gravel-bed braided rivers are prone to significant adjustment during overbank events. Extensive channel widening and thalweg shift may occur. Elevated bars may become more stable over time. Local channel incision may abandon and stabilise bar surfaces. Elevated surfaces may accumulate overbank fine-grained sediment, typically around vegetation. Floodplains formed by braid channel accretion and abandonment of sections of the braid belt are typically comprised of a wide range of geomorphic units. These areas are prone to reworking by lateral channel shift, avulsion and reoccupation of old braid channels.

The behavioural regime of laterally unconfined, medium-energy rivers

Laterally unconfined, medium-energy rivers are formed on moderate slopes and tend to be mixed-load systems. These sensitive rivers are prone to adjustment in vertical, lateral and wholesale dimensions. Bank-attached instream geomorphic units are observed more frequently than along high-energy rivers. These features are reworked less frequently than mid-channel bars. Composite banks are readily eroded and reworked, giving the channel significant capacity to adjust both laterally and vertically (i.e. both geometry and planform). Lateral migration results in a meandering channel planform (Figure 11.6). Along each bend, flow deflection results in deposition of point bars along convex banks and erosion of the concave bank. These rivers have significant potential for planform adjustment. Floodplains are formed by vertical or lateral accretion (within-channel and overbank deposits). They are also prone to wholesale adjustment via neck and chute cut-offs or avulsion of active meander belts. Floodplain formation and reworking generate a wide range of floodplain geomorphic units.

Low-flow-stage behaviour

At low flow stage, broken water is evident over riffles, while pools trap finer grained sediments in relatively still water.

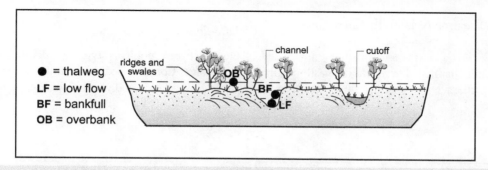

Figure 11.6 River behaviour in laterally unconfined, medium-energy rivers – active meandering river type. Helicoidal flow promotes deposition of scroll bars on the inner (convex) slope and erosion of the outer (concave) bank within asymmetrical channels of this mixed-load river. Ridge and swale topography demarcates phases of lateral migration, capped by overbank deposits (thereby creating composite floodplain sediments and associated banks). Cut-offs may be generated. Photographs: left, British Columbia (G. Brierley); right, Williams River, Alaska (N.D. Smith from http://lessen.museon.nl/generated/s900x600_1196c8e6add0dd0005daa11f5ae3bc8f.jpg).

The mobile bed is subject to recurrent shifts in character, composition, sorting and hydraulic unit diversity.

Bankfull-stage behaviour

Medium-energy meandering rivers have significant capacity to adjust in both the vertical and lateral dimensions at bankfull stage. The velocity (or shear stress) reversal hypothesis outlines a shift in flow dynamics at bankfull stage relative to low flow conditions. While pools are partially infilled at low flow stage, they tend to be scoured at bankfull stage. A regular pattern of scour and deposition reflects alternation of convergent and divergent flow along the channel, combined with secondary circulation currents. Surface flow convergence at the pool induces a descending secondary current which increases the bed shear stress and encourages scour, while surface flow divergence at the riffle produces convergence at the bed and thereby favours deposition. Once initiated, bed perturbations interact with flow to generate conditions necessary for the maintenance of riffle–pool sequences. Helicoidal flow in bends carries sediment up the convex slope of point bars while the thalweg is deflected to the concave bank, inducing scour and lateral movement of the channel. Where scour is greatest along the concave bank, a pool develops and an asymmetrical channel is formed. This self-sustaining lateral migration mechanism maintains the pool–riffle–point bar complex as patterns of erosion and deposition selectively transport and deposit material of differing calibre across the bed. In some instances, bankfull flows modify point bars as chute channels short-circuit the bend. These may be partially infilled by chute bar and ramp deposits, producing a compound point bar. Over time, scroll bars may develop adjacent to the point bar, initiating ridge and swale development on the floodplain.

Overbank-stage behaviour

Lateral migration is the dominant process at overbank stage. As the channel migrates, within-channel deposits form the basal component of the floodplain and are overlain by vertical accretion deposits laid down by overbank processes. Levees at channel margins may induce clear differentiation of channel-zone and floodplain processes and deposits. When flows go overbank, coarser sediments are deposited on the levee crest, while finer grained suspended-load materials accrete as flow energy is dissipated over the floodplain. Backswamp, wetland or floodplain ponds are formed in distal areas. Lateral accretion deposits are restricted to the proximal floodplain, where there is sufficient energy to rework part of the wide, alluvial plain. This results in a composite floodplain formed by two different sets of processes. Features such as palaeochannels, neck or chute cut-off channels, ridge and swale topography or flat floodplains provide a record of wholesale channel adjustment.

The behavioural regime of laterally unconfined, low-energy rivers

Interconnected multichannelled networks of anastomosing rivers commonly characterise laterally unconfined, low-energy rivers that form on very low slopes (Figure 11.7). Banks of these suspended-load rivers are comprised of vertically accreted silt and clay. Interconnected networks of narrow and deep, straight to sinuous channels are pinned in place by these cohesive, fine-grained banks, which may also have dense vegetation cover. Channels are unable to generate sufficient energy to promote bank erosion. These conditions limit the ability of channels to adjust laterally, such that they adopt a trench-like configuration, with a low width/depth ratio. The range of instream geomorphic units is limited because of the lack of bedload-calibre materials. Sculpted fine-grained geomorphic units are common. Areas between channels are characterised by extensive vertically accreted floodplains. Each anabranch may have its own planform. These moderately resilient rivers have localised capacity for adjustment.

Low-flow-stage behaviour

Flow between pools is typically discontinuous at low flow stage. Cohesive sediment limits hydraulic diversity, although vegetation and wood may create significant rough-

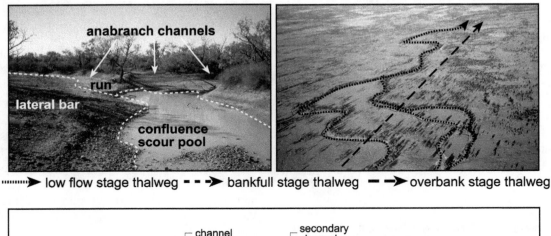

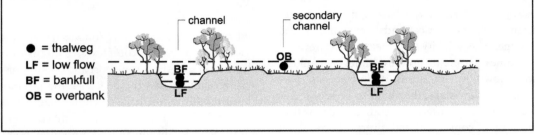

Figure 11.7 River behaviour in laterally unconfined, low-energy rivers with continuous channel(s) – anastomosing (anabranching) river type. Low flow stages deposit fine-grained sediments, partially infilling scoured pools. Erosion sculpts fine-grained instream geomorphic units at bankfull stage. Overbank flows inundate the floodplain, promoting vertical accretion. The channel banks of this suspended-load river are laterally stable as they are comprised of fine-grained cohesive sediment. Photographs: Cooper Creek, central Australia (K. Fryirs).

ness. Pools with standing water may partially infill with suspended-load sediments.

Bankfull-stage behaviour

Cohesive bank sediments limit adjustments to channel geometry in these suspended-load systems. There is little variability in geomorphic unit assemblage given the lack of bedload-calibre material. Pool scour at bankfull stage tends to occur at areas of flow convergence (e.g. at confluence zones between channels). Sections of planar bed between pools are subsequently accentuated with morphologies that resemble bars and shallow runs. During waning stages, fine-grained sediment drapes may accumulate via oblique accretion on channel banks and on bank-attached bars. Stepped banks and compound channel geometries may reflect erosional and/or depositional scenarios.

Overbank-stage behaviour

Floodplains are dominated by vertically accreted silt and clay forming relatively flat surfaces. In some cases, low levees and backswamps may form. Low width/depth ratio channels have little capacity to migrate. Changes in flow preference results in avulsion (channel abandonment) and differential rates of channel infilling.

The behavioural regime of laterally unconfined, low-energy rivers with discontinuous channels

In general, discontinuous watercourses are found in wide, unchannelised valleys where low-energy conditions promote the dissipation of floodwaters and the incremental accumulation of suspended-load materials via vertical accretion. The texture of the valley floor reflects sediment supply off adjacent hillslopes. Vegetation cover can induce significant resistance to shallow flows. Instream geomorphic units are seldom evident other than in discontinuous gullies, although swamps, floodouts, pools and ponds may be observed (see Figure 11.8). In their infilling (unchannelised) stage these rivers have a relatively simple geomorphic structure with little capacity for adjustment (i.e. they are moderately resilient). However, if subjected to incision they are prone to vertical, lateral and wholesale adjustment and are very sensitive.

Low-flow- and bankfull-stage behaviour

Analysis of river behaviour emphasises processes that form the valley fill along unchannelised rivers. These are more akin to floodplain processes and, hence, are interpreted as 'overbank'-stage behaviour. At low flow stage, flow is

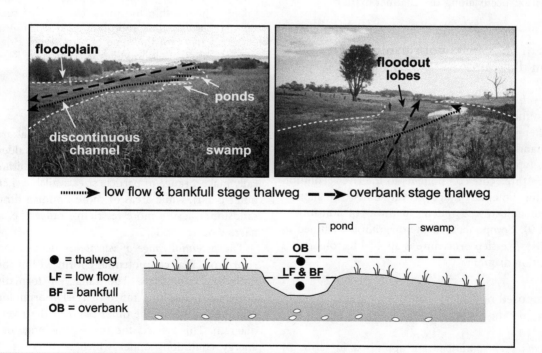

Figure 11.8 River behaviour in laterally unconfined, low-energy rivers with discontinuous channels – chain-of-ponds and floodout types. Unconfined flows dissipate energy over the valley floor, promoting vertical accretion of valley fills. At lower flow stages, virtually all flows are subsurface. Floodout deposits may occur as splay-like features immediately downstream of discontinuous channels. Photographs: left, Mulwaree River, NSW; right, Frogs Hollow Creek, NSW (K. Fryirs).

restricted to depressions or preferential drainage lines. Suspended-load deposits accrete in these depressions. Most flow is subsurface, retaining base flow to downstream reaches. Hydraulic diversity is limited.

Overbank-stage behaviour

At higher flow stages, a sheet of water may cover the entire valley floor. Overland flow generates suspended-load deposits. Any bedload materials are rapidly deposited as floodouts as flow energy is dissipated across the valley floor. If enough energy is created, ponds may scour along preferential drainage lines.

If these rivers incise an entrenched channel with a symmetrical or compound form is produced. Headcuts induce dramatic adjustments to river morphology. Concentration of energy within the channel produces an array of geomorphic units, including bank-attached and mid-channel features. If bank materials are relatively cohesive and resistant to change, the stepped channel cross-section may reflect erosion of flat-topped, elongate forms at channel margins (i.e. ledges). These are common along channelised fill rivers with a suspended-load transport regime. Where a sand substrate dominates, deposition adjacent to the bank produces benches. Both of these features are formed under high-flow-stage conditions, when erosion and deposition occur along the channel margin.

Analysis of river behaviour using the river evolution diagram

The *river evolution diagram* is a conceptual tool that can be used to summarise the range of river character and behaviour in different landscape settings. This tool builds upon an understanding of the 'natural' behavioural regime of a given river type.

There are three core components to the river evolution diagram: the potential range of variability, the natural capacity for adjustment and the pathway of adjustment (Figure 11.9). Components of the diagram are defined in Table 11.1. A five-step procedure is applied to construct a river evolution diagram.

Step 1. Imposed boundary conditions and the potential range of variability

Imposed boundary conditions are appraised in terms of valley setting, slope and lithology (Figure 11.9). Over geomorphic timeframes, these conditions are effectively set. Upstream catchment area, slope, valley confinement and sediment calibre determine the energy conditions under which rivers operate. Geologic setting influences landscape

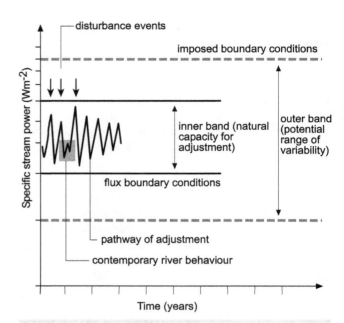

Figure 11.9 Components of the river evolution diagram. This conceptual scheme provides a framework to examine how rivers adjust over time. Energy settings are determined by imposed boundary conditions (outer band) and prevailing flux boundary conditions (i.e. flow and sediment regimes; inner band). When subjected to differing forms of disturbance events, the river adopts a pathway of adjustment (the jagged line within the inner band). This records the pattern and rate of morphological variability that is characteristic for that type of river. From Brierley and Fryirs (2005). © John Wiley & Sons, Ltd. Reproduced with permission.

relief and the range of material textures that are available to the river (i.e. whether it is bedrock-, boulder-, gravel-, sand- or mud-dominated). Areas of mixed lithology typically make a range of particle sizes available (hence a wide outer band), while areas of more uniform lithology (e.g. sandstone) have a more restrictive range (i.e. a relatively narrow outer band).

The *potential range of variability* defines the range of river types that can potentially form within the imposed boundary conditions. The range of formative stream powers and resulting range of river morphologies determine the width of the outer band of the river evolution diagram. This reflects the maximum range of formative energy conditions under which a range of river types operate for that specific landscape setting at that position in the catchment.

Stream power (see Chapter 5) is considered to provide the most appropriate measure with which to differentiate among variants of river settings (i.e. the *y*-axis of the river

Table 11.1   Definition of components of the river evolution diagram[a]

Component	Definition
Specific stream power (see Chapter 4)	Total stream power is calculated as the product of discharge acting in any given cross-section multiplied by channel slope. Specific stream power determines the energy acting on a given area (e.g. per unit channel width). Hence, as channel geometry adjusts, so does the use of available energy. Specific stream power is represented on the y-axis of the river evolution diagram using a logarithmic scale. Geomorphic work reflects the ability of a flow to induce adjustment in bed character, channel morphology, the assemblage of geomorphic units and channel planform without inducing change to a different river type.
Time	Represents the timeframe over which the full suite of behaviour occurs for a particular river type. Shown using a linear scale on the x-axis of the river evolution diagram.
Outer band	Reflects the *potential range of variability* in the types of rivers that can form under a certain set of *imposed boundary conditions* (i.e. valley setting, slope and lithology).
Inner band	Reflects the *natural capacity for adjustment* for a particular river type which represents the degree to which vertical, lateral and wholesale adjustments can occur for that type. The width of the inner band is defined by the *flux boundary conditions*, i.e. the range of flow and sediment fluxes and vegetation dynamics that dictate the potential extent of adjustment in the assemblage of geomorphic units, channel planform, channel morphology and bed character for that type.
Pathway of adjustment	Defined by the frequency and amplitude of system responses to disturbance events. The shape of the pathway reflects the variability in the trajectory and timeframe of recovery in response to disturbance events. This records the behavioural regime of a river. Some rivers may adjust among multiple states.
Disturbance event	Formative events that induce geomorphic adjustments to a river type. The size of the arrows represents the relative magnitude of the event that induced adjustment. The sequence of events may be a significant determinant of geomorphic response.
Contemporary river behaviour	Adjustments that take place under contemporary flux boundary conditions while maintaining the river type.

[a] From Brierley and Fryirs (2005). © John Wiley & Sons, Ltd. Reproduced with permission.

evolution diagram) as it reflects both the amount of energy that is available to be utilised in any given setting (total stream power) and it refers to the manner with which energy is used, as determined by channel capacity (and active channel width; i.e. specific stream power). Adjustments to channel geometry modify the use of energy, thereby altering the position of differing river settings (and associated channel configurations) within the outer bank on the river evolution diagram.

It is recognised explicitly that adjustments in other external variables may alter the width of the inner band (see below) or its position within the potential range of variability. For example, an influx of sediment may alter various attributes of river morphology, including channel capacity, thereby modifying formative specific stream power conditions. These mutual adjustments accentuate the underlying role of stream power as the most appropriate single determinant of river character and behaviour.

The placement of the outer bands of the potential range of variability is dependent on the energy conditions that fashion the range of river types that can occur in a particular valley setting. Stream power modelling is conducted to establish the highest and lowest values of formative stream powers that occur for this range of river types. These bands are normally added after Step 2 has been completed.

Step 2. Flux boundary conditions and the natural capacity for adjustment

The width of the inner band represents the contemporary range of *flux boundary conditions* within which the reach operates (Figure 11.9). Combinations of these factors, operating within the imposed boundary conditions, determine the range of river types and behavioural states that could be observed in that setting. The prevailing flux boundary conditions may be quite different to those experienced in the past. Hence, different types of rivers with differing characters and behavioural regimes may be observed within the same set of imposed boundary conditions.

The characteristic form for a given river type is not a static configuration or structure; rather, it reflects an array of potential adjustments among the assemblage of geomorphic units, channel geometry, channel planform and bed material organisation as determined by the contemporary range of flow, sediment and vegetation conditions. These considerations determine the *natural capacity for adjustment*, as shown by the width of the inner band on the river evolution diagram. The potential extent of adjustments is measured in terms of the range of formative specific stream powers that induce adjustments to various attributes of river morphology without resulting in river change. Rivers with significant natural capacity to adjust have wide inner bands. Those with limited natural capacity to adjust have narrow inner bands.

Behavioural attributes of the river evolution diagram outlined in this chapter are framed in relation to contemporary flux boundary conditions over timeframes in which a characteristic set of process–form associations has become established for a particular type of river. This timeframe may vary markedly from setting to setting and for different types of rivers. For some river types, the 'natural' behavioural regime may comprise differing states. In these instances, transitions between states in response to breaching of internal (intrinsic) threshold conditions are considered to be part of the natural capacity for adjustment for that type of river. Examples include cut-and-fill rivers, partly confined valleys prone to floodplain stripping, meandering rivers that adjust their slope following generation of cut-offs or rivers subjected to avulsion or changes in channel multiplicity. In general terms, the width of the inner band that conveys possible states varies with the ease of adjustment of the river. Sensitive rivers have wider bands than resilient rivers, reflecting the inherent range in the degrees of freedom within which rivers operate.

As each reach adjusts to disturbance events, the nature and extent of response may vary markedly. In terms of the behavioural regime of a river, the type and extent of adjustment do *not* result in the adoption of a different river character and behaviour (i.e. river change, as discussed in Chapter 12). The prevailing flux boundary conditions fashion the contemporary river type and its behavioural regime, as determined by its natural capacity for adjustment. The *natural capacity for adjustment* determines the range of *behaviour* that any particular type of river may experience, while the *potential range of variability* determines the range of river types that may be found in that landscape setting (i.e. within its imposed boundary conditions), thereby providing a measure of the possible states that the river could adopt if *change* occurred.

Natural rivers dynamically adjust so that their geomorphic structure and function operate within a range of variability that is appropriate for that type of river, and the range of flux boundary conditions under which that river type operates. Natural or expected river character and behaviour reflect the range of processes and associated forms that occur within the bounds determined by the inner band on the river evolution diagram.

The natural capacity for adjustment varies markedly for differing types of rivers, over differing timeframes. This reflects a combination of factors, such as:

1. *The variability of sediment mix at any given point along a river.* This may reflect local considerations that determine the relative balance of, say, gravel, sand and finer grained particles, or the influx of materials from upstream.
2. *The flow regime.* Some rivers are adjusted to relatively uniform flow conditions in which mean annual floods are the primary determinant of river form. In these situations, the inner band is relatively narrow. However, if the system is adjusted to significant flow variability, the inner band is wider.
3. *Riparian vegetation and wood.* These components of flow resistance vary markedly from setting to setting, potentially exerting a significant influence on the natural capacity for adjustment of certain types of rivers.
4. *System history.* In some instances, longer term climate-induced changes to the nature and pattern of sedimentation on the valley floor may impose constraints on contemporary system behaviour (e.g. gravel terraces or fine-grained cohesive banks that line river courses), thereby imposing a narrow band to the natural capacity for adjustment.

Step 3. Placing rivers within the potential range of variability

This step in construction of the river evolution diagram positions the river within the potential range of variability, based on prevailing energy conditions (Figure 11.9). If the contemporary river operates under relatively high energy conditions, the inner band is situated high in the potential range of variability. Alternatively, if contemporary energy levels are low (relative to the range of conditions that can be experienced under the imposed boundary conditions), the inner band is placed towards the bottom of the potential range of variability. The width of the inner band reflects the range of energy conditions experienced under prevailing flux boundary conditions.

Specific stream power modelling of multiple cross-sections within a particular reach is undertaken for the range of flow events up to and including the 1:100-yr flood. Estimates of the highest and lowest stream powers are used to position the upper and lower limits of the inner band respectively.

Stage 4. The pathway of adjustment

Responses to differing forms of disturbance must be appraised to assess the types and extent of adjustment that define the range of expected character and behaviour of a given river type. Collectively, these adjustments define the *pathway of adjustment* on the river evolution diagram (Figure 11.9). The behavioural regime of any given type of river, as defined by the natural capacity for adjustment, encompasses ongoing responses to alterations in flux boundary conditions. Reaches may operate at different positions within their natural capacity for adjustment as pulse disturbance events of differing magnitude and frequency alter water and sediment regimes and vegetation associations (Chapter 2). If a press disturbance breaches threshold conditions, positive feedback mechanisms may drive the system to a different state, possibly inducing a change in river type (Chapters 2 and 12). These considerations determine the pathway of adjustment of a reach, as marked by modifications to the arrangement and abundance of geomorphic units, adjustments to the organisation of material on the channel bed and local alterations to channel planform. Within the inner band of the river evolution diagram, system responses to disturbance events may be indicated by oscillation around a characteristic form or adjustments among various characteristic forms. A characteristic form retains key geomorphic attributes that reflect the fundamental process–form associations for the given river type.

The form of the pathway of adjustment summarises system response to disturbance events, indicating how any given river type is able to accommodate adjustments to flow and sediment transfer. Variability in specific stream power estimates integrates all forms of adjustment to determine the pathway shown on the river evolution diagram. In essence, these considerations describe the morphologic and behavioural adjustments to ongoing variability in the nature, extent and sequence of disturbance events on the one hand (i.e. impelling forces) and the capacity of the system to absorb prospective disruptions on the other hand (i.e. the effectiveness of response mechanisms as conditioned by resisting forces along the reach).

As noted above, river responses to disturbance events reflect reach sensitivity, measured as the ease with which the river is able to adjust its form. This provides a measure of the capacity of the system to accommodate the impacts of disturbance events via mutual adjustments, such that the river is able to sustain a characteristic form. Breaching of intrinsic thresholds may promote adjustments among various states that represent the behavioural regime for some river types. Disturbance events are indicated schematically on the river evolution diagram by arrows on the edge of the inner band (Figure 11.9). The frequency and sequence of disturbance events are conveyed by the spacing of arrows, while the size of the arrow indicates the relative magnitude of the event.

The form of the pathway of adjustment is defined by its amplitude, frequency and shape (Figure 11.10a). *Amplitude* reflects the extent of adjustment in response to a disturbance event. *Frequency* reflects the recurrence with which disturbance events drive geomorphic adjustments. The *shape* of the pathway of adjustment reflects the trajectory of response to disturbance events. Variants include progressive adjustments in a particular direction, oscillations around a mean condition or jumps between characteristic states. The spacing of disturbance events that drive adjustment varies in differing settings, influencing the river type and its sensitivity to adjustment. In behavioural terms, however, the collective response to disturbance events does not drive the system outside its natural capacity for adjustment.

The pathway of adjustment summarises system responses to sequences of disturbance events of varying magnitude and frequency. While it is possible to quantify the specific stream power conditions that are used to define the potential range of variability and the natural capacity for adjustment, it is not yet possible to model the pathways of adjustment in a simple way. Conceptual examples of differing forms and timeframes of system recovery that determine the shape of the pathway of adjustment are shown in Figure 11.10. The type and timeframe of response depend partly on whether the disturbance event induces adjustments that reinforce or counteract existing tendencies and whether specific stream power will likely increase or decrease in response. Recovery time may be highly variable, reflecting the condition of the system at the time of the impact, as influenced by the recent history of events, among many considerations. Disturbance responses may be instantaneous or delayed (i.e. lagged responses). Their consequences may be short lived or long lasting. Combinations of disturbance responses, and the resulting shape of the pathway of adjustment, can be simple (temporally uniform) or complex (temporally variable). The form of the pathway of adjustment represents an interpretative summary of river behaviour at low flow, bankfull and overbank stages.

If the geomorphic response is damped out, and the previous state is regained after a short recovery time, the pathway of adjustment has a jagged shape reflecting minor adjustments away from a characteristic form. This form of adjustment is exemplified by cut-off formation along an actively meandering river (Figure 11.10bA). Elsewhere, progressive adjustments may promote shifts to an alternative characteristic form, with an altered nature and/or level

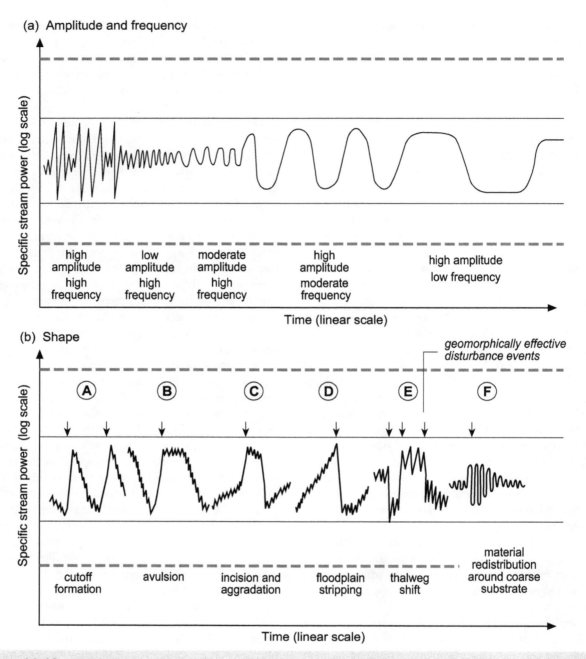

Figure 11.10   Components of the pathway of adjustment as used in the river evolution diagram. Significant variability in the form and rate of adjustment occurs for different types of rivers. Three components are considered in appraisal of pathways of adjustment, namely (a) amplitude, frequency and (b) shape. From Brierley and Fryirs (2005). © John Wiley & Sons, Ltd. Reproduced with permission.

of activity, but adjustments remain within the natural capacity for adjustment for that river type. In this case, steps along the pathway of adjustment record shifts among multiple characteristic states. Intervening flatter areas record minor modifications around one of these states. These types of rivers are prone to cyclical patterns of threshold-induced adjustments, such as avulsion (Figure 11.10bB), incision and aggradation (Figure 11.10bC) and floodplain stripping (Figure 11.10bD). Reaches that are prone to abrupt adjustments also have a cyclic pattern of adjustment with short recovery times. However, this pathway reflects recurrent (tight) oscillations around a characteristic form, as exemplified by thalweg shift in a braided river (Figure 11.10bE) or redistribution of bedload material around coarse substrate in a gorge (Figure 11.10bF).

Schematic applications of the pathway of adjustment for various river types are presented in Figure 11.11. The natural capacity for adjustment for a gorge may be relatively narrow, with adjustments maintaining a uniform state over timeframes up to 10^3 yr (Figure 11.11a). These deeply etched bedrock rivers are resistant to change and demonstrate very short periods of disturbance response, such that adjustments are barely discernible over the short to medium term (<10^2 yr). As the river has limited capacity to adjust, it is characterised by a low-amplitude, high-frequency pathway of adjustment within a narrow inner band.

Rivers in partly confined valley settings may be prone to floodplain stripping (Figure 11.11b). Although this type of river has relatively limited capacity for adjustment, and is considered to be resilient to change, it demonstrates stepped adjustments over timeframes of 10^3–10^4 yr. Such adjustments include channel expansion, floodplain building and floodplain reworking via stripping mechanisms. This is induced by the breaching of an energy threshold. The pathway of adjustment reflects different phases of response to disturbance events, as the river adjusts between periods of progressive floodplain aggradation and short periods of catastrophic erosion. During the aggradation phase, disturbance events tend to have a lower amplitude and lower frequency as periods of floodplain inundation decrease. Eventually, catastrophic events bring about floodplain stripping during a short phase of adjustment that is characterised by high-amplitude, moderate-frequency responses to disturbance.

Bedrock-based laterally unconfined rivers tend to act as transfer reaches, sustaining an approximate balance between sediment input and output with a relatively thin veneer of deposits over the valley floor. On the river evolution diagram, a low-sinuosity variant of this river with cohesive banks is characterised by low-amplitude, low-frequency adjustments over timeframes of 10^2–10^3 yr (Figure 11.11c). The river oscillates around a relatively stable form and configuration.

Braided rivers have significant capacity to adjust, with a wide inner band (Figure 11.11d). Frequent disturbance events induce recurrent reworking of bedload material via thalweg shift, flow stage adjustment and local adjustments to bed level over timeframes of 10^0–10^1 yr. The pathway of adjustment is characterised by low-amplitude, high-frequency responses to disturbance with short recovery times.

An active meandering sand-bed river in a laterally unconfined valley setting has a wide range in its natural capacity for adjustment (Figure 11.11e). These sensitive reaches have significant capacity to adjust both vertically and laterally. Progressive channel migration builds the meander belt over time. As sinuosity increases, the energy of the system decreases. Cut-off channels may induce phases of disturbance response as the channel readjusts its slope to the reduced sinuosity, typically over timeframes of 10^1–10^2 yr. The river is then subjected to progressive adjustments as the characteristic meandering form is maintained. Over longer timeframes, meandering sand-bed rivers may be prone to avulsion, as they adjust their course beyond the meander belt and sediments accumulate elsewhere on the valley floor. Following avulsion, the river re-establishes its meander belt via lateral migration and vertical accretion. Hence, this type of river is characterised by a stepped pathway of adjustment, with a wide range of disturbance responses of varying amplitude and frequency.

Low-energy alluvial rivers tend to be moderately resilient to adjustment. Although these rivers have a wide range in their natural capacity for adjustment that includes modifications to channel morphology and shifts in channel position on the valley floor, cohesive channel boundaries induce progressive rather than dramatic geomorphic adjustments. In the anastomosing example presented here, the pathway of adjustment is characterised by high-amplitude but low-frequency disturbance responses, as occasional avulsion events alter channel multiplicity (Figure 11.11f).

Cut-and-fill rivers have significant natural capacity for adjustment, as they oscillate between two characteristic states (Figure 11.11g). During the aggradation phase, discontinuous channels are quite resilient to adjustment. Eventually, however, exceedance of a threshold condition may promote dramatic incision and formation of a continuous channel. During the incision phase, the system responds more dramatically to disturbance events, as the energy and sensitivity of the system are enhanced. Over time, the channel infills, producing an intact valley floor once more. Typically, cut and fill cycles occur over timeframes of 10^2–10^3 yr. Responses to disturbance events vary during these different phases, with low-amplitude and low-frequency responses during the fill stage, but high-amplitude and high-frequency responses during the cut stage.

Step 5. Contemporary river behaviour

The final component in construction of the river evolution diagram entails determination of the *contemporary behaviour of the river* (see Figure 11.9). The contemporary river can sit anywhere on the pathway of adjustment for the river type. Appraisal of river behaviour is based on how the river adjusts its form in relation to contemporary flux boundary conditions. In some instances, former flow and sediment

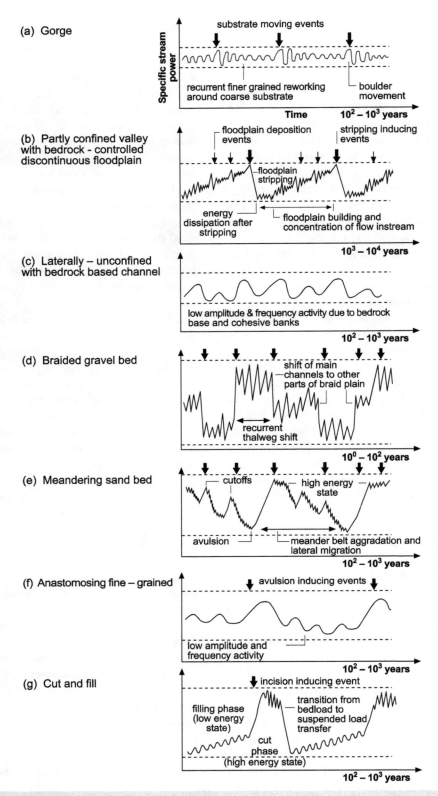

Figure 11.11 Pathways of adjustment for different types of rivers. This diagram, which only conveys the inner band of the river evolution diagram, indicates different pathways of adjustment in response to disturbance events of variable recurrence for different types of rivers. Profound differences in the ranges of variability and the timeframes of adjustment are indicated (see text for details). From Brierley and Fryirs (2005). © John Wiley & Sons, Ltd. Reproduced with permission.

conditions may impose constraints on the contemporary range of river character and behaviour. For example, it may take the system a considerable period of time to adjust to a major flood event that mobilised the coarsest bedload fraction if more frequent, lower magnitude events are unable to do so.

At any point in time, a river can operate anywhere within its natural capacity for adjustment. As each river type has a distinct set of process–form associations, its character and behaviour adjust to a given set of disturbance events through a certain range of responses. Ongoing interactions form and rework geomorphic units. The operation of flow and sediment fluxes and prevailing vegetation conditions shape the present character and behaviour of the river. Assessment of river behaviour is framed in terms of the period of time over which flux boundary conditions have remained relatively uniform such that a characteristic river form results, with a particular assemblage of geomorphic units, bed material organisation and channel planform.

Bringing it all together: examples of river evolution diagrams

Once the specific stream power modelling has been conducted for the river under consideration, the inner and outer bands are placed on a log–linear graph. The *y*-axis represents the range of specific stream powers under which different river types behave and the *x*-axis reflects the timeframe over which the river has remained in this characteristic state. Known dates/times of disturbance events can be added to this timeline. The pathway of adjustment is drawn as a conceptualisation of behaviour. This represents formative geomorphic adjustments for that type of river, indicating how adjustments to channel and floodplain forms increase or decrease specific stream power.

Examples of river evolution diagrams for rivers in differing valley settings are portrayed in Figure 11.12. A wide valley, with a relatively steep slope within a granitic catchment has a wide band. As the valley setting is laterally unconfined, there is considerable range in the energy conditions under which the river operates, and materials of differing calibre are available to be moved (Figure 11.12c). As such, a wide range of river morphologies and associated process domains may be adopted in this setting. A partly confined valley with a lower slope within a metasedimentary catchment will have a narrower band, as moderate energy conditions, valley confinement (i.e. less space to adjust) and the mixed texture of the sediment load produce a restricted range of river morphologies (Figure 11.12b). These situations contrast significantly with, say, a narrow, steep valley in a volcanic terrain, which is represented by a narrow band, as the confined valley setting and the uniform

sediment load impose particular river morphologies within a narrow range of high-energy conditions (Figure 11.12a). An influx of sediment in volcanic terrains can induce progressive geomorphic adjustments as large volumes of material are reworked. The position of different rivers within the imposed boundary conditions in Figure 11.12 reflects an energy gradient from high-energy variants on the left to low-energy variants on the right.

Predicting river responses to altered flux boundary conditions

The set of relationships between discharge acting on a given slope and the volume of sediment of a given calibre that is conveyed on the Lane balance (Chapter 5) presents an elegant qualitative framework with which to predict morphological adjustments to the bed of a river system. As what happens on the bed is a fundamental control upon bank processes, whether erosional or depositional, this is a critical guide to assessment of river character and behaviour and associated channel stability.

Theoretical and mathematical modelling applications provide quantitative insights into the direction, rate and extent of channel adjustments to altered boundary conditions. These relationships are derived primarily for uniform flow and sediment delivery scenarios (i.e. regime conditions). Channels respond to altered flow and/or sediment conditions by either eroding or depositing sediment. Channel responses vary in relation to prevailing morphology (e.g. sinuosity, degree of braiding) and the make-up of sediments and resisting elements on the valley floor. Based on empirical relations derived for stable channels under regime conditions, discharge Q can be related to channel width w, channel depth d, grain size D_{50} and slope s (Chapter 7). Typically, rivers are subjected to variable boundary conditions. As such, considerable caution must be heeded in predicting likely future adjustments based upon linear cause-and-effect reasoning. Due regard should be given to inherent uncertainties that underpin system-specific applications. Non-linear behaviour is common, often resulting in surprising outcomes. This presents significant challenges in interpretations and predictions of river behaviour for dynamic and evolving channels. While generalised relationships provide a helpful guide, placing reach-specific conditions in context of catchment-scale considerations is fundamental to meaningful prediction of likely future river adjustments. The best starting point for these analyses of river behaviour is to ask questions such as 'How can this river adjust?' 'Under what flow conditions does adjustment occur?' and 'What types of geomorphic units are being formed and reworked under these conditions?'

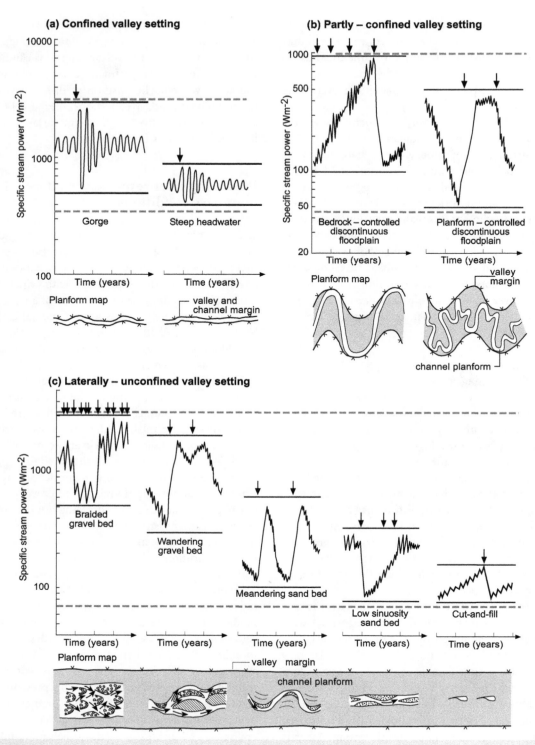

Figure 11.12 Schematic examples of the river evolution diagram in differing valley settings; (a) confined, (b) partly-confined, (c) laterally-unconfined. Note the variability in specific stream power estimates for these differing behavioural pathways of adjustment. See text for details. From Brierley and Fryirs (2005). © John Wiley & Sons, Ltd. Reproduced with permission.

Tips for reading the landscape to interpret river behaviour

Step 1. Identify individual landforms and their process–form associations

Appraise flow and energy conditions under which each geomorphic unit is formed and reworked (Chapters 8 and 9). This analysis relates impelling and resisting forces to available sediments to determine how materials are reworked within channel and floodplain compartments. Magnitude–frequency relations of formative processes are assessed. Remember, some instream and floodplain geomorphic units may be products of former flow conditions.

Step 2. Interpret river behaviour at the reach scale

Interpret reach-scale geomorphic adjustments over time-frames in which flux boundary conditions remain relatively uniform (i.e. geomorphic timescales; Chapter 2). Appraise behavioural attributes in relation to the assemblage of geomorphic units that make up a reach. A river's *capacity for adjustment* reflects the ways in which the channel can adjust in vertical, lateral and wholesale dimensions. Capacity for adjustment is interpreted by assessing the nature and frequency of adjustments to bed material organisation, bed/bank processes and channel geometry, assemblages of channel and floodplain geomorphic units and channel planform (including how the channel adjusts its position on the valley floor). Analysis of river sediments provides insight into the environments of deposition and reworking (Chapter 6). Interpreting boundaries between units (position, shape, definition) determines whether units were formed by a genetically linked set of processes, or whether reworking has altered or removed other geomorphic units. For example, inset units may be formed by a different set of processes relative to adjacent features. This may be demarcated by an erosional boundary, as shown by terrace risers.

Analysis of process–form associations of individual geomorphic units, and interactions between these features, enables interpretation of how the package/assemblage is formed and reworked under low flow, bankfull and overbank conditions. This aids assessment of controls upon the position and pattern of geomorphic units, determining what sequence of flow events brought about this assemblage of features.

Low-flow-stage behaviour involves analysing bed material organisation and sediment transport, assessing formative conditions for bedform generation and explaining depositional patterns (Chapter 6). In a sense, preservation of these features provides a record of the lack of subsequent events which have been able to rework these features (i.e. in many instances, these deposits provide insight into waning-stage conditions for the previous formative flow event that was able to deform the bed).

Analysis of *bankfull-stage behaviour* interprets formative processes and conditions that fashioned channel geometry and the assemblage of instream geomorphic units (Chapters 7 and 8 respectively). This includes assessment of changing flow alignment and energy conditions within the channel. The suite of erosional and depositional processes is summarised at the reach scale, differentiating among processes operating on the channel bed and bank, and whether resulting bar forms are primarily mid-channel or bank-attached features. It is critical to determine the aggradational/degradational balance of the channel bed and then to interpret erosional and/or depositional processes along the banks. From this, within-reach variability in channel geometry and associated channel–floodplain interactions are appraised. This entails assessment of whether channel boundaries are fixed (i.e. imposed or forced morphologies) or whether they are alluvial (and how readily they are reworked; i.e. Is this a bedload, mixed-load or suspended-load river?). Is bank composition consistent along the reach? How does sediment composition of the banks affect the width/depth ratio of the channel? What is the geomorphic role of vegetation and wood along the reach (Chapter 5)? Do ledge or bench features indicate channel expansion or contraction? Is there any evidence of bedrock or forcing elements (e.g. ancient alluvium or wood) that impose channel size and shape? How has human disturbance affected channel size and shape (Chapter 13)?

Overbank-stage behaviour interprets how the floodplain is formed and reworked and how the channel shifts over the valley floor. Analysis of floodplain geomorphic units provides insight into the behavioural regime of the river during overbank events (Chapter 9). This includes assessment of whether adjacent geomorphic units are genetically linked. From this, a determination is made as to how the floodplain formed and has been reworked and whether it is a product of the contemporary flow regime. A key interpretative guide in this assessment is determination of whether the contemporary river could have deposited the materials that make up the channel banks. Appraisal of formative flows should be linked to the alignment of flow at differing overbank-flow stages. Typically, the influence of the channel becomes less pronounced at higher flow stages, as flow more closely follows valley alignment. These analyses provide insight into the ability of the channel to move on the valley floor (type and frequency of adjustment), and the forms of lateral and/or wholesale adjustment.

Determination of the *natural range of variability* interprets whether the behavioural regime of the river reflects oscillations around a characteristic form or a set of characteristic forms (Chapter 2). This interpretation determines whether negative feedback mechanisms maintain the balance of slope, bed material size, channel size, etc. such that morphological attributes of the river self-correct over a given timeframe. Assessing the role of resistance factors and the timeframe over which these interactions occur is particularly important.

Step 3. Explain controls and impacts on river behaviour at the reach scale

Explanations of river behaviour entail assessment of how energy conditions and the mix of impelling (erosional) and resisting (depositional) forces mould available materials (sediment calibre and volume) under the range of flow conditions, thereby producing the characteristic morphology of the reach. Flow–sediment relationships fashion the aggradational–degradational balance, as noted by the Lane balance diagram (Chapter 5).

Efforts to relate geomorphic adjustment to the *discharge regime* determine how the climate setting affects river behaviour (Chapter 4). This includes analysis and flow variability, seasonality and the range of flood events. Is the stream perennial, intermittent or ephemeral? How often do bankfull stage and/or formative flows occur (magnitude–frequency relations)? How variable are flows (what is the coefficient of variation)? Is river behaviour fashioned primarily by regular or irregular events? How does the river use its energy at differing flow stages? Which surfaces are inundated under which flows, and what geomorphic work is done on each geomorphic surface – erosion and/or deposition? This entails analysis of the periodicity and duration with which flows exceed critical shear stress.

To relate geomorphic adjustment to the *sediment regime* (Chapter 6) it is important to determine whether the bed condition is imposed or not (e.g. bedrock; a historical artefact). Is it a bedload, mixed-load or suspended-load river? How much sediment is available for reworking? How frequently are materials on differing surfaces entrained and transported? How often is the coarsest material on the bed mobilised? How far it is likely to move? Is the reach supply or transport limited? Other important material properties include appraisal of the looseness of bedload material (packing, armour, paving, etc.) and differentiation of the mobile (active) fraction from pavement (lag) fractions. How readily do materials of different sizes move within channels of differing size (depth)? The range of bed material sizes on differing surfaces provides insight into the formative energy of geomorphic units and, hence, their range of behaviour. Therefore, it is important to assess the sedimentary make-up of different instream geomorphic units. What deposition and reworking processes are responsible for their position and morphology?

Resistance elements in the channel and on the floodplain must be assessed (Chapter 5). Are resisting elements 'forced' or are they primarily determined by adjustments in channel and boundary materials? How effective are these forms of resistance? How does the distribution of resistance elements affect patterns of erosion and deposition, and the resulting behavioural regime of the reach?

The river evolution diagram provides a visual summary of these relationships. Determination of the imposed boundary conditions is used to assess the potential range of variability of the reach. This measure of the total energy available to do geomorphic work in a valley sets the width and energy level of the outer band on the diagram. Flux boundary conditions under which the river operates influence the natural capacity for adjustment of the river type. This measure of the range of specific stream power conditions under which the river behaves sets the width of the inner band and its position within the outer band on the diagram. The width of the inner band reflects the capacity for river adjustment for the type of river under investigation. Finally, river behaviour is appraised to construct the pathway of adjustment. This depicts all forms of adjustment, the recurrence with which they occur and the types/role of disturbance events in triggering these adjustments.

Step 4. Explain how catchment-scale relationships affect river behaviour

River behaviour must be explained in relation to catchment-scale controls, framing each reach in its process domain (source, transfer or accumulation zone), alongside determination of the upstream catchment area, position in the landscape and the prevailing balance of impelling and resisting forces (i.e. slope–discharge relations for a given valley setting). Catchment-scale investigations assess how catchment size and shape, and associated distribution of tributaries, affect river behaviour along the trunk stream (Chapter 3). Process responses are framed in relation to the slope of the longitudinal profile and available energy, alongside the space within which the channel is able to move (i.e. patterns of valley setting and accommodation space). Valley morphology may exert a key control upon separation of channel and floodplain compartments and associated dissipation of flow energy. Transitions in river behaviour may (or may not) be evident upstream/downstream of tributary confluences. This reflects alterations to flow and sediment flux of the trunk stream. Hillslope–valley floor and reach–reach connectivity affect sediment input and resulting behaviour. Geologic and climatic disturbance events may profoundly alter the flow–sediment balance (e.g. volcanic eruption, cyclonic storm). Hence, appraisal of boundary conditions that fashion the behavioural regime of a river must incorporate assessment of the role of differing types, combinations and frequencies of disturbance events.

Regional geology sets the erodibility and erosivity of a landscape, as determined by its drainage pattern/density (landscape dissection) and availability of materials (calibre and volume). Climate setting and landscape configuration fashion flood history and effectiveness. Imposed boundary conditions reflect the geologic and topographic setting, including variability in valley morphology (slope and confinement). Human disturbance may have greatly altered these 'natural' conditions. Flow and sediment regimes and vegetation associations determine the behavioural regime of a river. Analysis of patterns and relationships along the longitudinal profile provides a powerful foundation to explain differences in behavioural regime in different parts of catchments.

Conclusion

Efforts to interpret river behaviour are framed in relation to the natural range of variability of any reach, viewing that reach in its catchment context. System dynamics are fashioned by ongoing flow and sediment fluxes and the history/impact of disturbance events. Adjustments around a characteristic form define the behavioural regime for a particular type of river. River change occurs when a reach adopts a different behavioural regime that is characterised by a different set of geomorphic units and associated process–form relationships (Chapter 12). In some instances, changes are irreversible; elsewhere they are not.

Different types of rivers have variable capacity to adjust in vertical, lateral and wholesale dimensions. The form and extent of geomorphic adjustment reflects bed material attributes (Chapter 6), bed and bank processes that fashion channel geometry (Chapter 7) and erosional and depositional processes that create and rework geomorphic units in channels and on floodplains (and associated channel planform attributes; Chapters 8 and 9).

The natural capacity for adjustment of any river is determined by the ease with which different forms of adjustment take place (i.e. sensitivity) and the range of disturbance events to which a river is subjected (and their magnitude–frequency relationships). Sensitivity is largely a product of the nature of available materials and the distribution of resistance elements along the valley floor. Magnitude–frequency relationships describe the geomorphic effectiveness of disturbance events in differing climatic settings. Some rivers are attuned to high-magnitude events that leave a persistent imprint upon the landscape. Others are progressively reworked by frequent, low-magnitude events.

River behaviour at low flow, bankfull and overbank flow stages is interpreted through analysis of the assemblage of instream and floodplain geomorphic units. Marked differences in behavioural regime are evident for rivers in different valley settings with differing energy conditions. Changes to boundary conditions may bring about a transition to another type of river with a different behavioural regime. These evolutionary considerations are addressed in Chapter 12.

Key messages from this chapter

- River behaviour reflects ongoing geomorphic adjustments that occur over timeframes in which flux boundary conditions remain relatively uniform. It is defined as adjustments to river morphology induced by a range of erosional and depositional processes by which water moulds, reworks and reshapes fluvial landforms, producing characteristic assemblages of geomorphic units at the reach scale.
- Geomorphic units are a unifying feature with which to interpret river behaviour for all types of rivers (erosional and depositional forms; channel and floodplain compartments).
- Rivers have the potential to adjust in vertical, lateral and wholesale dimensions. Vertical channel adjustments refer to the stability of the bed and include incision and aggradation processes. Lateral channel adjustments refer to processes of channel expansion and contraction. Wholesale channel adjustments refer to adjustments of the position of the channel on the valley floor and include the processes of lateral migration, avulsion, floodplain stripping and thalweg shift.
- The capacity for river adjustment is a measure of the range and extent of geomorphic adjustment that can occur for a given type of river (i.e. its natural range of variability). Rivers that can adjust in all three dimensions have greatest capacity to adjust.
- Rivers in confined valleys have limited capacity to adjust, while rivers in laterally unconfined valleys have considerable capacity for adjustment.
- River sensitivity is a measure of the ease with which a river can adjust and its proximity to a threshold condition. Rivers in laterally unconfined valleys with non-cohesive bed and bank materials are particularly sensitive to adjustment.
- Flood magnitude, frequency and duration induce markedly different geomorphic responses along different

types of rivers. The amount of geomorphic work performed and the geomorphic effectiveness of an event depend on the type of river and its capacity to adjust.

- For some rivers, recurrent bankfull-stage flows are the formative agents of geomorphic adjustment. In other instances, high-magnitude–low-frequency floods are the primary formative flows.
- River behaviour is interpreted at low flow stage, bankfull stage and overbank stage. Low-flow-stage behaviour involves analysing bed material organisation and sediment transport. Bankfull-stage behaviour involves analysing the assemblage of instream geomorphic units and the size and shape of the channel. Overbank-stage behaviour involves analysing how the floodplain is formed and reworked and how the channel shifts over the valley floor.
- The river evolution diagram provides a visual representation of how a river behaves. Measures of specific stream power are framed in relation to imposed and flux boundary conditions. Interpretations of the natural range of variability and pathway of adjustment highlight the various ways that the river behaves, and associated use of available energy.

CHAPTER TWELVE

River evolution

Introduction

River evolution is the study of river adjustment over time. Evolution is ongoing. Even if boundary conditions remain relatively constant, adjustments occur. Appraisal of the trajectory and rate of river evolution is required to assess whether ongoing adjustments are indicative of long-term trends or whether they mark a deviation in the evolutionary pathway of that river. Such insights guide interpretation of the likelihood that the direction, magnitude and rate of change will be sustained into the future. To perform these analyses, it is important to determine how components of a river system adjust and change over differing timeframes, and assess what the consequences of those changes are likely to be. Reconstructions of the past provide a means to forecast likely future river behaviour.

Instinctively, human attention is drawn to landscapes that are subject to change. Observations of bank erosion, river responses to flood events, anecdotal records of river adjustments or analyses of historical maps and aerial photographs provide compelling evidence of the nature and rate of river adjustments. Efforts to read the landscape must frame these insights in a broader context, examine their representativeness and isolate controls upon evolutionary trajectories. For example, do these adjustments reflect modifications around a characteristic state and associated equilibrium scenarios over a given timeframe? Are short-term adjustments indicative of longer term trends? Has the river been subjected to threshold-induced change? How has the balance of formative and reworking processes and controls changed over time? Is the river sensitive or resilient to disturbance? How are responses to disturbance manifest through the catchment, remembering that an erosional signal in one place is often matched by a depositional signal elsewhere?

Analysis of river behaviour in Chapter 11 highlighted how different types of rivers have differing capacities to adjust, such that they respond to differing forms of disturbance event in different ways. Attributes such as thalweg shift on braidplains, meander migration/translation, cutoff development or avulsion are characteristic behavioural traits for certain types of rivers. In some instances, alterations to the boundary conditions under which rivers operate may bring about river change, whereby the behavioural regime of the river is transformed, and the river is now characterised by a different set of process–form relationships. River evolution may occur in response to progressive adjustments, an instantaneous event (e.g. a major flood or an earthquake) or longer term changes to geologic and climatic boundary conditions. This distinction between behaviour and change is essentially a matter of timescale. All rivers change as they evolve over time. In essence, if the geomorphic structure of a river changes, so does everything else (i.e. process relationships and the balance of impelling and resisting forces at the reach scale (Chapter 5) encompass adjustments to bed material organisation (Chapter 6), assemblages of instream and floodplain geomorphic units (Chapters 8 and 9) and channel geometry (Chapter 7)).

River change can result from alterations to impelling forces, resisting forces, or both. Resulting adjustments modify the nature, intensity and distribution of erosional and depositional processes along a reach. In some instances, predictable transitions can occur. For example, a change from a wandering gravel-bed river to an active meandering river can occur as flux boundary conditions are altered to reduce sediment load and discharge, or vegetation cover is increased. However, just because a particular type of river in a given system responds to an event of a given magnitude in a certain way, does not mean that an equivalent type of river in an adjacent catchment will respond to a similar event in a consistent manner. Even if particular cause and effect relationships are well understood, some systems may demonstrate complex (or chaotic) responses to disturbance events (see Chapter 2). More importantly, no two systems are subjected to the same set of disturbance events. Each system has its own history and its own geography (configuration), with its own cumulative set of responses to disturbance events, and associated lagged and off-site

Geomorphic Analysis of River Systems: An Approach to Reading the Landscape, First Edition. Kirstie A. Fryirs and Gary J. Brierley.
© 2013 Kirstie A. Fryirs and Gary J. Brierley. Published 2013 by Blackwell Publishing Ltd.

responses. The trajectory of river change may be influenced by the co-occurrence of disturbance events, such as a large flood following vegetation clearance. Such concatenations may set the system on a trajectory of change that would not have occurred if the system had not been disturbed or if these disturbances had occurred independently. Also, similar outcomes may arise from different processes and causes (the principle of convergence or equifinality; see Chapter 2).

Geologic and climatic factors determine the environmental setting and the nature of disturbance events to which rivers are subjected. They set the imposed and flux boundary conditions that fashion the erodibility and erosivity of a landscape, and the resulting character, behaviour and pattern of river types. Stark contrasts can be drawn, for example, between a dry, low-relief landscape with negligible vegetation cover and a high-precipitation mountainous terrain with dense forest cover. Formative processes, rates of activity (magnitude–frequency relations) and evolutionary trajectories vary markedly in these differing settings. Hence, any consideration of river evolution must be framed in relation to these geologic and climatic controls. In this chapter these considerations are appraised for differing tectonic settings and morphoclimatic regions. Particular emphasis is placed upon how landscape setting influences the imposed boundary conditions (especially slope and valley width) that constrain the range of behaviour of rivers, and the flux boundary conditions (i.e. flow and sediment regimes) that determine the mix of erosional and depositional processes along any given reach. Critically, as noted from the Lane balance diagram, alteration to either the imposed or flux boundary conditions promotes evolutionary adjustments. Geologic factors set and alter the imposed boundary conditions under which rivers operate, through their influence on lithology, relief, slope, valley morphology and erosivity and/or erodibility of a landscape. For example, tectonic activity or volcanic events may disrupt the nature and configuration of a landscape. Climate considerations play two critical roles. First, they are key determinants of the type and effectiveness of geomorphic processes (flow and sediment interactions) that shape landscapes at any given place. Second, climatic factors mediate the role of ground cover, which affects hydrologic processes and landscape responses to geomorphic processes through its influence upon surface roughness and resistance. Alterations to flux boundary conditions drive adjustments to the flow–sediment balance, prospectively modifying the evolutionary trajectory of a system.

Evolutionary adjustment may take a mere moment in time (e.g. river responses to a volcanic eruption) or be lagged some time after a disturbance event. Elsewhere, landscapes may be stable or demonstrate progressive adjustment over time. Some rivers are adjusted to high coefficients of discharge variability (see Chapter 4), such that large floods are rare but not unusual – they are part of the 'formative process regime' for that particular setting. Other rivers are adjusted to smaller, more recurrent events. Many rivers flow on surfaces created by past events, or are still adjusting to past flow and sediment regimes. In these cases, geomorphic memory continues to exert a significant influence upon contemporary forms and the nature and effectiveness of processes. Understanding how contemporary processes relate to historical influences is a key challenge in efforts to read the landscape.

This chapter is structured as follows. First, timescales of river change are discussed. Second, pathways and rates of geomorphic evolution are summarised for different types of rivers. Third, geologic and climatic controls on river evolution are considered. Then, evolutionary responses to changes in boundary conditions are outlined, and the river evolution diagram presented in Chapter 11 is used to extend analysis of river behaviour to incorporate interpretations of the nature and capacity for river change for various types for rivers. Finally, tools to interpret river evolution by reading the landscape are reviewed.

Timescales of river adjustment

Timescale of river adjustment varies from place to place, dependent upon the range of adjustment of the system (its sensitivity/resilience), the range and sequence of disturbance events and the legacy of past impacts. Both sensitive and resilient systems are prone to disturbance – responses are more likely and/or recurrent in the former relative to the latter.

Analysis of river evolution frames system responses to disturbance events in relation to adjustments over geologic and geomorphic time (see Chapter 2). Geologic controls set the imposed boundary conditions within which rivers operate. Over timeframes of millions of years, tectonic setting exerts a primary control upon topography, determining slope and valley settings that influence river morphology and behaviour. Over geomorphic time, rivers adjust to climatically fashioned flux boundary conditions (flow variability, sediment availability and vegetation cover) over hundreds or thousands of years. Any disruption to flux boundary conditions may affect the evolutionary trajectory of a river. The key consideration here is whether the reach is able to accommodate adjustments while it continues to operate as the same type of river (i.e. it operates within its behavioural regime) or whether these altered conditions bring about a transition in process–form relationships (i.e. river change occurs).

As noted in Chapter 2, river responses to disturbance events range from gradualist (uniformitarian) adjustments through to catastrophic change. A continuum of responses to disturbance events may be discerned:

- *No response may be detected*, as systems absorb the impacts of disturbance. Stable rivers can tolerate considerable variation in controlling factors and forcing processes. For example, gorges are resilient to adjustment or change. Alluvial systems with inherent resilience induced by the cohesive nature of valley floor deposits, or the mediating influence of riparian vegetation and wood, may demonstrate limited adjustment over thousands of years. In these cases, responses to disturbance events are short-lived or intransitive, and change does not occur.
- *Part of progressive change.* Rivers may respond rapidly at first after disruption, but in a uniform direction thereafter, such that change occurs gradually over a long period. For example, progressive denudation results in gradual reduction of relief over time, as gravitationally induced processes transfer sediments from source to sink. This results in long-term changes to slope and, hence, river type. Progressive adjustments are often observed following ramp or pulse disturbance events, so long as threshold conditions are not breached.
- *Change may be instantaneous* as breaching of intrinsic or extrinsic threshold conditions prompts the transition to a new state or even a new type of river. These effects tend to be long lasting or persistent.
- *Change may be lagged.* Off-site impacts of major disturbances may induce a lagged response in downstream reaches (e.g. conveyance of a sediment slug). The subsequent history of disturbance events affects the nature/rate of response and prospects for recovery.

Efforts to read the landscape seek to unravel variability in forms, rates and consequences of adjustments within any given system over differing timeframes. Pathways and rates of adjustment and evolution vary markedly for different types of rivers.

Pathways and rates of river evolution

Evolutionary pathways and rates of adjustment of rivers vary in differing geologic and climatic settings. This reflects differing ways in which boundary conditions and disturbance events affect flow–sediment interactions along a river. Alternatively, disturbance events may affect the surfaces upon which these processes are acting (e.g. role of fire, human disturbance – see Chapter 13). Evolutionary adjustments are likely to be most marked for those systems that have the greatest capacity to adjust and change. Hence, the nature and rate of evolution tend to be most pronounced in freely adjusting alluvial settings. These rivers have the greatest range in their degrees of freedom, such that pronounced disturbance events may trigger adjustments in channel planform, channel geometry (bed and bank processes), assemblages of channel and floodplain geomorphic units, and bed material organisation. The mix of water, sediment and vegetation conditions, as such, influences likely pathways of river adjustment for rivers in differing settings. Characteristic examples of evolutionary pathways are presented for rivers in differing valley settings below.

Likely evolutionary pathways of rivers in confined valley settings

Rivers in confined valley settings have limited capacity for adjustment. Their morphologies are largely imposed and are comprised largely of an array of imposed (bedrock) erosional forms. Steep headwater rivers progressively rework assemblages of slope-induced erosional geomorphic units as channels cut into bedrock via incisional processes over timeframes of thousands of years (Figure 12.1a). In contrast, gorges are stable and resilient systems over timeframes of hundreds or thousands of years. However, progressive incision and lateral valley expansion eventually create space along the valley floor for floodplain pockets to develop in partly confined valleys (Figure 12.1a). These transitions reflect changes to imposed boundary conditions.

Likely evolutionary pathways of rivers in partly confined valley settings

Just as gorges progressively widen to partly confined valleys with bedrock-controlled floodplain pockets over thousands of years, so sustained widening of these valleys eventually promotes the transition to partly confined valleys with planform-controlled floodplain pockets (Figure 12.1a). Increased valley width and reduced valley floor slope or changes in material texture, in turn, may result in a transition in the type of planform-controlled floodplain pockets that are observed, say from a low-sinuosity variant to a meandering planform variant (Figure 12.1a).

Likely evolutionary pathways of rivers in laterally unconfined valley settings

Variants of channel planform were conveyed along a continuum in Chapter 10. Adjustments to flow–sediment relations (i.e. flux boundary conditions) may bring about a transition to adjacent types of rivers along this continuum. Reduced energy conditions induced by lower flow and/or sediment availability may transform a braided river into a wandering gravel-bed river, and vice versa (Figure 12.1b). In turn, reduced energy conditions induced by lower flow

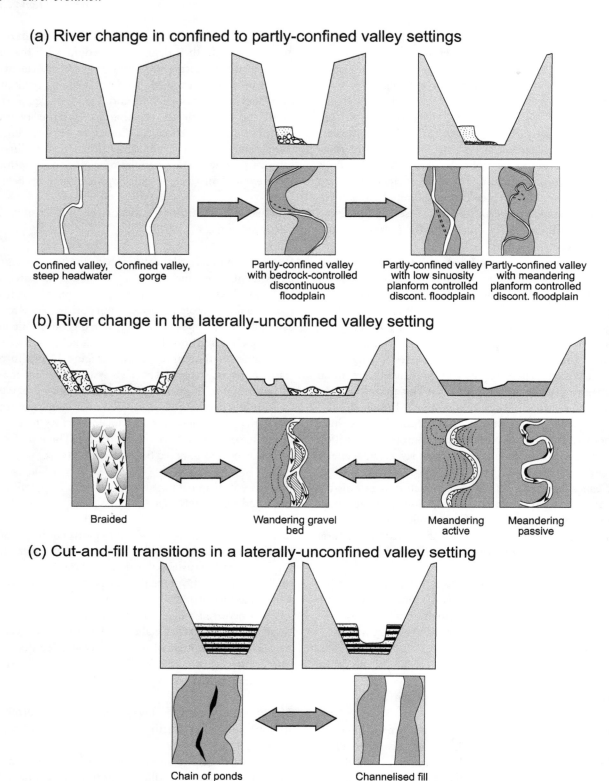

(a) River change in confined to partly-confined valley settings

Confined valley, steep headwater | Confined valley, gorge

Partly-confined valley with bedrock-controlled discontinuous floodplain

Partly-confined valley with low sinuosity planform controlled discont. floodplain | Partly-confined valley with meandering planform controlled discont. floodplain

(b) River change in the laterally-unconfined valley setting

Braided

Wandering gravel bed

Meandering active | Meandering passive

(c) Cut-and-fill transitions in a laterally-unconfined valley setting

Chain of ponds

Channelised fill

Figure 12.1   Likely evolutionary transitions in differing valley settings. River change reflects alterations to imposed or flux boundary conditions. (a) Confined to partly confined transitions require changes to imposed boundary conditions via valley incision and sidewall retreat and lowering of valley slope. These changes are irreversible. (b) Laterally unconfined river transitions require changes to flux boundary conditions, including decreases in stream power, sediment calibre and load. These transitions are reversible. (c) Cut-and-fill river transitions require changes to flux boundary conditions, including increases in stream power, sediment load and fill slope. These transitions are reversible.

and/or sediment availability may transform a wandering gravel-bed river into an active meandering river, and vice versa (Figure 12.1b). Alternatively, increase in sediment load (bedload fraction) may transform a passive meandering (suspended-load) river into an active meandering (mixed-load) river, and vice versa (Figure 12.1b). Various stages of evolutionary adjustments may be discerned along a discontinuous watercourse, reflecting cut and fill phases (Figure 12.1c). However, should certain circumstances eventuate, the river may maintain a continuous watercourse.

The examples outlined in Figure 12.1 convey progressive evolutionary adjustments. In essence, the types of rivers that are found in an adjacent position along the longitudinal profile (i.e. an energy gradient) are likely to present the next step or phase in the evolutionary adjustment of a river. This may reflect conditions of decreasing energy associated with progressive landscape denudation, or increasing energy associated with uplift (i.e. steeper slope conditions). This line of reasoning, whereby juxtaposed river types along slope-induced environmental gradients provide guidance into likely evolutionary adjustments, is a direct parallel to Walther's law of the correlation of facies (Chapter 6): adjacent sedimentary deposits in contemporary landscapes are used to guide inferences into stacked depositional units within basin fills.

Geologic and climatic controls are the primary determinants of imposed and flux boundary conditions, and the associated suites of disturbance events to which rivers are subjected. Although these geologic and climatic considerations act in tandem, they are considered separately below for simplicity.

Geologic controls upon river evolution

Geologic setting determines the imposed boundary conditions within which rivers adjust and evolve. The nature and movement of tectonic plates is a primary determinant of the distribution and relief of terrestrial and oceanic surfaces. The nature and position of mountain belts and depositional basins is determined largely by the distribution of plates and geologic processes that occur at different types of plate boundaries. Landscape relief and topography are fashioned by the balance of endogenetic processes (i.e. geologic processes that are internal to the Earth) and exogenetic processes (i.e. geomorphic processes that erode and deposit materials at the Earth's surface). The nature, frequency and consequences of geologic disruption and disturbance events vary markedly in different tectonic settings. This is determined largely by position relative to a plate margin and the nature of tectonic activity at that margin. Vertical and lateral displacement along fault-lines

is common in some settings. Contorted strata of folded rocks attest to the incredible forces at play. Faulting, folding and tilting generate distinctive topographic controls upon slope, valley morphology and drainage patterns (Chapter 3). Volcanic activities and subsidence modify relief and availability of materials.

Tectonic setting frames the long-term landscape and dynamic context of river systems. Simplified schematic representations of primary tectonic settings are presented in Figure 12.2. Tectonic motion generates three primary types of plate boundaries: collisional (constructional) (Figure 12.2a and b), pull-apart (Figure 12.2c and d) and lateral displacement (Figure 12.2e). The rate of uplift, subsidence or lateral movement of a plate affects the relative stability and nature of landscape adjustment. Uplifting rivers incise into their beds, creating narrow valleys. Subsiding rivers aggrade, creating expansive floodplains. Rivers in low-relief, plate-centre settings are characterised by long-term stability, as they slowly denude and rework the landscape (Figure 12.2f).

Long-term changes to the position of plate boundaries affect the nature of constructional and reworking processes at any given locality (Figure 12.3). These geological foundations determine patterns of lithological and structural variability, affecting the erodibility and erosivity of landscapes.

The convergence of continental plates generates major mountain chains. Deeply incised bedrock channels in headwater settings contrast starkly with transport-limited braided rivers, low-relief rivers atop uplifted plateau landscapes or deeply incised gorges at plateau margins (Figure 12.4a–c respectively). Uplift of supply-limited plateau landscapes may create deeply entrenched, superimposed drainage networks. For example, Figure 12.4d shows the planform of a meandering river that previously formed on a relatively flat alluvial plain that has been retained as the landscape was uplifted, creating a deeply etched, bedrock-controlled gorge in which river character and behaviour are imposed. Differing forms of constructional landscapes are generated through subduction of dense but relatively thin oceanic plate beneath a continental plate. Recurrent phases of tectonic activity produce basin and range topography comprised of mountain ranges, volcanic chains and intervening basins, exerting a dominant imprint upon contemporary drainage networks. The imprint of landscape setting upon river character, behaviour and evolution is clearly evident in pull-apart basins. This tectonic setting is characterised by striking alignment of lakes and straight, bedrock-controlled river systems. In some instances, basins that pulled apart in the past may retain a dominant imprint upon contemporary landscapes, forming escarpments (Figure 12.4e) and rift valleys (Figure 12.4f). Alternatively, lack of tectonic activity is a primary determinant of river

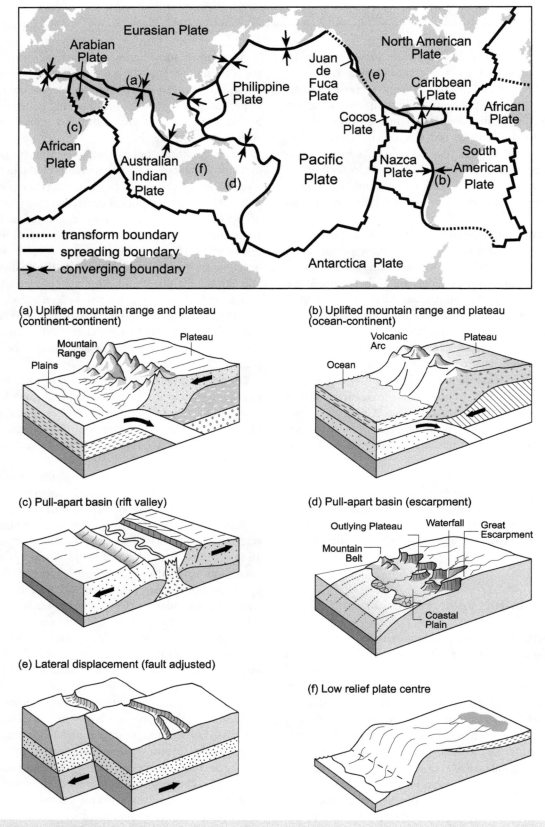

Figure 12.2 Plate tectonic settings. Landscape settings and river types vary with position relative to plate boundaries. Uplifted mountainous terrains occur at convergence zones, whether continent–continent interactions that generate high-elevation mountain and plateau topography (e.g. Himalayas and Qinghai–Tibetan Plateau (a)) or subduction zones characterised by ocean–continent collision (e.g. Andes (b)). Rift valleys and escarpments are created in pull-apart basins (e.g. eastern Africa (c)) and the Great Escarpment in eastern Australia (d) (modified from Ollier and Pain (1997)). Lateral displacement occurs when plates slide past each other along a fault-line, realigning drainage patterns (e.g. San Andreas Fault, western USA (e)). Low-relief, low-lying areas in plate centres are tectonically stable (e.g. central Australia (f)).

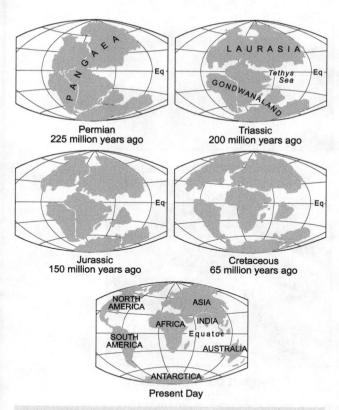

Permian
225 million years ago

Triassic
200 million years ago

Jurassic
150 million years ago

Cretaceous
65 million years ago

Present Day

Figure 12.3 Plate movements over the past 225 million years. Progressive adjustments that created the present configuration of continents reflect the breakup of Pangaea and subsequently Laurasia and Gondwanaland. This induced significant but variable disturbance to river systems in differing settings in the form of volcanoes, earthquakes, fault adjustment and uplift. Plate movement also brings about long-term shifts in climate regime. For example, Antarctica was once part of Gondwanaland and had rainforests and a warm climate. These considerations fashion the imposed boundary conditions within which rivers operate today.

processes and forms in plate-centre landscapes. These low-relief, low-erosion settings often have profound stability and antiquity (Figure 12.4g).

Long-term changes to plate tectonic boundaries ensure that any given landscape setting has likely been subjected to differing forms and phases of tectonic activity (Figure 12.3). Geologic adjustments are sometimes imprinted atop each other. Elsewhere, the imprint of past events has been virtually erased, though metamorphosis of rocks may provide insights into former conditions. Importantly, tectonic setting not only fashions the relief and erodibility of a landscape, it also affects the climate and, hence, the erosivity of that landscape.

Climatic influences on river evolution

Spatial and temporal variability in climate are genetically linked to geologic considerations, as mountain belts and other topographic factors influence temperature and precipitation regimes and the movement of weather systems. The distribution of landmasses and latitudinal factors fashion continental or maritime climate conditions and solar radiation effects. Topographic and climatic conditions can be combined to differentiate morphoclimatic regions (Chapter 4; Figure 12.5).

Climatic controls upon river evolution are manifest in two primary ways. *Direct* influences reflect hydrologic considerations and thermal conditions, expressed primarily by the flow regime. This drives the flux boundary conditions under which rivers operate. *Indirect* influences are manifest primarily through climatic influences upon ground cover (and rainfall–runoff associations) and resistance factors (i.e. surface roughness). Any alteration to these relationships affects the flux boundary conditions under which rivers operate. Adjustments to the flow and sediment balance may alter the evolutionary trajectory of a river. In many settings, past climatic conditions continue to exert an influence upon the effectiveness of contemporary geomorphic processes (i.e. climatic memory).

Direct and indirect impacts of climate variability vary markedly in differing morphoclimatic regions. Some tropical humid regions are characterised by high temperatures and high precipitation throughout the year, and have rainforest vegetation associations. Rivers in these regions are attuned to recurrent high flow conditions and considerable roughness on valley floors, but interannual variability in flow is limited (i.e. the coefficient of variation for discharge is low; Chapter 4). Tropical humid areas with prominent dry and monsoonal seasons are characterised by savanna vegetation. Although seasonal variability in flow and geomorphic activity is pronounced, interannual variability is limited. Rivers in these areas are especially sensitive to the effectiveness of the monsoon. Mid-latitude regions are dominated by arid and semi-arid climates. Desert and steppe landscapes have limited vegetation cover. Pronounced, highly effective geomorphic activity occurs during short storms. Desert environments with limited sediment availability are characterised by etched/sculpted bedrock rivers. Other deserts have ephemeral rivers with high sand availability, resulting in high width/depth channels because of the non-cohesive, non-vegetated nature of bank materials.

Humid-temperate rivers have perennial flow. Vegetation cover exerts a primary influence upon process–form relationships. Warmer humid regions are not subjected to severe winter conditions, but summers can be hot and dry. Vegetation cover may be relatively sparse and shrub-like

(a) Uplifting continental plate, Waimakariri River, New Zealand

(b) Low-relief plateau, Qinghai

(c) Incised river on edge of plateau, Huang He

(d) Superimposed drainage, San Juan River, Goosenecks State Park, Utah

(e) Pull-apart escarpment, Great Escarpment, eastern Australia

(f) Pull-apart basin, rift valley, Iceland

(g) Low-relief plate centre, Central Australia

Figure 12.4   Rivers in different tectonic settings. (a) A braided river in an uplifted continental plate, Waimakariri River, New Zealand; photograph: G. Brierley. (b) A passive meandering-anabranching river in an uplifted, low-relief, plateau terrain, Qinghai–Tibet Plateau; photograph: G. Brierley. (c) Incised gorge at the margins of the uplifting Qinghai–Tibet Plateau, upper Yellow River, China; photograph: G. Brierley. (d) Superimposed drainage has produced bedrock-controlled gorges in a supply-limited plateau terrain, Goosenecks State Park, Utah, USA; photograph: http://www.geology.wisc.edu/~maher/air/air04.htm. (e) Escarpment and gorge country on a pull-apart plate boundary, Great Escarpment, eastern Australia; photograph: R. Ferguson. (f) Anastomosing river in a pull-apart rift valley of Iceland; photograph: E. Hafsteinsdottir. (g) Anastomosing river in a low-relief, plate-centre location, Channel Country, central Australia; photograph: G. Nanson.

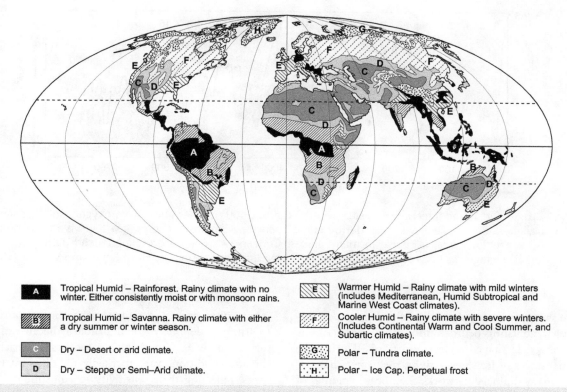

A	Tropical Humid – Rainforest. Rainy climate with no winter. Either consistently moist or with monsoon rains.
B	Tropical Humid – Savanna. Rainy climate with either a dry summer or winter season.
C	Dry – Desert or arid climate.
D	Dry – Steppe or Semi–Arid climate.

E	Warmer Humid – Rainy climate with mild winters (includes Mediterranean, Humid Subtropical and Marine West Coast climates).
F	Cooler Humid – Rainy climate with severe winters. (Includes Continental Warm and Cool Summer, and Subartic climates).
G	Polar – Tundra climate.
H	Polar – Ice Cap. Perpetual frost

Figure 12.5 Morphoclimatic regions. Latitude and altitude are key determinants of the relationship between relief and climate type. These factors, in turn, affect the flux boundary conditions that fashion river behaviour and/or change.

in Mediterranean areas, but is much more substantive in subtropical regions. There is marked variability in runoff generation and geomorphic effectiveness of floods in this morphoclimatic zone. Some areas have extremely high coefficients of variation in discharge, with significant inter-annual variability in flood events. Often, river systems are attuned to extremely high, but infrequent, flows. Mediterranean rivers have seasonal discharge and variable ground cover. Ephemeral streams are subjected to irregular reworking by flash floods. Discontinuous watercourses are prominent. Cooler humid regions have severe winters and continental climates, with significant areas of boreal forest. Rivers freeze in winter, and there is extensive permafrost in northerly latitudes. Profound adjustments may occur during spring melt. Polar regions are dry and cold, and bedrock-dominated rivers are relatively inactive.

Landscape history and climate setting bring about marked variability in flora and fauna across the globe. Faunal interactions with rivers can affect the nature, rate and effectiveness of geomorphic processes. A wide range of ecosystem engineers is evident. Ants and worms induce bioturbation in soils, impacting upon sediment supply and transfer on hillslopes. Beaver dams exert a direct impact upon channels. Hippopotamus tracks may induce channel

realignment. Wombat burrows may locally enhance rates of bank erosion. Changes to these faunal interactions may alter the evolutionary trajectory of the river. Similarly, any factor that alters vegetation cover (and associated resistance/roughness) can have a significant affect upon the evolutionary trajectory of a river. For example, the geomorphic role of fire varies markedly in differing morphoclimatic regions. Savanna and Mediterranean areas are especially prone to fire events that clear ground cover, resulting in pulsed flow and sediment inputs into river systems.

Climate is a key driver of river change. It fashions the sequence of disturbance events that bring about geomorphic adjustments, influencing system dynamics and the behavioural regime of any given reach. In some instances, floods or droughts may bring about transitions to a different type of river. Impacts upon the flow regime, and changes to ground cover, alter the rate of sediment movement in river systems, thereby affecting both sides of the Lane balance diagram.

Long-term changes during the Quaternary period have been induced by glacial–interglacial cycles (Figure 12.6). At the coldest part of the last glacial maximum (15 000–18 000 yr ago), ice covered one-third of the land area of the Earth to an average depth of 2–3 km, but in places up to

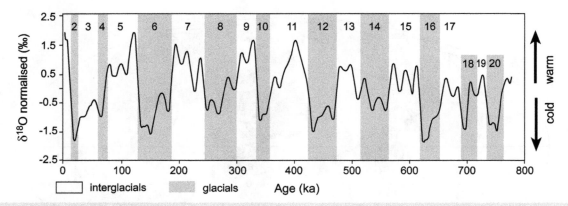

Figure 12.6 Glacial and interglacial cycles over the last 800 000 yr, determined from oxygen isotope stages. Ice sheets and glaciers cover extensive areas during glacial periods. Climate is cooler, sea level is lower, rainfall is reduced and vegetation cover is sparse in many places. Converse conditions are experienced during interglacial periods. Marked changes to flow–sediment flux occur during transitional phases. From Lowe and Walker (1997).

4 km. Ice sheets created sculpted/denuded landscapes, creating slowly adjusting bedrock-dominated rivers. Alpine glaciers carved U-shaped valleys and fiords. During the recessional stages of ice sheet activity, meltwater channels realigned many drainage networks. Significant volumes of glacially reworked materials drape many landscapes, providing large sediment stores that can be reworked by river systems. Hence, there are marked differences in the historical imprint upon contemporary landscapes in glaciated and non-glaciated settings.

Glacial cycles also brought about significant falls in sea level (up to 120 m). This exerted a profound impact upon patterns and rates of sedimentation in lowland basins as base level adjusted. Deep canyons were carved into what are now parts of the continental shelf. These effects were propagated upstream, leaving terraces at valley margins. Subsequent sea level rise during interglacial periods created drowned valleys and ria coastlines. Floodplain, terrace and marine sediments in infilled lowland valleys and estuaries retain records of multiple phases of sea level rise and fall.

Longer term glacial–interglacial cycles also brought about major river changes in arid morphoclimatic zones, altering the distribution and extent of monsoonal climatic influences. As climate changes, so too does the vegetation cover. Hence, geomorphic adjustments reflect alterations to both impelling forces (the flow regime) and resisting forces (ground cover) (see Chapter 5).

Geomorphic responses to climate change are markedly variable in different parts of the world. The impact of climate change is not simply a measure of the direction or extent of change. Temperature changes from −20 °C to −30 °C may not induce a marked difference in process response, but transition from −5 °C to +5 °C certainly does. Similarly, change in annual precipitation from 9000 to 10 000 mm a^{-1} is unlikely to induce marked variability in

geomorphic process activity, but changes from 500 to 1500 mm a^{-1} definitely would, primarily because of altered vegetation cover. Geomorphic responses to variability in climatic conditions vary markedly for different types of rivers, reflecting their sensitivity to adjustment (Chapter 11). They also vary dependent upon the condition of the system at the time of the disturbance event (especially its resistance). In many instances, contemporary landscapes have been fashioned largely by conditions from the past.

Landscape memory: imprint of past geologic and climatic conditions upon contemporary river processes, forms and evolutionary trajectory

Contemporary rivers flow upon, and rework, surfaces created by past events. Hence, historical influences may exert a primary influence upon the distribution, rate and effectiveness of erosional and depositional processes. This imprint from the past varies markedly in differing settings. As noted in Chapter 2, landscape memory is fashioned primarily by past geologic and climatic conditions, or events. Some landscapes also retain a prominent memory of former anthropogenic activities (discussed in Chapter 13).

Geologic controls determine the relief, topography and erodibility of a landscape (see Figure 12.7a and b). The influence of elevation upon potential energy manifests itself as impelling forces (and associated kinetic energy) driven largely by slope (i.e. erosivity). This exerts a primary control upon the effectiveness of erosional processes and the resulting degree of landscape dissection. Geologic factors also influence the nature and extent of accommodation space and associated patterns of sediment stores in landscapes. Valley setting, in turn, affects

(a) Tectonically uplifting landscapes, Southern Alps - South Island, New Zealand

(b) Tectonically stable and ancient landscapes, Channel Country, Central Australia

(c) Glacially-carved valley, Haizi Shan plateau

(d) Paraglacial sediment stores, Bridge River, British Columbia

(e) Low-relief, bedrock dominated, supply limited systems, Canadian Shield

(f) Low-relief, wide valleys with fine grained floodplains formed under palaeoclimates, Macquarie River, NSW

Figure 12.7 Landscape memory. Examples of geologic memory include controls exerted by tectonic setting upon the erodibility and erosivity of landscapes. Stark contrasts are evident in uplifting plate-margin mountainous areas on the South Island of New Zealand (a) and the ancient landscapes of plate-centre locations such as in the Channel Country, central Australia (b) (photograph: G. Nanson). Climatic memory may manifest itself as deeply carved glacial troughs, such as in the Haizi Shan (c) (photograph: J. Harbor; http://web.ics.purdue.edu). Alternatively, paraglacial sediment stores may influence the contemporary form and sediment dynamics of rivers, such as along the Bridge River, British Columbia (d) (photograph: G. Brierley). Some landscapes were largely stripped of sediments by ice sheets, creating low-relief, entrenched, bedrock-dominated, supply-limited rivers, such as rivers on the Canadian Shield (e) (photograph: Bill Van Geest, http://www.nrdc.org/land/forests/boreal/page3.asp). Finally, former wetter phases may have formed much larger river systems into which contemporary channels are set, as exemplified by the Macquarie River, Australia (f) (photograph: K. Fryirs).

channel–floodplain relationships, thereby influencing the contemporary capacity for adjustment of rivers.

The contemporary climate regime is a primary determinant of the flux boundary conditions under which rivers operate, affecting discharge and flow energy and vegetation and/or ground cover which resist erosion processes. Critically, these relationships have changed over time. The impact of these changes is especially pronounced in those parts of the world affected by Pleistocene glacial activity. Glaciers carved deep and narrow valleys in mountain areas, constraining the range of geomorphic behaviour of contemporary channels in these settings (Figure 12.7c). Many downstream areas were draped with glacially reworked materials. In some instances these vast (paraglacial) sediment stores that reflect former climatic conditions continue to influence contemporary river behaviour (Figure 12.7d). The distribution of these sediment stores is influenced largely by geologic controls upon the accommodation space in landscapes, such as wider sections of valleys that store glacio-fluvial, glacio-lacustrine and alluvial fan materials. In many other settings, ice sheets stripped surface materials from vast areas, limiting contemporary rates of sediment supply across largely denuded areas (Figure 12.7e). Another form of climatic memory is that associated with floodplain deposits of underfit streams (i.e. contemporary channels are too small to have formed the valleys within which they presently flow; Figure 12.7f). These inherited forms influence contemporary river morphology and associated patterns and rates of sediment erosion, transport and deposition. In this instance, climatic memory directly reflects geologic memory, as past geologic controls induced the accommodation space along palaeovalleys within which contemporary rivers flow.

Landscapes retain a selective memory of past events. Sometimes a sharp erosional boundary reflects a major disjunct in time, highlighting the removal or erasure of a significant part of the record (Chapter 6). Indeed, some landscapes may retain a very limited history of past events. Elsewhere, especially in long-term depositional basins in accretionary environments, a remarkable long-term record may be preserved (much of which is buried subsurface). Hence, different parts of a landscape retain variable records of past activity. Ultimately, changes to boundary conditions drive river evolution.

River responses to altered boundary conditions

The Lane balance diagram provides a simplified basis with which to interpret primary controls upon river evolution. Essentially, if the bed stability of a river changes, so will the geomorphology. In other words, the balance becomes unsettled and adjustments ensue. The two key considerations here are the amount of water acting on a given slope and the volume and texture of sediment delivered to the channel. As noted above, geology and climate are the primary determinants of these factors. The tectonic setting determines the rate of uplift (i.e. relief and sediment generation, and erosion rate), while lithology determines the breakdown size of weathered/eroded materials. Uplift or subsidence also alters the slope upon which geomorphic processes are acting. Climatic factors determine the flow regime and the amount of water available to do work in river systems. Evolution is driven by changes to these various controls.

Davisian notions of landscape adjustment infer that rivers evolve as slopes decrease and valley floors widen over geologic time, prior to uplift kick-starting the cycle once more (Chapter 2). Such continuity in boundary conditions, and even the direction of change, is seldom observed in reality, as invariably something happens to disrupt these patterns over timeframes of millions or tens of millions of years. Disturbance events may alter the flow–sediment balance along a river, whereby changes to geologic and climatic conditions induce adjustments in process relationships along valley floors, and resulting river morphologies. Various examples of river evolutionary adjustments in response to altered boundary conditions, disturbance events and flow–sediment fluxes are outlined below.

River responses to tectonic uplift and displacement along fault-lines

Uplift of a fault block, or even an entire plateau landmass within a plate, induces rejuvenation, whereby rivers are made young again and incise into underlying bedrock. If the rate of bed incision is unable to keep up with the rate of uplift, convex bulges are created along longitudinal profiles. These areas are characterised by waterfalls and/or oversteepened sections of the bed profile (see Chapter 3). In some instances, knickpoint erosion may instigate river capture, wherein flow that was previously part of a separate basin is realigned and captured as a headward-cutting channel eats through the drainage divide over time (see Figure 12.8). An underfit stream now flows within the abandoned valley (i.e. the stream is much smaller than the river that created the valley itself). Elsewhere, stepped longitudinal profiles with multiple waterfalls reflect the recurrence of uplift events and the hardness of bedrock layers through which knickpoint retreat occurs. The pulsed nature of bed incision and knickpoint retreat in tectonically active settings is often accompanied by dramatic influxes of sediment from hillslope failures, some of which dam the river with variable longevity. Alternatively, lateral displacement along fault-lines during earthquake events

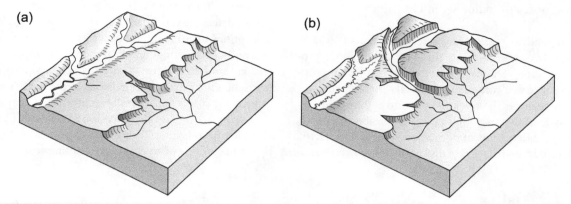

Figure 12.8 River capture. (a) In this example, knickpoint retreat in an escarpment-dominated landscape results in extension of river valleys into adjacent catchments. (b) River capture changes drainage direction and catchment area–discharge relationships. The remnants of the captured system often become underfit as their headwater catchments no longer supply water and sediment to the lower system.

(a) Laterally displaced fault block, San Andreas Fault, California North America

(b) Vertically uplifted fault block, Tachia River, Taiwan

Figure 12.9 Lateral and vertical fault displacement along rivers. (a) Lateral offsets realign drainage patterns along the San Andreas Fault, USA (photograph: Robert E. Wallace (Ed) The San Andreas Fault System, California. US Geological Survey Professional Paper 1515, Chapter 2 online Photograph Album, Figure 2.20. http://www.geologycafe.com/california/pp1515/chapter2Album.html). (b) Vertical offsets change bed slope along rivers, producing bedrock steps, knickpoints and base level adjustments. The Tachia River, Taiwan, was uplifted during an earthquake in 1999.

can realign and/or reconfigure river systems. This can occur in a lateral dimension (Figure 12.9a) or vertical dimension (Figure 12.9b).

River responses to long-term changes in valley setting

Rivers are products of the valleys in which they flow. Long-term changes to valley morphology reflect geologic controls (see Figure 12.2c). For example, progressive knickpoint retreat along trunk and tributary rivers at the plate margin creates series of dissected gorges in escarpment-dominated landscapes at the margins of pull-apart basins. These valleys cut backwards and incise far more rapidly than they widen. Changes to valley floor slope and valley width over millions of years induce transitions from a gorge to a partly confined valley with bedrock-controlled discontinuous floodplains (Figure 12.1b) and subsequently to a partly confined valley with planform-controlled discontinuous floodplains (Figure 12.1c).

River responses to major sediment inputs

Rivers respond to marked increases in sediment load by aggrading. In some instances this may bring about profound landscape responses. For example, volcanic eruptions can drape vast volumes of material across a landscape, transforming incised bedrock streams into highly sediment-charged systems that may infill valleys to considerable depth, promoting the development of braided rivers (Figure 12.10). These localised and irregular disturbance events are rela-

tively spatially constrained (i.e. they occur in semi-predictable places, determined primarily by tectonic setting). Volcanic disruptions to river systems occur primarily in subduction and pull-apart settings and in response to hot spot activity (i.e. areas of thin crust through which molten materials from the upper mantle are released at the Earth's surface).

Volcanic events are generally recurrent (i.e. they occur at the same place on repeated occasions, and resulting materials build up over time). Landscape responses are fashioned by the magnitude of an eruption, resulting sediment inputs

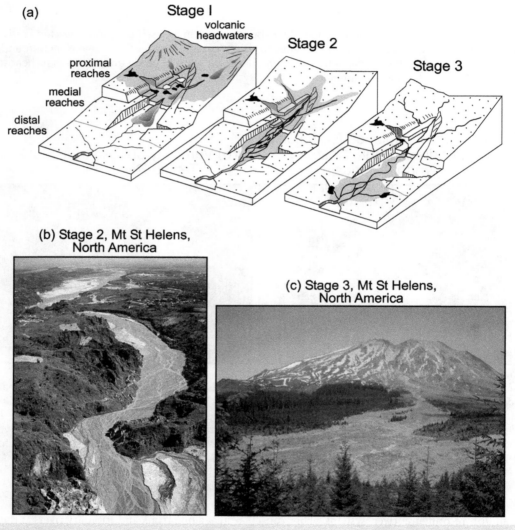

Figure 12.10 Volcanically induced river forms. Evolution of the Taupo River following the Taupo eruption 1.8 ka (a). Pyroclastic and debris flows covered the flanks of the volcano and adjoining valleys (Stage 1). Lakes developed in side valleys that become dammed. Rills and channels rework large volumes of sediment deposited (Stage 2). Incision and knickpoint retreat greatly increase sediment supply to downstream areas, potentially producing lahars. Sediment reworking along the main valley axes forms braided rivers (Stage 3). Incision into sediments produces large terraces. Reprinted from Sedimentary Geology, 220 (3–4), Manville, V., Segschneider, B., Newton, E., White, J.D.L., Houghton, B.F. and Wilson, C.J.N., Environmental impact of the 1.8 ka Taupo eruption, New Zealand: Landscape responses to a large-scale explosive rhyolite eruption, 318–336, © 2009, with permission from Elsevier. The 1980 Mt St Helens eruption, USA, produced morphologies similar to Stage 3 landscapes (b, c). Photographs: left, http://earthobservatory.nasa.gov/Features/Lahars/lahars_2b.php; right, http://www.ngdc.noaa.gov/hazardimages/picture/show/943 © NOAA/NGDC, NOAA National Geophysical Data Center.

and the interval between events (i.e. the length of time over which sediment reworking occurs). In general terms, volcanic landscapes that have not experienced an eruption for a significant period tend to become deeply etched bedrock-controlled systems. These rivers are resilient to change during flood events. However, eruptions bring about dramatic transformations, altering all attributes of the river. Lahars and debris flow deposits line valley floors. Aggradation induces braided rivers with an array of mid-channel depositional geomorphic units. Typically, these are short- to medium-term adjustments post-eruption, as the river progressively adapts to prevailing flow–sediment conditions by incising into its bed (i.e. flow conditions remain relatively consistent over time, while the rate of sediment production is not maintained). Incision and reworking promote a transition back to increasingly imposed river morphologies. Downstream transfer of materials accentuates bed incision and the deeply etched character of the landscape.

The imprint of volcanic events brings about a range of localised and off-site impacts. Tephra deposits may create a significant drape of materials over vast areas. Rivers subsequently flow within very light, low-density, highly porous materials, such that coarse bed material is readily conveyed within the channel (often as suspended load). In other settings, ignimbrite flows may infill valleys and create plateau-like landscapes with caps of extremely resistant materials. Valley incision and headward retreat subsequently demarcate these materials as knickpoints and waterfalls along longitudinal profiles.

Long-term erosion of volcanic landscapes can create inverted relief. This occurs when lava flows infill valleys, flattening out the ground surface (Figure 12.11). As the thicker basaltic materials are often more resistant to erosion than the surrounding country rock, long-term progressive erosion may result in basalt-peaked caps derived from materials previously deposited on valley floors as the high points in these landscapes.

While volcanic events induce massive sediment inputs into riverscapes over irregular but infrequent timescales, more recurrent but much smaller sediment inputs occur in response to landslides and associated hillslope instability events (Figure 12.12). A range of outcomes may occur, dependent upon the amount of sediment input, the size of the valley and the capacity of the river to rework these deposits. In extreme instances the valley may become blocked, forming a dam and lake. This alters the base level of the trunk stream, resulting in aggradation and delta growth within the lake. Downstream, the channel responds to reduced sediment loads by incising. Eventually the dam may break. This results initially in extensive flooding and erosion of downstream reaches. Subsequently, the massive influx of deposits induces aggradation as a sediment slug

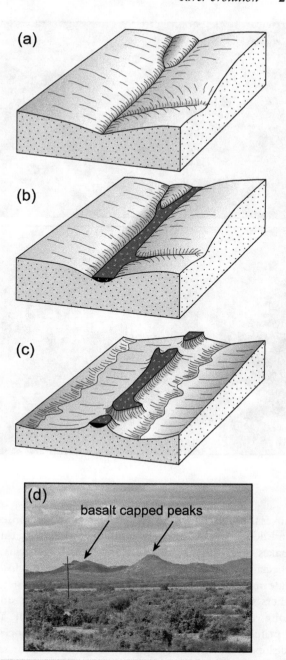

Figure 12.11 Inverted relief. In this example, lava flow fills a valley with materials that are more resistant to erosion and weathering than the surrounding country rock (a, b). Over millennia, the valley margins and drainage divides are lowered, such that the volcanic materials become perched as at the highest surface in the landscape (i.e. the basement rocks that line the new valley floor are much younger than the lava flow) (c). From Cundari and Ollier (1970). © Taylor and Francis. Reproduced with permission. (d) Photograph of El Capitan near Cobar, western NSW, Australia, showing the lava capped hills. Photograph: Coddley flickr.

Figure 12.12 Impact of landslides upon a trunk stream. In extreme circumstances, landslides may completely block river valleys, forming lakes upstream of the blockage. Dam breaks can have catastrophic consequences for downstream river morphology. These photographs are from landslide-blocked valleys in Sichuan Province, China. Photographs courtesy of Zhaoyin Wang, Tsinghua University. (a) The Tangjiashan quake lake is removed by scouring the landslide dam and releasing the lake water. (b) Landslide dams create lakes upstream of the landslide along Jiuzhai Creek. (c) Landslide dams created a knickpoint along Jiuzhai Creek.

moves through the system. Extreme landscape responses to landslide events are especially pronounced following earthquakes or extreme storms (cyclones). Such scenarios are especially pronounced in steep, dissected terrains close to plate margins in regions with (sub)tropical climates. In others settings, hillslope-derived materials may be stored along valley floors for a considerable period of time. This is primarily determined by valley width, and associated hillslope–valley floor connectivity and the space for sediment storage (Chapter 14). If these deposits are not accessible to the channel, they may have a negligible impact upon river behaviour and change.

River responses to climate change (flow regime and ground cover changes)

Impacts of climate change and variability may be manifest through localised extreme events, semi-regular, systematic changes (e.g. glacial–interglacial cycles) or long-term adjustments associated with the movement of tectonic plates (the distribution of land masses is a primary control upon weather patterns; Figure 12.3). Of primary concern here is

the impact of changing boundary conditions and disturbance events upon the way in which rivers operate, and their evolution. Examples of system responses to climate-induced alterations to the flow–sediment balance and ground cover are outlined below.

Climate change induces marked variability in the character, behaviour and evolution of river systems in glaciated and non-glaciated landscapes. Phases of glacial activity in mountainous terrain induce extensive erosion and sculpting of landscapes. The mountains themselves are etched and denuded, while valleys are carved. Stripped surficial materials and bedrock are broken down and conveyed considerable distances from source. As a consequence, the boundary conditions upon which rivers operate are transformed.

Transitional climatic phases at the ends of ice ages are periods of intensive geomorphic activity. This period is referred to as the paraglacial interval. Melting glaciers and ice sheets result in pronounced discharge variability. Hillslopes are unstable, as previously supporting ice has melted, and vegetation cover is negligible. This results in extensive sediment movement, aggrading valley floors and the formation of large alluvial fans. Steep slopes, abundant bedload-calibre material and fluctuating discharge result in braided river systems. Extensive braidplains (or valley sandar) are evident at the margins of many contemporary glaciers or ice sheets. Over time, discharge is reduced and streams incise into their beds, creating extensive terrace sequences (Figure 12.7d). Rivers retain extensive sediment loads, and braided channel planforms extend well beyond the mountain front. Amelioration of climatic conditions over thousands of years results in less variable discharges, diminishing sediment loads and increases in vegetation cover on hillslopes and valley floors. Rivers respond by changing to wandering gravel-bed or active meandering systems (Figure 12.1b).

In some instances, post-glacial climate changes may generate some truly epic landscapes, inducing profound alterations to river systems. This is exemplified by breaching of ice-dammed lakes, which release vast volumes of water in truly catastrophic flows (termed jokulhlaups). These floods may etch and sculpt vast terrains, fashioning future drainage networks and resulting river morphologies (Figure 12.13). Elsewhere, streams beneath glaciers or significant

(a) Scablands of the Columbia River, eastern Washington, USA.

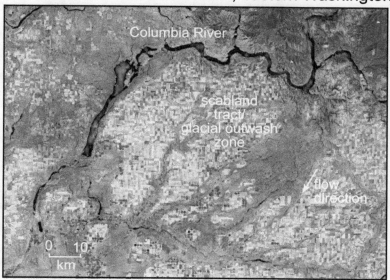

(b) Dry falls in the scablands of eastern Washington, USA.

Figure 12.13 Impact of breached ice dams upon evolutionary trajectories of rivers. Ice dams may trap vast volumes of water. Failure of these dams results in catastrophic flood events that may etch landscapes and realign drainage patterns over significant areas, resulting in underfit streams. The great Missoula floods created the Channel Scablands of the upper Columbia River, Washington State, USA. The scablands show up as dark-grey scars on the landscape, as the soil has been stripped and the bedrock (basalt) eroded into characteristic channels of scabland morphology. (a) © Google Earth 2011, based on Delinger and O'Connell (2010). (b) Photographer: Ikiwana, from http://www.uvm.edu/~geomorph/gallery/.

meltwater flow can realign drainage networks (a form of river capture). Many landscapes retain a significant climatic memory from these post-glacial events.

Although non-glaciated terrains are not subjected to paraglacial sedimentation and breaching of ice-dammed lakes, dramatic landscape responses to changing climatic conditions may occur in these settings. For example, former river courses in some desert landscapes have been draped by sand dunes in response to drier climates and reduced vegetation cover during glacial periods (Figure 12.14). Some non-glaciated landscapes have been subjected to progressive drying over hundreds of thousands of years. Marked reductions in channel geometry, along with notable

decreases in sinuosity and bed material size, result in rivers that are clearly undersized for the valleys within which they flow (i.e. these rivers are underfit; see Figure 12.15).

Figure 12.16 conveys marked transitions in the post-glacial history of a river in a non-glaciated landscape. In this evolutionary sequence, sparsely vegetated slopes and fluctuating discharge conditions induced a braided river planform at the last glacial maximum. Climate amelioration and vegetation growth brought about dramatic transformation of the flow–sediment balance, whereby the energy of the system was diminished to such an extent that the river became a discontinuous watercourse with a fine-grained swamp that accumulates suspended-load deposits.

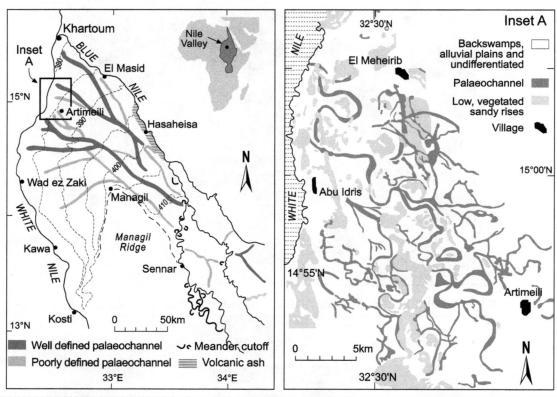

Figure 12.14   Climate-induced transition from river courses to dune fields. Many of the late Pleistocene and Holocene palaeochannels of the White and Blue Nile, North Africa that are preserved in the floodplains sit atop a former lake bed and are now overlain by desert sand dunes. At the last glacial maximum, floods were up to 4–5 m higher than today. White Nile levels were high around 14.7–13.1 ka, 9.7–9.0 ka, 7.9–7.6 ka, 6.3 ka and 3.2–2.8 ka. These wetter intervals are associated with global warming that began 15 000 ka, leading to a northward shift of the intertropical convergence zone into northern Sudan at intervals during the early to mid Holocene. Millennial-scale fluctuations are linked to changes in summer monsoon intensity linked to sea surface temperature changes in the South Atlantic and Indian oceans. Some of the centennial-scale changes may be related to variations in solar activity. Decadal fluctuations are closely correlated with El Niño–southern oscillation events. Incision by the Blue Nile and main Nile has caused progressive incision in the White Nile amounting to at least 4 m since the terminal Pleistocene some 15 ka and at least 2 m over the past 8 ka. The Blue Nile seems to have cut down at least 10 m since 15 ka and at least 4 m since 8 ka. Progressive channel incision drained the formerly swampy floodplains making these areas available for settlement by Neolithic farmers. Figures and text reproduced from Global and Planetary Change, 69 (1–2), Williams, M.A.J., Late Pleistocene and Holocene environments in the Nile basin, 1–15, © 2009, with permission from Elsevier.

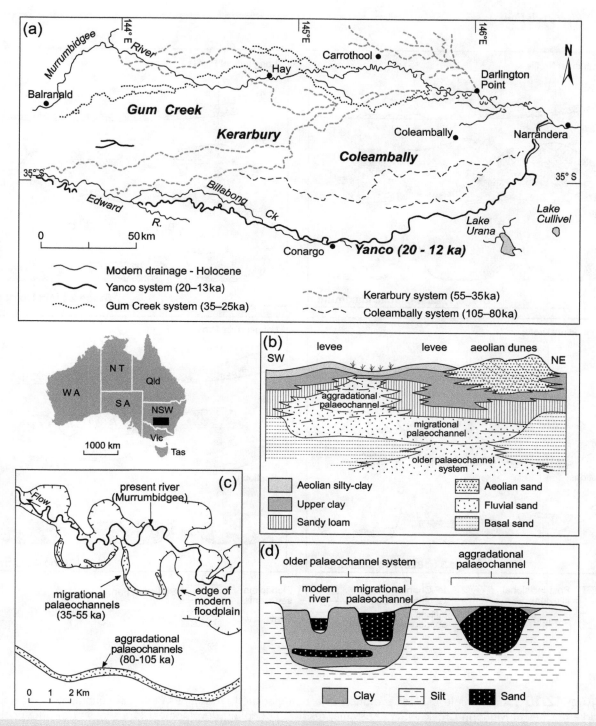

Figure 12.15 Late Quaternary evolution of the Murrumbidgee River. The Riverine Plain of southwestern NSW records a history of fluvial (in)activity that spans the last 105 ka (a). Aggradational palaeochannels dating from 80 to 105 ka were laterally accreting mixed-load rivers (b). They terminated with an aggradational phase comprised of bedload vertical accretion (b). Migrational palaeochannels dating from 35 to 55 ka had a high sinuosity with ridges and swales (c). Former bankfull discharges were four to eight times higher than the present day during oxygen isotope stage (OIS) 3 and 5 (see Figure 12.6) between 100 and 80 ka (OIS 5) and between 55 and 25 ka (OIS 3) (d). Since the last glacial maximum from 20 to 12 ka, discharges were high and channels carried around four times the present bankfull discharge. The Holocene saw the transition to the modern drainage network and flow regime characterised by smaller, more sinuous channels (c). This sequence of palaeochannel activity reflects the drying of the continent and increases in aridity over the last several hundred thousand years. This transition from mixed- to suspended-load rivers is characterised by planform adjustments that reduce the channel slope by increasing stream length (i.e. sinuosity), creating channels with cohesive boundaries that are less prone to adjustment (i.e. evolutionary adjustment from an active meandering to a passive meandering river). Modified from Page *et al.* (2009). © Taylor and Francis. Reproduced with permission.

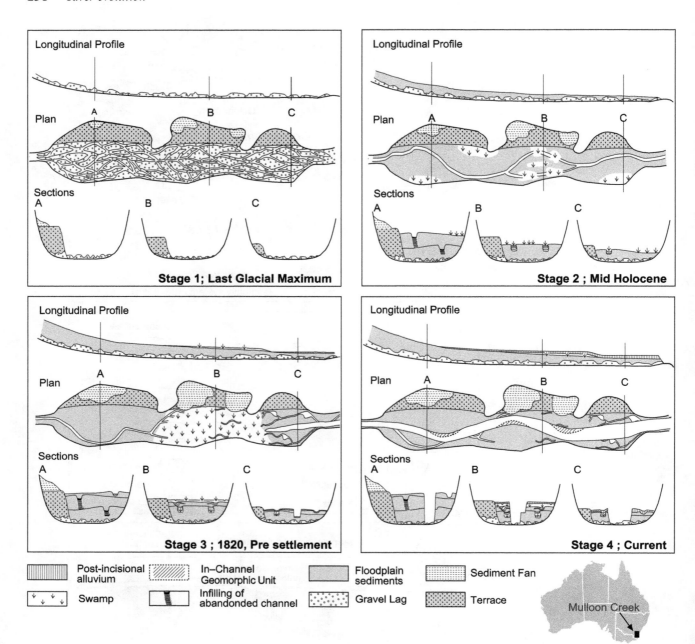

Figure 12.16 An example of river evolution in a non-glaciated landscape. Valley confinement along upper–middle reaches of Mulloon Creek, NSW, Australia, restricts floodplain development to a series of distinct pockets. These pockets comprise a downstream-thinning wedge of sediment packages. Stage 1 depicts the maximum extent of sediment reworking inferred to have occurred at the last glacial maximum. Some time before 12 500 yr ago the bedload-dominated braided river was transformed into a suspended-load system. By mid-Holocene, vertically accreted swampy valley fill deposits extended down-pocket as sediment lobes with perched channels (Stage 2). Sedimentation at the head and centre of the swamp starved the downstream areas of sediment, forming a wedge-shaped deposit that decreases in thickness down-pocket (Stage 3). A channel extended through the pocket shortly after European settlement (Stage 4). Today, a wedge-shaped unit of post-incisional alluvium thickens down-pocket, commencing in mid-pocket atop the swamp facies. From Johnston and Brierley (2006). © SAGE Publications. Reproduced with permission.

These valley floor deposits have subsequently been incised to create a continuous channel.

Glacial–interglacial cycles induce significant sea level change (eustasy). This may bring about geomorphic adjustment along the lower course of rivers. Sea levels may be lowered by 120–150 m during glacial maxima, essentially extending river courses onto what is now the continental shelf. The nature of geomorphic adjustment varies for differing fluvial–marine interactions (i.e. whether a delta or estuary is present) and the nature/extent of the continental shelf itself. Profound adjustments are noted along the lowland plains of large rivers, where incised valleys and fills develop significant terrace sequences (Figure 12.17). These terraces, in turn, constrain subsequent channel responses during periods of rising sea levels. Alternatively, the profound weight of accumulated deposits along the lowland plains of rivers, or in inland-draining (endorheic) basins, may induce subsidence via isostatic adjustment. Given the very low slope of these settings, avulsion may be experienced along these low-energy, suspended-load rivers.

Ongoing climate changes associated with global warming are bringing about marked geomorphic transitions for some rivers. For example, melting permafrost has increased discharges and the erosive potential of many rivers that drain into polar regions. Impacts of ice flows following spring melt have been accentuated. This exemplifies regionally specific patterns and trends in the evolutionary adjustment of rivers.

Finally, the impacts of climate changes upon rivers must be related to the magnitude–frequency relations of formative events, especially the geomorphic effectiveness of extreme floods. As noted previously, there is significant variability in response in differing landscape and climatic settings. This reflects the sensitivity/resilience of a river, and the extent to which the river is attuned to seasonal and interannual variability in discharge. In some instances, extreme floods may exert a profound imprint or memory upon the system, whereby the river is subsequently unable to adjust its boundaries. Depending upon the condition of the system at the time of the event, and associated availability of sediment, flows may be highly erosive or highly depositional. Either way, transformation of channel boundaries exerts a significant influence upon the subsequent evolutionary adjustments of the river. These various pathways and rates of geomorphic evolution are meaningfully captured using the river evolution diagram.

Linking river evolution to the natural capacity for adjustment: adding river change to the river evolution diagram

The river evolution diagram introduced in Chapter 11 can be used to evaluate pathways of river evolution in relation to the behavioural regime for any given type of river. In general terms, the greater the range of variability in geomorphic behaviour demonstrated by a river (i.e. the greater the degrees of freedom and capacity for adjustment), the greater the likelihood that evolutionary adjustments and geomorphic change will occur over shorter timeframes. Conversely, the more resilient the river the longer the timeframe for discernible geomorphic adjustment. These determinations reflect the imposed boundary conditions within which a river operates, as shown by the outer band of the river evolution diagram. The width of the outer band increases from confined through partly confined to laterally unconfined settings, as the potential range of variability increases. Rivers can more readily adopt differing morphologies and behavioural attributes if there is space for the channel to adjust on the valley floor.

A similar degree of variability is evident in the width of the inner band on these figures. This reflects the natural capacity for adjustment as determined by flux boundary conditions. The width of this inner band represents the range of states that the river can adopt while still being considered to be the same type of river (i.e. retaining a consistent set of core geomorphic attributes that reflect the character and behaviour of that river type). In a sense, this is a measure of the sensitivity of the river, as it records the ease with which the river is able to adjust. As indicated for the potential range of variability, the width of the inner band is greatest in laterally unconfined settings.

River responses to disturbance events are indicated by the arrows shown at the top of the inner band on the river evolution diagram. The spacing of the arrows indicates their frequency, while the size of an arrow indicates its magnitude. In most instances, disturbance events promote river adjustments but the reach remains within the inner band (i.e. perturbations fall within the natural capacity for adjustment). River adjustment within the inner band may breach intrinsic threshold conditions, marking a shift in the way energy is used (either concentrated or dispersed). Typically, this reflects an adjustment in the character or distribution of resisting forces (e.g. bed resistance, form resistance, resistance induced by riparian vegetation or wood). These internal adjustments alter the assemblage of erosional and depositional landforms on the valley floor, yet fall within the behavioural regime of the river.

In other instances, changes to the prevailing flux boundary conditions and/or severe disturbances may bring about changes to the formative processes that fashion river morphology (i.e. river change has occurred). This scenario is highlighted by the shift in the position of the inner band. In these instances, altered stream power relationships reflect differing energy use in relation to prevailing flux boundary conditions. Reaches now operate within a different inner band on the river evolution diagram, with altered

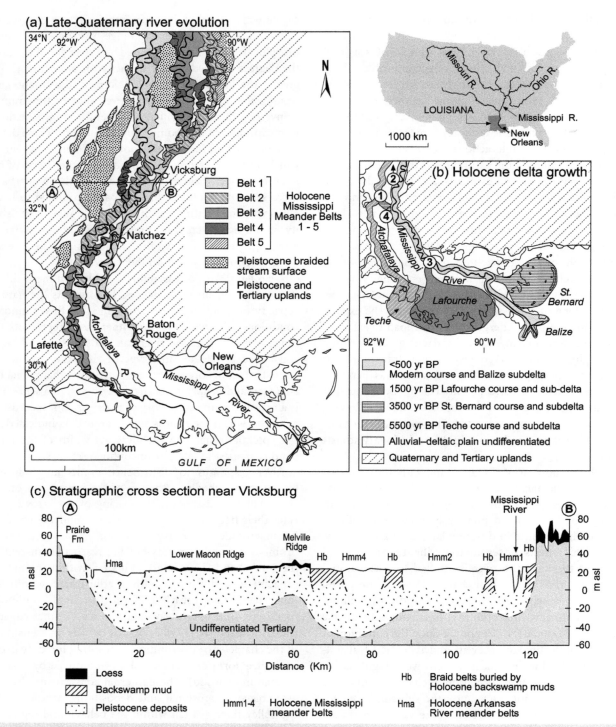

Figure 12.17 Example of river responses to eustatic sea level changes along a lowland plain. Aggradation of the lowland plain of the lower Mississippi River in the period since sea level lowstand at the last glacial maximum has brought about a series of dramatic evolutionary adjustments. (a, c) Up to five meander belts have accumulated above former braided river deposits. (b) At least three phases of avulsion repositioned areas of delta growth. Modified from Aslan *et al.* (2006) © Society for Sedimentary Geology and Rittenour *et al.* (2007) © Geological Society of America. Reproduced with permission.

energy conditions. The shape of the pathway for adjustment, shown by the jagged line within the inner band, has a different form for the new river type, depicting a change in process–form associations along the valley floor, such that there is a change in river morphology. The new configuration represents a different type of river, with a different appearance (character) and set of formative processes (behaviour). Inevitably, there may be some overlap in the position of former and contemporary bands, and some geomorphic units may be evident in both situations. However, the assemblage of geomorphic units in the two bands differs, reflecting a change to river character and behaviour.

The shift in the position of the inner band can be induced by a press disturbance that exceeds an extrinsic threshold. This usually reflects alteration to flux boundary conditions, as modified flow and sediment transfer regimes (i.e. impelling forces) drive river change. In this case, the time that is required for recovery following perturbation is longer than the recurrence interval of disturbance events. Effectively, the previous configuration of the river was unable to cope with changes to the magnitude and rate of applied stress. Rare floods of extreme magnitude, or sequences of moderate-magnitude events that occur over a short interval of time, may breach extrinsic threshold conditions, transforming river character and behaviour.

Dependent on the subsequent set of process–form associations adopted by the river, the natural capacity for adjustment may widen or contract as the new type of river adjusts to different flux boundary conditions. The position of the inner band within the potential range of variability (the outer band) indicates whether the change in river type marks a transition to a higher energy state (an upwards adjustment) or a lower energy state (downward adjustment). Changes to the amplitude, frequency and shape of the pathway of adjustment within the inner band indicate how the river responds to pulse disturbances of varying magnitude and frequency.

In some cases, change may occur during a threshold-breaching flood. In this instance the two inner bands are located adjacent to each other and the date of the disturbance event is noted. In other cases change may be lagged or occur progressively over time. In these instances, the space between the two inner bands is widened to depict whether change occurred over years or decades. In more complex situations, transitional river types are depicted on the diagram. For simplicity, only a major shift between one river type and another is shown in the examples outlined below.

Figures 12.18–12.23 build upon the interpretation of river behaviour for various river types presented in Chapter 11, using documented examples of river evolution from the literature. Emphasis is placed upon the nature of evolu-

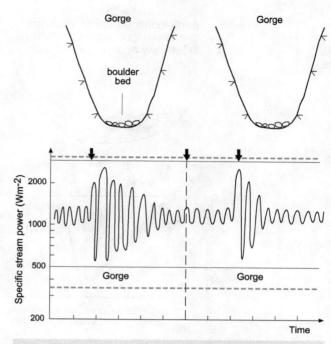

Figure 12.18 River evolution diagram for a gorge. Resilience is high and the system responds quickly to disturbance events, with negligible adjustments to river morphology.

tionary changes to the river, timeframes over which these changes occur and evolutionary trajectory.

Figure 12.18 represents an imposed river configuration such as a gorge. Disturbance events that have the capacity to induce changes in other settings are unable to bring about significant geomorphic adjustments along confined rivers, as the inherent resilience of the system is too strong. Perturbations to the flow and sediment regime are accommodated by instream adjustments to hydraulic resistance, such as the nature and distribution of bedforms, dissipating flow energy. Adjustments to river character and behaviour are negligible and the river type remains the same. Millions of years of valley widening may allow for out-of-channel deposition and generation of floodplain pockets, but the assemblage of erosional and depositional geomorphic units along the reach is likely to remain consistent over tens of thousands of years (at least).

A different pattern of responses to changes in external stimuli may be experienced in partly confined valley settings, where the potential range of variability is somewhat broader than in confined valleys (Figure 12.19). This enables a greater range of possible river morphologies to develop. Antecedent controls and prevailing flux boundary conditions shape the contemporary configuration of the river. A bedrock-controlled discontinuous floodplain river has negligible capacity for adjustment because of the

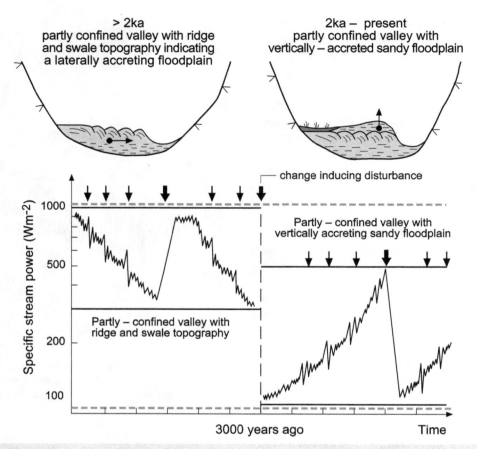

Figure 12.19 River evolution diagram for partly confined rivers. Lateral accretion of the Tuross River (South Coast, New South Wales, Australia) pushed the channel against the valley margin (i.e. the channel attained its maximum sinuosity). Since then, the river has been transformed into a lower energy system characterised by vertical accretion processes (Ferguson and Brierley, 1999a,b). From Brierley and Fryirs (2005). © John Wiley & Sons, Ltd. Reproduced with permission.

bedrock-imposed setting. Valley widening over tens of thousands of years results in progressive transition to a planform-controlled situation. The example demonstrates potential adjustments in this situation, as there is greater capacity for adjustment because of the greater degrees of freedom. Local areas of the channel are able to adjust their planform within the partly confined valley. For example, lateral migration may form ridges and floodchannels within the vertically accreted silty floodplain. In this instance, the natural capacity for adjustment has shifted to a lower energy river type. This is indicated on the river evolution diagram by a downward shift in the position of the inner band (the natural capacity for adjustment) within the outer band (the potential range of variability). In addition, the range of river behaviour has been reduced (i.e. the width of the inner band has narrowed; note the logarithmic scale).

Rivers are more sensitive to change in laterally unconfined valley settings relative to partly confined and confined valleys (i.e. the potential range of variability and the natural capacity for adjustment are greatest in laterally unconfined valley settings). Changes are shown from a braided configuration to a meandering mixed-load system (Figure 12.20), from a mixed-load meandering to a suspended-load meandering river (Figure 12.21), from a gravel-bed braided to a low-sinuosity sand-bed river (Figure 12.22) and from a braided to a fine-grained discontinuous watercourse (Figure 12.23). These changes reflect alterations to both the impelling forces that promote change (i.e. less variability in flow, less coarse-sized material on the valley floor, etc.) and internal system adjustments that modify the pattern and extent of resistance. A major shift in the assemblage of geomorphic units ensues, resulting in altered patterns of mid-channel and bank-attached geomorphic units, and

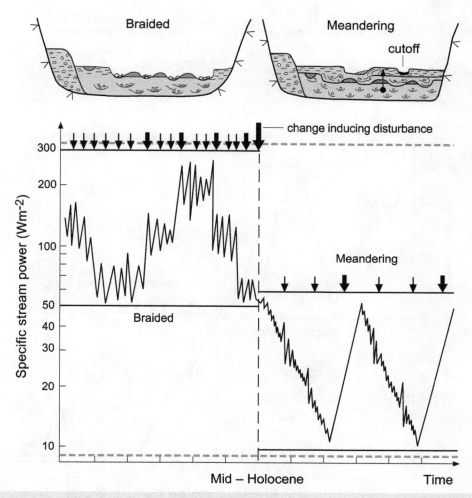

Figure 12.20 River evolution diagram for braided–meandering transition in a laterally unconfined valley setting. Many rivers subjected to high sediment loads in formerly glaciated landscapes were characterised by braided planforms. As sediment availability decreased, vegetation became established and flow regimes became less variable in the Holocene, energy diminished and braided rivers were transformed into meandering planforms. These active meandering systems operated under lower energy conditions than their braided counterparts. From Brierley and Fryirs (2005). © John Wiley & Sons, Ltd. Reproduced with permission.

processes of floodplain formation and reworking. Channel geometry and bedform assemblages are transformed as well. Critically, these adjustments occur over much shorter timeframes than those indicated for the examples shown in confined and partly confined settings in Figures 12.18 and 12.19.

The four transitions in river character and behaviour shown for laterally unconfined valley settings in Figures 12.20–12.23 all show a downward shift in the natural capacity for adjustment within the potential range of variability. This reflects the adoption of a lower energy river type within the same landscape setting. This transition is especially pronounced in Figure 12.23. In some instances, an increase in resistance increases the capacity of the system to trap finer grained materials, thereby aiding the transi-

tion to a single-channelled or discontinuous channel configuration. Increased stability enhances prospects for vegetation development on the valley floor. As a result, the natural capacity for adjustment is narrower, reflecting a reduction in the range of behaviour. Changes to energy relationships reflect the consumption of energy, altering the pathway of adjustment. For example, the transition from braided to meandering configurations shown in Figure 12.20 is marked by a switch from tight chaotic oscillations reflecting recurrent reworking of materials on the channel bed to a jagged shape that reflects the occasional formation of cut-offs and subsequent readjustment of channel geometry, planform and slope.

Post-glacial adjustments to flow and sediment fluxes commonly induced changes from a braided to a meandering

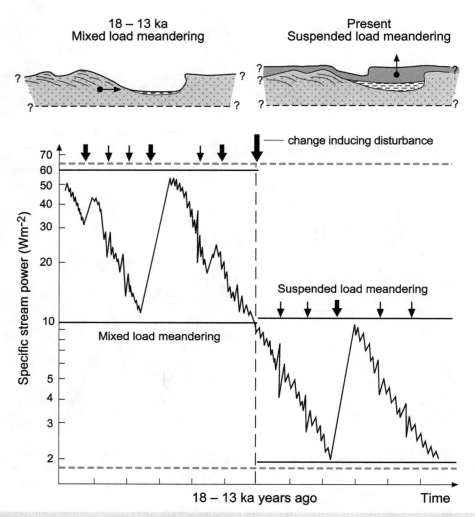

Figure 12.21 River evolution diagram showing transition from a mixed-load river to a suspended-load river in a laterally unconfined valley setting. Rivers on the wide plains of western New South Wales, Australia have adjusted to much lower energy regimes in the period since the last glacial maximum. This has been marked by a transition from mixed-load meandering rivers that were prone to progressive lateral migration to suspended-load systems that are characterised predominantly by vertical accretion. This diagram is based on phases and timescales of river evolution along the Murrumbidgee River shown in Figure 12.15. From Brierley and Fryirs (2005). © John Wiley & Sons, Ltd. Reproduced with permission.

channel planform (Figure 12.20). In the early post-glacial interval, abundant sediment, highly variable flows and negligible vegetation cover promoted the development of braided rivers. A wide range of mid-channel bars and shifting channels of varying size characterised these bedload-dominated systems. Progressive reduction in sediment availability in the post-glacial era, along with reduced variability in discharge and progressive encroachment of vegetation onto the valley floor, brought about the transformation of many of these braided rivers into mixed-load meandering systems by the mid-Holocene. These rivers are now characterised by laterally migrating single channels with point bars and associated instream geomorphic units,

and an array of laterally and vertically accreted floodplain forms.

The impacts of long-term drying upon the planform and geometry of rivers in non-glaciated environments shown in Figure 12.15 are reconstructed using the river evolution diagram in Figure 12.21. This shows the transformation from a mixed-load laterally migrating channel into a slowly migrating suspended-load river with a much smaller channel capacity. This transition reflects a decline in fluvial activity driven by changes to the discharge regime.

Different pathways and rates of adjustment may be experienced by different types of rivers subjected to similar climatically induced changes to prevailing flow and sedi-

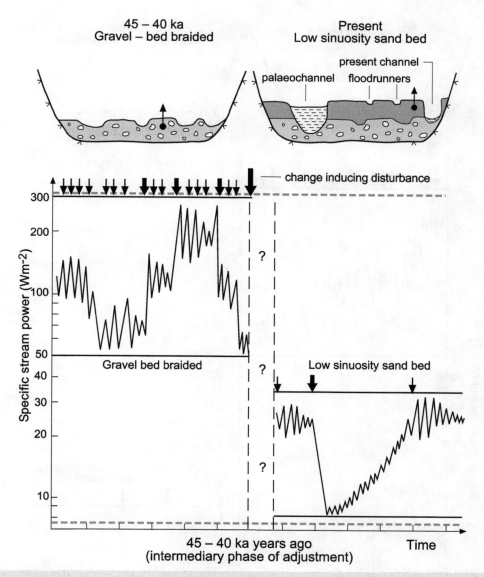

Figure 12.22   River evolution diagram showing transition from a gravel-bed braided to a low-sinuosity sand-bed river in a laterally unconfined valley setting. Profound changes to river morphology have characterised the late Quaternary evolution of the Hawkesbury–Nepean River, New South Wales, Australia. This marks a transition from a gravel-bed meandering system to a low-sinuosity sand-bed river that is effectively inset within a gravel braidplain (based on Nanson *et al.* (2003)). From Brierley and Fryirs (2005). © John Wiley & Sons, Ltd. Reproduced with permission.

ment fluxes. In Figure 12.22, a stable low-sinuosity sand-bed river with a vertically accreted floodplain has replaced a gravel-bed braided river. Figure 12.23 builds upon the example shown in Figure 12.16, showing the transition from a braided configuration at the last glacial maximum to the development of a fine-grained discontinuous watercourse in a cut-and-fill river.

A range of tools and approaches used to analyse and interpret river evolution is outlined in the following section.

Reading the landscape to interpret river evolution

Interpretations of river evolution by reading the landscape can be complemented by sediment analysis and use of dating techniques, process measurements, appraisal of historical records and modelling applications (see Table 12.1). In some instances, ergodic reasoning (space for time substitution) can be used to construct evolutionary sequences

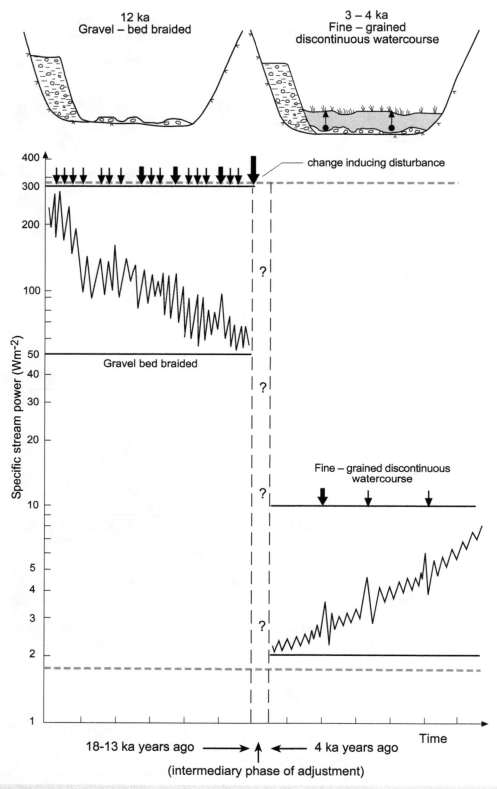

Figure 12.23 River evolution diagram showing the transition from a gravel-bed braided river to a fine-grained discontinuous watercourse in a laterally unconfined valley setting. A marked reduction in energy conditions and river morphology has characterised the Holocene evolution of Mulloon Creek, New South Wales, Australia. Amelioration of climate and encroachment of vegetation onto the valley floor promoted the accretion of fine-grained materials and the development of discontinuous watercourses atop former braided-river deposits (see Figure 12.16). From Brierley and Fryirs (2005). © John Wiley & Sons, Ltd. Reproduced with permission.

Table 12.1 Sources of evidence used to detect river change

Direct observations	Reading the landscape. Local anecdotal knowledge. Ergodic reasoning (space for time substitution) based upon reconstructed evolutionary time slices.
Process measurements	Instrument records (rarely continuous) – typically applied over intervals from minutes to years (e.g. discharge, sediment flux (bedload/suspended load), bank erosion, bed aggradation/degradation, meander migration, floodplain sedimentation). Ground surveys – such as repeated field surveys of cross-sections. *The representativeness of process measurements must be carefully appraised, in spatial and temporal terms. It is often difficult to measure rates of activity during formative (high-magnitude) events.*
Historical records	Historical records, such as explorers and survey notes, paintings, bridge plans, hydrographic surveys, newspaper articles, etc. Archival maps, such as portion plans. Vertical aerial photographs, remote sensing and satellite imagery. Historical photographs. *Inevitably, the availability of these records is something of a lottery. Most records are incomplete. Often, multiple disturbance events occur within available time intervals.*
Sedimentary evidence	Long-term records of river changes are largely derived from complex and generally incomplete sedimentary records. This entails analysis of: • Surface forms and palaeochannels. • Subsurface forms using techniques such as ground-penetrating radar. • Exploratory sedimentary data (bore hole, core-log and trench data). • Bed material calibre, bedforms, palaeocurrent indicators and the architecture of depositional units. For example, facies types provide insight into depositional process and flow energy (Chapter 6). • Geometry and assemblage of depositional units, and their bounding surfaces (erosional or depositional). • Slackwater deposits provide evidence of flood stage. *Inevitably records are incomplete, and are often inaccessible. Chronologic controls are often lacking.*
Dating techniques	1. *Relative methods* • Relative height. • Organic remains. • Artefacts (archaeological remains, especially pottery shards). 2. *Absolute methods*

Dendrochronology	Up to 10 000 yr
Lichenometry	Up to 10 000 yr
Radiocarbon (C-14 and accelerated mass spectrometry (AMS))	Up to 55 000 yr
Lead-210 (Pb-210)	<250 yr; optimum 5–100 yr
Caesium-137 (Cs-137)	Post 1952
Cosmogenic dating	<1000 yr to millions of years, but generally <500 000 yr
Uranium series dating	Typically 5000–350 000 yr, but may be as recent at 100 yr
Thermoluminescence (TL) and optically stimulated luminescence (OSL)	Typically around 1000–300 000 yr but may be as recent as 100 yr
Electron spin resonance (ESR)	Up to 300 000 yr
Potassium–argon dating (K–Ar)	Up to millions of years
Tephrachronology	Up to millions of years
Isothermal plateau fission track dating	<140 yr
Palaeomagnetism	1000–10 000 000 yr

(Chapter 2). Assessment of river evolution at any given locality must be framed in its spatial context (within-catchment position and in relation to regional patterns/trends), alongside broader scale geologic and climatic considerations (i.e. tectonic setting and records of climate change). Typically, topographic maps, geology maps, remotely sensed images and resources such as Google Earth® are analysed prior to going into the field. Controls upon the contemporary character and behaviour of the river (Chapters 10 and 11) must be assessed before meaningful interpretations of evolutionary adjustments can be performed. This entails analysis of river forms and processes in relation to geologic and climatic controls upon imposed and flux boundary conditions, and the associated range of disturbance events to which the river is subjected. Questions asked in these preliminary investigations include:

- What is the landscape setting – geology, climate, vegetation cover? Is this a glaciated landscape, a desert, a meltwater channel, an urban stream, a tropical rainforest, the flanks of a volcano? How does the setting impact upon the erodibility/erosivity of this landscape, and associated flow–sediment fluxes?
- How does position in the landscape/catchment, and associated slope, catchment area and valley width affect the nature and effectiveness of erosional and depositional processes (i.e. is this a source, transfer or accumulation zone)?
- How is the reach affected by downstream or upstream controls? How connected are hillslopes to the valley floor?

Building upon these geographic relationships, field analyses of river evolution interpret the range and pattern of geomorphic units observed in a given setting. Analysis of the sedimentary record involves interpretation of the internal structure and characteristics of sedimentary sequences for a given landform (Chapter 6). Spatial relationships between landforms provide a basis to interpret depositional and erosional histories at the reach scale. By interpreting the sequences of sediments preserved in basin fills, stages or phases of evolutionary adjustments can be differentiated and formative events can be appraised.

Inevitably, any landscape retains an incomplete record of past activities and events. Bare bedrock in confined valleys and supply-limited landscapes is indicative of erosional surfaces. Cosmogenic dating techniques can be used to determine exposure dates of differing surfaces, from which erosion rates can be determined. Reworking of deposits provides a partial preservation record in partly confined valley settings. More complete depositional sequences are evident in laterally unconfined settings and transport-limited landscapes where basin fills may record activities over long timescales. Much of the record may be buried. Terraces and floodplains often preserve records of

deposition and reworking that extend back over thousands or tens of thousands of years. Insight into reworking events can be gleaned from erosional surfaces (discontinuities or unconformities) in the sediment record. Are boundaries onlapping (depositional) or erosional (e.g. local scour or floodplain reworking)? Do they indicate changes in river behaviour (e.g. change in type of floodplain deposit)? Are depositional sequences in bank exposures consistent with deposits laid down by the contemporary river, or are they indicative of change? Disjuncts (unconformities) in the depositional record are indicative of erosive events.

Linking sediment sequences to their chronology is vital in determining phases and rates of activity. Assessment of the preservation potential of deposits provides guidance into what may be missing (erasure) and the record of events that may have been obliterated by later erosion. Juxtaposition of units often represents a hiatus and/or change in process relationships. When combined with dating techniques, phases of river evolution can be interpreted and rates of change determined. Dating tools can be used to generate age estimates of depositional features, providing insight into the time they were laid down (or reworked), the timeframe of disjuncture between eroded units and the period of time that has been lost from the depositional record. Ideally, the erosional/depositional history in one reach is related directly to evolutionary adjustments in upstream and downstream reaches.

These interpretations can be supplemented by process measurements to assess the rate and effectiveness of geomorphic process activity. From this, magnitude–frequency relationships can be derived to assess how much work is likely to be performed for an event of a given magnitude. These relationships are extremely important in deriving rating curves that estimate sediment transport (Chapter 6) or formative flows that fashion channel geometry (Chapter 7). A range of logistical problems besets field-based measurement of geomorphic processes. First and foremost, the representativeness of the data (in space and time) must be assessed. How accurate/precise are the data themselves? How reliably can they be extrapolated to other situations? In many instances, measurement techniques may disturb the observed processes. As yet, many processes and phenomena cannot be observed or measured directly or even indirectly. Real-time or lapse observations and measurements may be extremely helpful in interpreting frequent low-magnitude events, but instruments are often destroyed in catastrophic high-magnitude events. Ironically, these events may well be the primary agents of landscape adjustment. All too often, the timescale of human observation is much shorter than that of the phenomenon under study. There are remarkably few, sustained programmes of longer term (decadal) process measurement. As such, it is difficult to discern magnitude–frequency relationships in a comprehensive manner.

In some instances, stages of landscape evolution can be appraised through reasoning by analogy (ergodic reasoning), which is the recognition of similarity among different things. Comparative frameworks can be used to relate states (or stages) of evolutionary adjustment in different areas that have a similar landscape configuration (i.e. equivalent features are produced by the same set of processes under an equivalent set of conditions). This is referred to as space for time substitution. Time slices can be used to interpret the pathway of adjustment that is likely to be experienced for reaches of the same river type. The reliability of predictions is dependent on the similarity of the places that are being compared and the range and rate of processes and disturbance events to which they are subjected. Similar outcomes may arise from different processes and causes (the principle of convergence or equifinality). A common origin or equivalent causality is a prerequisite for effective comparison.

Increasingly, geomorphologists simulate real-world understanding as a basis to interpret process understandings, identify key controls upon process–form linkages, assess rates of activity and predict evolutionary trajectories through the use of physical and numerical models and experimental procedures (e.g. flume studies). This provides an important platform to assess understandings of real-world situations. Hypotheses and future scenarios can be tested.

While modelling provides a critical basis to assess magnitude–frequency relationships for individual processes, it is difficult to 'scale up' processes and interactions in a way that meaningfully captures landscape-scale dynamics at the catchment scale. Models cannot generally take account of the intrinsically random or chaotic disturbances that drive landscape change, or their non-linear and complex responses. Concerns arise about the selection of input parameters and the transferability of insights from one system to another. Hence, significant questions remain about the representativeness and replicability of modelled output to real-world situations. Field verification provides the critical test of our understanding. Tools such as reading the landscape are required to meaningfully adapt findings from modelling applications to real-world conditions, circumstances and situations, linking field interpretations to theoretical understandings.

Tips for reading the landscape to interpret river evolution

Step 1. Identify individual landforms and their process–form associations

Critical insight into river evolution can be gained by assessing whether the landforms that make up any given reach are products of the contemporary behavioural regime of the river or they reflect former conditions. Palaeolandforms provide a 'glimpse' into the past. For example, the morphology, position and sedimentary structure of terraces and palaeochannels can guide insight into past flow and sediment regimes. This aids determination of whether the bed is aggrading or degrading, former channel dimensions and planform, and the sediment-transporting regime (e.g. bedload, mixed-load or suspended-load river).

Step 2. Interpret river change at the reach scale

The key to analysing river change is to determine whether a wholesale change in river character and behaviour has occurred such that a new river type occurs. Critically, what has changed, and why? Analysis of the pattern/position of landforms can aid interpretations into likely sequences of events. It is also important to determine the timeframe over which change has occurred (last year, decade, century, 1000 yr, 10 000 yr, million years) and to assess stages/phases of evolution as a series of evolutionary time slices. Analysis of changes to the geomorphic unit assemblage provides insight into the altered process regime, indicating how erosional or depositional processes, and their relative balance/effectiveness have changed over time. For example, terraces, inset floodplains and knickpoints are indicative of changing geomorphic conditions and associated phases of activity. Similarly, differing packages of units may be evident across the valley floor (e.g. ridge and swale topography, cut-offs, avulsion). Beyond this, any indications of adjustments to channel geometry and associated bank erosion/deposition processes must be unraveled, determining whether these attributes fit with the contemporary flow and sediment regime. Is the size and shape of the channel a function of the current flow and sediment regime or a remnant from the past? Is there any indication of changes to channel planform? Interpretation of erosional or depositional boundaries between geomorphic units provides insight into the reworking of these features and phases/sequences of events that characterise reach evolution.

Step 3. Explain controls on river change at the reach scale

It is often exceedingly difficult to isolate the impacts of past events upon river evolution. Some events or phases of geomorphic impacts leave a dominant imprint upon the contemporary landscape, essentially overriding (overprinting) the geomorphic signal of previous events. While the records of these events may persist, they may erase signals of previous activity. In some instances, relatively trivial events may be selectively preserved, whereas impacts of other formative events may have been entirely erased. Erasure creates a disjunct in time in the features that are preserved along the valley floor. However, accumulation zones may retain a near-continuous record of sediment preservation over time.

In simple terms, whatever fashioned the valley controls the river. In some instances, valley morphology may have been shaped by past glacial activity. Elsewhere it may have been superimposed following tectonic uplift, or it may be produced by river capture. These long-term landscape controls fashion contemporary process relationships on valley floors.

Evidence used to determine the evolutionary sequence can be used to interpret changes to the boundary conditions and associated disturbance events that triggered the history of adjustments. How and why has channel geometry changed? Have alterations to channel alignment or enlargement/contraction modified the use of energy along the reach, changing the mix of bed and/or bank processes and channel–floodplain linkages? Has the assemblage of instream geomorphic units changed (e.g. have transitions occurred between mid-channel and bank-attached features, reflecting alteration of the erosional/depositional balance of the reach)? From this, it is important to ask why formative processes are different. What factors altered the flow–sediment balance? Has available energy increased or decreased over time? Have resistance elements been altered (e.g. vegetation cover, wood loading)? Floodplain sediments may provide an indication of past depositional environments. Some floodplains were formed by different sets of processes that operated under very different conditions to those that occur today. Is a hiatus evident – is the contemporary floodplain forming in the same way it did in the past (e.g. transitions from lateral to vertical accretion)? Are formative and reworking processes consistent over time? If not, why not? Is there any evidence of palaeo-forms or transitions in floodplain type?

The Lane balance diagram can be used to relate river changes to altered imposed and flux boundary conditions. Is there any evidence that the rate of change over time has been altered? Was evolution progressive or did the exceedance of threshold conditions bring about dramatic/rapid change? What was the role of catastrophic events? Alternatively, was change lagged after the disturbance (Chapter 2)? What kinds of disturbance events brought about evolutionary adjustments to the river? Did geologic controls such as uplift, faulting, folding and tilting induced by earthquakes, volcanic activity or subsidence events bring about these adjustments, or were climatic factors responsible (e.g. cyclones, intense local storms, drought)? Are there any indicators of drivers of river evolution, such as flood debris (slackwater deposits), volcanic ash, fault scarp, mega landslides, earthquakes, etc.?

These analyses must be framed in relation to river sensitivity. Marked variability in the type, ease and rate of adjustments is evident for bedload, mixed-load and suspended-load rivers. Similarly, reaction and relaxation times after disturbance vary.

Step 4. Explain how catchment-scale relationships affect river evolution

Putting it all together at the landscape scale entails explanation of catchment-wide river responses to changes in imposed and flux boundary conditions, and associated disturbance events that drive evolution. Evolutionary assessments must appraise what is going on at any given location in relation to what is going on elsewhere in the system. In some instances, contemporary landscapes reflect lagged responses to disturbance events elsewhere in the catchment. Response gradients record how linkages and connectivity transmit upstream signals of disturbance response to downstream reaches. In essence, reading the landscape entails linking these spatial and temporal considerations in any given system, determining how what happened in the past or elsewhere in the system affects what is observed in any given reach today.

Pathways and rates of river evolution are determined primarily by geologic and climatic controls upon landscape setting (Figures 12.2 and 12.5), the memory or imprint of the past (Figure 12.7) and the combination of disturbance events to which a river is subjected. Constructing evolutionary sequences is a system-specific exercise. Each river must be placed in its *geologic/tectonic* and *climatic* context to interpret drivers of river evolution. This entails assessment of adjustments to boundary conditions and the nature/effectiveness of disturbance events over time. River location must be appraised in its tectonic context, assessing position relative to plate margins and the type/rate of activity at that margin. The erodibility and erosivity of a landscape exert key controls upon river types and their evolutionary adjustment. Long-term topographic changes that affect relief, slope, drainage pattern, drainage density, valley width, etc. must be assessed to interpret how imposed boundary conditions may have changed. In some instances, underfit streams may have developed, wherein the contemporary river flows within a valley created by much larger flows. This may reflect river capture or the influence of meltwater channels. Other forms of geologic imprint (memory) include responses to disturbance events such as earthquakes or volcanic eruptions.

Inevitably, geologic controls must be viewed in relation to changing climatic conditions and associated variability in flow, sediment and vegetation interactions (i.e. flux boundary conditions). Increases or decreases in discharge may alter the available energy of formative flows. However, associated changes to vegetation cover and surface roughness alter the effectiveness of these flows to do geomorphic work. Non-linear responses are common. Appraisal of the effectiveness of disturbance events and associated magnitude–frequency relations is a key consideration in determination of evolutionary adjustments. In some instances a dominant imprint from the past may influence contemporary process–form interactions (i.e. climatic memory). For example, sediment availability fashioned by paraglacial processes is a primary determinant of river character and behaviour in many glaciated landscapes. The key issue here is determining how changes to climatic factors have influenced the natural range of behaviour and evolutionary trajectory of the river, assessing how and why the river has responded to changes in flux boundary conditions in the way in which it has. Perhaps the key question to address here is whether the river is well adjusted to its contemporary setting (slope, discharge regime, etc.) or whether certain attributes are products of former geologic and climatic considerations.

Evolutionary considerations must be framed in relation to within-catchment position, appraising the ways in which disturbance responses are conveyed through a landscape, and their consequences. Changes to base level may induce significant evolutionary adjustments, whether as a product of progressive knickpoint retreat, the role of resistant lithologies, landslide-induced dams (however temporary) or sea level changes. Also, alterations to tributary–trunk stream relationships may bring about adjustments to river character and behaviour.

Comparison of system-specific evolutionary histories is required to appraise the transferability of insights from one situation to another. Are inferred evolutionary records consistent from locality to locality within a catchment and across a region? If so, this can be related to broad-scale climate variability. Is the direction and rate of change local or regional; is it systematic, or is it subject to local-scale variability in controls? Informed appraisals from the regional record can be used to generate a more complete picture of phases or stages of river evolution and their timing. Catchment-specific applications should be related to regional trends to see whether scenarios are similar or different from adjacent catchments. These records should be analysed in relation to climate patterns, histories of flood events and responses to known disturbance events. The challenge here lies in unravelling these various spatial and temporal considerations.

Conclusion

Long-term river evolution is fashioned largely by tectonic setting and geologic history. This determines the imposed boundary conditions within which contemporary processes operate. Inset within this, climatic controls determine the mix of water, sediment and vegetation interactions that occur in any given landscape. Changes to flux boundary conditions drive the evolutionary trajectory of a river, prospectively inducing river change to a different river type. Many landscapes are products of recent adjustments. Elsewhere, landscapes may reflect great antiquity, such that rivers retain the imprint of antecedent geologic or climatic conditions. Geomorphological interpretation of river evolution unravels how a river has adjusted and changed over various timeframes and the range of disturbances (causes) that induced these changes. However, reading the landscape does not end there! In the next chapter, various forms of human disturbance that may modify river character, behaviour and evolution are outlined.

Key messages from this chapter

- River evolution is the study of river adjustment and change over time. Insights into the evolutionary trajectory of a river can be used to determine causes of geomorphic adjustment and/or change, providing a basis to forecast likely future trends.
- River change is defined as a wholesale adjustment in river character and behaviour such that a new river type occurs. River change can occur in response to altered flux and/or imposed boundary conditions.
- Evolutionary adjustments occur in response to disturbance events over geologic and geomorphic timeframes. Responses can be progressive, instantaneous or lagged.
- Geologic setting determines the imposed boundary conditions within which rivers adjust and evolve. It determines the relief, topography and erodibility of a landscape. Tectonic setting is a key control on the topography of the landscape. Uplifted plate margins produce steep, highly erosive, sediment-charged rivers. Pull-apart margins may contain escarpment-dominated rivers or rift valleys. Plate-centre locations contain low-lying, low-energy rivers with broad open plains. Tectonic adjustments can induce river entrenchment, capture and underfitness.
- Climate variability drives fluxes of water, sediment and vegetation along rivers (flux boundary conditions). Climate changes impact upon catchment hydrology and the flow regime of rivers. Alterations to flow–sediment balances may reflect event-specific, seasonal, interannual/decadal or much longer term (e.g. glacial–interglacial) trends. Responses to climate change vary markedly in differing morphoclimatic regions.
- Constructing evolutionary sequences is a system-specific exercise.

- The sedimentary record and historical documents provide critical sources of information with which to guide interpretations of river evolution. Interpretation of the position and assemblage of geomorphic units provides insight into river adjustments over differing timeframes (i.e. primary depositional forms versus secondary reworking; differentiation of channel versus floodplain versus terrace compartments; relationship between river behaviour and change). Boundaries between geomorphic units provide insight into reworking of these features and evolution.
- The river evolution diagram can be used to conceptualise the pathway of river evolution in response to imposed and flux boundary conditions and sequences of disturbance events.

CHAPTER THIRTEEN

Human impacts on river systems

Introduction

Change is an integral part of all river systems. Disturbance is ongoing, as rivers adjust to altered boundary conditions and associated flow and sediment fluxes. Human disturbance modifies the boundary conditions under which processes operate, prospectively altering the pattern, rate and consequences of river adjustments. The nature of human activities at a particular place is just as important as this innate diversity in environmental setting. Everything is contextual. Most rivers now operate under fundamentally different conditions to those that existed prior to human disturbance.

Human impacts do not directly alter the fundamental hydraulic and geomorphic processes such as the mechanics of sediment transport, erosion, and deposition along rivers. However, human disturbance modifies the spatial distribution (pattern, extent and linkages) and rate (accentuated/accelerated or decelerated/suppressed) of these processes, often inducing profound changes to river morphology, whether advertently or otherwise. Some processes and landscape responses now happen more often in more places than they did prior to human disturbance. The converse situation also occurs. Other processes may no longer occur in areas where they once were common, or they may occur less frequently within a smaller geographic range.

Understanding of contemporary river forms and processes, tied to interpretations of natural variability and longer term evolution, provide a basis to assess river responses to differing forms of human disturbance. From this, predictions of likely future character, behaviour and condition can be made. Appraisal of human impacts upon landscapes and ecosystems must be framed in relation to natural variability and evolutionary trajectory of a given system, asking the question 'What adjustments are likely to have taken place in the absence of human disturbance?' Efforts to unravel human impacts from natural variability must appraise whether the mix of processes, and their consequences, has been altered. Detailed analyses of river evolution are required

to isolate the imprint of human disturbance from natural variability and evolutionary tendencies of a river (see Chapter 12). Catchment-specific appraisals are required to unravel the cumulative, layer-upon-layer responses of river systems to multiple forms of disturbance.

River responses to human disturbance vary markedly across the planet, affected by factors such as environmental setting, population pressure (today and in the past) and level of economic/industrial development. Differing forms of human disturbance vary in terms of their spatial and temporal distribution and extent, their intensity and their recurrence (i.e. whether they are one-off events or sustained impacts). As differing environmental settings present variable opportunities for human exploitation, there is marked spatial variability in the extent and intensity of differing forms of human disturbance. For example, bedrock rivers have limited agricultural potential but they may present opportunities for dam construction. In contrast, many alluvial reaches that have significant capacity for geomorphic adjustment also have considerable agricultural potential. Indeed, virtually all readily accessible alluvial reaches have been 'developed' for human purposes.

Human-induced changes to the boundary conditions within which rivers operate bring about non-uniform responses to the nature and rate of landscape adjustments. Timeframes of river adjustment, and the character/extent of human impacts, vary markedly from system to system. Typically, different reaches within a catchment are at differing stages of adjustment to differing forms of human and natural disturbance. In many instances, responses may represent a legacy from past events or off-site impacts triggered from elsewhere in the system. Hence, catchment-scale investigations are required to analyse the changing nature of biophysical fluxes and the strength of linkages between different landscape compartments. The key issue in assessment of human impacts on river systems is determination of whether human-induced disturbance has unsettled the 'natural' balance of processes at any given place, and the consequences of this unbalancing.

Geomorphic Analysis of River Systems: An Approach to Reading the Landscape, First Edition. Kirstie A. Fryirs and Gary J. Brierley.
© 2013 Kirstie A. Fryirs and Gary J. Brierley. Published 2013 by Blackwell Publishing Ltd.

Process responses to altered flow and sediment fluxes and resisting forces along valley floors are manifest through adjustments to the pattern and effectiveness of erosional and depositional processes, prospectively altering channel geometry/planform and the assemblage of geomorphic units along a reach. Fundamentally, human impacts upon river systems are brought about by alterations to:

1. The flow regime, whether induced by altered runoff conditions, groundwater relationships or flow regulation.
2. The sediment regime, whether as a result of altered sediment delivery to channels or the ease with which sediments are conveyed through the system (exhaustion/ starvation versus oversupply).
3. The distribution of resistance elements on the valley floor such as vegetation type and coverage.

Impacts of flow regulation have been profound across the world. Efforts to ensure water supply, power generation, flood control and irrigation programmes have altered the natural variability in flow for virtually all large rivers. Access to potable water has been a key determinant of the sites of human settlement and land use since the dawn of civilisation. Humans now use over 50% of readily available runoff. In most instances, wherever significant opportunities for water development exist, they have already been exploited. There is now more water in storage facilities than flowing in rivers. Modifications to flow regimes have altered the sediment-transporting capacities of rivers, altering the balance of erosional and depositional processes along a river. Pronounced off-site implications often ensue.

Human activities have modified many other components of the hydrologic cycle. Land use change has greatly modified evapotranspiration and runoff relationships. Typically, deforestation increases stream flow, while afforestation decreases stream flow. Many swamps have been drained, increasing landscape connectivity and reducing groundwater stores, thereby impacting upon base flow conditions along rivers. Increased areas of impermeable surfaces result in shorter duration, more-peaked flood events. Often, these flows are 'controlled' along channelised and artificially leveed rivers, with marked off-site implications.

Human impacts upon sediment transfer relationships along rivers have been just as profound as these hydrologic adjustments. Deforestation, swamp drainage and other forms of land use change have greatly increased sediment availability. Shallow landslides, soil wash and gullying processes occur much more frequently when native vegetation cover is removed. Ploughing breaks up soil materials, resulting in threefold increases in the supply of fine-grained sediment. Land use changes have locally increased rates of floodplain sedimentation by at least an order of magnitude. Although soil erosion has brought about profound increases in sediment availability, the flux of sediment reaching the world's coasts has been markedly reduced because of sediment retention in reservoirs. Globally, more than 50% of basin-scale sediment flux in regulated basins is potentially trapped in artificial impoundments.

Environmental factors influence the pattern and intensity of human disturbance. For example, agricultural potential is partly determined by geologic controls upon relief and soil type, and climatic influences on water availability. Variability in river response to human disturbance in different parts of the world reflects:

1. Inherent natural variability of river systems. Some systems are subjected to relatively minor 'natural' disturbances; others are forever adjusting to ongoing disturbance events.
2. The forms, scale and intensity of human disturbance, both past and present, upon a given river system.
3. The relative sensitivity to disturbance of a river system. Some rivers are just waiting to be nudged 'over the edge', whereby exceedance of a threshold condition results in the transition to a different type of river. Others are remarkably resilient to change.

This chapter is structured as follows. First, a summary history of human interactions with river systems is presented. This is followed by an appraisal of direct and indirect human impacts upon rivers. Finally, river responses to human disturbance are framed in relation to recovery notions, adding a further layer of complexity to the river evolution diagram.

Historical overview of human impacts upon river systems

Water is a remarkable resource. It is renewable, readily stored and transported, and is reusable. Secure and reliable access to water of an appropriate quality and quantity is a fundamental prerequisite for human well-being. Throughout history, access to water resources has been at the forefront of human endeavours. Many human activities are intimately tied to river systems, whether for human and agricultural practices, navigation and trade purposes, or industrial uses. As a consequence, human activities have profoundly altered river systems, both advertently and inadvertently, over the last 5000 yr. The intensity of development of land and water resources has been particularly pronounced over the past 500 yr. This reflects population growth, technological advancements and the cumulative nature of impacts. Recognition of the profound significance of human impacts upon river systems, along with societal desire and capacity to engage with river repair, has brought about significant growth in conservation and rehabilitation initiatives in recent decades.

Human activities are influenced, in part, by geologic and climatic factors that fashion environmental attributes of any particular setting. Just as landscapes in differing settings have variable potential for human exploitation, differing land uses have variable impacts in terms of their intensity, extent and consequences. Past human actions affect the contemporary behaviour of river systems. This is termed *anthropogenic memory*. For example, in many parts of the world materials previously mobilised and restored by human activities (termed legacy sediments) are the primary sediment stores that are being reworked by contemporary landscape-forming processes. Often, rates of sediment movement may diminish over time in response to depletion of sediment sources (i.e. exhaustion). Human disturbance may enhance or suppress rates of geomorphic activity.

Lakes and river systems provided reliable freshwater sources for prehistoric societies. These flat, accessible and resource-rich lands provided ideal opportunities for human endeavour, supplying access to food resources through various plants, animals and fish. Clearance of ground cover through the use of fire was an early indirect form of human disturbance to landscapes and ecosystems. However, low population densities limited impacts to relatively small areas, and nomadic hunter–gatherers simply moved to alternative locations when resources in a given place had been used. Similarly, impacts of fire were relatively localised, though this likely brought about pulsed sedimentation events.

The dawn of civilisation emerged from marshland areas on floodplains, as humans manipulated flows and environmental conditions to support more sedentary lifestyles. The emergence of hydraulic civilisations started several thousand years ago along rivers such as the Tigris–Euphrates, Yangtze, Nile, Indus and Ganges. This marked a notable transition in both human relationships to the environment and to social organisation in these cradles of civilisation. Concerns for steady water supply for domestic and agricultural use brought about notable transitions in both governance arrangements with which to manage society and technological developments with which to manage water resources. Wetlands were drained, ditches were dug and small dams constructed to control flow for irrigation purposes. Environmental problems such as siltation and salinisation brought about the demise of some hydraulic civilisations.

Awareness of the impacts of human disturbance upon erosion and sedimentation extends back over thousands of years. This reflected notable landscape responses to clear-ance of vegetation from hillslopes and valley floors. These activities were near ubiquitous in some instances. In many instances this induced incision and greatly increased rates of sedimentation. Centuries later, vegetation clearance and land use change were often accompanied by desnagging – removal of trees and wood from channel courses to aid navigation and the conveyance of flow. Somewhat ironically, survival of remnant forests often reflected the hunting aspirations of feudal lords; several of these areas remain as parklands, woodlands and reserves to this day.

The next major phase in human relationships to river systems occurred in the early days of the Industrial Revolution in northern Europe and northeastern North America. Development of navigable channels for trade and timber transport, water mills and weirs were accompanied by channelisation and dredging activities, along with local use of diversions and in-channel structures. This era saw massive expansion in the use of rivers for industry (mills, factories, cooling, etc.), power generation and irrigation. Large-scale dam, canal and river diversion schemes were accompanied by significant channelisation initiatives for navigation of industrial rivers. The rapid growth of urban areas required the development of water transfer projects to supply freshwater for domestic use. Dams and reservoirs were constructed in nearby hills and mountains, along with canals or pipes to transfer water. Increasing use of bricks, steel and concrete to protect infrastructure and agricultural lands saw massive expansion in efforts to control, stabilise and train rivers. At the same time, agriculture became increasingly mechanised, ploughing vast areas of land on a more regular basis. Accelerated erosion transferred large volumes of fine-grained materials to river systems.

Progressive disruption of river systems, alongside limited awareness of the inherent dangers of living on floodplains and overconfidence in these actions, brought about a major shift in management actions in the late nineteenth and early twentieth centuries in efforts to rectify problems. Artificial levees constructed to control floods became even higher. Greater use was made of structural revetments and canals. Major water resource projects and multiple-use river projects were developed. Subsequently, large dams and river diversions have regulated the flow of virtually all major rivers.

Enhanced awareness of the impacts of human disturbance to landscapes and river systems has brought about various reforms in land and water management. In some parts of the world, regeneration of land cover or forest regrowth has altered water, sediment and nutrient fluxes, bringing about secondary adjustments to river forms and processes. In recent decades there has been an attitudinal shift from flood control to flood defence to flood management programmes in some areas. As most opportunities for large-scale dam and irrigation programmes have already been developed, emphasis in water management programmes has shifted to the demand side of the water use equation, rather than focusing solely upon concerns for water supply. Low impact schemes are increasingly popular, such as small-scale dams in 'run of river' water use programmes. Moves towards sustainable river management are increasingly incorporated within integrated river plans.

Inevitably, approaches to river management vary markedly in differing parts of the world. These are largely societal choices and questions of priority, reflecting river condition in any given region, pressures placed upon rivers and the capacity of society to apply and maintain rehabilitation treatments. Environmental values have been recognised in the re-regulation of flows, typically framed through efforts to mimic the natural flow regime. In some instances channel maintenance flows have been applied to rework the channel bed and adjust channel geometry. Mitigation, enhancement and rehabilitation techniques increasingly apply soft and/ or environmentally sensitive engineering techniques, including the use of hybrid and bioengineering revetments. Whenever possible, these options are tied to riparian vegetation management and resnagging programmes.

In analysing these various human impacts upon river systems, differentiation can be made between direct and indirect forms of disturbance.

Direct and indirect forms of human disturbance to rivers

Human modifications to biophysical attributes of river systems can be direct or indirect (Table 13.1). While most direct modifications are intended, indirect modifications are inadvertent. *Direct* modifications reflect resource development activities (e.g. water supply, power generation, gravel extraction) or structural engineering works designed to alleviate the effects of flooding. Clearance of riparian vegetation cover and removal of wood have generally accompanied these activities. *Indirect* human impacts refer to adjustments brought about as secondary responses to landscape changes which modify the discharge and/or sediment load of the river. Changes to ground cover modify the nature and balance of flow–sediment fluxes. The scale, extent and rate of ground cover change exert a significant influence upon geomorphic response to human disturbance. In many instances, changes have considerable lagged and off-site impacts. It is often very difficult to differentiate river responses to direct human disturbance at the reach scale from indirect human impacts at the catchment scale. Ultimately, the history of river adjustments reflects cumulative responses to disturbance, whether natural or human induced.

Table 13.1 Forms of human disturbance to river courses

Direct channel changes	Indirect catchment changes
River regulation • Water storage in reservoirs behind dams • Water diversion schemes (e.g. for irrigation)	*Land use changes* • Changes to ground cover, including forest clearance, afforestation and changes in agricultural practice (e.g. conversion of grazing land to arable land and emplacement of agricultural drains and irrigation channels)
Channel modifications • River engineering. Channelisation programmes include flood control works (levee and stopbank construction), bed/bank stabilisation structures and channel realignment • Sand/gravel extraction and dredging programmes • Clearance of riparian vegetation and removal of wood	• Urbanisation and building/infrastructure construction, including stormwater systems • Mining activity • Road construction *Climate change* • Regional variability in rainfall, precipitation, runoff patterns, vegetation cover, etc., and associated changes in land use

Indirect human impacts upon river systems

Indirect human impacts upon river systems are mediated through altered boundary conditions that affect flow and sediment flux, thereby adjusting river processes and forms. These landscape modifications affect river character and behaviour in ways that were unplanned and/or unforeseen. Impacts of inadvertent human disturbance are often delayed until well after the original activity has ceased. Spatial and temporal lags have varying intensity, reflecting catchment-specific conditions. In this section, river responses to ground cover and land use changes are followed by brief discussion of the impacts of water abstraction, urbanisation, mining and other indirect factors.

Changes to ground cover and land use changes

River responses to indirect human disturbance are contingent upon the nature of the landscape itself, as well as the nature of human activities. Broad-scale agricultural transformations of landscapes were initiated in the Middle East around 10 000 yr ago. Phases of agricultural intensification resulted in severe environmental degradation. For example,

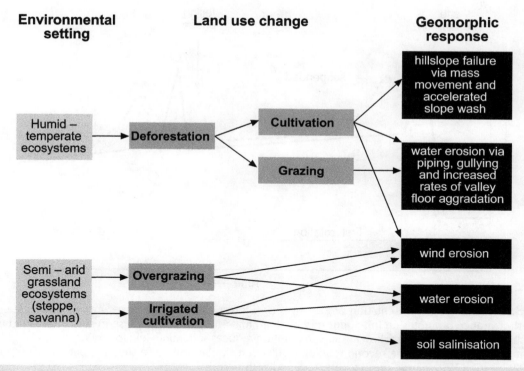

Figure 13.1 Geomorphic responses to land use change in differing environmental settings. Modified from Starkel (1987).

deforestation induced dramatic erosion of much of the Mediterranean region around 5000 yr ago. The nature of ground cover – its density, structure, extent of canopy cover, root networks and a host of other factors – influences resisting forces that impede erosion. In general terms, runoff and sediment yields are high from cultivated and heavily grazed rangeland and relatively low for forests and ungrazed rangeland. Any changes that eliminate or reduce vegetative cover are likely to increase sediment discharge proportionately more than water discharge.

Simplified pathways of geomorphic responses to land use change in different environmental settings are presented in Figure 13.1. Grassland cultivation in drier lands with sparse vegetation cover often enhanced rates of wind erosion and/or desertification. Deflation, in turn, enhanced water erosion. Collectively, this increased prospects for salinisation, thereby inhibiting prospects for landscape recovery. Development of soil crusts promoted rapid runoff. Pulsed sediment movement occurred once crusts were broken. Prospects for recovery may be limited in these settings.

Significant human-induced changes to forests have occurred for at least the last 5000 yr. In historic times, humans have reduced global forest cover to about half its maximum Holocene extent, and eliminated all but a frac-

tion of the world's aboriginal forests. Today, forests still cover around a third of the Earth's land surface. Initial forest clearance likely occurred in a piecemeal manner, with phases of forest regeneration in accord with changes in population density and settlement history. Subsequent forest clearance was more rapid and extensive. Changes to forest cover impact upon both the hydrologic regime and sediment delivery mechanisms. Fire and burning practices increase runoff rates and sediment loads. Deforestation increases stream flow, while afforestation decreases stream flow. Reduction in forest cover increases water yield, while establishment of forest cover on sparsely vegetated land decreases water yield. Alterations to infiltration capacity modify flood hydrographs by shortening lag times and increasing flood peaks (Chapter 3). Increased rates of runoff, along with more concentrated and peaked flows, increased the capacity for sediment transfer. Hillslope instability is greatly increased once forests are cleared, enhancing landslide and debris flow activity. Gully development via headcuts increases sediment supply to downstream reaches. Removal of protective forest cover increases the sensitivity of soils to erosion and reduces the rainfall threshold for erosion initiation. Forest clearance may increase river sediment loads by an order of magnitude or more.

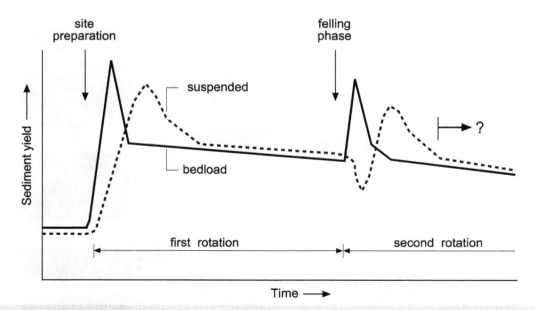

Figure 13.2   Schematic representation of the impacts of forest clearance upon sediment yields in steepland settings. Forest clearance promotes an initial pulse in bedload materials, while the pulse in suspended load is delayed. Re-establishment of ground cover following site preparation reduces sediment yield following the initial pulsed event. Subsequent forest clearance in the second rotation results in a similar, but more subdued pattern of response. From Leeks (1992). © John Wiley and Sons, Ltd. Reproduced with permission.

A schematic representation of increasing sediment yield in response to forest clearance is presented in Figure 13.2. Clearcutting of steep hillslopes markedly increases sediment loads. Much of the pre-logging drainage pattern is obliterated by mass movement and debris torrents. Runoff and erosion rates are greatly increased. Once sediments have been mobilised from hillslopes, the capacity for subsequent erosion and sediment transfer is determined by the rate of sediment regeneration. This is influenced largely by the lithology and the weathering regime. Replanting of logged areas stabilises hillslopes and reduces runoff, erosion and sediment yield. Sediment yields typically decline during the second rotation phase. Selective logging practices markedly reduce these impacts.

Average sediment supply rates from immature forests (i.e. stands <20 yr old) are four times greater than from mature forests. Landslide rates from roads are roughly 45 times the rate from mature forests, and landslide rates from clearcut areas are roughly four times that of mature forests. Gully initiation and expansion may extend drainage networks. Accentuated runoff and erosion on hillslopes may accelerate rates of bank erosion along tributaries, increasing rates of floodplain sedimentation in downstream reaches by an order of magnitude or greater. Sediment overloading accentuates the tendency for downstream channels to become wider, shallower, less sinuous and more braided. If sediment supply declines sufficiently, yet flows

remain effective, channel entrenchment may ensue, possibly leading to the formation of inset floodplains.

Development of agricultural systems in the New World was more extensive, widespread and synchronous than equivalent endeavours in the Old World. Clearance of riparian forests and removal of wood from channels was a priority of pioneer settlers in New World settings, as these were the most fertile and well-watered lands. Technological advances facilitated remarkable increases in the intensity of these activities, bringing about rapid landscape and river changes within the first generation after colonisation. Adjustments along some sensitive rivers were so profound that threshold conditions were breached and river change occurred, whereby the river adjusted to a completely different set of river-forming processes under altered channel/catchment boundary conditions. Many of these rivers have not subsequently recovered.

Forests have naturally regenerated or been planted in some parts of the world, altering water budgets and associated sediment loads. In many places, improved conservation practices and reforestation initiatives have drastically reduced rates of soil erosion and transfer of fine-grained sediment to rivers. In some instances, secondary responses to afforestation include channel narrowing as streams incise. Elsewhere, influxes of exotic vegetation have smothered channel beds, increasing resistance to flow and promoting aggradation.

Impacts of water abstraction (and return flows)

Pervasive impacts of water abstraction for irrigation and other human uses are evident along nearly all major rivers. Water abstraction reduces flow and its competence, thereby altering the flow–sediment balance of a river. Reductions to the sediment transporting capacity of the trunk stream may promote build-up of sediment at tributary confluences. Alternatively, return flows from drains and/or outfalls may accentuate sediment transfer, as countless agricultural drains and ditches locally increase flow conveyance. Groundwater abstraction may also affect the hydrologic regime of river systems.

Urbanisation

Impacts of urbanisation are more localised but more pronounced than other forms of land use change. Impervious surfaces and efficient stormwater systems increase the area of low or zero infiltration capacity and increase flow transmission in channels. This increases the volume of runoff for a given rainfall, resulting in a flashier runoff regime with shorter lag times and higher peak discharges. Greater flow velocity and competence may accelerate rates of erosion and increase channel slope, increasing deposition in downstream areas. Geomorphic consequences reflect the form, extent and intensity of modifications to streams and the ground cover, the type of urban development, and the physical and climatic setting of the city.

Impacts of urbanisation upon sediment yield can be considered as a two-phase response (Figure 13.3). Exposure of large amounts of soil during the construction phase induces extensive erosion, increasing sediment concentration and yield by one and two orders of magnitude respectively. Sediment yields decline after the construction phase as sediment availability is reduced in response to the increased area of impermeable surfaces. Infiltration excess and saturated overland flow become more pronounced, and stormwater drains extend the drainage network, conveying increased peak flows from urbanised catchments. However, sediment yields are typically lower than in forested basins. Typically, urban streams are channelised to minimise erosion. As a consequence, their off-site impact in downstream reaches may be pronounced, characterised by channel enlargement and changes to the assemblage of geomorphic units. Recent innovations in water-sensitive urban design and low-impact design measures have reduced these impacts.

Mining

Mineral extraction for fossil fuels (e.g. coal, lignite and peat), metals (e.g. gold, silver, lead, zinc and copper) and aggregates (e.g. alluvial sand and gravel) has induced a profound impact on river systems in various parts of the world. These activities disrupt the hydrologic regime (through vegetation removal and drainage modification), accelerate hillslope erosion and increase sediment delivery to rivers. Passive dispersal refers to the transport of mining waste along with the indigenous load without major disruption to river morphology. In contrast, active transformation refers to the movement of sediment slugs, often bringing about changes to river character and behaviour through aggradation, channel widening and the adoption of a multichannel planform. Fine-grained fractions are deposited some distance from source on floodplain surfaces or in the accumulating basin. As mining-induced supply diminishes, channel degradation produces a narrow single channel, with a series of terraces. Contamination by metalliferous fine-grained materials and other toxic substances can exert an indirect impact on river forms and processes over extensive lag periods. For example, toxic waste products may retard the development of riparian vegetation development for centuries. In extreme situations, large open-cut mines may remove entire hills or even mountains, sometimes infilling intervening valleys.

Other forms of indirect impacts

Alongside land use change, climate change is probably the primary form of indirect human impact upon river systems. These impacts are pervasive, but trends vary markedly from region to region. This reflects altered precipitation and temperature patterns, snowpacks and flow regimes, and secondary adjustments to vegetation cover. In some places, climate change has increased opportunities for agricultural development and intensification, accentuating impacts of land use change. Other indirect forms of human disturbance upon river geomorphology include stock grazing and trampling, boat-induced wave action and alterations to ecological conditions (e.g. removal of beaver dams, incursions of exotic vegetation).

Direct human impacts upon river processes and forms

In contrast to inadvertent secondary consequences of indirect human impacts upon rivers, direct human impacts refer to planned activities that purposefully modify river character and behaviour. Many direct human impacts are site- or reach-specific forms of disturbance that induce an immediate geomorphic response. This section analyses geomorphic impacts of dams and inter-basin transfers, channelisation programmes, removal of riparian vegetation and wood, sand/gravel extraction and impacts of rehabilitation schemes.

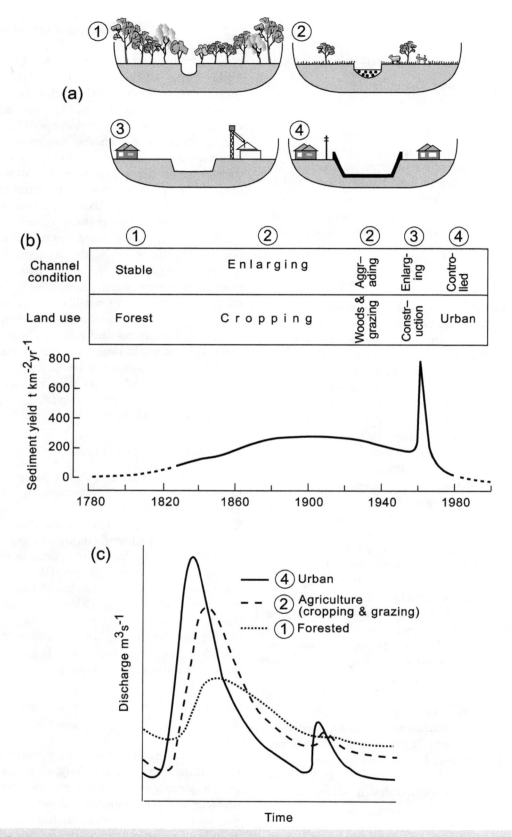

Figure 13.3 Geomorphic responses to forest clearance and subsequent transition from agricultural to urban land use. Discharge and sediment yields are conveyed for four stages of land use change, alongside channel morphological responses. (a) Initially channels enlarge, but they are subsequently stabilised by engineering structures. Modified from Roberts (1989). © John Wiley and Sons, Ltd. Reproduced with permission. (b) There is a sharp peak in sediment yield during the construction phase. From Wolman (1967). © John Wiley & Sons, Ltd. Reproduced with permission. (c) Although impermeable surfaces reduce sediment yield, runoff rates are increased, with more peaked flows, prospectively increasing downstream (off-site) impacts.

Dams and reservoirs

Measures used to control and regulate river flow, and associated concerns for water supply for agricultural (irrigation), commercial/industrial or residential purposes, have been the primary forms of direct human impact upon river systems. The extent of these programmes is staggering. More than 400 000 km² of land has been inundated by reservoirs behind the world's largest dams. The global volume of freshwater trapped in reservoirs now exceeds the volume of flow along rivers. Dams and/or inter-basin transfers have fragmented nearly all of the world's large rivers. Although dams are a point-source form of disturbance, they induce considerable off-site effects because they alter the longitudinal connectivity and base-level conditions of the river. Some of the most profound responses have occurred at the coastline, as morphodynamic interactions between fluvial and coastal processes are modified.

Dams have been constructed for more than 5000 yr. The pace of construction quickened dramatically after 1950, with more than 200 large dams completed each year. A recent decline in the rate of dam development and associated water transfer projects across much of the Western world reflects the lack of remaining reasonable opportunities. There are now more than 45 000 large dams (>15 m high or 5–15 m high if reservoir volume is greater than $3 \times 10^6 \, m^3$). Many dams are more than 300 m high. The Three Gorges Dam along the Yangtze River is 185 m high and 2309 m long. It impounds $39.3 \times 10^9 \, m^3$ of water, with its reservoir extending upstream for 600 km. The river discharges 530×10^6 t of silt per year into the reservoir.

Individual dams commonly form part of integrative water supply or hydro-electricity programmes, such as inter-basin transfer schemes, wherein water storage and flow is accentuated in some systems, but diminished elsewhere. Typically, dams act as both water supply facilities and flood control impoundments. Flow regulation reduces the stochastic, unpredictable nature of flows. The extremes of flow are reduced, substantially lowering flow maxima (i.e. peak flows) but increasing flow minima (i.e. base flow).

Disruption to water and sediment transfer impacts directly on river structure and function both upstream and downstream of a dam (Figure 13.4). The reduction in channel gradient following elevation of base level upstream of dams reduces the transporting capacity of flow as it enters the reservoir. Reservoirs also create higher base levels for tributaries upstream of the dam, inducing sediment deposition along their lower courses. Delta growth upstream of the backwater limit reduces the water storage capacity of the reservoir. Although aggradation takes place rapidly initially, its upstream extent may be limited or long delayed, dependent upon sediment supply conditions. Reservoirs are excellent sediment traps. They commonly retain more than 90 % of the total load and the entire coarse fraction (i.e. all bedload sediment and all or part of the suspended load).

Downstream impacts of altered flow regimes and sediment reductions vary widely, depending on the amount of reservoir storage, the dam operations and the location of the dam relative to sediment sources. Downstream impacts include slope change, alterations to bed material texture and adjustments to channel geometry and planform of both the trunk stream and tributaries. Rivers respond to reduced sediment availability by adjusting their sediment transport capacity. This is primarily achieved by coarsening of the bed rather than alterations to slope. Sediment starvation immediately downstream of dams induces bed and bank erosion. Water releases from dams have the energy to move sediment, but little or no sediment load is available to them. This 'hungry water' expends its energy by eroding the channel bed and banks. Bed incision and channel narrowing may reduce bankfull cross-sectional area by over 50 % over timeframes ranging from 10 to over 500 yr. Downstream-progressing degradation along the trunk stream can induce upstream-progressing degradation and entrenchment along tributaries, promoting accelerated deposition at tributary confluences. Channel widening often follows bed incision and headward extension. Vegetation encroachment may greatly increase channel stability.

The effects of river regulation tend to diminish with distance downstream, as non-regulated tributaries make an increasing contribution to the flow. In some instances, the pattern and rate of morphodynamic interactions may be altered a considerable distance from the control structure. This is exemplified by accelerated erosion and shoreline recession at the coastal interface. These are system-specific scenarios. Since many dams were constructed in the twentieth century, it may be a century or more before the lagged nature of river adjustment processes is fully realised, especially in downstream reaches.

Channelisation

Ever since the Roman era humans have endeavoured to train channels to control the impacts of floods in efforts to protect infrastructure. Among many goals, channelisation programmes strive to improve drainage, prevent erosion and maintain navigational arteries. Typically, these endeavours set out to stabilise and regulate a channel by fixing it in place with a particular size and configuration. Systematic channelisation programmes began in earnest in the seventeenth century. Since then, hundreds of thousands of river kilometres have been channelised, simplifying the geomorphic structure of rivers and altering flow interactions and sediment movement. Swampy areas have been extensively

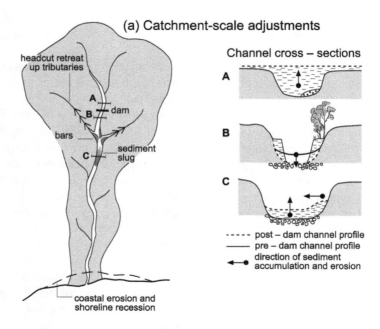

(a) Catchment-scale adjustments

headcut retreat
up tributaries

A
B — dam

bars

C — sediment
slug

coastal erosion and
shoreline recession

Channel cross – sections

A

B

C

- - - - post – dam channel profile
——— pre – dam channel profile
◄●— direction of sediment
accumulation and erosion

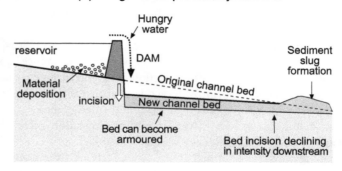

(b) Longitudinal profile adjustments

Hungry
water

reservoir   DAM

Material
deposition   Original channel bed

incision   New channel bed

Bed can become
armoured

Sediment
slug
formation

Bed incision declining
in intensity downstream

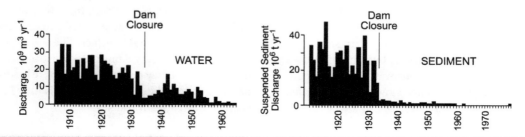

(c) Effects on water and sediment discharge of the Hoover Dam, Colorado

Figure 13.4 Geomorphic impacts of dams. (a) Dam construction traps sediment in a delta, creating an accumulation zone at the entrance to the reservoir (point A). Suspended-load sediments drape the former channel at this point, which now lies beneath the reservoir. At point B, immediately downstream of the dam, reduced bedload and increased erosive potential of 'hungry' water have induced bed incision following dam closure. A slot channel has been produced, and the bed has become armoured. Inset floodplains that line the compound channel have been colonised by dense, rapidly growing weeds. The original floodplain is increasingly decoupled from the channel because of changes to the flow regime and morphological adjustments. The channel at point B has become an area of net sediment loss following dam closure. Sediments released from this zone have accumulated downstream at point C, where the channel has contracted through the formation of lateral bars. As accelerated rates of bedload sediment supply cannot be sustained from upstream because of the armouring effect at point B, reworking of sediments is likely and effects are progressively propagated further downstream. Off-site impacts of dam construction may include incision of tributary streams and altered morphodynamics at the coastline (typically shoreline recession). From Brierley and Fryirs (2005). © John Wiley & Sons, Ltd. Reproduced with permission. (b) Sediment accumulation behind the dam starves downstream reaches of bedload material. Hungry water incises into the bed downstream of the dam, reworking sediment stores that may progress downstream as a sediment slug. (c) Closure of the Hoover Dam on the Colorado River, USA, reduced sediment yield up to 95 %. From Meade and Parker (1985). © US Geological Survey. Reproduced with permission.

drained for agricultural purposes. Concrete lining and pipes are commonly used in urban and peri-urban areas.

Channel straightening/realignment is typically applied for flood protection to evacuate flow more rapidly (Figure 13.5). Artificial cut-off and realignment programmes may increase the efficiency with which the channel is able to convey flow and sediment in the short term, initially enhancing prospects for flood control and navigability. This steepens the gradient, as flow now follows a shorter path. Flow velocity and transport capacity are increased. Degradation ensues, progressing upstream as a headcut. Depending upon the composition of the bed and bank, erosion may increase sediment load to the reach downstream, ultimately flattening its slope and promoting aggradation. Hence, these measures are often accompanied by bank protection and stabilisation measures.

Resectioning refers to increases in channel capacity (widening and/or deepening) in efforts to increase flow conveyance to reduce overbank flooding. Widening reduces velocity and unit stream power, thereby lowering sediment transporting capacity. This promotes deposition in the form of benches along channel margins. Dredging is often required to maintain channel dimensions.

Levee, floodwall and stopbank construction increase channel capacity by raising the banks in efforts to provide protection against floodwaters and to maintain channels. This reduces floodplain inundation and sedimentation rates. As embankments and artificial levees accentuate peak flows, they induce off-site impacts in downstream reaches. In some instances, deeper flows may promote bed incision. Floodwaters may become trapped behind levees in extreme events.

Channel stabilisation and bank protection works use structures such as paving, gabions, steel piles, subaqueous mattressing, dykes and jetties to control bank erosion. These structures alter channel width and roughness components, with secondary implications for bed incision and subsequent sediment release, thereby adjusting channel bed slope. This may promote sedimentation adjacent to the bank, potentially increasing flooding if channel capacity is reduced. Bridge crossings are another form of localised direct disturbance. Bridge pylons and embankments concentrate flow and accentuate scour. Culverts under roads and crossings locally concentrate flow, prospectively inducing scour downstream of the concreted section of channel, while acting as a base level for the upstream channel.

Dredging refers to removal of sediment from the bed to deepen the channel, thereby maintaining navigable channels. This is especially prevalent along the thalweg in lowland reaches. Dredging may promote degradation through lowering of base level. Upstream-migrating headcuts contribute additional sediment to the dredged reach. Deepening may also promote bank collapse and upstream-

progressing degradation within tributaries. Clearing and desnagging refer to removal of obstructions from the river in efforts to aid flood passage and navigation capacity. This decreases resistance and increases flow velocity, thereby promoting bed degradation, subsequent widening and marked increase in-channel capacity.

Weirs and lock emplacements are channel-spanning features that are used to regulate slope for navigation. These features alter bed slope, reducing conveyance of sediment. They also modify river structure, promoting elongate pools in place of hydraulic diversity. These point-source disturbances have localised off-site consequences.

Since channelisation involves manipulation of one or more of the dependent hydraulic variables of slope, depth, width and roughness, feedback effects are initiated which promote adjustments towards a new characteristic state. Geomorphic response times following the emplacement of river engineering works depend on the types of works installed and the extent to which they alter flow and stream power, sediment supply and vegetation cover. It may take hundreds of years to attain this new characteristic state.

In contrast to dam construction, which essentially represents a point disturbance with off-site impacts, channelisation activities are applied over varying lengths of river. Catchment-wide applications may be implemented in urban settings or along major navigational arteries. Indeed, once channelisation begins, secondary instability and/or channel adjustments elsewhere in the system often prompt extension of the channelisation programme.

Removal of riparian vegetation and wood

Riparian vegetation and wood are amongst the most important resistance elements along some river courses. Variability in geomorphic response to vegetation clearance and wood removal reflects the role played by riparian vegetation and wood as determinants of process–form associations for differing types of river. In addition, the inherent capacity of a river to adjust, the position of the reach within its catchment and the sequence of driving factors (i.e. floods) that promote change influence the effectiveness of vegetation and wood as resistance elements along river channels. Responses are most pronounced in those settings where vegetation and wood exert greatest influence on river morphology, namely sand-bed alluvial rivers (Figure 13.6). Once the inherent resilience of valley floors is breached and incision is triggered in these settings, channel expansion ensues. The capacity of wood to increase roughness and stabilise instream sediment may significantly enhance geomorphic river recovery following disturbance. Hence, wood emplacement is often viewed alongside riparian revegetation programmes as fundamental components of river rehabilitation practice.

(a)

Natural channel

Channelised stream

Diverse array of geomorphic units. Channel is able to adjust on valley floor. Sorted gravels. High channel-floodplain connectivity.

Near – continuous riffles. Battered and reinforced banks with artificial leves. Unsorted gravels, uniform channel bed morphology. Enlarged channel with no channel-floodplain connectivity.

Pool environment

Low water velocities. Abundant instream wood and natural undercut banks.

High channel velocities. No instream wood. Uniform bank morphology.

Riffle environment

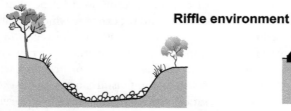

Sufficient water depth to maintain connectivity between pools and riffles during dry periods. Range of water surface morphologies.

Insufficient depth of flow or no flow during dry periods. Relatively uniform water surface morphology.

Longitudinal profile

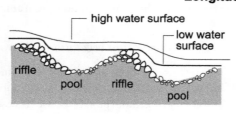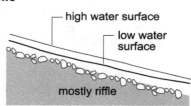

(b)

Figure 13.5 Geomorphic effects of channelisation. (a) Flow and sediment conveyance are greatly enhanced in more uniform, straighter channels. (b) The Ishikari River, near Sapporo, Japan, was channelised to drain a swamp and provide flood protection for development of agricultural land through construction of a hydraulically efficient, low-sinuosity, enlarged channel. (a) Modified from Corning (1975). (b) Photographs provided by Tomomi Marutani.

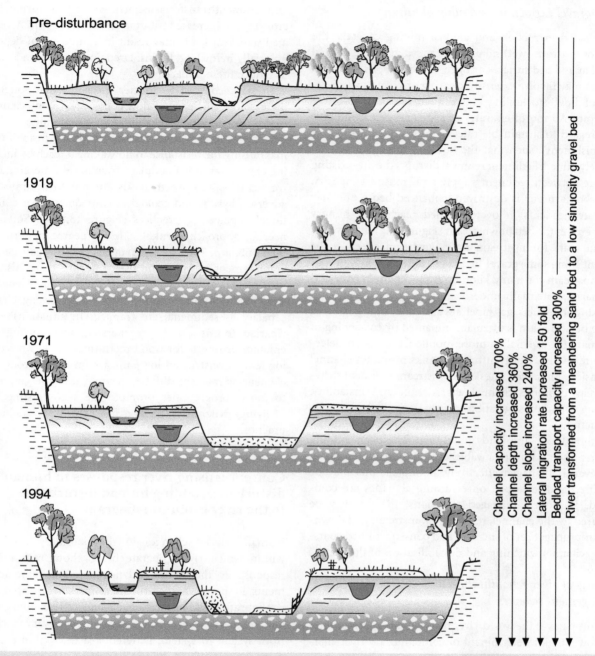

Figure 13.6 Dramatic geomorphic responses to clearance of riparian vegetation and removal of wood. Based on Brooks *et al.* (2003). At the time of European settlement, the Cann River, Victoria, Australia, was a low-capacity sinuous channel flowing within a rainforest. Partial clearance of riparian vegetation had occurred by 1919, but a desnagging programme in the 1960s induced channel metamorphosis. The near-instantaneous reduction of vegetative roughness elements lowered threshold conditions that determine bed level stability and critical bank height, such that the channel became highly sensitive to change. Flood events that brought about minor perturbations under intact vegetation conditions were much more geomorphologically effective under altered boundary conditions. Exceedance of threshold conditions brought about fundamental shifts in river character and behaviour via incision, straightening and channel expansion. The progressively enlarging channel increasingly concentrated flow energy at flood stage. Channel capacity increased by 700%, channel depth increased by 360%, channel slope increased by 240%, and there was a 150-fold increase in the rate of lateral channel migration. The channel became increasingly decoupled from its floodplain. From Brierley and Fryirs (2005). © John Wiley & Sons, Ltd. Reproduced with permission.

Sand/gravel extraction and alluvial mining

Gravel/sand extraction and alluvial mining can take the form of instream (wet) mining, where sediment is extracted from bar and bed surfaces or open floodplain pits. Instream mining may involve extensive clearing of vegetation, diversion of flow, stockpiling of sediment and excavation of deep pits. All too often, rates of sediment extraction give little regard to sustainable rates of bedload transport (i.e. replenishment), such that the bed and floodplain are effectively mined. Sediment removal disrupts the pre-existing balance between sediment supply and transport capacity. Once bed armour is destroyed, enhanced bed scour may generate headcuts in oversteepened reaches, as hungry water erodes the bed downstream (Figure 13.7). It may take several years for upstream or downstream effects to be evident along some gravel-bed rivers. Headcuts may propagate upstream for many kilometres on the main river and tributaries, potentially undermining bridges and weirs and exposing aqueducts, gas pipelines and other utilities buried in the bed. Incision is often accompanied by coarsening of bed material, as smaller, more mobile fractions are selectively reworked. Undercutting of banks promotes channel expansion. Enhanced rates of downstream sediment delivery may promote channel aggradation and instability, altering riparian vegetation associations and hydraulic interactions.

Removal of sand and gravel via floodplain mining also represents non-renewable exploitation of resources. However, if managed effectively, floodplain pits may be stabilised and left open once mining activities are completed. In poorly managed situations, the pit may be captured by the channel, resulting in upstream and downstream propagation of incision and consequent bed coarsening, channel widening and destabilisation of the banks.

Impacts of river rehabilitation and management schemes

In many parts of the world, river rehabilitation programmes are now a major determinant of contemporary river character and behaviour. In the past, river management practices emphasised concerns for river stability in efforts to 'train' or 'improve' river courses by pinning the channel in place (Figure 13.8a–d). Endeavours were typically applied in a piecemeal manner, with little consideration given to basin-wide perspectives or off-site impacts. As a consequence, many activities not only failed to achieve their intended goals; they also had unforeseen, undesirable effects. Indeed, some engineering works have promoted or enhanced river instability, inducing local increases in bedload, uncontrolled aggradation and channel widening. Although engineering works generally reduce sediment flux, local areas may experience accelerated rates of erosion

and sediment transfer associated with bed scouring, bank erosion, and increased tributary sediment supply. Accentuated sediment loads may result in a build-up of deposits, especially in lowland basins. Dredging is often undertaken to reduce these impacts.

In stark contrast to former measures that sought to increase the efficiency of flow conveyance along channels, many contemporary rehabilitation programmes seek to redress secondary problems of bed/bank instability by maximising the resistance to flow along a reach by increasing channel and/or floodplain roughness. Emerging environmentally sensitive methods attempt to minimise the adverse physical and ecological consequences of conventional engineering practices (Figure 13.8e–j). Soft engineering approaches entail a lesser degree of structural manipulation, such as riparian vegetation management, emplacement of wood and use of flexible materials (geotextiles). These measures increase the structural heterogeneity and roughness along a reach, decreasing channel capacity as sediments are trapped. In situations where riparian forests can be regenerated, fast-growing trees enhance prospects for wood recruitment as key pieces for log jams. Constructed log jams are important roughness elements in many rehabilitation strategies. Emerging 'space to move' programmes promote the self-adjusting basis of living, dynamic rivers as a platform for rehabilitation practice.

Conceptualising river responses to human disturbance: adding human disturbance to the river evolution diagram

Human-induced changes to the boundary conditions within which rivers operate bring about non-uniform responses to the nature and rate of geomorphic adjustments. Responses to disturbance reflect catchment-specific configuration and history. Timeframes of river adjustment, and the character/extent of human impacts, vary markedly from system to system. Different reaches within a catchment are typically at differing stages of adjustment to differing forms of human and natural disturbance. Individual forms of human disturbance seldom occur in isolation from others. In general, impacts of indirect human disturbance are often delayed until well after the original activity has ceased. Responses to direct changes are usually more rapid. Recovery time following disturbance depends upon the extent of displacement, the subsequent flow regime and the availability of sediment to drive recovery processes. In systems with large buffering capacity and/or with large thresholds to overcome, there may be considerable time lags between perturbation and morphological response. Some landscapes are capable of withstanding external

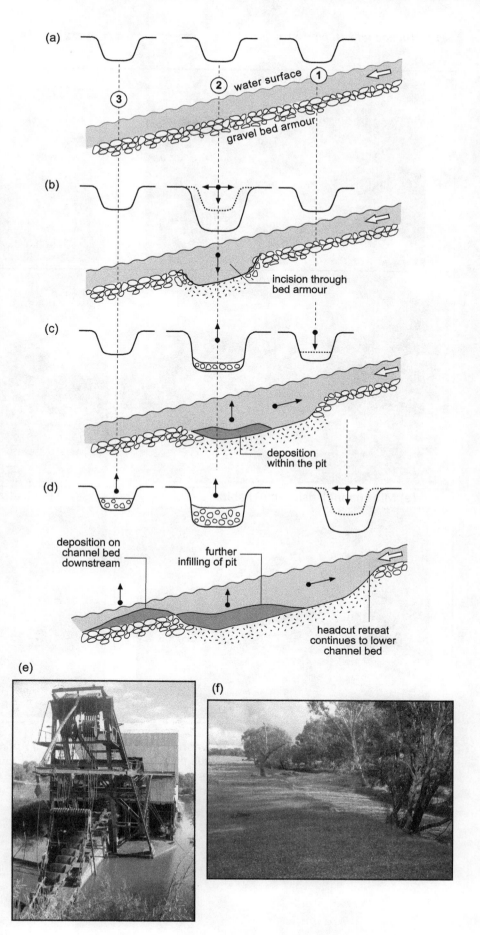

Figure 13.7 Geomorphic impacts of instream gravel mining. (a) In the pre-extraction condition, the sediment load and the force available to transport sediment are continuous through the reach. (b) Excavation of an instream pit breaks the bed armour and instigates a headcut at the upstream end of the pit. Initially, the pit traps sediment, interrupting the transport of sediment through the reach. Downstream, the river retains the capacity to transport sediment but has no sediment load. (c) Headward extension of the headcut acts to maintain bed surface slope. Hungry water erodes the downstream end of the pit, as incision expands both upstream and downstream. (d) Sediments released following the upstream progression of the headcut and associated channel expansion partially infill the incised and expanded trench of downstream zones in the form of bars and benches. This results in a compound channel form. Modified from Landscape and Urban Planning, 28 (2–3), Kondolf, G.M., Geomorphic and environmental effects of instream gravel mining, 225–243, © 1994, with permission from Elsevier. (e) The Eldorado Dredge in the Reedy Creek in Victoria, Australia churned over $27 \times 10^6 \, m^3$ along 18 km of river between 1936 and 1954 to extract tin and gold. (f) A sediment slug was produced that has infilled downstream channels and inundated the floodplain. This slug is moving at $50–150 \, m \, yr^{-1}$ and is threatening the heritage-listed Ovens River. Photographs: K. Fryirs.

Labels within figure:

(a) water surface, 1, 2, 3, gravel bed armour

(b) incision through bed armour

(c) deposition within the pit

(d) deposition on channel bed downstream, further infilling of pit, headcut retreat continues to lower channel bed

(e)

(f)

Figure 13.8 Changing approaches to river rehabilitation practice. (a–d) Approaches to river management along the Upper Hunter catchment, NSW, Australia that were used in the 1950s–1960s. Photographs: NSW Department of Lands; reported in Spink *et al.* (2009). (e–j) These invasive hard engineering structures have subsequently been replaced by less-invasive vegetation and wood (soft engineering) practices. Photographs: Keating *et al.* (2008).

(g) Mid channel engineered log jam

(h) Riparian revegetation

(i) Stock exclusion and revegetation

(j) Riparian revegetation

Figure 13.8 *(Continued)*

disturbance, while others are subjected to dramatic and irreversible changes

Contemporary river processes and forms reflect cumulative responses to disturbance impacts, and their interconnected consequences. It is often very difficult to isolate the consequences of individual forms of disturbance, as perturbations build upon each other, making it difficult to isolate specific cause-and-effect relationships and predict future consequences. Regardless of the nature, extent and direction of human disturbance, each system has its own collective memory. Disturbance responses are accentuated at the bottom end of catchments, where the cumulative effects of upstream changes are manifest.

Appraisal of the geomorphic consequences of human-induced changes to river courses must be framed in the context of natural patterns and rates of adjustment. Different types of rivers have distinct behavioural regimes and associated propensity for adjustment. Human disturbance may transform, negate/accelerate or induce little change to these behavioural regimes.

River response to human disturbance can be built into the river evolution diagram, as shown in Figure 13.9. In this figure, Zone A represents the natural capacity for adjustment within which a range of river behaviour is evident. If direct or indirect human disturbance occurs, the capacity for adjustment can expand or contract depending on whether the range of behaviour is accentuated or suppressed. An expanded capacity for adjustment, termed the contemporary capacity for adjustment, is depicted in Zones B and C. If human disturbance expands the range of behaviour for that type of river, adjustments move away from a natural range of states towards an altered range of states.

In general terms, when rivers are subjected to relatively low levels of impact spread over a considerable length of time, they progressively adjust while maintaining a roughly equivalent state. In systems in which the natural behavioural regime and morphological character of rivers fluctuate among multiple states, gradual and low-impact forms of human disturbance may increase the periodicity with which adjustments among these various states take place and the capacity for adjustment expands. However, these adjustments are unlikely to push the system to a new state that falls outside the contemporary capacity for adjustment. In this instance, although rates of change are modified, adjustments tend to be localised and reversible, with only modest or localised adjustments to their geomorphic configuration. These sorts of adjustments tend to occur in relatively resilient systems.

When reversible geomorphic change occurs, a fundamental shift in the type of river does *not* occur. This is represented by a shift from Zone A to Zone B in Figure 13.9. During these changes, the key defining attributes of the type of river do not change (i.e. the key geomorphic

units remain unaltered). However, other structural and functional attributes of the river are considered to be out of balance. Although the key geomorphic structure of the river may remain the same, the rate of adjustment may be affected and the range of adjustment may be altered. Hence, the potential exists for the river to operate outside its natural capacity for adjustment, but ongoing adjustments are reversible.

In contrast, profound human disturbance over a short period of time may breach threshold conditions, pushing the system outside its long-term range of behaviour, and river change may ensue. This transition may take the form of a relatively simple, one-step transformation, or disturbance may set in train progressive adjustments. Regardless, changes from the pre-disturbance condition are likely to be irreversible over hundreds of years, if ever. These types of responses tend to occur along sensitive reaches that are vulnerable to disturbance. Irreversible change to a new type of river is represented in Figure 13.9 by a shift from Zone A to Zone C if change is induced from a natural state, or a shift from Zone B to Zone C if change is induced by continued/sustained human disturbance. In this case, the key defining attributes of the type of river have changed (i.e. the key geomorphic units have changed) and other structural and functional attributes have been altered, forming a different type of river.

Various examples of river evolution diagrams that incorporate responses to human disturbance are presented in Figures 13.10–13.13. Figure 13.10 conveys the impact of a dam upon a relatively resilient river. Although the river type remains the same, the capacity for adjustment of the river has been suppressed, as the range of flux boundary conditions has been reduced. This is represented on the river evolution diagram by the narrowing of the inner band on the right-hand side of the figure.

Figure 13.11 conveys dramatic geomorphic responses to clearance of riparian vegetation and removal of wood along a sensitive sand-bed river. Incision and channel expansion brought about irreversible geomorphic change from a meandering channel to a low-sinuosity gravel-bed river. This is depicted by a shift to higher specific stream conditions with a different pathway of adjustment moving from left to right in the figure.

Direct human disturbance induced by channelisation of a formerly meandering river is depicted on the river evolution diagram in Figure 13.12. The limited range of behaviour of the former fine-grained meandering (swampy) system was transformed into a low-sinuosity channel with a greatly enhanced cross-sectional area. The river is also regulated. These geomorphic changes have increased the capacity for adjustment to a wider band on the right-hand side of this river evolution diagram, but the pathway of adjustment is restricted by flow regulation. However,

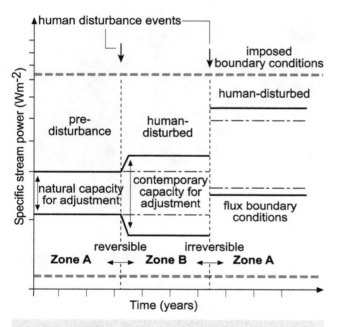

Figure 13.9 Adding human disturbance to the river evolution diagram. Human disturbance often induces an additional layer of complexity to that associated with 'natural' disturbance events. This is noted on the river evolution diagram by an expansion in the capacity for adjustment if changes are reversible (Zone B). If changes are irreversible, the capacity for adjustment may be expanded and the position of the inner band may be shifted, such that a different type of river is adopted (Zone C). In some instances, human disturbance may suppress the capacity for adjustment of a river. From Brierley and Fryirs (2005). © John Wiley & Sons, Ltd. Reproduced with permission.

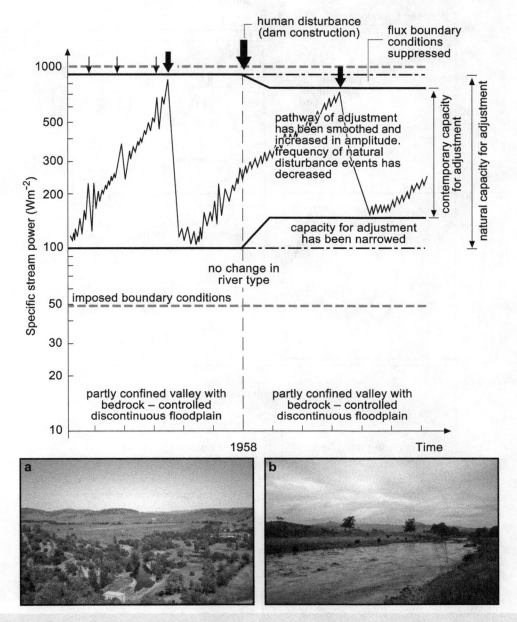

Figure 13.10 River evolution diagram showing geomorphic responses to dam construction along a relatively resilient river. Construction of Glenbawn Dam along the upper Hunter River, New South Wales, Australia, did not bring about a change in river type, but the capacity for adjustment of the river was narrowed. Prior to 1958, this system operated as a partly confined river with bedrock-controlled floodplain pockets. The channel comprised an array of gravel point bars, bedrock pools and gravel riffles, with discontinuous pockets of floodplain on the insides of bends. Dam construction and the modified flow regime exerted relatively minor changes to the geomorphic structure of this resilient bedrock-controlled river. The river retains the key geomorphic attributes of the same type of river that was evident prior to dam construction (i.e. changes have been reversible in geomorphic terms, as noted on the right-hand side of the figure). However, the dam has had a range of secondary geomorphic impacts. The dam traps bedload material supplied from the upper catchment. Downstream of the dam, degradation and armouring have occurred. The flow regime has been altered, as water releases from the dam maintain base flow conditions for irrigation purposes. Peak flows have been reduced and the seasonality of flow has been altered. The capacity for adjustment of the river has been suppressed, as the range of flux boundary conditions has been reduced. This is represented on the river evolution diagram by the narrowing of the inner band on the right-hand side of the figure. Similarly, the amplitude and frequency of the pathway of adjustment have been reduced, reflecting the lower geomorphic effectiveness of flood events. Photograph (a) shows the river downstream of the dam, while photograph (b) demonstrates the maintenance of base flow conditions following dam construction. Photographs: K. Fryirs. From Brierley and Fryirs (2005). © John Wiley & Sons, Ltd. Reproduced with permission.

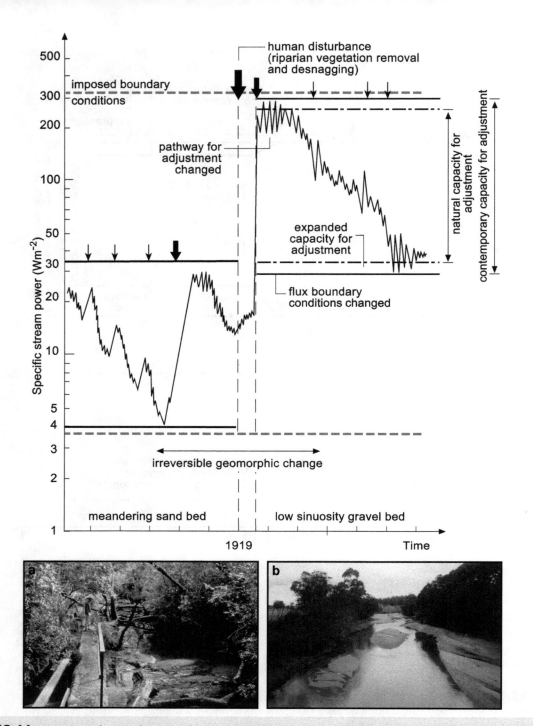

Figure 13.11 River evolution diagram showing geomorphic responses to clearance of riparian vegetation and removal of wood for a sensitive sand-bed river. Removal of riparian vegetation and desnagging operations induced river change along the Cann River, Victoria, Australia, in the mid-late twentieth century. Prior to European settlement of the area, the river operated as a low-capacity, slowly meandering sand-bed channel with a high loading of wood and rainforest vegetation associations on the floodplain (see Figure 13.6). Every few thousand years the river was subjected to avulsion, as indicated by the natural capacity for adjustment on the left-hand side of the figure. Following human disturbance, this sensitive river was primed to respond to flood events, such that river character and behaviour were fundamentally altered and the system was transformed into a low-sinuosity sand-bed river (depicted on the right-hand side of the figure). Channel incision and lateral expansion have created a low-sinuosity trench that is largely decoupled from the floodplain. Rates of sediment transfer are several orders of magnitude higher than prior to disturbance. The capacity for adjustment of the new river system is much greater than its predecessor. The pathway of adjustment has been altered to reflect the change in river behaviour. As the energy of the system has increased significantly within this enlarged channel, the new river type sits higher within the potential range of variability. Based on sediment supply and transport rates in the contemporary system, it is estimated that it would take many thousands of years for the system to recover to its pre-disturbance state. The photographs show (a) the adjacent Thurra River, which remains in an intact condition, (b) the Cann River today. Photographs: A. Brooks and K. Fryirs. From Brierley and Fryirs (2005). © John Wiley & Sons, Ltd. Reproduced with permission.

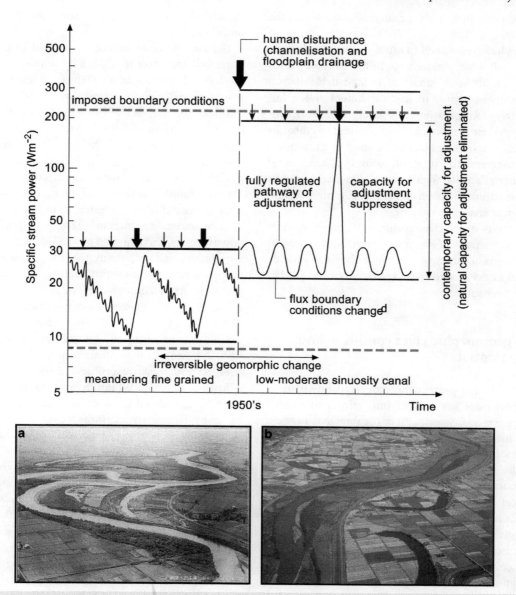

Figure 13.12 River evolution diagram showing geomorphic adjustments to channelisation and floodplain drainage on the Ishikari River, Hokkaido, Japan. Prior to channelisation, this river system was a low-energy, fine-grained meandering river with a marshland floodplain (left-hand side of the figure). Large wetlands and cut-offs occurred on the floodplains. After the Second World War, the city of Sapporo expanded significantly, and additional land along the Ishikari River was required for development. The marshlands were drained and resurfaced with fill, and an extensive channelisation scheme was undertaken. A canal was dredged and lined with concrete bricks. Meander bends were cut off and plugged, significantly shortening the river, to convey flood flows as efficiently as possible to the sea. An extensive network of flood-control structures and canals was emplaced, some utilising the old channel network. The meandering fine-grained Ishikari River was irreversibly altered and retains little in the way of its inherent geomorphic diversity. The energy of the enlarged low-sinuosity channel (canal) has likely increased, but the capacity for adjustment has been severely constrained (right-hand side of the figure). Water quantity and sediment supply are stringently controlled through the use of reservoirs and weirs, producing a regularly fluctuating, artificial pathway of adjustment. Other than localised bank erosion, little geomorphic adjustment is allowed to occur. Infrequent high-magnitude events flood areas beyond the channel zone. Photographs: Tomomi Marutani. From Brierley and Fryirs (2005). © John Wiley & Sons, Ltd. Reproduced with permission.

extreme floods may induce accentuated impacts within the larger channel.

Most rivers have responded to multiple forms of human disturbance, both direct and indirect. In the example shown in Figure 13.13, an initial phase of upland deforestation increased sediment supply to mid-catchment rivers, but a braided river configuration was retained. Subsequent impacts of gravel extraction and emplacement of embankments induced incision and transformation to a meandering gravel-bed river with a different pathway of adjustment in a lower energy setting. Significant channel adjustments via migration and/or avulsion can still occur, but irreversible geomorphic change is evident.

These examples of the river evolution diagram not only highlight evolutionary trajectories of rivers and responses to differing forms of human disturbance, they also provide a platform to assess prospective river recovery and predict likely future river futures.

Assessing geomorphic river condition and recovery potential

Understanding of the present trajectory and rate of river adjustment provides key insight into efforts to predict likely future scenarios. Assessment of the state that a system is adjusting towards and the timeframe over which it will achieve that state are contingent upon availability of sediment and the future flow regime, and the way in which process interactions are fashioned by human disturbance and historical lagged and off-site impacts. A catchment perspective is required to examine the changing nature of biophysical fluxes and the strength of linkages between different landscape compartments (see Chapter 14). Understanding of contemporary river forms and processes, tied to interpretations of longer term river evolution, provides a basis to assess river responses to differing forms of disturbance. From this, predictions of likely future character, behaviour and condition can be made in the context of whether river recovery will, or will not, occur.

Geomorphic river condition is defined as the contemporary physical state of the river (Figure 13.14). It is a measure of the capacity of the river to perform functions that are expected for that type of river within the setting that it occupies. The key issue to address in making this determination asks: is the river operating as expected for its type, or do anomalous processes and forms indicate that the physical state of the river is compromised in some way? Determination of the condition of a reach requires a clear understanding of the 'expected' behaviour of that river type based on its position in the catchment and its environmental setting. A solid understanding of river evolution is required to assess the natural variability of the river, and to explain whether human disturbance has altered the behavioural regime.

Geomorphic river recovery is defined as the ability (or potential) of a river to change its condition over the next 50–100 yr. For example, if the river was left alone, would its condition deteriorate or improve? This requires insight into how the river is likely to adjust in the future given its current condition and the impact of limiting factors and pressures that operate in the system. *Limiting factors* are internal to the system and may include changes to sediment availability (e.g. passage of sediment slugs or sediment starvation), runoff relations and vegetation cover. *Pressures* refer to factors that are external to the system, such as climate variability, human changes to landscape forms and processes, and a myriad of socio-economic factors (e.g. population, land use). Analysis of limiting factors and pressures is a catchment-specific exercise. Each reach must be placed in its catchment context, interpreting lagged and off-site impacts in the conveyance of disturbance responses.

Notions of geomorphic river recovery encapsulate a sense of how a river has adjusted from its 'natural' condition following human disturbance, whether change has been reversible or irreversible and what state that river is adjusting towards. While changes to river morphology must be considered to be irreversible (in practical terms) in many river systems, some rivers have proven to be remarkably resilient to change, while others have started on a pathway towards recovery.

Recovery is a natural process that reflects the self-healing capacity of river systems. Here, *geomorphic river recovery is defined as the post-human disturbance trajectory of change towards an improved condition.* Assessing the pathway of geomorphic river recovery is a predictive process.

Recovery rarely reflects an orderly, progressive and systematic process. Different components of a system adjust in different ways and at variable rates, such that individual reaches undergo transitions between different states at different times. Multiple potential trajectories are likely. These reflect the condition of a reach and prospective responses to future disturbance events. These considerations must be viewed alongside prevailing, system-specific driving factors and time lags. Sensitive reaches are more prone to a wider range of prospective trajectories than their resilient counterparts.

The river recovery diagram provides a framework to appraise river responses to disturbance and prospects for recovery (Figure 13.15). The vertical axis on the left conveys a *degradation* pathway, starting from an intact state at the top, with progressively more degraded conditions down this axis. The axes on the right represent the potential recovery pathways of a reach. The position when initial signs of recovery are noted is represented by the turning point. The *restoration* pathway reflects a system that shows

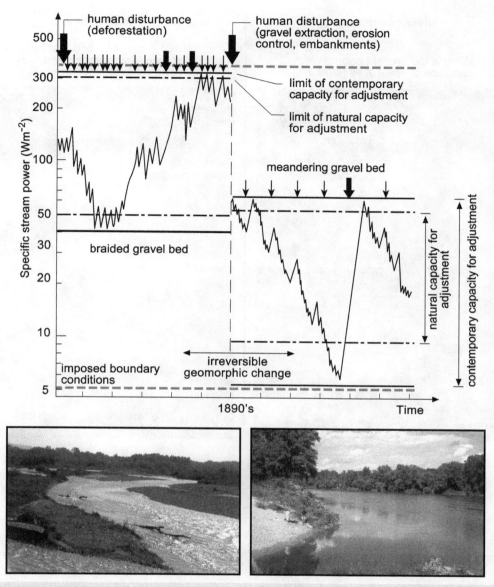

Figure 13.13 River evolution diagram showing geomorphic responses to indirect and direct human disturbances. Rivers in the French Prealps had already been subjected to significant indirect responses to forest clearance by the late nineteenth century, but they retained a braided river configuration. The capacity for adjustment of these rivers was high, as indicated on the left-hand side of the figure. Wide and shallow channels transported and stored significant volumes of gravel, and floodplains were subjected to regular flooding. Subsequent phases of rural depopulation brought about significant direct management actions, such as construction of channel embankments and instream gravel extraction. These programmes were accompanied by afforestation and erosion-control management strategies in the upper catchment, including construction of artificial reservoirs. This altered the yearly water fluxes, reduced peak discharges and decreased seasonal flows. As a consequence, sediment supply decreased and incision occurred downstream. A single channel formed within the previous braidplain, marked by transformation to a meandering gravel-bed river (right-hand side of the figure). Subsequent encroachment of vegetation into this alluvial corridor has led to channel constriction and the formation of inset floodplain surfaces. As the energy of the system has decreased over time, the contemporary sinuous single-thread pattern sits at a lower position within the potential range of variability on the river evolution diagram, and has a different pathway of adjustment than the former braided river configuration. Management strategies now aim to reinstigate a braided river system in parts of these catchments through artificial injection of gravel and removal of artificial sediment storage reservoirs from the upper catchment. Based on Bravard *et al.* (1999), Piegay *et al.* (2000) and Piegay and Schumm (2003). Photographs: G. Brierley. From Brierley and Fryirs (2005). © John Wiley & Sons, Ltd. Reproduced with permission.

River condition

What is the present geomorphic
condition of River X vs River Y?

River recovery

What is the likelihood that River X
will recovery towards River Y over
the next 50-100 years?

Figure 13.14 Defining river condition and river recovery potential. Geomorphic river condition is defined as the contemporary physical state of the river. River recovery potential is defined as the ability (or potential) of a river to change its condition over the next 50–100 yr. Photographs from Bega catchment: K. Fryirs.

signs of returning towards primary attributes of the intact state. These reaches have experienced *reversible* change from their intact condition (i.e. adjustments that have occurred are part of the behavioural regime for that type of river). The *creation* pathway reflects recovery towards a different condition. These reaches have experienced *irreversible change*, so it is no longer realistic or expected that the degraded river will return towards the pre-disturbance condition over the next 50–100 yr.

In some parts of the world a pre-disturbance condition continues to provide a useful benchmark with which to appraise human impacts upon river systems. Elsewhere, the history and impacts of human disturbance have been so profound that a pre-disturbance condition no longer provides a meaningful basis with which to consider prospective river futures. In these cases, it is more realistic to frame analyses of river recovery in relation to prevailing boundary conditions (i.e. flow, sediment and vegetation interactions). In these instances, the state at the top of the degradation pathway reflects a period when catchment

boundary conditions have reached the contemporary stage of 'development'. Stages of adjustment can be determined for reaches of the same river type.

Five examples of evolutionary trajectories for rivers affected by human disturbance, and interpretations of recovery prospects, are conceptualised in Figure 13.16. Example 1 is an intact river. The reach sits at the top of the degradation pathway, as it has not experienced human-induced deterioration in condition. The river is adjusting within its natural capacity for adjustment, reflecting the behavioural regime for this type of river, rather than experiencing a shift in state (i.e. river change). Example 2 is a reach where human disturbance has prompted deterioration away from an intact condition and the river has moved down the degradation pathway. However, this example continues to operate as the same type of river that was evident prior to disturbance. Although the behavioural regime of the river has been altered, there has not been an irreversible change in river character and behaviour. Limiting factors and pressures will determine whether this

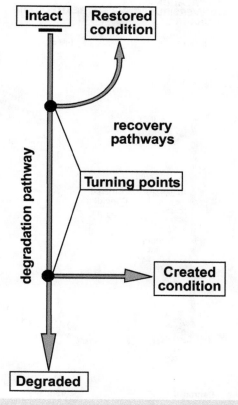

Figure 13.15 The river recovery diagram. From Brierley and Fryirs (2005). © John Wiley & Sons, Ltd. Reproduced with permission.

The character, configuration, connectivity and evolutionary trajectory of each river system induce non-synchroneity in the timing, pattern and rates of biophysical responses to differing forms of disturbance reflect. These considerations, along with appraisal of pressures and limiting factors, and lagged and off-site responses, require catchment-specific information with which to predict likely future river character and behaviour.

Tips for reading the landscape to interpret human impacts on river systems

Step 1. Identify individual landforms and their process–form associations

Framing contemporary river forms and processes in relation to interpretations of natural variability and long-term evolution provides a basis to consider river responses to human disturbance. The initial step in these analyses entails assessment of whether the package and positioning of a geomorphic unit along the river is as 'expected' for a 'natural' variant of that river, or whether the assemblage has been altered by human disturbance. For example, is there any indication of anomalous geomorphic units that do not 'belong' along this type of river? Alternatively, are key geomorphic units modified or absent? Are all geomorphic units located in their 'expected' positions, or are they misplaced? Have the magnitude–frequency and process–form associations of geomorphic units been altered along the river?

Step 2. Interpret how human disturbance has altered river character and behaviour at the reach scale

Human disturbance may alter the availability of energy in a reach, or the way in which that energy is used (i.e. alterations to the distribution and effectiveness of resisting elements). Alterations to the flow regime, sediment regime or resistance elements on the valley floor may change the rate and pattern of erosional and depositional processes, often resulting in changes to river morphology. First, the types of human disturbances occurring in the system need to be determined (past and present). This entails assessment of the nature, scale, extent, pattern, intensity and recurrence of differing forms of direct and indirect human disturbance. For example, how has land use changed? How has vegetation cover and the loading of instream wood changed? Have more imposed morphologies been applied (e.g. channelisation, levees, stop banks)? Second, system responses to differing forms of direct and indirect human disturbance must be interpreted. Typical questions to be addressed include:

system moves along a restoration or creation pathway. The third example represents rivers where human disturbance has brought about threshold-induced adjustments to river character and behaviour, such that the river now operates as a different river type (i.e. irreversible change has occurred). The degradation pathway of the pre-disturbance river has been severed. Given its poor condition, the reach sits low on this new degradation pathway, with low recovery potential. The river has shifted onto a trajectory to a new river type (the creation pathway). The fourth example represents rivers that have been subjected to irreversible change to a new river type, but now show signs of recovery. Given the poor condition of the reach, it sits low on the degradation pathway, limiting prospects for recovery along a restoration pathway. The trajectory of change is along the creation pathway, whereby the biophysical characteristics of the river differ from those that have occurred in the recent evolution of the reach. However, the behavioural regime of the river is now improving. Examples of this type of recovery occur along urban and regulated rivers. Example 5 reflects enhanced recovery mechanisms along a recovery pathway, whereby human interventions have improved the behavioural regime of the river.

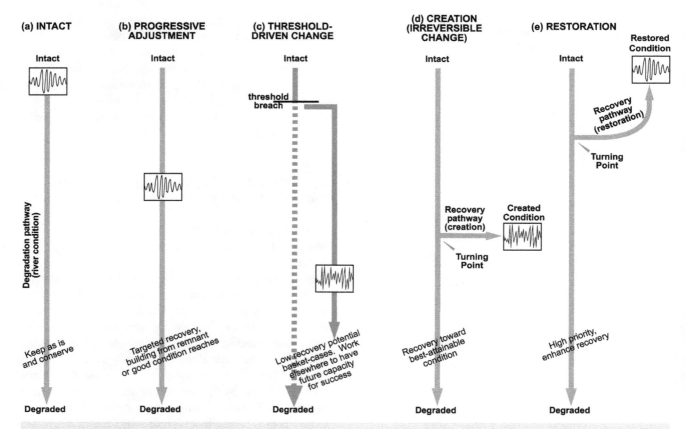

Figure 13.16   Use of the river recovery diagram to demonstrate trajectories of river adjustment and the application of enhanced recovery mechanisms: (a) intact, (b) turning point, (c) irreversible, threshold-drive degradation (d) irreversible change, created condition, (e) restored recovery trajectory. From Brierley and Fryirs (2008). © Island Press, Washington, DC. Reproduced with permission.

- Has channel geometry changed? Is the bed stable? Gas critical bank height been exceeded? Has the channel widened, contracted, or retained near-consistent dimensions? Has channel–floodplain connectivity been altered, with associated adjustments to floodplain-forming processes?
- Has human disturbance simplified or broadened the range of geomorphic units?
- Are channels smoother or rougher than they were prior to human disturbance? How do altered resistance elements affect sediment conveyance and the distribution of erosional and depositional processes (aggradational/degradational balance) of the reach?
- Has human disturbance induced a wholesale change in river type?

Step 3. Explain how human disturbance has altered river character and behaviour at the reach scale

To explain to what degree, and how or why human activities have impacted on a reach, the contemporary character and behaviour of a reach must be related to its evolutionary sequence and trajectory. Typical questions to be addressed include:

- How has the nature, scale, extent and intensity of direct human influences affected the range of behaviour and evolutionary trajectory of the river? Has the intensity of impacts increased or decreased over time? Do the various forms of human disturbance work in the same direction (reinforce and accentuate change through positive feedback mechanisms), or do they counteract each other via negative feedback mechanisms?
- Do adjustments merely reflect alterations to the rate of process activity (whether accelerated or decreased), or has the range and pattern of geomorphic processes been significantly modified by human disturbance? Have human activities set out to counterbalance negative impacts?
- How is the cumulative legacy of multiple human impacts imprinted upon the landscape?

- To what degree is the character and behaviour of the river a product of contemporary human disturbances relative to human disturbances from the past?
- How have advertent river management plans and activities impacted upon the river, and over what timeframe?

Step 4. Explain how catchment-scale relationships affect river responses to human disturbance

Interpreting river responses to human disturbance is a catchment-specific exercise. In some places the impacts have been relatively localised, whereas in others the off-site impacts have occurred after a significant lag time. Catchment-scale analyses of system responses to cumulative human impacts, framed in relation to the natural variability and longer term evolutionary adjustment, provide an information base with which to forecast a range of potential, future evolutionary trajectories for the river. This enables determination of whether changes are reversible or irreversible over particular timescales and whether river recovery is possible.

Questions to be addressed in assessing catchment-scale responses to human disturbance include:

- Where have impacts occurred in the catchment relative to the reach under investigation?
- How has human disturbance altered the evolutionary trajectory of the reach?
- Has the range of behaviour for this system been expanded or suppressed as a result of human disturbance? Has this been localised or is it ubiquitous?
- How sensitive to human disturbance are different parts of the landscape? How are disturbance responses transmitted through the catchment (i.e. what is the response gradient)? Has human disturbance altered landscape connectivity?
- What is the geomorphic condition of the reach and what is the likelihood that condition will improve over the next 50–100 years? What is the recovery potential of the reach given catchment limiting factors and pressures?

Conclusion

The pattern and extent of human-induced modification to rivers vary markedly across the world. Many modifications are purposeful or direct. Indirect changes have been more extensive and pervasive. Regardless of underlying causes, whether natural, purposeful or unintended/accidental, all river systems are subject to disturbance events that promote adjustments to their behavioural regime. In some instances, change occurs. River responses to human disturbance have gathered momentum over time, especially since the nineteenth century. The construction of structures such as dams, levees and concrete-lined trapezoidal channels, and activities such as sand/gravel extraction, have induced enormous damage to river structure and function. Many channels have been homogenised and effectively separated from their floodplains. Intended modifications have resulted in a range of unintentional consequences, such as changes to flow and sediment transfer regimes, patterns and rates of erosion and sedimentation, hydraulic resistance and flow velocity.

Analysis of evolutionary trajectories and lagged responses are required to unravel the impacts of human disturbance from 'natural' variability in any given system. The type and extent of human impacts varies markedly, ranging from site-specific works along a particular reach (e.g. bridge construction or emplacement of a stormwater outlet) to catchment-wide changes in ground cover. Catchment-specific attributes, and variability in the character, extent, history and rate of human induced disturbance, ensure that cumulative changes are system specific.

Key messages from this chapter

- Understanding contemporary river forms and processes, tied to interpretations of natural variability and long-term evolution, provides the baseline information against which to consider river responses to human disturbance. Human disturbance modifies the flow regime, sediment regime and resistance on the valley floor, altering the spatial distribution, rate and effectiveness of geomorphic processes. In many instances this instigates changes to river morphology.
- Differing environmental settings, demographic pressures and historical considerations ensure that different forms of human disturbance vary in their spatial and temporal distribution, intensity and recurrence in different parts of the world.
- Direct human impacts occur at a site and include river regulation (dam building), channel modification, sediment extraction, removal of vegetation and wood, etc. Dam construction modifies flow regimes and induces sedimentation in reservoirs. Hungry water results in bed incision downstream. Channelisation reduces geomorphic heterogeneity and roughness while increasing channel slope and hydraulic efficiency. Removal of vegetation and wood reduces the resistance of channels to erosion. Sand/gravel extraction and alluvial mining remove sediment from the system, resulting in bed incision.

- Indirect human impacts occur as secondary responses to landscape changes that occur elsewhere in a catchment and alter the flow and/or sediment regimes. Examples include land use changes such as forest clearance and afforestation, urbanisation, mining etc. Changes to ground cover impact on sediment and water yield. Urbanisation tends to increase water yield and decrease sediment yield. Mining tends to elevate sediment loads in rivers, producing sediment slugs. River rehabilitation has left a legacy of 'works' along river courses.

- Various forms of human disturbance in a catchment result in cumulative and off-site responses. The timing of adjustment may be lagged behind the disturbance or occur near instantaneously, depending on the type of disturbance and its intensity.

- Human disturbance affects the physical condition and recovery potential of rivers. River condition is defined as the contemporary physical state of the river. It measures the capacity for the river to perform functions that are expected for that river type within the setting that it occupies. Geomorphic recovery potential is defined as the ability of a river to improve its condition over the next 50–100 yr. River recovery potential is contingent on the impact of catchment-scale limiting factors and pressures on a reach. Restoration or creation trajectories reflect whether irreversible change has occurred as a result of human disturbance.

- Human disturbance alters the range of variability of processes occurring along a river. The rate and extent of adjustment often exceeds the natural range of variability. On the river evolution diagram this is represented as an expansion of the capacity for adjustment. In other cases, behaviour may be suppressed (e.g. via dam construction). In this case human disturbance reduces the range of variability.

CHAPTER FOURTEEN

Sediment flux at the catchment scale: source-to-sink relationships

Introduction

Erosion and deposition are natural processes that shape and rework landscapes. Problems only arise when they impact upon human society. Hence, prediction of magnitude–frequency relationships that fashion sediment flux, and associated implications across a catchment, are important considerations in risk and hazard management. Clearly, avoidance of areas subjected to rapid and unpredictable sediment movement is the most effective management strategy. However, avoidance is no longer a feasible or realistic option in densely settled parts of the world. As a consequence, many areas are prone to sediment disasters. Vulnerability is especially pronounced in areas that are subjected to earthquakes and volcanoes and/or extreme climatic events such as typhoons. Local variability in sediment flux reflects sediment availability, the history/sequence of disturbance events and the rate at which sediment production generates new stores of material that can be reworked. These considerations, alongside catchment connectivity, are key components in the analysis of catchment-scale sediment budgets.

This chapter is structured as follows. First, sediment flux is conceptualised at the catchment scale through analysis of sediment inputs, outputs and stores, outlining how quantification of sediment delivery underpins construction of sediment budgets. A cross-scalar approach to analysis of sediment cascades is developed. This entails analysis of landscape setting and memory as controls on the erodibility/erosivity of the contemporary landscape, and the accommodation space in which sediments are stored and reworked. This sets the contemporary conditions under which sediment cascades operate. Catchment-scale (dis)connectivity affects the internal dynamics of the sediment cascade, both spatially and temporally, providing a basis to examine cumulative responses to disturbance, whether natural or human induced. Within this landscape context, reach sensitivity (capacity for adjustment and history of change) determines the balance of erosion and

deposition and the output and storage functions of a reach. Process-form associations of geomorphic units determine the magnitude–frequency relationships with which different stores are formed and reworked. Finally, sediment flux is analysed across three scales: global, landscape setting and catchment-scale analyses. Examples are used to show how analysis of catchment-scale sediment flux and connectivity can be used to assess river recovery in different landscape settings.

Conceptualising sediment flux through catchments

As noted in Chapter 3, river systems act as conveyor belts that move sediments from source zones (hillslopes in headwater areas, where net erosion occurs) through transfer zones in mid catchment, where erosion and deposition are approximately in balance, to accumulation (sink) zones (i.e. lowland plains, oceans or inland basins). Analysis of sediment budgets entails identification of sediment sources and erosion rates (whether primary erosion or reworking of sediment stores), measurement of sediment storage/accumulation (i.e. deposition and restorage of materials within landscape compartments) and determination of sediment output at the basin outlet. Changes to relationships between sediment sources, storage elements and pathways of sediment movement may bring about dramatic changes to landscape form.

In summary terms, a sediment budget provides an account of sediment movement from the headwaters to the mouth of a catchment, over a given timeframe. It is measured as:

$$O - I \pm \Delta S = 0$$

where O is sediment output, I is sediment input and ΔS refers to the change in sediment storage. If sediment inputs exceed outputs, then net deposition (storage) has occurred. If outputs exceed inputs, then net erosion has occurred.

Geomorphic Analysis of River Systems: An Approach to Reading the Landscape, First Edition. Kirstie A. Fryirs and Gary J. Brierley.
© 2013 Kirstie A. Fryirs and Gary J. Brierley. Published 2013 by Blackwell Publishing Ltd.

In general, sediment budgets are constructed at the (sub) catchment scale, reflecting the primacy of the catchment as the fundamental unit of landscape analysis. However, sediment budgets may be a useful tool at smaller scales, such as reach-scale analyses of sediment transfer along a river or plot-scale analyses of soil erosion on agricultural fields.

Sediment yield refers to the quantity of sediment that reaches the basin outlet. This 'output' component of a sediment budget reflects denudation and erosion rates, and the connectivity of the system (i.e. how effectively sediments are transferred to the outlet). The sediment delivery ratio defines the proportion of sediment leaving an area, relative to the amount of sediment eroded in that area (i.e. the efficiency of sediment throughput). A catchment with a sediment delivery ratio of 5% is inefficient, as only 5% of the sediment eroded from a catchment has exited at the river mouth. Derivation of a catchment-scale sediment budget requires that the volumes of sediment eroded/sourced from a catchment and exiting a catchment can be calculated. On its own, this measure only provides a coarse guide to the efficiency of sediment transfer. It fails to recognise the spatial and temporal variability of sediment movement through landscapes, and provides no basis for understanding transport processes.

Significant basin-to-basin variability in sediment cascades reflects spatial and temporal controls upon sediment availability, landscape (dis)connectivity and the history of erosion/reworking events. *Sediment sources* can be either colluvial or alluvial. The colluvial system involves hillslope sediment production via processes such as sheetwash, gullying, landslides, etc. Materials contributed to the valley floor become part of the alluvial system, where they may be reworked by channels. Processes such as bank erosion, incision and floodplain reworking contribute alluvial sediment to the channel network.

The nature and pattern of *sediment storage* within a catchment reflect spatial variability in accommodation space (i.e. places where sediments may be stored) and the effectiveness of geomorphic processes that rework these sediments (i.e. the magnitude–frequency relationships of reworking processes, and sequences of events, that determine whether sediments in stores are eroded or additional materials are added to that store). Sediment storage occurs in colluvial storage units such as soil mantles, colluvial footslopes, etc., or in alluvial storage units such as floodplains, terraces, bars, etc. Residence time, which refers to the amount of time sediment remains in storage, varies significantly depending on the type of storage unit (Figure 14.1). Residence time varies with height above the channel, the range of reworking processes and the nature of the sediment (i.e. the ease with which materials can be reworked). Sediment *stores* are transient features made up of readily mobile sediments such as mid-channel bars. Residence time is short (months or years), as these features are reworked regularly. Sediment *sinks* trap materials for extended periods of time (hundreds or thousands of years; e.g. terraces). In general terms, the residence time of sediment stores tends to increase with distance downstream. Indeed, as sediments are buried and the thickness of the sediment fill increases, subsidence may trap and compress sediments within depositional basins.

Sediment transport is an episodic process. Different grain-size fractions move via different mechanisms. Bedload movement is characterised as a jerky conveyor belt (Chapter 6). The time a grain spends in transport is significantly shorter than the time spent in storage. The suspended load makes up a much larger proportion of the alluvial sediment budget. In some instances, the solution load may be very significant. The manner and extent to which sediments are conveyed through river systems once they are mobilised reflects the connectivity or coupling of a system.

Sediment budget flow diagrams summarise source-to-sink relationships that characterise catchment sediment cascades. Three examples shown in Figure 14.2 exemplify significant basin-to-basin variability in the nature of sediment inputs and outputs and patterns of sediment storage over time. These relationships reflect three key process interactions: generation of sediments, efficiency of their conveyance and space in the landscape for the restorage of materials. As a simple guide, if there are limited sediment stores or sinks in a landscape, sediment conveyance through that part of the system may be close to 100%. While sediment budgets and associated flow diagrams are very useful for determining how a catchment sediment cascade is operating, it must be recognised that these are spatially and temporally clumped products. To truly understand the dynamics of catchment-scale sediment flux, and changes over time, catchment-scale linkages and (dis)connectivity in water and sediment transfer must be examined in relation to the ways in which materials are reworked and restored in landscapes. However, in all three examples shown in Figure 14.2, the vast majority of eroded materials is restored within these landscapes, such that resulting sediment delivery ratios are <20%.

Techniques used to construct a sediment budget

Any sediment budget must give careful consideration to the spatial and temporal scales at which it is to be applied. Typically, a trade-off must be made between scale and precision and the level of accuracy that is sought. Local-scale studies tend to be field intensive, involving in-depth investigation into interactions among landscape compo-

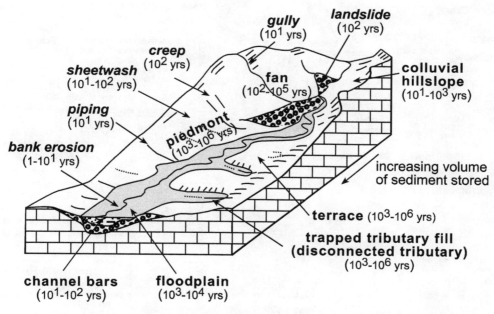

gully
(10¹ yrs)

landslide
(10² yrs)

creep
(10² yrs)

sheetwash
(10¹-10² yrs)

fan
(10²-10⁵ yrs)

**colluvial
hillslope**
(10¹-10³ yrs)

piping
(10¹ yrs)

piedmont
(10³-10⁶ yrs)

bank erosion
(1-10¹ yrs)

increasing volume
of sediment stored

terrace (10³-10⁶ yrs)

**trapped tributary fill
(disconnected tributary)**
(10³-10⁶ yrs)

channel bars
(10¹-10² yrs)

floodplain
(10³-10⁴ yrs)

xxx = sediment source processes and their recurrence of reworking
xxx = sediment stores and sinks, and their residence time

Figure 14.1 Timeframes of sediment (re)generation for differing sediment sources and residence times for sediment stores in river systems. Differing colluvial and alluvial landforms operate as sediment sources and stores/sinks over variable timeframes, ranging from years to many thousands of years. The recurrence with which sediments are sourced or stored is largely dependent on position in a catchment and the recurrence of geomorphically effective disturbance events.

nents that make up the system. In contrast, at national or global scales sediment budgets tend to consider sediment yields (outputs) rather than identification of sources and stores. Given the stochastic nature of the forcing elements that determine sediment movement and storage, timescale of analysis is a critical consideration in the derivation of sediment budgets. Longer term perspectives may incorporate changes in climate, land use and sea level, whereas shorter, contemporary analyses can measure system responses to a given event (or sequence of events), such as impacts of floods, fire or land use alterations. From this, magnitude-frequency relationships can be derived.

Generic procedures used to construct a sediment budget are summarised in Figure 14.3. Depending upon the spatial and temporal scale of investigation, a range of techniques can be employed at each stage of analysis. Sediment movement in landscapes is rarely monitored. It is often very difficult to measure sediment movement at formative flow stages, and extrapolation from rating curves is typically applied (see Chapter 6). Seasonal changes may be an important driver of sediment flux (e.g. tropical wet–dry seasonality, or impacts of snowmelt). Sediment flux at any given time is greatly influenced by the amount of sediment that is available to be moved at that time, especially materi-

als in readily accessible stores. This has major implications when extrapolating data over longer time periods.

Measuring sediment storage involves identifying and calculating the volumes of sediment sourced from various areas, transported along the channel network and stored in various landforms throughout a catchment. To accurately calculate the volumes of material being sourced, transported or stored at any one particular time, or to track changes over time, involves identifying changes in the size and assemblage of landforms over time. Techniques used to derive and analyse sediment budgets range from simple qualitative conceptualisations used to examine process interactions to in-depth quantitative analyses that derive volumetric information about the rates at which sediment is entering and leaving landscape components. Numerical and physical modelling procedures are commonly applied. Most budgets use a combination of desk, field and analytical techniques.

Advances in computer-based technology, remote sensing imagery and geographic information science (GIS) have aided the development of more detailed and complex sediment budgets, increasing what can be achieved in terms of speed, accuracy and scale. GIS allows large datasets to be overlaid with greater spatial accuracy. Repeat surveys using

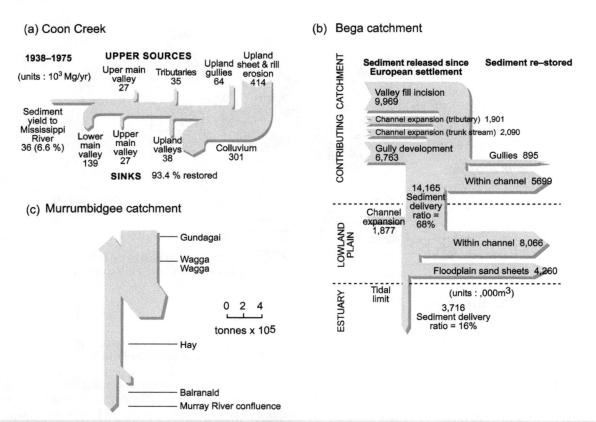

Figure 14.2 Examples of sediment budget diagrams. The thickness of arrows represents the proportion of sediment sourced, transferred or restored in landforms at different parts of the catchment. Constrictions in the 'transport' arrow reflect disconnectivity of sediment flux and associated sediment storage. (a) Hillslope sources were the primary contributor to the sediment budget of Coon Creek, Wisconsin, USA, a tributary of the Mississippi River, from 1938 to 1975. Units are in 10^3 Mg yr^{-1}. Upland sheet and rill erosion accounted for 75% of material input into the budget. Of this material, over 70% was restored as colluvium (i.e. hillslope deposits). Collectively, almost 94% of the material sourced from hillslopes and reworked along valley floors was restored within the system, yielding just 6.6% of these materials to the Mississippi River over this period. Reprinted from Geomorphology, 108 (1–2), Trimble, S., Fluvial processes, morphology and sediment budgets in the Coon Creek Basin, WI, USA, 1975–1993, 8–23, © 2009, with permission from Elsevier. (b) Alluvial sediment budget for Bega catchment, NSW, Australia (× 10^3 m^3). Hillslopes are largely disconnected from valley floors in this landscape, so the post-European settlement sediment budget is dominated by reworking of valley floor (alluvial) deposits. Incision of valley fills is the dominant sediment source. Around one-third of the materials eroded from the upper catchment are restored within the channel. Sediment conveyance to the lower catchment is relatively efficient. Channel widening along the lowland plain adds materials to the sediment budget. However, within-channel deposition and accumulation of sand sheets on the floodplain have been the dominant processes in this part of the catchment in the period since European settlement. Sediment storage in this area results in just 16% of the total sediment inputs being transferred to the basin outlet. Reprinted from Geomorphology, 38 (3–4), Fryirs, K. and Brierley, G.J., Variability in sediment delivery and storage along river courses in Bega catchment, NSW, Australia: implications for geomorphic river recovery, 237–265, © 2001, with permission from Elsevier. (c) Annual suspended-load sediment budget (t yr^{-1}) for the Murrumbidgee River, NSW, Australia. Suspended-sediment sources are dominated by materials that are sourced in the eastern highlands and gully networks in tributaries. Little in the way of inputs is made to the system once the river enters the western plains. Most suspended sediment is restored on the expansive Riverine Plain (see Figure 12.15). Only a small proportion of the sediment reaches the Murray River. From Olive *et al.* (1994).

advanced remote-sensing technologies such as light detection and ranging (LiDAR) can be applied to generate precise digital elevation models from which fine-resolution sediment budgets can be derived. Alternatively, repeat aerial photographs can be used to assess changes in river morphology, and associated sediment flux, of bedload or mixed-load rivers.

Fieldwork provides important contextual information with which to ground modelling applications. Most sediment budgets have a fieldwork component. Changes to

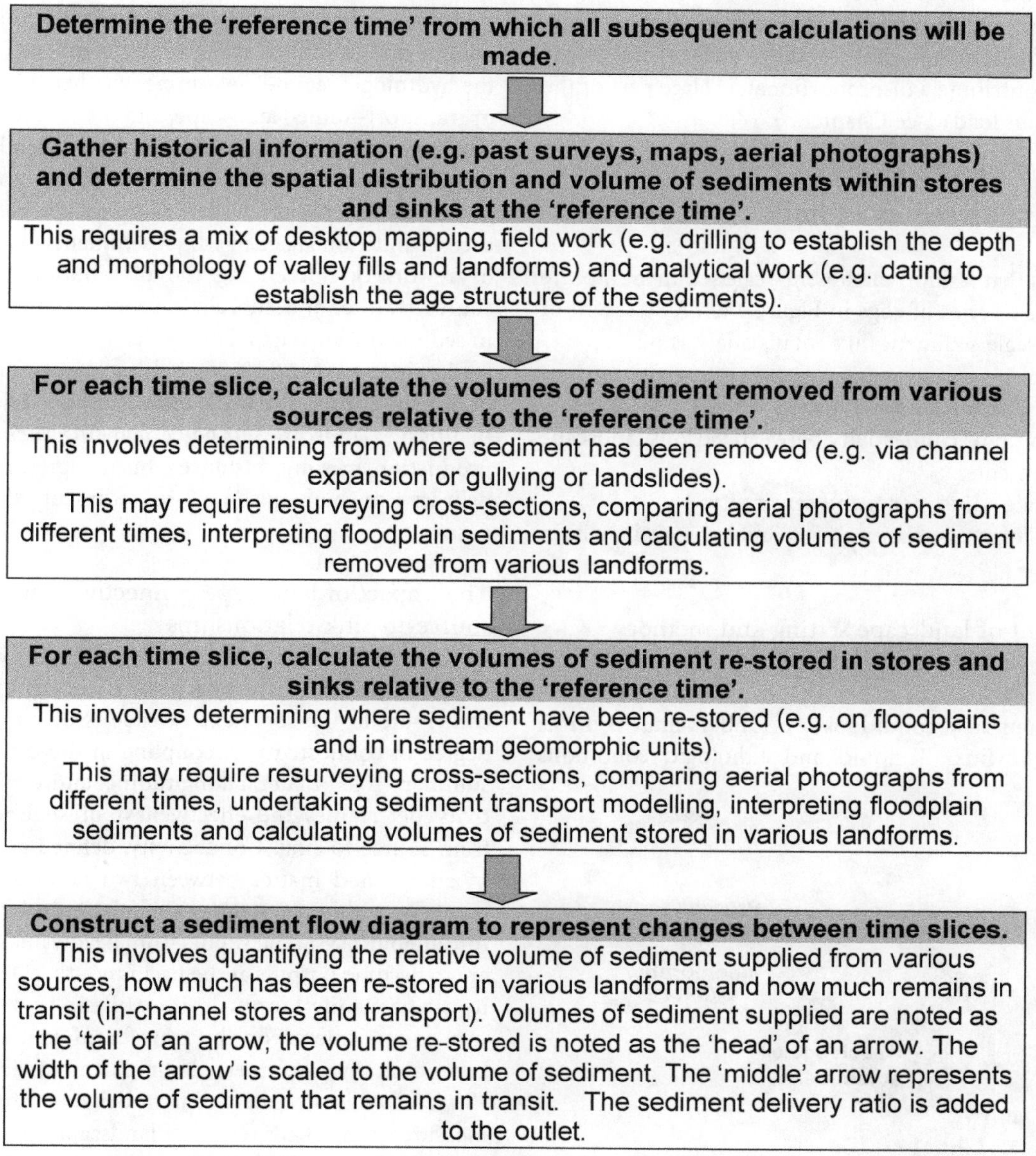

Determine the 'reference time' from which all subsequent calculations will be made.

Gather historical information (e.g. past surveys, maps, aerial photographs) and determine the spatial distribution and volume of sediments within stores and sinks at the 'reference time'. This requires a mix of desktop mapping, field work (e.g. drilling to establish the depth and morphology of valley fills and landforms) and analytical work (e.g. dating to establish the age structure of the sediments).

For each time slice, calculate the volumes of sediment removed from various sources relative to the 'reference time'. This involves determining from where sediment has been removed (e.g. via channel expansion or gullying or landslides). This may require resurveying cross-sections, comparing aerial photographs from different times, interpreting floodplain sediments and calculating volumes of sediment removed from various landforms.

For each time slice, calculate the volumes of sediment re-stored in stores and sinks relative to the 'reference time'. This involves determining where sediment have been re-stored (e.g. on floodplains and in instream geomorphic units). This may require resurveying cross-sections, comparing aerial photographs from different times, undertaking sediment transport modelling, interpreting floodplain sediments and calculating volumes of sediment stored in various landforms.

Construct a sediment flow diagram to represent changes between time slices. This involves quantifying the relative volume of sediment supplied from various sources, how much has been re-stored in various landforms and how much remains in transit (in-channel stores and transport). Volumes of sediment supplied are noted as the 'tail' of an arrow, the volume re-stored is noted as the 'head' of an arrow. The width of the 'arrow' is scaled to the volume of sediment. The 'middle' arrow represents the volume of sediment that remains in transit. The sediment delivery ratio is added to the outlet.

Figure 14.3 Flow chart of generic tasks undertaken to construct a sediment budget.

sediment storage and/or measurement of sediment transport rate can be achieved through recurrent resurveying of channel cross-sections, continuous monitoring of bed (e.g. scour chain analysis) and bank mobility (e.g. erosion pins), measurement of bedload or suspended-load transport and sediment analyses. Concerns may arise regarding reliability and representativeness of these data. However, more ground can be covered with higher accuracy using computer-based techniques. Sequential aerial photographs, satellite images and historical maps aid identification of erosion and deposition rates for differing geomorphic surfaces.

Palaeo-records can be used to interpret how sediment fluxes have changed over time, isolating responses to different forcing mechanisms such as climatic changes or extreme events. Application of dating techniques allows more accurate interpretation of the timescales and rates of sediment deposition and reworking. Various sediment fingerprinting techniques can be employed to trace sources and pathways of sediment movement and accumulation. Fallout radionuclides such as lead-210 and caesium-137 can be used to trace recent sediment flux (since the 1950s). Similarly, sediment-bound contaminants can be used to unravel deposition rates extending back to around the time of the industrial revolution, especially in areas where more recent contamination has occurred. Longer term records can be obtained by analysing the mineralogy of sediments stored in floodplains using X-ray fluorescence and X-ray diffractometer technology. When coupled with dating techniques, these sequences can reveal changes in sediment sources over time. This enables determination of geomorphically effective tributaries (i.e. those areas of the catchment that contribute a disproportionately large part of the total sediment load) (see Chapter 3).

Controls upon sediment flux

Interactions that fashion catchment-scale sediment flux are visualised as a series of cogs in Figure 14.4. Assessment of catchment-scale sediment flux entails analysis of:

- landscape setting and memory;
- the strength of connectivity between various landscape compartments;
- reach-scale adjustments and sensitivity;
- process–form associations of geomorphic units within a reach.

The imprint of landscape setting and memory on source-to-sink relationships

Imposed boundary conditions are fundamental controls on sediment flux. Tectonic and lithologic conditions

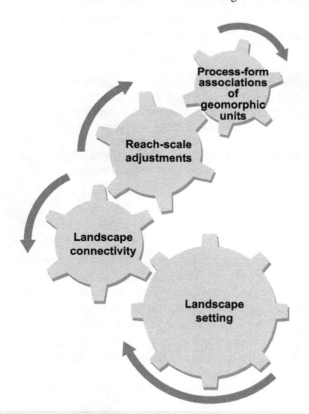

Figure 14.4 Conceptualisation of controls upon sediment flux at the catchment scale.

fashion the erosivity/erodibility of a landscape and the calibre of sediment that is made available to a river. The volume and calibre of available sediment, and relations to the hydrologic regime, determine whether this is a supply- or transport-limited landscape, and the resulting pattern of bedrock-controlled and alluvial reaches. Weathering rates and landscape dissection are key determinants of sediment generation. Slope and valley morphology determine the accommodation space in which sediments are stored and/ or reworked, influencing the distribution of sediment source, transfer or accumulation zones. Marked differences in sediment transport relationships are evident along bedload, mixed-load and suspended-load rivers. The influence of human disturbance upon sediment flux varies markedly in differing landscape settings, enhancing rates of sediment production in some instances, but suppressing sediment delivery elsewhere. In some cases sediment exhaustion can occur.

The impact of landscape connectivity on source-to-sink relationships

The extent to which various parts of a catchment actively contribute to the sediment cascade is determined by the degree of connectivity or coupling in the catchment. The summary role of lateral, longitudinal and vertical connectivity determines the effectiveness of sediment transfer from source to sink. Connectivity, defined as the transfer of energy and matter between two landscape compartments or within a system as a whole, must be maintained throughout a system if inputs from headwater source zones are to become outputs at the basin mouth. (Dis)connectivity can be manifest through either physical contact between two compartments or the transfer of material between two physically disconnected compartments (Figure 14.5). Viewed in this way, the sediment delivery ratio provides a measure of the effectiveness of landscape connectivity.

Hillslopes are the primary source of materials moved through river systems. In many instances, however, reworked hillslope materials that now make up alluvial depositional sequences on the valley floor are the dominant contemporary source of sediments moved by the river. In a sense, these alluvial sediment stores can be considered secondary sediment sources.

A range of gravity-induced mechanisms moves sediments on hillslopes, including rockfalls, landslides, debris flows/avalanches, earth flows, soil creep and gully mass-movement complexes. The nature of sediment movement varies markedly for these different processes, in terms of the amount and calibre of sediment moved, the distances moved and the frequency/recurrence with which movement takes place. In some instances large volumes of sediment are moved considerable distances; elsewhere, small

(a) Hillslope - channel coupling

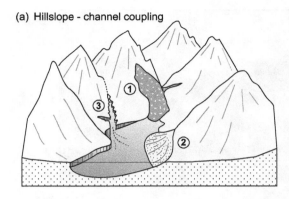

(b) Hillslope - channel decoupling

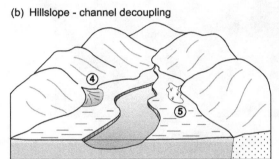

Figure 14.5 Hillslope–valley floor (de)coupling. (a) In coupled landscapes, hillslope materials are efficiently transferred to the valley floor, where they are prone to be readily reworked. Landslide deposits, fans and gully networks (numbered 1–3 respectively) transfer materials to channels. (b) In decoupled landscapes, sediments released from hillslopes at valley margins may be stored further down-slope (i.e. within the catena) or they may be stored atop extensive floodplains (numbers 4 and 5 respectively).

incremental movements of sediment are the dominant hillslope mechanism. Of key concern here is whether sediments are moved within the hillslope compartment itself (i.e. restored within the catena; Figure 14.5a) or whether sediments are transferred to the valley floor (Figure 14.5b). As slope flattens and valley floors widen in downstream parts of catchments, there is greater accommodation space and increased capacity for sediment storage. As a result, sediment transfer from hillslopes and along the valley floor is increasingly spatially interrupted in downstream parts of catchments. Floodplain pockets are longer term stores (sinks) than their instream counterparts. Terraces have even longer residence times than the contemporary floodplain, further acting to decouple the system.

Various landforms may impede sediment conveyance, either within landscape compartments or within the catchment as a whole. Buffers, barriers and blankets disrupt longitudinal, lateral and vertical linkages respectively

(Figure 14.6). These blockages can be natural or human induced.

Buffers disrupt longitudinal and lateral linkages within catchments, preventing sediment from entering the channel network. Fans, piedmont zones and floodplain pockets may disconnect sediment transfer from hillslopes to the channel. Other buffers include features such as intact valley fills and floodouts that have discontinuous or absent water courses or low-slope alluvial floodplains. Elevated floodplain or terraces may block tributary confluences, disconnecting material supply from lower order drainage lines to the primary channel network. In some instances, channel incision may decouple channel and floodplain processes. Human disturbance may greatly affect these relationships. Ground cover may act as an important control upon the effectiveness of buffers. Many direct management actions seek to limit sediment delivery from hillslopes. Channelisation programmes disconnect channels from their floodplains. Systematic drainage of swamps in cut-and-fill landscapes transforms former sediment accumulation zones into source zones.

Barriers impede downstream conveyance of sediment once it has reached the channel network. They most commonly disrupt longitudinal linkages through their effect on base level or bed profile. For example, bedrock steps or wood may locally reduce slopes by introducing a local base-level control. Sediments are trapped as they backfill areas immediately upstream of the step, inducing local discontinuity in sediment transfer. Similarly, a valley constriction can act as a barrier that initiates valley backfilling and sediment storage in floodplains or instream units. Sediment slugs may act as plugs to sediment movement along channels. Similarly, overwidened channels may not have the competence to carry sediments made available to them, acting as barriers to downstream sediment conveyance. Breaching of barriers results in pulsed sediment movement and the formation of transient sediment stores. These 'natural' circumstances have been greatly accentuated by human activities. The impacts of dams are especially profound. Large dams are exceptionally efficient traps for bedload and suspended-load fractions. Hydrologic and geomorphic impacts of countless smaller dams may also be pronounced. Human impacts on roughness elements within channels also disrupt the effectiveness of barriers, as do secondary (indirect) responses such as the generation of sediment slugs following deforestation.

Finally, *blankets* are features that disrupt vertical linkages in landscapes by smothering other landforms. As such, they protect these sediments from reworking, temporarily removing them from the sediment cascade. Blankets can occur instream or on floodplains. They include features such as floodplain sand sheets or fine-grained materials that infill the interstices of gravel bars. Bed armour acts as

Figure 14.6 Buffers, barriers and blankets. These landforms impede lateral, longitudinal and vertical connectivity of sediment flux in landscapes. They promote sediment storage, thereby reducing the efficiency of sediment transfer through a catchment. *Buffers.* (a) Valley-bottom swamps, Wingecarribee swamp, NSW. Photograph: K. Fryirs. (b) Alluvial fans in steep terrain, Tibet. Photograph: G. Brierley. (c) Trapped tributary fill, Macdonald River, NSW. Photograph: K. Fryirs. (d) Broad, open alluvial plains, North Coast, NSW. Photograph: R. Ferguson. *Barriers.* (e) Sediment slug, Waiapu River, New Zealand. Photograph K. Fryirs. (f) Dam, Gordon River, Tasmania. Photograph: K. Fryirs. *Blankets.* (g) Floodplain sediment sheet, King River, Tasmania. Photograph: K. Fryirs. (g) Fine-grained materials in the interstices of gravels, King River, Tasmania. Photograph: K. Fryirs.

a blanket which prevents the reworking of subsurface sediments. Blankets are most commonly found along alluvial rivers where instream sediment stores and floodplain sinks are common. Human alteration of landscape and channel boundaries greatly modifies these relationships. This is especially pronounced in urban areas, where impermeable surfaces are dominant. However, any factor that modifies 'natural' rates of sediment generation and the efficiency of conveyance may affect the effectiveness of blankets. For example, vast volumes of material reworked by mining activities commonly create a significant drape over the valley floor. In contrast, concrete-lined channels completely inhibit surface–subsurface process interactions.

Reach sensitivity

Different types of river store and rework sediments in different ways. Confined rivers tend to act as efficient sediment transfer zones, whereas alluvial valleys tend to act as transfer or accumulation zones within which a range of river types can form. The sensitivity and capacity for geomorphic adjustment, and associated sediment flux, vary for these differing river types (Chapter 11; Figure 14.4). Human disturbance may alter these relationships. Similarly, reach sensitivity can change over time depending on the type and severity of disturbance and the condition of the reach at the time of the disturbance. For example, bed level lowering and headcut retreat (Chapters 2 and 4) can alter the process zone distribution in a catchment and the sensitivity of reaches to sediment storage, transfer or accumulation (Figure 14.7). Once bed incision and channel expansion are initiated, former sediment accumulation zones can be transformed into sediment sources, potentially releasing significant volumes of alluvial materials. These reaches become sensitive to adjustment and sediment release. Alternatively, channelisation or the formation of blockages that disconnect reaches from disturbance influences may desensitise a reach, transforming its geomorphic structure and function. These ongoing adjustments induce profound spatial and temporal variability in sediment flux.

Process–form associations at the geomorphic unit scale and impacts upon material reworking

At the finest scale of resolution (the smallest cog in Figure 14.4) the recurrence with which geomorphic units are formed and reworked (i.e. their residence time) determines the extent to which sediments are stored and/or transported within a reach/compartment. Roughness of different surfaces, the balance of impelling and resisting forces and the degree of sediment organisation (e.g. packing, armouring) determine the extent to which flows of various magnitude and frequency can rework and transport sediment. Any reduction in roughness may increase the effectiveness with which flows move sediment. Hence, these relationships are extremely sensitive to human impacts upon valley floors. In extreme instances, management actions strive to 'fix' channel boundaries using concrete or oversized boulders. Elsewhere, indirect responses to human disturbance may include generation of loose, more-malleable channel boundaries, if pulses of bedload materials are flushed along the channel. The role of vegetation and wood as resistance elements is extremely important at this scale. While roughness may be greatly reduced following human disturbance to riparian vegetation cover, rapid spread of dense exotic vegetation may greatly increase within-reach roughness. Such responses are often indirect consequences of human actions. They create entirely different boundary conditions for flow–sediment interactions.

Synthesising catchment-specific controls upon spatial and temporal variability in sediment flux

Sediment availability and flux are spatially and temporally contingent. They reflect the configuration of the system and the history of system responses to disturbance events. Rates of sediment supply from upstream and the ability of a reach to trap sediments are key considerations for analysis of river recovery. Catchment-specific configuration and landscape history determine how within- and between-compartment connectivity affects sediment flux in any given system at any given time. The conveyor belt operates very efficiently in systems with a high sediment delivery ratio, effectively transporting sediments unimpeded to the mouth. Under other sets of conditions, the conveyor belt breaks down. Hence, sediment availability and the spatial and temporal patterns of landscape (dis)connectivity affect how off-site responses are manifest in the system and the timeframe over which they will occur. The downstream pattern of river types determines how effectively the system conveys (or restores) sediments from reach to reach.

The type and distribution of buffers, barriers and blankets dictate the strength of coupling between landscape compartments. The effectiveness of these agents of landscape disconnectivity reflects catchment-specific configuration, landscape history and system responses to disturbance events. For example, landslide interactions with valley floors are conveyed schematically in Figure 14.8. Extreme landslides may reconfigure drainage networks, realigning headwater streams and their valley morphology. Should the landslide materials become channelised along the trunk stream, they may develop an extensive linear form that is reworked by subsequent flows. In laterally connected landscapes, large landslide events may disrupt longitudinal transfer of sediments, as temporary dams disrupt the slope of the longitudinal profile and associated base

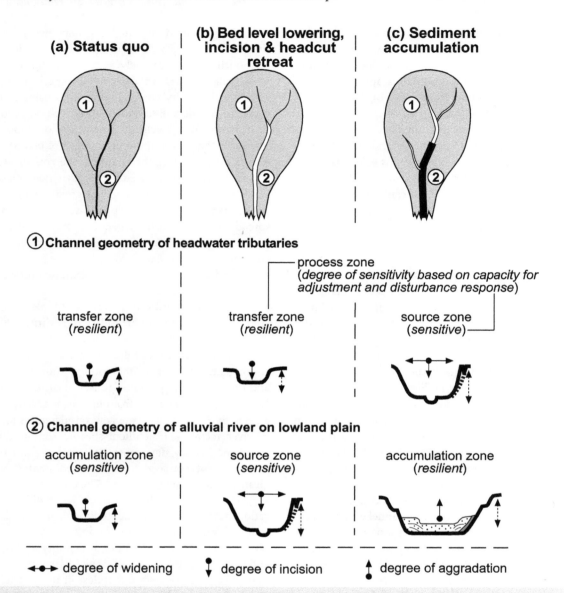

Figure 14.7 The changing nature of process zones and reach sensitivity as a result of disturbance. In this figure bed level lowering and headcut retreat alter the process zone distribution of a catchment. Associated changes to reach sensitivity are noted. (a) In the status quo situation (undisturbed), headwater tributaries act as transfer zones and the lowland plain acts as an accumulation zone. (b) Following bed level lowering, reaches along the lowland plain become sensitive to adjustment. Channel incision and expansion release sediment, transforming the reach into a source zone. Headward retreat and network extension induce incision and expansion of tributary channels, releasing additional sediment. These reaches now act as sensitive source zones. (c) Sediment released from upstream source zones is trapped along the lowland plain, switching this reach back to an accumulation zone.

level. Breaching of these dams may result in dramatic consequences downstream, as large volumes of sediment are unevenly dispersed. Such activity is especially pronounced following earthquake activity. These point impacts result in very pulsed sediment flux. If landslides fall into water bodies such as glacial lakes, materials may be stored for extensive periods (i.e. until the next reworking process,

such as a glacial advance, reactivates these materials). Finally, in decoupled settings, materials mobilised on hillslopes are re-stored down-slope or in fans at the base of hillslopes.

Tributary confluence zones may disrupt patterns and rates of sediment flux along the trunk stream (see Chapter 3). Landscape configuration describes the frequency with

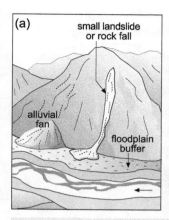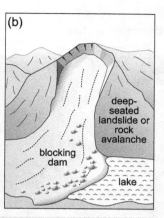

Figure 14.8 Conceptual representation of land-slide impacts upon trunk streams. (a) Landslides are too small to reach the drainage network. The ratio of landslide runout to slope length is insufficient, and valley floor buffer width is wide, resulting in no sediment delivery to the channel. (b) Landslides are large enough to form dams, thus disrupting or obliterating the drainage system. Long-term blockage will force backwater sedimentation and trapping behind land-slide dams or, in extreme cases, drainage reversal. Landslide size in relation to cross-valley length scale is sufficient to overwhelm the channel and form a stable landslide-dammed lake. From Korup (2005). © John Wiley and Sons, Ltd. Reproduced with permission.

which tributary streams join the trunk stream, and resulting patterns/rates of sediment flux. If several sediment-charged streams join the trunk stream within a relatively short distance, aggradation occurs. Some tributary systems may be disconnected from the trunk stream, as sediments are trapped behind floodplain pockets, at confluences or in trapped tributary fills. Geomorphically effective tributaries disrupt the balance of erosional and depositional processes along the trunk stream.

Similarly, pinch points as valleys narrow play a major role in the distribution of flow energy in river systems, promoting sediment storage in upstream reaches. Reaches that accentuate rates of sediment transfer can be described as boosters. These include mid-catchment gorges or other forms of local valley constriction, such as channelised or artificially straightened reaches that enhance the capacity of the system to convey sediment downstream. Pronounced decrease in flow energy downstream of these reaches results in dramatic transition to depositional zones and associated sediment stores.

Sediment movement is not uniform over time. Rather, pulses in sediment reflect stochastic inputs and transfer mechanisms. Each landscape compartment is subjected to a suite of processes that has its own magnitude–frequency

domain. The effectiveness of a sediment cascade reflects the synchroneity of process relationships across differing compartments of the landscape. This reflects the nature/timing of hillslope processes and the nature/timing of reworking processes on the valley floor. The magnitude and frequency of events that trigger and deliver sediment to valley floors may differ markedly from those events that actively transport sediment along tributaries and trunk streams. Therefore, phased (dis)connectivity operates over differing temporal scales.

The degree to which a catchment sediment cascade is connected or disconnected can be viewed at various levels of temporal resolution, ranging from floods of various magnitude and frequency, through to a long-term response to disturbance (e.g. land use change). The timescale over which the sediment cascade operates is dependent upon the frequency with which geomorphically effective flows are able to breach blockages within a catchment. The breaching capacity of buffers, barriers and blankets is measured as the magnitude, frequency and sequencing of events required to remove a blockage and promote sediment movement (see Figure 14.8).

The area that directly contributes to, or transports sediment along, the channel network is referred to as the *effective catchment area*. This provides a measure of the degree to which the catchment is longitudinally, laterally and vertically connected. The position of buffers, barriers or blankets determines the extent to which a catchment is (dis)-connected. *Effective timescales* relate to the timeframe over which sediment movement within a catchment is connected or disconnected. The magnitude–frequency characteristics of perturbations required to breach a blockage, and the propagation time for change to be manifest in the system, determine effective timescales of (dis)connectivity. Hence, responses to disturbance may be manifest throughout the system or absorbed elsewhere within the system.

Combining these two concepts, sediment flux at the catchment scale can be considered as a series of switches which determine which parts of the landscape contribute to the sedimentary cascade over different time intervals. Figure 14.9 conveys a conceptual framework by which these interactions may be assessed. In this conceptualisation, hillslopes and channels are buffered by alluvial fans in the upper tributary systems, and floodplains and terraces occur along the lower trunk stream. Fine-grained materials blanket gravel bars at various positions along the tributary networks. A valley constriction along the trunk stream acts as a barrier that disrupts the transfer of sediment from the upper catchment to the lower catchment. Sediments supplied to the lower catchment form a sediment slug along lowland reaches.

The pattern of buffers, barriers and blankets characterised in Figure 14.9 influences the timeframe over which

Effective timescale

| Frequent, low magnitude | Less frequent, moderate magnitude | Infrequent, high magnitude |

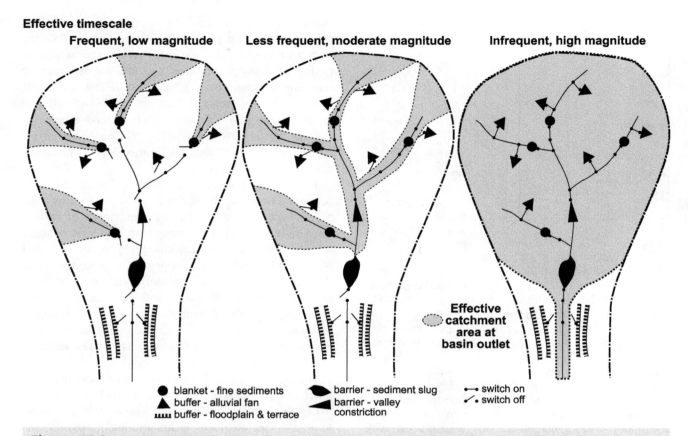

Effective catchment area at basin outlet

● blanket - fine sediments
▲ buffer - alluvial fan
⊔⊔⊔ buffer - floodplain & terrace

⬛ barrier - sediment slug
◀ barrier - valley constriction

•→ switch on
↗• switch off

Figure 14.9   Magnitude–frequency relationships that fashion the operation of buffers, barriers and blankets that control effective timeframes and effective catchment areas of sediment flux. Changes to the effective catchment area and effective timescales of (dis)connectivity are depicted as a series of switches that are active or inactive under certain flow conditions. Reprinted from Catena, 70 (1), Fryirs, K., Brierley, G., Preston, N. and Kasai, M., Buffers, barriers and blankets: The (dis)connectivity of catchment-scale sediment cascades, 49–67, © 2007, with permission from Elsevier.

sediments are reworked in different landscape compartments. At low flow stages associated with frequent, low-magnitude energy inputs, the capacity for hillslope erosion and fluvial sediment reworking is limited. Landscape disconnectivity is significant and the effective catchment area is low. At this stage, none of the buffers, barriers or blankets is breached and sediment cascading in the channel network is limited. As flow stage increases within the channel network, more readily reworked instream barriers and blankets are breached. There is sufficient energy to initiate reworking of interstitial fines that form instream blankets, and sediments are conveyed through the valley constriction, connecting the upstream channel network to the lowland plain. As connectivity increases, the effective catchment area increases. Other buffers and barriers are breached by high-magnitude–low-frequency events. Reworking of alluvial fans connects the hillslopes to the channel network. The sediment slug along the lowland plain is reworked, contributing sediment to the river mouth. However, unless an extreme event occurs, the floodplains and terraces along

the lowland plain are not reworked and maintain their buffering capacity. Indeed, sediments derived from upstream may be deposited on the floodplain surface, forming a blanket. During high-magnitude events, connectivity within the system is at its highest and the effective catchment area the largest. All switches may be turned on.

The distribution and types of buffers, barriers and blankets may change over time, altering the operation of sediment cascades through their impacts upon the effective catchment area and effective timescales of (dis)connectivity. For example, the large sediment slug that acts as a barrier along the lowland plain may be removed, reconnecting the lowland plain to the river mouth. Alternatively, small sediment slugs in tributaries may form barriers that subsequently disconnect the upper catchment from the lowland plain. Changes to the pattern of blockages in a catchment affect the extent of connectivity during the next effective flow. Disconnection of sediment transfer by upstream barriers, and hence depletion of sediment supply, may produce hungry rivers that promote incision. Any factor that lowers base level promotes bed degradation and associated upstream progression of headcuts along the trunk stream and tributaries. Alternatively, excess sediment loading from upstream may overload reaches downstream. The connectivity of reaches and associated tributary–trunk stream linkages are key determinants of the pattern, extent and consequences of human disturbance to river courses. In highly connected landscapes, the cumulative effects of alterations to forms of (dis)connectivity in upper parts of a catchment are manifest relatively quickly. In contrast, in highly disconnected landscapes, changes to the nature of (dis)connectivity in one part of a catchment are absorbed or suppressed in the system and geomorphic changes are not propagated through the catchment. Hence, there is marked variability in sediment flux, and associated responses to human disturbance, in differing landscape settings.

Analysis of sediment flux across various scales

The conceptualisation of sediment flux presented in Figure 14.4 highlights the primacy of landscape setting as the key determinant of sediment generation and erodibility/erosivity relationships. Geologic and climatic controls result in marked variability in rates of geomorphic activity and their consequences across the world. This section presents a summary overview of flow–sediment relationships at the global scale. This is followed by a comparison of sediment flux in two tectonic settings. Finally, various examples of the impact of human disturbance upon catchment-scale sediment flux relationships are outlined.

Sediment flux at the global scale

Estimated global water discharge to the sea is $40\,000\,km^3\,yr^{-1}$. Although changes to land surface vegetation cover have increased runoff by around $200\,km^3\,yr^{-1}$, reservoirs trap around $170\,km^3\,yr^{-1}$. Marked variability in flow is evident for the major rivers across the world (Figure 14.10). Flow is greatest in tropical areas of the world, where monsoonal climates generate significant seasonal discharge. Seasonality of flow varies markedly in different hydrologic regions (Figure 14.11). Some rivers experience runoff all year round (perennial rivers), while others have marked winter–spring flows associated with snowmelt and others have marked dry seasons during summer. Extended periods of drought occur in many regions. This variability impacts on the generation of runoff and the magnitude–frequency relationships of hydrologic and geomorphic events experienced in different settings. In general terms, interannual variability in peak discharge correlates with mean annual runoff, such that more arid basins have much greater inter-annual variability than drainage basins with greater precipitation and runoff totals (Table 14.1). These discharge relationships are a key control on the operation of the sediment cascade in different climatic and landscape settings.

The global pattern of sediment yield to the ocean reflects factors such as tectonic setting (topographic controls), lithology, precipitation and discharge, and human influences (e.g. flow regulation trapping sediment in reservoirs, accelerated erosion due to deforestation; Figure 14.12). Suspended sediment is the primary component of global sediment flux. These materials are readily transported to river mouths, whereas bedload material tends to be stored within catchment. The 'natural' (pre-anthropogenic disturbance) global flux of suspended sediment to the oceans is estimated to be $14 \times 10^9\,t\,yr^{-1}$. Bedload contributions increase this total to $15.5 \times 10^9\,t\,yr^{-1}$ (i.e. bedload comprises around 10% of the total). Almost 60% of global sediment delivery is derived from catchments that drain high mountainous terrain (>5000 m asl). Low mountains (1000–3000 m asl) that comprise large areas of continental landmass contribute the second largest amount.

Marked difference in sediment yield per catchment area is evident when comparing large catchments that drain passive margin settings relative to small catchments that drain active tectonic margins (Figure 14.12). All rivers with large sediment loads originate in mountainous terrain. Although catchments along passive plate margins tend to have larger areas, with the majority of sediment derived from the small mountainous headwaters, yield per unit catchment area is relatively low. Yields are also low in polar regions, where rates of sediment generation and frequency of formative flows are low. In contrast, rivers that drain active tectonic margins tend to have relatively small catchment areas but transfer significant volumes of sediment to the ocean. Sediment yields per unit catchment area are among the highest rates in the world.

Human activities have impacted significantly upon patterns and rates of sediment flux. Land clearance has markedly increased sediment loads. However, dam construction has greatly reduced sediment flux to the coast because of sediment retention in reservoirs. Taking out the latter effect, it is estimated that deforestation would increase sediment flux to $16.2 \times 10^9\,t\,yr^{-1}$ of suspended sediment, or $17.8 \times 10^9\,t\,yr^{-1}$ with bedload included. However, the effect of reservoirs and dams has actually decreased sediment yield to the oceans relative to pre-disturbance times by 10% (i.e. $12.6 \times 10^9\,t\,yr^{-1}$). Large dams trap 20% of the global sediment load and small reservoirs a further 6%. However, there are large regional differences in the pattern of post-disturbance sediment flux from the continents. Most of the increase in sediment yield has occurred in lower lying areas (<3000 m asl), where disturbance has

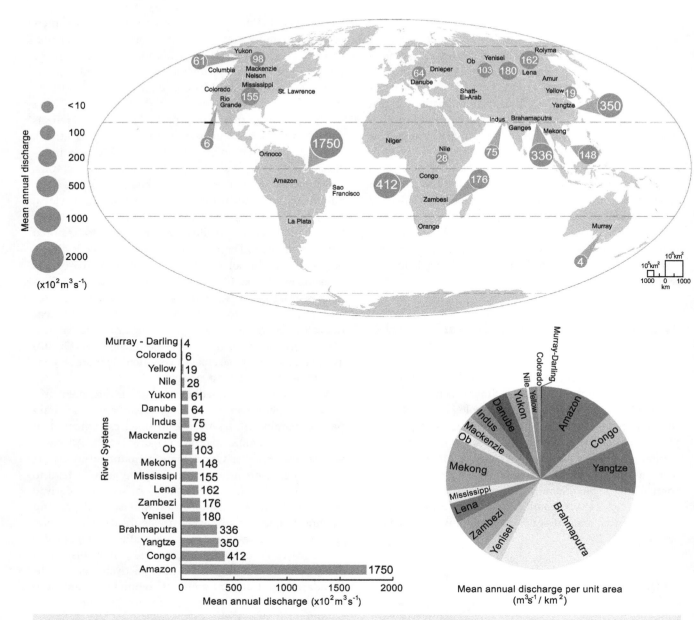

Figure 14.10 Global variability in mean annual discharge and mean annual discharge per unit area. In volumetric terms, the highest discharge is generated by the Amazon Basin. However, the highest yield is generated by the Brahmaputra, which has the highest mean annual discharge per square kilometre of catchment area in the world. Data from Wohl (2007).

been greatest. Sediment yield in mountain and high mountain areas (>3000 m asl) has decreased due to a 53% retention of sediment in reservoirs and dams.

Sediment flux in differing tectonic settings

Figure 14.13 shows how (dis)connectivity and (de)coupling notions can be used to explain spatial and temporal variability in river dynamics, evolution and sediment cas-

cading processes, and how this information can be used to inform assessments of river recovery in differing tectonic settings. In highly coupled systems, longitudinal, lateral and vertical linkages are strong, such that a significant proportion of the catchment actively contributes sediment to the cascade. The potential for geomorphic river recovery is enhanced in these settings. Uplifting, dissected terrains have high drainage densities within connected landscape configurations (Figure 14.13a). High relief promotes both

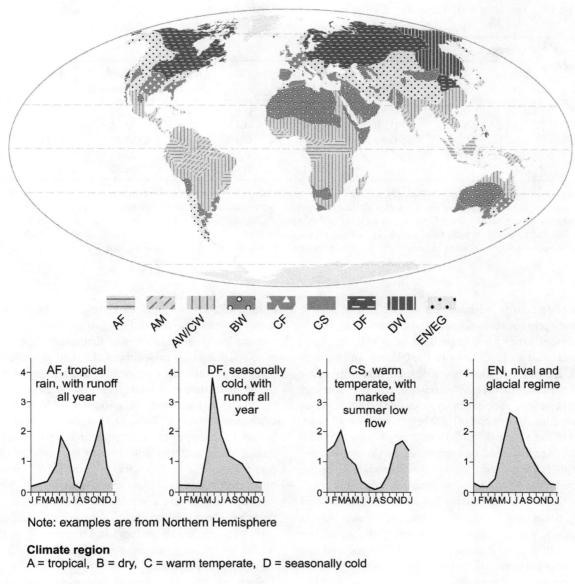

Note: examples are from Northern Hemisphere

Climate region
A = tropical, B = dry, C = warm temperate, D = seasonally cold

Flow regime
F = appreciable runoff all year, W = marked winter low flow, S = marked summer low flow, EN/EG = nival and glacial

Figure 14.11 Seasonal flow variability for climate regions across the globe. In general terms, *interannual variability* in peak discharge correlates with mean annual runoff, such that more arid basins have much greater interannual variability than drainage basins with greater precipitation and runoff totals. From Burt (1996). © John Wiley & Sons, Ltd. Reproduced with permission.

the generation of large volumes of sediment and efficient conveyance of hillslope-derived materials. Sediment from hillslopes is delivered directly to the channel. High-energy conditions on valley floors enhance rates of sediment reworking along the channel network, efficiently conveying all but the coarsest material. While disconnectivity does occur in this landscape setting, it is typically short-lived.

Hence, effective catchment areas are high and a large proportion of the catchment area is connected to the sediment cascade. Response to disturbance occurs quickly, so recovery prospects are high.

This scenario contrasts starkly with disconnected landscape settings, where sediment reworking along river courses is limited. For example, in low-relief,

Table 14.1   Coefficient of variation C_v for world and Australian streams[a] (from Finlayson and McMahon (1988))

Catchment area (km²)	Average C_v								
	World	Australia	Southern Africa	Northern Africa	Asia	North America	South America	Europe	South Pacific
0–10³	0.45	0.59	0.81	—	0.47	0.31	0.39	0.30	0.26
10³–10⁴	0.48	0.88	0.78	0.54	0.45	0.39	0.33	0.27	0.22
10⁴–10⁵	0.37	0.98	0.70	0.37	0.30	0.38	0.34	0.31	—
>10⁵	0.33	1.12	0.54	0.25	0.28	0.35	0.41	0.25	—
All	0.43	0.70	0.78	0.31	0.38	0.35	0.35	0.29	0.25

[a]Australia and southern Africa are hydrologically distinct from the rest of the world because the C_v of annual runoff and the standard deviation of logarithms to base 10 of peak discharges for all basin areas is the highest in the world, and the ratio of the peak discharge for the 1 % annual exceedance probability flood to the mean annual flood is generally higher than elsewhere. Australian and southern African streams are generally characterised by high flood variability or steep annual series flood frequency waves. Hence, the largest floods are not necessarily larger than elsewhere in terms of magnitude, but the difference between the largest and smallest floods is the greatest in the world.

low-drainage-density landscapes at the margins of passive plates, sediment source areas are relatively small and extensive sediment stores form impediments to sediment conveyance (Figure 14.13b). Longitudinal, lateral and vertical linkages are weak or disconnected, such that only some areas of the catchment actively contribute to the sediment cascade. Long lag times may be experienced between phases of geomorphic change and off-site response. In some cases, changes to the nature of (dis)connectivity in one part of a catchment are absorbed or suppressed in the system and geomorphic changes are not propagated through the catchment. Under these conditions, sediment supply is a primary limiting factor to river recovery following disturbance and recovery times may be extremely long. Alluvial sediment stores may be significantly reworked along the channel, but slow rates of sediment generation on hillslopes, and the inefficient conveyance of these materials to valley floors, inhibit prospects for recovery.

Examples of human disturbance upon sediment flux relationships at the catchment scale

Appraisals of geomorphic river recovery potential entail catchment-scale analyses of response gradients and trajectories of change. The character, behaviour and evolutionary traits of each catchment, with its specific configuration and landscape history, determine the capacity and rate of recovery. Factors that influence these relationships include the nature and extent of disturbance, the inherent sensitivity of the river type and the operation of biophysical fluxes (both now and into the future) as determined by landscape connectivity.

Catchment-specific sediment flux is contrasted for three examples in Boxes 14.1–14.3. Although the nature of

human disturbance (and its timing) has been similar in these catchments, the differing configurations of these systems has resulted in variable landscape responses, with marked differences in sediment flux and associated prospects for geomorphic river recovery. Responses to the same forms of disturbance were profound in Bega catchment (Box 14.1), but localised in the Hunter and in Twin Streams catchments (Boxes 14.2 and 14.3). Although there are similarities in the types of rivers present in these three catchments, their patterns and history of disturbance have resulted in variable responses. In some places discontinuous watercourses have been sensitive to change (Bega and Twin Streams), whereas in other places (Hunter) these river types have been relatively resilient to adjustment. Sediment calibre (controlled by geology) and imprints from the past (valley morphology, the position of terraces, etc.) control these variable responses.

Connectivity and sediment flux relationships in these catchments are quite different. The Bega catchment was transformed from a highly disconnected system that was buffered by significant discontinuous watercourses and limited hillslope–channel coupling, to a system that evacuated sediment to the lowland plain very efficiently. The system is now highly connected to its lowland plain, but not the coast. Twin Streams catchment is also highly connected, but in this instance hillslopes are well coupled to valley floors and sediments have been efficiently evacuated out of the catchment to the estuary. In contrast, limited capacity for adjustment and significant buffering of hilllopes and valley floors has resulted in the maintenance of a largely disconnected system in the upper Hunter catchment.

As a direct result of connectivity relationships, the same type of river has variable recovery potential in the three

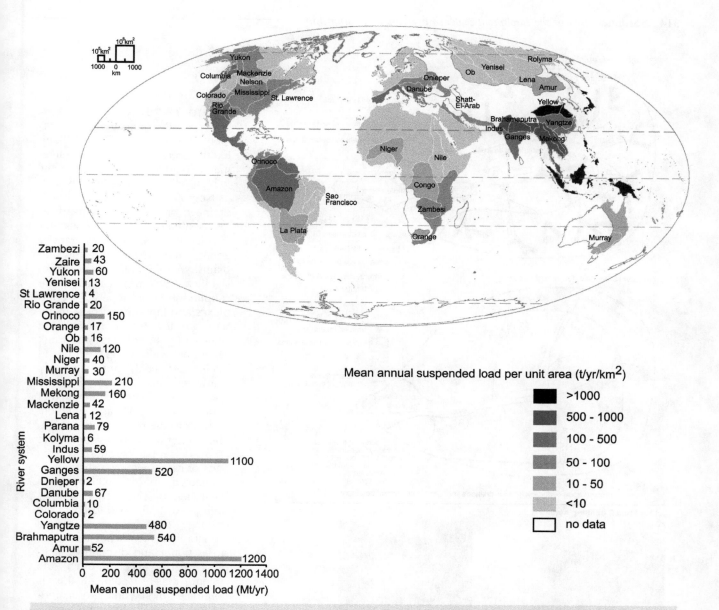

Figure 14.12 Global distribution of suspended-sediment load. Asia is the largest total continental contributor of fluvial sediment to oceans, followed by South America and North America. The lowest sediment yield occurs in polar regions, while tropical and warm temperate regions have the highest sediment yield, accounting for two-thirds of the global sediment delivery. Almost 60% of global sediment delivery is derived from catchments that drain high mountainous terrain (>5000 m asl). Low mountains (1000–3000 m asl) that comprise large areas of continental landmass contribute the second largest amount.

Although the Amazon has the largest sediment load in the world (1200 Mt yr^{-1}), its yield is only 190 t km^{-2} yr^{-1}. In contrast, rivers that drain active tectonic margins tend to have relatively small catchment areas but transfer similar volumes of sediment to the ocean. Per unit catchment area, they are some of the most sediment-charged rivers in the world. For example, the Waiapu River in New Zealand has a sediment load of 28 Mt yr^{-1} but a staggering yield of 20000 t km^{-2} yr^{-1}. The lowest yielding catchments in the world are found in passive margin settings with very low terrain (e.g. parts of Australia) or in polar regions.

The extent and type of human impact on sediment yield to the coast varies significantly across the globe. Africa and North America have seen the largest reduction in sediment yield to the coast (39% and 19% respectively), with Asia having a 13% reduction. For example, the Nile and Colorado rivers deliver almost no sediment to the ocean, and many other rivers, such as the Zambezi, Mississippi and Indus, have experienced marked decreases in sediment yield because of a range of activities such as dredging and extraction, and the construction of dams. Low sediment yields in northern Europe and England at least partially reflect channel management and increased vegetation cover. In contrast, the Brahmaputra has experienced an increase in sediment yield due to deforestation in Nepal. Recent construction of large dams along the Yangtze and Yellow rivers has altered these global-scale comparisons. In Indonesia and tropical regions, sediment yield has increase by 45% and 24% respectively, largely due to deforestation. Data from Syvitski et al. (2005), Milliman and Meade (1983), Milliman and Syvitski (1992) and Summerfield and Hulton (1994).

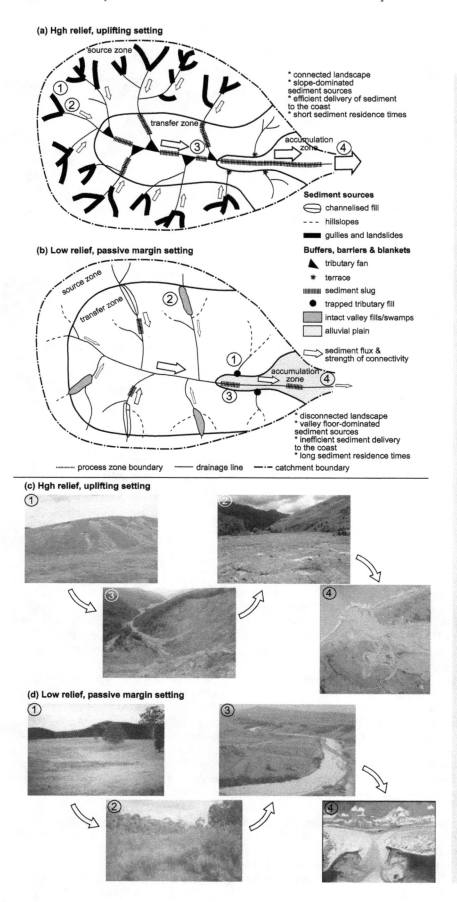

Figure 14.13 Generalised patterns of (dis)connectivity in sediment flux in high-relief, uplifting terrains versus low-relief, passive-margin settings. (a, c) Catchments in high-relief, uplifting terrains such as New Zealand tend to have high drainage densities. Steep headwater (sediment source) zones are transitional to very efficient transfer zones and expansive lowland plains. Within the source zones, significant volumes of material are supplied from hillslope gully complexes and landslides. These materials are transferred as sediment slugs along the tributary and trunk stream network. Along the lowland plains, sediments are effectively evacuated from the catchment. Overall, the degree of catchment connectivity is high and sediment delivery to the river mouth is high. The timeframe of geomorphic recovery is relatively short, as large parts of catchments are recurrently connected. (b, d) Catchments in low-relief, passive-margin settings such as Australia tend to have low drainage densities and subdued topography. Significant sediment stores buffer and disconnect sediment conveyance through the system. Recent sediment flux is dominated by reworking of materials from valley floor stores. Overall, there is significant disconnectivity and sediment delivery ratios at the basin mouth are low. At the continental scale, only 20% of the total load to Australian rivers is exported to the coast. Around 90% of the river sediment load is generated from only 20% of the contributing catchment area; that is, effective catchment areas are small. Once disturbed, timeframes for recovery may be very long (i.e. centuries or millennia) because of sediment exhaustion. Modified from Catena, 70(1), Fryirs, K., Brierley, G., Preston, N. and Kasai, M., Buffers, barriers and blankets: The (dis)connectivity of catchment-scale sediment cascades, 49–67, © 2007, with permission from Elsevier.

Box 14.1   Bega catchment, NSW, Australia

Types of rivers

This escarpment-dominated granitic landscape has numerous discontinuous watercourses at the base of the escarpment, extensive lengths of partly confined valleys and a short alluvial river heading into a bottleneck reach adjacent to the coast.

Nature of human disturbance

The catchment was largely cleared for agriculture from the mid-nineteenth century. The Bega Valley has one of Australia's largest dairy industries.

Capacity for adjustment of rivers

Because of bedrock control, confined and partly confined valleys have limited capacity to adjust and are resilient to change. Most discontinuous watercourses have become continuously channelised, as they were highly sensitive to human disturbance. The lowland plain river has adjusted significantly through channel expansion and storage of a sediment slug.

Patterns of rivers

Rivers that drain from the escarpment have continuous channels throughout. Those that drain directly from the escarpment once had intact valley fills at the base of the escarpment that have subsequently become incised.

Connectivity and response gradients

Hillslopes and channels are disconnected and the majority of sediment supply results from reworking of valley floor materials. Some systems have always been connected, but where valley fills have been incised, connectivity has been markedly enhanced (i.e. buffers have been removed). The mid-catchment bedrock reaches have acted as boosters that have efficiently delivered sediment to the lowland plain where it remains trapped as a sediment slug. Sediment delivery to the coast has been inefficient, as large sediment stores have been added to the long-term sediment sink along the lowland plain. The upper catchment is largely exhausted of sediment. The process zone distribution of the catchment has been altered.

Evolutionary trajectories

Most of the bedrock-controlled rivers function as they have throughout the Holocene. Intact valley fills remain as conservation priorities. The prospects for recovery are limited in channelised fills, as sediment supply is limited. Recovery along the lowland plain is limited, as the sediment slug smothers this reach.

Summarised from:

Brooks and Brierley (1997), Fryirs and Brierley (1998, 1999, 2000, 2001), Brierley and Fryirs (1998, 1999, 2000, 2005, 2009), Brierley *et al.* (1999, 2002) and Fryirs (2002, 2003).

---

different catchments. Given that Bega catchment is now sediment-supply limited, river recovery potential is low and will take many hundreds of years. In Twin Streams catchment, significant sediment regeneration and connectivity results in high recovery potential. In the Hunter catchment, sediment disconnectivity is high, but rivers have not been profoundly altered and remain relatively resilient to change. As such, they have high recovery potential.

Analysis of the configuration and dynamics of each catchment provides significant insight into how river systems respond to disturbance. Such catchment-specific information is valuable for forecasting future evolutionary scenarios and management applications. Landscapes are emergent and contingent. Each catchment has its own set of process–form linkages and responds to disturbance in its own way.

Tips for reading the landscape to interpret catchment-scale sediment flux

Step 1. Identify individual landforms and their sediment flux relationships at the local scale

Over time, rivers convey sediment from their headwaters to the basin outlet. However, there is significant variability in the way that sediment erosion, transport and deposition occur at different positions in catchments. Local-scale variability in sediment flux is analysed by determining answers to the following questions:

- What types of sediment storage units occur along a reach (i.e. geomorphic units)?
- What geomorphic units are stores and which are sinks?

Box 14.2   Upper Hunter catchment, NSW, Australia

Types of rivers

This catchment has different geologies either side of the Hunter–Mooki fault. On the east, rivers are dominantly confined and partly confined valleys. On the western side, accommodation space allows laterally unconfined rivers to form. A number of discontinuous watercourses once occurred. A unique river type, an entrenched tortuous meandering fine-grained river, flows within this low-lying area.

Nature of human disturbance

The catchment was largely cleared for agriculture from the mid-nineteenth century. Significant river works were installed from the 1950s onwards. Wine growing, horse studs and mining are the dominant land uses today.

Capacity for adjustment of rivers

Rivers in confined and partly confined valleys have limited capacity to adjust and are considered resilient. Rivers in the laterally unconfined valley setting are underfit and have limited capacity to adjust given the fine-grained nature of the materials that make up the banks. These rivers are also considered resilient. Marked local-scale variability in adjustment and sensitivity to disturbance along the trunk stream reflects the inherited sinuosity induced by terrace confinement.

Patterns of rivers

In the east, confined and partly confined rivers extend along the majority of the river length. In the west, rivers have very short confined headwaters and extensive sections of laterally unconfined valleys.

Connectivity and response gradients

On the eastern side, rivers are highly connected and efficient conveyors of sediment (except along the upper Hunter trunk stream, where Glenbawn Dam disconnects sediment transfer). On the western side, significant buffers are present, disconnecting hillslopes from the channel network. However, incision of discontinuous watercourses has increased longitudinal connectivity over time.

Evolutionary trajectories

Because adjustments have been localised and changes to landscape connectivity have been minor, overall geomorphic adjustments have been limited. No areas of pronounced channel change have been detected and major off-site impacts have not occurred. Hence, there is reasonable potential for geomorphic recovery.

Summarised from:

Hoyle et al. (2008), Fryirs *et al.* (2007a,b, 2009), Fryirs and Brierley (2010), Brierley *et al.* (2010), Spink *et al.* (2009, 2010) and Jain *et al.* (2008).

Box 14.3   Twin Streams catchment, Auckland, New Zealand

Types of rivers

This catchment contains short and relatively steep confined and partly confined streams that drain from the flanks of a dissected volcano. Many lower order tributaries contain wetlands and swamps. Other sediment stores are restricted to a short alluvial section adjacent to the estuary.

Nature of human disturbance

This peri-urban catchment has experienced phases of forest clearance and agricultural development over the last 150 yr.

Capacity for adjustment of rivers

There is very limited capacity for adjustment in confined and partly confined valleys that dominate the catchment. Along more alluvial reaches, banks are cohesive, given the fine-grained nature of the sediments. Discontinuous watercourses in tributaries are sensitive to adjustment once disturbed.

Patterns of rivers

The concave-up longitudinal profile results in a pattern of steep headwaters with a rapid transition to partly con-fined valleys and a very short lowland plain/estuary. Many of the smaller tributaries contain discontinuous watercourses downstream of steep headwaters.

Connectivity and response gradients

Other than the discontinuous watercourses along tributaries, these are highly connected systems. High hillslope–channel coupling occurs across the majority of the catchment, especially in headwater areas.

Evolutionary trajectories

Large volumes of sediment were released from headwaters after forest (kauri) clearance. This resulted in stream incision and local expansion of channels. Given high connectivity, the vast majority of sediments were flushed through to the coast. Sediments have subsequently been dispersed within Waitemata harbour. River recovery potential is high because of the bedrock-controlled character of the streams. Protection of remaining intact discontinuous watercourses is a high conservation priority.

Summarised from:

Gregory *et al.* (2008) and Reid *et al.* (2008, 2010).

- What are the materials properties of these units?
- Are the units vegetated or not? Does this provide insight into timeframes of reworking?
- What processes form and rework these materials, and over what timeframes?
- Is there any evidence that the pattern of stores and sinks has changed over time (whether in response to natural factors or human disturbance)?

Step 2. Interpret sediment flux at the reach scale

Some reaches store sediments for various periods of time, some transfer sediment such that inputs and outputs are roughly balanced, while others evacuate sediments. Reach-scale variability in sediment flux is analysed by determining patterns of sediment storage and assessing their ease/frequency of reworking. Key questions to address include:

- Is this a source, transfer or accumulation zone?
- If alluvial, is this a bedload, mixed-load or suspended-load river?
- Which units along the reach act as short-term storage features (stores) relative to long-term storage features (sinks)?
- What is the balance of erosion and deposition processes, and the behavioural regime, of the reach? Is there any evidence of aggradation (sand sheets/slugs/waves) or degradation (headcuts or bed erosion), suggesting sediment accumulation or erosion respectively?
- What is the sediment input/output ratio for the reach under investigation?
- How frequently are different sediment (grain-size) fractions moved on differing surfaces?

Step 3. Explain controls and impacts on sediment flux at the reach scale

To explain how various reaches operate requires an understanding of how the river responds to the prevailing sediment flux. This is directly linked to how the river behaves and the associated pattern of erosional and depositional processes that form geomorphic units along a river. Some questions to ask include:

- How do prevailing slope and discharge conditions affect flow energy in the reach, the erosional–depositional balance, patterns of sediment stores/sinks and their ease of reworking and the behavioural regime of the river?
- Is there any evidence for river change, whether natural or human induced?

- How does the sediment regime of the reach respond to a range of flow conditions? Is sediment available to be moved?
- Is the reach sediment-supply or -transport limited?
- How readily and frequently are sediment stores and sinks reworked? What is their residence time? Are sediment stores and sinks redistributed and/or replenished over time?
- How effectively does the channel transport the sediment available to it? What is the relationship between bed material size and volume relative to available energy?
- How have disturbance events (e.g. flood history, vegetation associations, drought impacts) affected sediment mobility?
- How have human activities altered the availability and mobility of sediments?

Step 4. Explain how catchment-scale relationships affect sediment flux

Sediment budgets document and quantify sediment source, transfer and accumulation relationships (i.e. source-to-sink relationships) over a given timeframe. These catchment-specific relationships reflect landscape setting, connectivity and system responses to human disturbance. Analysis of catchment-scale sediment budgets entails determination of process domains – patterns of source, transfer and accumulation zones along the trunk stream and primary tributaries:

- Where are sediments eroded from? Are the primary sources colluvial or alluvial valley floor sediments? What is the volume of sediment being eroded?
- What types of sediment stores and sinks occur in the catchment? Where are sediment stores and sinks located in the catchment and what are their volumes?
- How often are sediment stores and sinks reworked (i.e. what are their residence times)? If reworked, how far are they transported?
- Where are sediments accumulating in the catchment? Are these accumulations stores or sinks?
- How readily are sediment sources replenished? Is exhaustion of sediment sources an issue? How have flow and sediment availability, and the relationship between them, changed over time?

Reach-scale behaviour must be appraised in relation to what is happening upstream or downstream, framing each reach in its catchment context to determine how sediment flux will impact on future evolutionary trajectories and where sediments will come from to aid river recovery. This entails analysis of the response gradient, which assesses how reaches have responded to disturbance (i.e. their capacity for adjustment, sensitivity and the nature/frequency of disturbance events) and the ways in which reaches interact at the catchment scale (i.e. how geomorphic responses in any given reach are propagated through the catchment, impacting upon prevailing flux boundary conditions at any given place/time). Critically, changes to these relationships over time must be considered. These relationships are fashioned by geologic, climatic and anthropogenic controls on sediment flux. Interactions among these controls determine internal sediment dynamics and sediment yield at the reach and catchment scales. Typical questions to address include:

- How does topographic setting and lithology affect sediment availability and calibre? How does the discharge regime affect the transfer of sediment through the catchment?
- What is the pattern of energy that is available to erode, transport and deposit sediments along the river course?
- How do catchment shape and size, and associated longitudinal and lateral linkages, vary from subcatchment to subcatchment? Which tributaries are geomorphically significant and why?
- Has the pattern of process zones changed over time? Is there evidence for changes to the pattern of sediment source, transfer and accumulation zones over time? Which geomorphic unit interpretations support these assertions? Over what timeframes have adjustments occurred, and why?
- Has available accommodation space, erodibility/erosivity, drainage pattern/density and connectivity/disconnectivity of the catchment changed over time? Are sediments deposited under past climates still stored (or activated) in the system (e.g. paraglacial sediments, legacy sediments)?
- Is the catchment coupled or decoupled? Are hillslopes connected to channels? Are tributaries and trunk streams connected? Have dams disconnected reaches? Are channels and floodplains connected? What blockages affect the sediment cascade of the catchment? What types of buffers, barriers and blankets occur and where are they distributed in the catchment?
- What controls the nature, distribution and effectiveness of these blockages? Under what flow conditions are these blockages removed or formed? Over what timeframes and under what types of flow conditions are switches turned on and off? What are the effective catchment area and effective timescales of sediment conveyance in the catchment?
- How have the pattern, functionality and effectiveness of buffers, barriers and blankets changed over time?

- How have human activities altered sediment flux in the catchment? How have human activities altered the degree of connectivity of sediment flux in the catchment?
- What lagged and off-site responses are evident, and over what timeframe? Is there any evidence that disturbance responses in one reach have breached threshold conditions elsewhere in the catchment?
- How have alterations to sediment flux affected the balance of erosion and deposition and resulting adjustments to channel geometry, assemblages of instream geomorphic units, and process–form relationships on floodplains throughout the catchment?
- Based on assessment of the evolutionary trajectory, what future states of adjustment are envisaged?

Conclusion

Sediment budgets are an integral part of geomorphic investigations into sediment movement through landscapes. Identification and quantification of sediment availability within a catchment (i.e. the calibre and volume of sediments in differing stores, and their accessibility/ease of reworking), alongside conceptualisations of landscape connectivity (i.e. the efficiency of sediment conveyance through landscapes), are key considerations in efforts to forecast sediment flux. Recent conceptual and technological advances in sediment budgeting aid interpretations of landform development and responses to anthropogenic disturbance. Analysis of sediment budgets provides critical insights into how landscapes look and operate, aiding selection and prioritisation of management applications.

Key messages from this chapter

- River systems act as conveyor belts that move sediment from source zones through transfer zones to accumulation (sink) zones. Sediment budgets provide an account of sediment movement over a given timeframe. Sediment yield refers to the quantity of sediment that reaches a basin outlet. It is often measured as the sediment delivery ratio, which records the proportion of sediment eroded from a catchment that reaches the basin outlet. Sediment flow diagrams summarise source-to-sink relationships by quantifying sediment sources, transport, stores/sinks and yield.
- Sediment sources can be colluvial (on hillslopes) or alluvial (on the valley floor). Sediment transport is an episodic process. Sediment storage occurs where accommodation space is made available for sediment to be deposited. Residence time refers to the amount of time

sediment remains in storage. Sediment stores are transient landforms that hold sediment for short periods of time (e.g. bar storage over months or years). Sediment sinks are more persistent landforms that hold sediment for extended periods of time (e.g. floodplain and terrace sinks over hundreds or thousands of years).

- Catchment-scale sediment flux can be visualised as a series of cogs that interact across a range of spatial and temporal scales. Controls on sediment flux include landscape setting and associated geologic and climatic controls, the degree of landscape connectivity, reach adjustments and roughness elements.
- Connectivity is defined as the transfer of energy and matter between two landscape compartments. Blockages (called buffers, barriers and blankets) can disrupt the sediment cascade. The area that actively contributes sediment to the conveyor belt (i.e. is connected) is called the effective catchment area. Coupled landscapes have few blockages and high effective catchment areas. Decoupled or disconnected landscapes have significant blockages and small effective catchment areas.
- Sediment movement is not uniform over time. The magnitude and frequency of events that move sediment on hillslopes, floodplains and within-channel features vary markedly.
- Effective timescales relate to the timeframe over which the catchment is connected and active sediment transport occurs.

- The operation of catchment sediment cascades can be considered as a series of switches that turn off and on over various timeframes, connecting or disconnecting various areas from the sediment conveyor belt.
- The degree to which a catchment is connected or disconnected determines how off-site impacts of disturbance are manifest through a catchment. Highly disconnected systems tend to absorb or suppress change, while cumulative impacts are readily manifest and change is efficiently propagated through highly connected landscapes.
- There is significant variability in the sediment yield of the world's largest rivers. The pattern is largely controlled by tectonic setting, whereby steep, tectonically active rivers have significantly higher sediment yields than large rivers that drain low-lying plate-centre locations.
- Sediment generation and landscape connectivity vary markedly in different tectonic settings and under the influence of various forms of human disturbance (especially land use change and flow regulation).
- Each reach must be placed in its catchment context to determine how it will adjust, change and respond to disturbance (i.e. lagged and off-site responses).

CHAPTER FIFTEEN

The usefulness of river geomorphology: reading the landscape in practice

Introduction

Once triggered, a quest for geomorphic endeavours is never fully sated. There are always more adventures to be had, experiences to be gained and landscape puzzles to be unravelled. Although perspectives change and grow, and the repository that is a geomorphologist's mind has an increasing range of experiences to draw upon, there are always new lessons to be learnt. An open-ended spirit of enquiry is required to learn from landscapes themselves. Indeed, landscapes are a source of inspiration. Analysis is never boring, as no two situations are ever exactly the same. This may prove to be challenging, but it is also great fun! This spirit of enquiry and endeavour underpins the perspective from which this book has been written.

Understanding of landscape processes, forms and evolution provides a critical template with which to frame a host of management applications. If the geomorphic structure and function of a landscape changes, so does everything else! For example, habitat availability/viability and measures of ecosystem functionality are altered, with a myriad of direct and indirect implications for biodiversity values. Landscape considerations, among many factors, influence water quality and turbidity. Critically, this is a dynamic, interlinked template, wherein changes to one part of a landscape impact elsewhere within that system.

Most rehabilitation activities manipulate the physical structure of a river (its geomorphology) in attempts to enhance ecological values, improve water quality and attain a particular aesthetic. Management efforts are unlikely to achieve their intended outcomes unless they build upon an appropriate understanding of the geomorphic structure, function and evolutionary trajectory of the system under investigation, framing responses to human disturbance in relation to natural variability. Such understandings respect the inherent diversity, variability and complexity of any given river system, promoting programmes that 'work with nature' by describing and explaining the contemporary physical state of the system and identifying causes of adjustment. Placing site/reach information in a catchment context allows assessment of off-site limiting factors and pressures on future trajectories of change and recovery potential. Such efforts build upon the principles outlined in this book. This entails reading the landscape and seeing and conceptualising river systems as dynamic wholes rather than static collections of parts. Cross-scalar, system-specific insights are required to frame coherent management applications that address environmental values in relation to human needs, and associated concerns for resource development, risk/hazard management, infrastructure protection (geotechnical engineering), etc.

Geomorphological processes determine the structure of a river system, providing an *integrative physical template* with which to assess habitat associations and *linkages of biophysical processes* in landscapes (Chapter 1). For example, changes to the geomorphic character and behaviour of rivers influence the availability of habitat for flora and fauna. Secondary adjustments to biotic and chemical interactions impact upon the thermal regime of a river, the production, processing and retention of nutrients and organic matter and their role in food-web processes. Other measures of aquatic ecosystem functioning may also be affected, such as water quality, pH, etc. For example, flow–sediment interactions are primary determinants of the distribution and retention of coarse particulate organic matter along rivers. Altered hydraulic conditions may change the stream's species assemblage and associated predator–prey relationships. Rather than consider elements from ecology, geomorphology, hydrology or aquatic geochemistry in isolation, *integrative river science* builds upon holistic cross-disciplinary analyses of river systems. The emergence of notions such as riverscape, ecohydrology, ecohydraulics and geodiversity is testimony to the adoption of more holistic approaches to enquiry. Effective approaches to river management address concerns for the key drivers and relationships that determine the integrity and functionality of any given system. In framing management initiatives to maintain ecosystem integrity, measures must target those elements of *ecological resilience* that are vulnerable or under stress, striving to enhance the self-sustaining capacity of the system.

The critical role of geomorphic enquiry in informing cross-disciplinary applications builds upon catchment-specific analyses of landscape diversity. These understandings must be framed in relation to broader contextual principles and theories. Reading the landscape and field-based investigations are critical tools with which to guide these interpretations. Many other complementary skills underpin useful geomorphology, especially spatial analysis and relationship skills derived from mapping, remote sensing and GIS applications, modelling skills, and training in related disciplines such as geology, hydrology, ecology, chemistry, engineering, Quaternary science, etc. Synthesis is fundamental, and teamwork is essential in most applications. No one can do everything! Even greater importance must be given to genuine collaboration with the users of geomorphic understandings. Working hand in hand with environmental managers, decision-makers, stakeholders, social scientists and politicians is fundamental to truly informed applications. Increasingly, economic cases for environmental applications recognise the importance of 'working with nature' as a basis for cost-effective practices. Working with the business sector will play an ever-increasing role in the emergence of 'useful geomorphology'.

Authentic, precautionary approaches to river management seek to protect and enhance the values of a given *place*. Biophysical and human relationships that shape the way rehabilitation and management are undertaken, and associated institutional arrangements, must be framed in the context of what is realistically achievable in any given catchment. Place-based values are the cornerstone of environmental protection. The inherent diversity, variability and complexity of landscapes emphasise concerns for the distinctive and the unique, alongside normal, typical or representative sites. These attributes and associations can only be captured through system-specific understandings.

In drawing this book to a close, emphasis is placed upon three primary principles that guide the application of geomorphic understandings of river systems:

1. Respect diversity by reading the landscape.
2. Appreciate system dynamics and evolution by reading the landscape.
3. Bring together spatial and temporal considerations to read the landscape in efforts to know your catchment.

Respect diversity

The scale at which efforts to read the landscape are undertaken determines what is seen. A head bowed over a gravel bar measuring the sizes of pebbles provides a very different perspective to that derived by viewing the landscape from the highest local point, or an aerial overview.

A reliable information base is required to determine the diversity of a given system and to identify the unique, rare and characteristic attributes of that system. Analyses of diversity require the adoption of an open-ended approach to enquiry. Questions about the types of rivers found in any given place must sit at the core of this approach. River diversity reflects a continuum of environmental conditions. There is no magic number of variants of river types. Recognising why certain rivers are found where they are, and understanding controls on their behaviour, are critical attributes in efforts to manage river systems.

Key principles to remember in respecting river diversity include:

- Gather and use information that is appropriate for the system under consideration.
- Recognise that there is significant diversity in river forms and processes at a range of spatial and temporal scales.
- Do not place undue emphasis on what the river looks like at the expense of interpreting its range of behaviour and adjustment.
- Recognise unique and distinctive attributes as well as characteristic forms and processes in the continuum of river diversity.

Understand system dynamics and evolution

Efforts to 'work with nature' are a key consideration in the design and implementation of sustainable and cost-effective river rehabilitation measures. Appropriate understanding of system dynamics and adjustment under prevailing flow and sediment fluxes is required to inform this process. As differing types of rivers adjust in different ways, appropriate management actions are framed in relation to the behavioural regime that is expected for any given type of river, rather than striving to 'impose' a particular structure and function. The emergence of 'erodible corridor', 'space to move' and 'channel migration zone' programmes reflects the recognition by management agencies of benefits gained by allowing channels to self-adjust, rather than trying to 'fight' river behaviour.

Meaningful differentiation can be made between river behaviour and river change. Importantly, guiding images for rehabilitation programmes should reflect the natural range of behaviour of that type of river, rather than a notional fixed state. Given their variable capacity for adjustment, some reaches are inherently more sensitive to physical disturbance than others. Identifying where and under what conditions river change may (or may not) occur is critical for determining where conservation or rehabilitation strategies should be implemented.

Realistic river rehabilitation plans build upon insights into the functionality of the system. The key issue here is whether the reach still behaves and adjusts today as it did in the past (i.e. adjustments lies within the natural behavioural regime for that type of river), or whether fundamental change to a different type of river, with a differing behavioural regime, has taken place. Understanding *causes* of system adjustment is required to determine the type and level of intervention that is needed to rehabilitate river systems. Framing contemporary processes in their evolutionary context helps in making these interpretations.

Key principles to remember when developing an understanding of system dynamics and evolution include:

- Recognise that different components of rivers adjust over geologic, geomorphic and engineering timescales.
- Interpret how rivers adjust at low flow, bankfull and overbank stage to define the natural range of variability.
- Recognise that just as the range of behaviour varies for different types of rivers, so too does the sensitivity to adjustment and change. Some rivers are subjected to threshold-driven adjustments, while others gradually adjust and may be considered resilient to change over longer timeframes.
- Frame interpretations of the contemporary behavioural regime in their evolutionary context, determining how rivers have changed over time.
- Relate river responses to human disturbance to the natural range of variability, determining the extent to which river behaviour has been altered, or change induced.
- Recognise that river responses to human disturbance vary from place to place. In some instances, human disturbance increases the range of system behaviour. In other instances, human disturbance decreases (suppresses) the range of system behaviour. The key consideration here is whether river responses to human disturbance have induced irreversible change in river type.
- Determine to what degree geologic, climatic and anthropogenic conditions have left an imprint on the system and how this affects contemporary river behaviour.
- Recognise that historical imprints, off-site (lagged) disturbance responses and complex responses (amongst other factors) ensure that rivers seldom operate as simple cause-and-effect systems and that system-specific responses determine the range of potential future pathway(s) of adjustment.

Know your catchment

Each catchment has its own history and its own boundary conditions and is subject to a system-specific set of contin-

gencies and disturbance events that shape contemporary forms and processes. Therefore, understanding catchment-specific patterns of differing river types, their connectivity and their adjustment in response to human disturbance are key considerations in the development of place-based approaches to river management. Unravelling system-specific evolutionary histories provides critical insight with which to guide interpretation of contemporary river adjustments and prospective future trajectories of adjustment. Catchment-scale analyses are required to identify potential off-site impacts, lag times and complex responses. This requires understanding of reach-scale sensitivity to adjustment and threshold responses, and the way in which these responses are mediated through catchments (i.e. response gradients).

Non-synchroneity in the timing, pattern and rates of geomorphic responses to differing forms of disturbance, and the associated river changes reflect the character and configuration of each river system. Given the catchment specific patterns of river types and their linkages, predictions of reach- and catchment-scale responses to disturbance events, and the recovery potential of the system, must be considered in the context of river history and an understanding of spatial linkages of physical processes in the catchment of concern. Downstream patterns of rivers, and associated measures of landscape (dis)connectivity, determine the degree to which disturbance in one part of a system will be expressed or absorbed elsewhere, and the timeframe over which this occurs.

As such, rehabilitation planning is a catchment-specific exercise that builds upon a body of knowledge on system character and behaviour at a range of spatial and temporal scales. Catchment-scale investigations aid efforts to identify and treat the causes, rather than the symptoms, of degradational processes, determining whether degradational influences are site specific or reflect off-site impacts induced by disturbance events elsewhere in the catchment. Meaningful management applications frame analyses of 'what is expected' at the reach scale (i.e. the range of behaviour for that type of river) in relation to the downstream pattern of reaches.

Key principles to remember when getting to know your catchment include:

- A hierarchical approach to analysis of river forms and function aids efforts to know your catchment. How does the system fit together? A building-block approach to analysis of rivers provides an integrated set of knowledge from site to reach to catchment scales. Understanding forces and resistance elements and bed material transport at fine scales helps explain how channels adjust and how instream and floodplain geomorphic units are formed and reworked. Understanding of process–form associations of geomorphic units at the reach scale allows interpretation of channel planform, river behaviour and river evolution. Piecing together the patterns of rivers at the catchment scale, and analysis of system responses to geologic, climatic and anthropogenic controls allows interpretation of sediment flux. Such insights are required to explain the present-day condition of a river, to interpret how alterations to one part of a system may affect other parts (lagged and off-site responses to disturbance), and to predict likely future river adjustments. This provides a basis to ensure that management actions address the underlying causes rather than the symptoms of change.

- Know what types of river you have, how they behave and how they have evolved.
- Place each reach in its catchment context. Identify the downstream pattern of rivers and why they form where they do.
- Determine connectivity relationships that fashion the sediment cascade. Appraise the effect of (dis)connectivity on the manifestation of off-site and/or cumulative responses to disturbance, recognising that there may be considerable time lags of adjustment.
- Forecast how catchment boundary conditions will affect river recovery potential and future trajectories of adjustment.
- Recognise that it is imperative to frame management actions in relation to the character and behaviour of any given reach, appropriately placing contemporary

attributes in their spatial (within-catchment) and temporal (evolutionary) context.

Closing comment: how the book should be used

Despite the uniformity of the underlying physics that shapes river behaviour, river systems demonstrate a remarkable array of biophysical interactions and evolutionary trajectories. Complex arrays of processes and forms are largely the result of system-inherent, dynamic genesis and development. Researchers and practitioners have developed a sophisticated understanding of the primary controls upon this diversity, variability and complexity. Each catchment has its own configuration and history of disturbance events. Inevitably, it is one thing to have these insights, but quite another to consider how this understanding is used!

Effective approaches to reading the landscape build upon careful observation, measurement, interrogation, interpretation and monitoring in efforts to *describe* and *explain* what is happening and why. The heart and soul of this process is captured by the expression, 'thinking like an ecosystem'. When used appropriately, these understandings provide a platform to *predict* environmental futures, thereby providing a proactive platform for management applications.

References

Aslan, A., Autin, W.J. and Blum, M.D. (2006) Causes of river avulsion: Insights from the Late Holocene avulsion history of the Mississippi River, USA. *Journal of Sedimentary Research* 75, 650–664.

Asselman, N.E.M. (1999) Suspended sediment dynamics in a large drainage basin: the River Rhine. *Hydrological Processes* 13, 1437–1450.

Baker, V.R. (1978) Palaeoflood hydrology and extraordinary flood events. *Journal of Hydrology* 96, 79–99.

Benda, L., Andras. K. and Miller, D. (2004) Confluence effects in rivers: interactions of basin scale, network geometry and disturbance regimes. *Water Resources Research* 40, W05402. DOI: 10.1029/2003WR002583.

Bertoldi, W., Gurnell, A.M., Surian, N., Tockner, K., Zanoni, L., Ziliani, L. and Zolezzi, G. (2003) Understanding references processes: Linkages between river flows, sediment dynamics and vegetated landforms along Tagliamento River, Italy. *River Research and Applications*, 25, 501–516.

Beven, K.J. (2001) *Rainfall–Runoff Modelling: The Primer.* John Wiley and Sons, Ltd, Chichester.

Bravard, J.P., Landon, N., Peiry, J.L. and Piégay, H. (1999) Principles of engineering geomorphology for managing channel erosion and bedload transport, examples from French rivers. *Geomorphology* 31, 291–311.

Brice, J.C. (1960) Index for description of channel braiding. *Bulletin of the Geological Society of America* 71, 1833.

Brierley, G.J. (1996) Channel morphology and element assemblages: a constructivist approach to facies modelling. In Carling, P. and Dawson, M. (eds) *Advances in Fluvial Dynamics and Stratigraphy.* Wiley Interscience, Chichester, pp. 263–298.

Brierley G.J. and Fryirs, K. (1998) A fluvial sediment budget for upper Wolumla Creek, South Coast, New South Wales, Australia. *Australian Geographer* 29(1), 107–124.

Brierley G.J. and Fryirs, K. (1999) Tributary–trunk stream relations in a cut-and-fill landscape: a case study from Wolumla catchment, N.S.W., Australia. *Geomorphology* 28, 61–73.

Brierley G.J. and Fryirs, K. (2000) River Styles in Bega catchment, NSW, Australia: implications for river rehabilitation. *Environmental Management* 25(6), 661–679.

Brierley, G.J. and Fryirs, K.A. (2005) *Geomorphology and River Management: Applications of the River Styles Framework.* Blackwell Publications, Oxford, UK.

Brierley, G.J. and Fryirs, K.A. (eds) (2008) *River Futures: An Integrative Scientific Approach to River Repair.* Island Press, Washington, DC.

Brierley, G.J. and Fryirs, K. (2009) Don't fight the site: geomorphic considerations in catchment-scale river rehabilitation planning. *Environmental Management* 43(6), 1201–1218.

Brierley, G.J., Cohen, T., Fryirs, K. and Brooks, A. (1999) Post-European changes to the fluvial geomorphology of Bega catchment, Australia: implications for river ecology. *Freshwater Biology* 41, 839–848.

Brierley, G., Fryirs, K., Outhet, D. and Massey, C. (2002) Application of the River Styles framework as a basis for river management in New South Wales, Australia. *Applied Geography* 22, 91–122.

Brierley, G.J., Reid, H., Fryirs, K. and Trahan, N. (2010) What are we monitoring and why? Using geomorphic principles to frame eco-hydrological assessments of river condition. *Science of the Total Environment* 408, 2025–2033.

Brooks, A.P. and Brierley, G.J. (1997) Geomorphic responses of lower Bega River to catchment disturbance, 1851–1926. *Geomorphology* 18, 291–304.

Brooks, A.P., Brierley, G.J. and Millar, R.G. (2003) The long-term control of vegetation and woody debris on channel and floodplain evolution: insights from a paired catchment study in southeastern Australia. *Geomorphology* 51, 7–29.

Burt, T.P. (1996) The hydrology of headwater catchments. In: Petts, G. and Calow, P. (eds) *River Flows and Channel Forms.* Blackwell Publishing, Oxford, pp. 6–31.

Chang, H.H. (1988) *Fluvial Processes in River Engineering.* John Wiley and Sons, Inc., New York.

Chorley, R.J., Schumm, S.A. and Sugden, D.E. (1984) *Geomorphology.* Methuen and Co. Ltd., New York.

Chow V.T. (1959) *Open Channel Hydraulics.* McGraw Hill, New York.

Church, M. (1992) Channel morphology and typology. In: Callow, P. and Petts, G.E. (eds) *The Rivers Handbook.* Blackwell, Oxford, UK, pp. 126–143.

Cohen, T.J. and Brierley, G.J. (1997) Channel instability in a forested catchment: a case study from Jones Creek, East Gippsland, Australia. *Geomorphology* 32(1–2), 109–128.

Corning, R.V. (1975) Channelisation: shortcut to nowhere. *Virginia Wildlife* (February), 6–8.

Costa, J.E. and O'Connor, J.E. (1995) Geomorphically effective floods. In: Costa, J.E., Miller, A.J., Potter, K.W. and Wilcock, P.R. (eds) *Natural and Anthropogenic Influences in Fluvial Geomorphology*, Geophysical Monograph 89. American Geophysical Union, Washington, DC, pp. 45–56.

Cundari, A. and Ollier, C.D. (1970) Inverted relief due to lava flows along valleys. *Australian Geographer* 11, 291–293.

Delinger, R.P. and O'Connell, D.R.H. (2010) Simulations of cataclysmic outburst floods from Pleistocene Glacial Lake Missoula. *Geological Society of America Bulletin* 122, 678–689.

Dunne, T. and Leopold, L.B. (1978) *Water in Environmental Planning*. W.H. Freeman, New York.

Eaton, B.C., Millar, R.G. and Davidson, S. (2010) Channel patterns: braided, anabranching, and single-thread. *Geomorphology* 120, 353–364.

Ferguson, R.J. and Brierley, G.J. (1999a) Downstream changes in valley confinement as a control on floodplain morphology, lower Tuross River, New South Wales: A constructivist approach to floodplain analysis. In: Miller, A.J. and Gupta, A. (eds) *Varieties of Fluvial Form*. John Wiley & Sons, Ltd, Chichester, pp. 377–407.

Ferguson, R.J. and Brierley, G.J. (1999b) Levee morphology and sedimentology along lower Tuross River, south-eastern Australia. *Sedimentology* 46, 627–648.

Fryirs, K. (2002) Antecedent landscape controls on river character, behaviour and evolution at the base of the escarpment in Bega catchment, South Coast, New South Wales, Australia. *Zeitshrift für Geomorphologie* 46(4), 475–504.

Finlayson, B.L. and McMahon, T.A. (1988) Australia vs the world: a comparative analysis of streamflow characteristics. In: Warner, R.J. (ed.) *Fluvial Geomorphology of Australia*. Academic Press, Sydney, pp. 17–40.

Fryirs, K. (2003) Guiding principles for assessing geomorphic river condition: application of a framework in the Bega catchment, South Coast, New South Wales, Australia. *Catena* 53, 17–52.

Fryirs, K. and Brierley, G.J. (1998) The character and age structure of valley fills in upper Wolumla Creek catchment, South Coast, New South Wales, Australia. *Earth Surface Processes and Landforms* 23, 271–287.

Fryirs, K. and Brierley, G.J. (1999) Slope channel decoupling in Wolumla catchment, South Coast, New South Wales, Australia: the changing nature of sediment sources since European settlement. *Catena* 35, 41–63.

Fryirs, K. and Brierley, G.J. (2000) A geomorphic approach for the identification of river recovery potential. *Physical Geography* 21(3), 244–277.

Fryirs, K. and Brierley, G.J. (2001) Variability in sediment delivery and storage along river courses in Bega catchment, NSW, Australia: implications for geomorphic river recovery. *Geomorphology* 38, 237–265.

Fryirs, K. and Brierley, G.J. (2010) Antecedent controls on river character and behaviour in partly-confined valley settings: upper Hunter catchment, NSW, Australia. *Geomorphology* 117, 106–120.

Fryirs, K., Brierley, G.J., Preston, N.J. and Kasai, M. (2007a) Buffers, barriers and blankets: the (dis)connectivity of catchment-scale sediment cascades. *Catena* 70, 49–67.

Fryirs, K., Brierley, G.J., Preston, N.J. and Spencer, J. (2007b) Catchment-scale (dis)connectivity in sediment flux in the upper Hunter catchment, New South Wales, Australia. *Geomorphology* 84, 297–316.

Fryirs, K., Spink, A. and Brierley, G. (2009) Post-European settlement response gradients of river sensitivity and recovery across the upper Hunter catchment, Australia. *Earth Surface Processes and Landforms* 34, 897–918.

Gilvear, D., Winterbottom, S. and Sichingabula, H. (2000) Character of channel planform change and meander development: Luangwa River, Zambia. *Earth Surface Processes and Landforms* 25, 421–436.

Gomez, B., Naff, R.L. and Hubbell, D.W. (1989) Temporal variations in bedload transport rates associated with the migration of bedforms. *Earth Surface Processes and Landforms* 14, 135–156.

Gregory, C., Reid, H. and Brierley, G. (2008) River Recovery in an urban catchment: Twin Streams catchment, Auckland, New Zealand. *Physical Geography* 29(3), 222–246.

Gurnell, A.M., Petts, G.E., Hannah, D.M., Smith, B.P.G., Edwards, P.J., Kollmann, J., Ward, J.V. and Tockner, K. (2000) Wood storage within the active zone of a large European gravel-bed river. *Geomorphology*, 34, 55–72.

Hickin, E.J. and Nanson, G.C. (1975) The character of channel migration on the Beatton River, north-east British Columbia, Canada. *Bulletin of the Geological Society of America* 86, 487–494.

Hickin, E.J. and Nanson, G.C. (1984) Lateral migration rates of river bends. *Journal of Hydraulic Engineering* 110(11), 1157–1167.

Hjulstrom, F. (1935) Studies of the morphological activity of rivers as illustrated by the River Fryis. *Bulletin of the Geological Institute of Uppsala* 25, 221–527.

Howard, A.D. (1967) Drainage analysis in geologic interpretation; a summation. *The American Association of Petroleum Geologists Bulletin* 51(11), 2246–2259.

Hoyle, J., Brooks, A.P., Brierley, G.J., Fryirs, K. and Lander, J. (2008) Variability in the nature and timing of channel response to typical human disturbance along the upper Hunter River, New South Wales, Australia. *Earth Surface Processes and Landforms* 33(6), 868–889.

Jain, V., Fryirs, K. and Brierley, G. (2008) Where do floodplains begin? The role of total stream power and longitudinal profile form on floodplain initiation processes. *Geological Society of America Bulletin* 120, 127–141.

Johnston, P. and Brierley, G. (2006) Late Quaternary river evolution of floodplain pockets along Mulloon Creek, New South Wales, Australia. *The Holocene* 16(5), 661–674.

Keating, D., Spink, A., Brooks, A., Sanders, M., Miller, C., Schmidt, T., Kyle, G., Fryirs, K. and Leishman, M. (2008) *The UHRRI Rehabilitation and Research Project 2003–2007*. Macquarie University. ISBN 978-1-74138-298-3.

Knighton, A.D. (1974) Variation in width–depth discharge relations and some implications for hydraulic geometry. *Bulletin of the Geological Society of America* 85, 416–426.

Knighton, D. (1998) *Fluvial Forms and Processes: A New Perspective*. Arnold, London, UK.

Kondolf, G.M. (1994) Geomorphic and environmental effects of instream gravel mining. *Landscape and Urban Planning* 28, 225–243.

Korup, O. (2005) Large landslides and their effect on sediment flux in South Westland, New Zealand. *Earth Surface Processes and Landforms* 30, 305–323.

Lane, E.W. (1955) The design of stable channels. *Transactions of the American Society of Civil Engineers* 120, 1234–1260.

Leeks, G.J.L. (1992) Impact of plantation forestry on sediment transport processes. In: Billi, P., Hey, R.D., Thorne, C.R. and Taconi, P. (eds) *Dynamics of Gravel-bed Rivers*. John Wiley and Sons, Ltd, Chichester, pp. 651–670.

Leopold, L.B. (2004) Geomorphology: a sliver off the corpus of science. *Annual Review of Earth and Planetary Sciences* 32, 1–12.

Long, Y.Q. and Xiong, G.S. (1981) Sediment measurement in the Yellow River. In: Walling, D. and Tacconi, P. (eds) *Erosion and Sediment Transport Measurement*, IAHS Publication, No. 133. IAHS Press, Wallingford, pp. 275–285.

Lowe, J.J. and Walker, M.J.C. (1997) *Reconstructing Quaternary Environments*. Addison Wesley Longman, London.

Manville, V., Segschneider, B., Newton, E., White, J.D.L., Houghton, B.F. and Wilson, C.J.N. (2009) Environmental impact of the 1.8 ka Taupo eruption, New Zealand: landscape responses to a large-scale explosive rhyolite eruption. *Sedimentary Geology* 220, 318–336.

Miall, A.D. (1985) Architectural element analysis: a new method of facies analysis applied to fluvial deposits. *Earth Surface Reviews* 22, 261–308.

Meade, R.H. and Parker, R.S. (1985) National water summary 1984 – hydrologic events, selected water-quality trends, and ground-water resources. *US Geological Survey Water-Supply Paper 2275*, 49–60 DJVU. US Geological Survey.

Miller, M.C., McCave, I.N. and Komar, P.D. (1977) Thresholds of sediment motion under unidirectional currents. *Sedimentology* 24, 507–527.

Milliman, J.D. and Meade, R.H. (1983) World-wide delivery of sediment to the oceans. *Journal of Geology* 91(1), 1–21.

Milliman, J.D. and Syvitski, J.P.M. (1992) Geomorphic/tectonic control of sediment discharge to the ocean: the importance of small mountainous rivers. *The Journal of Geology* 100, 525–544.

Montgomery, D.R. and Buffington, J.M. (1997) Channel-reach morphology in mountain drainage basins. *Geological Society of America Bulletin* 109(5), 596–611.

Montgomery, D.R., Abbe, T.B., Buffington, J.M., Peterson, N.P., Schmidt, K.M. and Stock, J.D. (1996) Distribution of bedrock and alluvial channels in forested mountain drainage basins. *Nature* 381, 597–589.

Moog, D.B. and Whiting, P.J. (1998) Annual hysteresis in annual bed load rating curves. *Water Resources Research* 34(9), 2393–2399.

Nanson, G.C. (1986) Episodes of vertical accretion and catastrophic stripping: a model of disequilibrium floodplain development. *Geological Society of America Bulletin* 97, 1467–1475.

Nanson, G.C. and Young, R.W. (1981a) Downstream reduction of rural channel size with contrasting urban effects in small coastal stream of south eastern Australia. *Journal of Hydrology* 52, 239–255.

Nanson G.C. and Young D.M. (1981b) Overbank deposition and floodplain formation on small coastal streams of New South Wales. *Zeitshrift für Geomorphologie* 25(3), 332–347.

Nanson, G.C., Cohen, T.C., Doyle, C.J. and Price, D.M. (2003) Alluvial evidence of major late-Quaternary climate and flow-regime changes on the coastal rivers of New South Wales, Australia. In: Gregory, K. and Benito, G. (eds) *Palaeohydrology: Understanding Global Change*. John Wiley and Sons, Ltd, Chichester, pp. 233–258.

Ollier, C.D. and Pain, C.F. (1997) Equating the basal unconformity with the palaeoplain: a model for passive margins. *Geomorphology* 19(1–2), 1–15.

Olive, L.J., Olley, J.M., Murray, A.S. and Wallbrink, P.J. (1994) Spatial variation in suspended transport in the Murrumbidgee River, New South Wales, Australia. In: Olive, L.J., Loughran, R.J. and Kesby, J.A. (eds) *Variability in Stream Erosion and Sediment Transport*, IAHS Publication No. 224. IAHS Press, Wallingford, pp. 241–249.

Page, K.J. and Nanson, G.C. (1982) Concave bank benches and associated floodplain formation. *Earth Surface Processes and Landforms* 7, 529–543.

Page, K.J., Kemp, J. and Nanson G.C. (2009) Late Quaternary evolution of Riverine Plain palaeochannels, southeastern Australia. *Australian Journal of Earth Sciences* 56, S19–S33.

Parker, G. (2008) Transport of gravel and sediment mixtures. In: Garcia, M.H. (ed) *Sedimentation Engineering: Processes, Measurements Modelling and Practice*, ASCE Manuals and Reports on Engineering Practice No. 110. American Society of Civil Engineers, Virginia, USA, pp. 165–243.

Patton, P.C. and Schumm, S.A. (1975) Gully erosion, northwestern Colorado. *Geology* 3, 88–90.

Phillips, J.D. (2007) The perfect landscape. *Geomorphology* 84(3–4), 159–169.

Piegay, H. and Schumm, S.A. (2003) Systems approaches in fluvial geomorphology. In: Kondolf, G.M. and Piegay, H. (eds) *Tools in Fluvial Geomorphology*. John Wiley and Sons, Ltd, Chichester, pp. 103–132.

Piegay, H., Bornette, G., Citterio, A., Herouin, E. and Moulin, B. (2000) Channel instability as a control on silting dynamics and vegetation patterns within the perifluvial aquatic zones. *Hydrological Processes* 14(16–17), 3011–3029.

Reid, H.E., Brierley, G.J. and Boothroyd, I.K.G. (2010) Influence of bed heterogeneity and habitat type on macroinvertebrate uptake in peri-urban streams. *International Journal of Sediment Research* 25(3), 203–220.

Reid, H., Gregory, C. and Brierley, G. (2008) Measures of physical heterogeneity in appraisal of geomorphic river condition for urban streams: Twin Streams catchment, Auckland, New Zealand. *Physical Geography* 29(3), 247–274.

Reinfelds, I. and Nanson, G.C. (2004) Torrents of terror: the August 1998 storm and the magnitude, frequency and impact of major floods in the Illawarra region of New South Wales. *Australian Geographical Studies* 39(3), 335–352.

Rice, S. and Church, M. (1998) Grain size along two gravel-bed rivers: statistical variation, spatial pattern and sedimentary links. *Earth Surface Processes and Landforms* 23, 345–363.

Rittenour, T., Blum, M.D. and Goble, R.J. (2007) Fluvial evolution of the lower Mississippi river valley during the last 100 ky glacial cycle: response to glaciations and sea-level change. *Geological Society of America Bulletin* 119(5–6), 586–608.

Ritter, D.F., Kochel, R.C. and Miller, J.R. (1978) *Process Geomorphology*. McGraw-Hill, New York.

Robert, A. (2003) *River Processes: An Introduction to Fluvial Dynamics*. Arnold, London.

Roberts, C.R. (1989) Flood frequency and urban-induced channel change; some British examples. In: Beven, K. and Carling, P. (eds) *Floods; Hydrological, Sedimentological and Geomorphological Implications*. John Wiley and Sons, Ltd, Chichester, pp. 57–82.

Schumm, S.A. (1968) *River adjustment to altered hydrologic regimen*. Murrumbidgee River and palaeochannels, Australia. United States Geological Survey Professional Paper, 598.

Schumm, S.A. (1969) River metamorphosis. *Proceedings of the American Society of Civil Engineers* 95, 255–273.

Schumm S.A. (1977) *The Fluvial System*. John Wiley and Sons, Inc., New York.

Schumm, S.A. (1985) Patterns of alluvial rivers. *Annual Review of Earth and Planetary Sciences* 13, 5–27.

Schumm, S.A., Harvey, M.D. and Watson, C.C. (1984) *Incised Channels: Morphology, Dynamics and Control*. Water Resources Publications, Littleton, CO.

Selby, M.J. (1983) *Hillslope Materials and Processes* (2nd edition). Oxford University Press, Oxford.

Simons, D.B. and Richardson, E.V. (1966) Resistance to flow in alluvial channels. *United States Geological Survey Professional Paper*, P 0442-J, J1–J61.

Sklar, L. and Dietrich, W.E. (1998) River longitudinal profiles and bedrock incision models: stream power and the influence of sediment supply. In: Tinkler, K. and Wohl, E.E. (eds) *Rivers over Rock: Fluvial Processes in Bedrock Channels*, AGU Geophysical Monograph Series, vol. 107. AGU, Washington, DC, pp. 237–260.

Spink, A., Fryirs, K. and Brierley, G.J. (2009) The relationship between geomorphic river adjustment and management actions over the last 50 years in the upper Hunter catchment, NSW, Australia. *River Research and Applications* 25, 904–928.

Spink, A., Hillman, M., Fryirs, K., Brierley, G.J. and Lloyd, K. (2010) Has river rehabilitation begun? Social perspectives on river management in the upper Hunter catchment. *Geoforum* 41, 399–409.

Starkel, L. (1987) Man as a cause of sedimentologic changes in the Holocene. *Striae* 26, 5–12.

Summerfield, M.A. and Hulton, N.J. (1994) Natural controls of fluvial denudation rates in major world drainage basins. *Journal of Geophysical Research* 99(B7), 13,871–13,883.

Surian, N., Mao, L., Giacomin, M. and Ziliani, L. (2009) Morphological effects of different channel forming discharges in a gravel-bed river. *Earth Surface Processes and Landforms* 34(8), 1093–1107.

Syvitski, J.P.M., Vorosmarty, C.J., Kettner, A.J. and Green, P. (2005) Impact of humans on the flux of terrestrial sediment to the global coastal ocean. *Science* 308, 376–380.

Tarboton, D.G. (2003) *Rainfall–Runoff Processes: A Workbook to Accompany the Rainfall–Runoff Process Web Module*. http://www.engineering.usu.edu/dtarb/rrp.html. Utah State University, USA.

Thomson, J., Taylor, M.P., Fryirs, K.A. and Brierley, G.J. (2001) A geomorphological framework for river characterisation and habitat assessment. *Aquatic Conservation: Marine and Freshwater Ecosystems* 11, 373–389.

Thorne, C.R. (1999) *Stream Reconnaissance Handbook: Geomorphological Investigation and Analysis of River Channels*. John Wiley and Sons, Ltd, Chichester.

Tockner, K., Ward, J.V., Arscott, D.B., Edwards, P.J., Kollmann, J., Gurnell, A.M., Petts, G.E. and Maiolini, B. (2003) The Tagliamento River: A model ecosystem of European importance. *Aquatic Sciences*, 65, 239–253.

Trimble, S. (2009) Fluvial processes, morphology and sediment budgets in the Coon Creek Basin, WI, USA, 1975–1993. *Geomorphology* 108, 8–23.

Walling, D.E. and Webb, B.W. (1982) Sediment availability and the prediction of storm-period sediment yields. In: Walling, D.E. (ed.) *Recent Developments in the Explanation and Prediction of Erosion and Sediment Yield*, IAHS Publication No. 137. IAHS Press, Wallingford, pp. 327–337.

Williams, G.P. (1983) Palaeohydrological methods and some examples from Swedish fluvial environments *Geografiska Annaler A* 65, 227–243.

Williams, M. (2009) Late Pleistocene and Holocene environments in the Nile basin. *Global and Planetary Change* 69, 1–15.

Wohl, E.E. (2007) Hydrology and discharge. In: Gupta, A. (ed.) *Large Rivers: Geomorphology and Management*. John Wiley and Sons, Ltd, Chichester, pp. 29–44.

Wolman, M.G. (1967) A cycle of sedimentation and erosion in urban river channels. *Geografiska Annaler A* 49(2–4), 385–395.

Selected readings

Books in fluvial geomorphology

Baker, V.R., Kochel,C. and Patton, P.C. (1998) *Flood Geomorphology*. John Wiley and Sons, Ltd, Chichester. 491 pp.

Bridge, J.S. (2003) *Rivers and Floodplains: Forms, Processes and Sedimentary Record*. Blackwell Publishing, Oxford. 489 pp.

Brierley, G.J. and Fryirs, K.A. (2005) *Geomorphology and River Management: Applications of the River Styles Framework*. Blackwell Publications, Oxford. 398 pp.

Brierley, G.J. and Fryirs, K.A. (eds) (2008) *River Futures: An Integrative Scientific Approach to River Repair*. Island Press, Washington, DC. 304 pp.

Chang, H.H. (1988) *Fluvial Processes in River Engineering*. John Wiley and Sons, Inc., New York. 432 pp.

Downs, P.W. and Gregory, K.J. (2004) *River Channel Management: Towards Sustainable Catchment Hydrosystems*. Arnold, London, UK. 395 pp.

Gurnell, A.M. and Petts, G.E. (eds) (1996) *Changing River Channels*. John Wiley and Sons, Ltd, Chichester. 422 pp.

Huggett, R.J. (2003) *Fundamentals of Geomorphology*. Routledge, London.

Knighton, D. (1998) *Fluvial Forms and Processes : A New Perspective*. Arnold, London. 383 pp.

Kondolf, G.M. and Piegay, H. (eds) (2003) *Tools in Fluvial Geomorphology*. John Wiley and Sons, Ltd, Chichester. 687 pp.

Leopold, L.B., Wolman, M.G. and Millar, J.P. (1964) *Fluvial Processes in Geomorphology*. Dover Publications, New York.

Miller, A. and Gupta, A. (eds) (1999) *Varieties of Fluvial Form*. John Wiley and Sons, Ltd, Chichester.

Morisawa, M. (1968) *Streams: Their Dynamics and Morphology*. McGraw Hill, New York. 175 pp.

Petts, G.E. and Amoros, C. (1996) *Fluvial Hydrosystems*. Springer, New York. 322 pp.

Petts, G.E. and Calow, P. (eds) (1996) *River Flows and Channel Forms*. Blackwell Science, Oxford. 262 pp.

Rhoads, B.L. and Thorn, C.E. (eds) (1996) *The Scientific Nature of Geomorphology*. John Wiley and Sons, Ltd, Chichester. 481 pp.

Schumm S.A. (1977) *The Fluvial System*. John Wiley and Sons, Inc., New York.

Schumm, S.A. (1991) *To Interpret the Earth: Ten Ways to be Wrong*. Cambridge University Press, Cambridge.

Schumm, S.A., Mosley, M.P. and Weaver, W.E. (1987) *Experimental Fluvial Geomorphology*. John Wiley and Sons, Ltd, Chichester. 413 pp.

Thomas, D.S.G. and Allison, R.J. (1993) *Landscape Sensitivity*. John Wiley and Sons, Ltd, Chichester. 374 pp.

Thorne, C.R., Hey, R.D. and Newson, M.D. (eds) (1997) *Applied Fluvial Geomorphology for River Engineering and Management*. John Wiley and Sons, Ltd, Chichester.

Tinkler, K.J. and Wohl, E.E. (eds) (1998) *Rivers over Rock: Fluvial Processes in Bedrock Channels*. Geophysical Monograph Series vol. 107. American Geophysical Union, Washington, DC.

Wohl, E.E. (2004) *Disconnected Rivers: Linking Rivers to Landscapes*. Yale University Press, New Haven. 301 pp.

Chapter 1 Geomorphic analysis of river systems: an approach to reading the landscape

Baker V.R. and Twidale C.R. (1991) The re-enchantment of geomorphology. *Geomorphology* 4, 73–100.

Phillips, J.D. (2007) The perfect landscape. *Geomorphology* 84(3–4), 159–169.

Preston, N.P., Brierley, G.J. and Fryirs, K. (2011) The geographic basis of geomorphic enquiry. *Geography Compass* 5(1), 21–34.

Slaymaker, O. (2009) The future of geomorphology. *Geography Compass* 3–1, 329–349.

Chapter 2 Concepts in river geomorphology

Chappell, J. (1983) Thresholds and lags in geomorphologic changes. *Australian Geographer* 15, 357–366.

Church, M. (2002) Geomorphic thresholds in riverine landscapes. *Freshwater Biology* 47, 541–557.

Chorley, R.J. and Kennedy, B.A. (1971) *Physical Geography: A Systems Approach*. Prentice Hall, London.

Lane, S.N. and Richards, K.S. (1997) Linking river channel form and process: time, space and causality revisited. *Earth Surface Processes and Landforms* 22, 249–260.

Phillips, J.D. (2003) Sources of nonlinearity and complexity in geomorphic systems. *Progress in Physical Geography* 27(1), 1–23.

Phillips, J.D. (2011) Emergence and pseudo-equilibrium in geomorphology. *Geomorphology* 132(3–4), 319–326.

Schumm, S.A. (1973) Geomorphic thresholds and the complex response of drainage systems. In: Morisawa, M. (ed.). *Fluvial Geomorphology*. State University of New York, Binghampton, pp. 299–310.

Schumm, S.A. and Lichty, R.W. (1965) Time, space and causality in geomorphology. *American Journal of Science* 263, 110–119.

Wolman, M.G. and Gerson, R. (1978) Relative scales of time and effectiveness of climate in watershed geomorphology. *Earth Surface Processes and Landforms* 3, 189–208.

Wolman, M.G. and Miller, J.P. (1960) Magnitude and frequency of forces in geomorphic processes. *Journal of Geology* 68, 54–74.

Chapter 3 Catchment-scale controls on river geomorphology

Begin, Z.B., Meyer, D.F. and Schumm, S.A. (1981) Development of longitudinal profiles of alluvial channels in response to base-level lowering. *Earth Surface Processes and Landforms* 6(1), 49–68.

Brierley, G.J. and Fryirs, K. (1999) Tributary–trunk stream relations in a cut-and-fill landscape: a case study from Wolumla catchment, N.S.W., Australia. *Geomorphology* 28, 61–73.

Chorley, R.J. (1969) The drainage basin as the fundamental geomorphic unit. In: Chorley, R.J. (ed.) *Water, Earth, and Man*. Methuen and Co. Ltd, Canada.

Hack, J.T. (1957) Studies of longitudinal stream profiles in Virginia and Maryland. *United States Geological Survey Professional Paper*, 294B, 45–97.

Horton, R.E. (1945) Erosional development of streams and their drainage basins: hydrophysical approach to quantitative morphology. *Bulletin of the Geological Society of America* 56, 275–370.

Langbein, W.B. (1964) Profiles of rivers of uniform discharge. *United States Geological Survey Professional Paper* 501B, 119–122.

Mackin, J.H. (1948) Concept of a graded river. *Geological Society of America Bulletin* 5, 463–511.

Poole, G.C. (2002) Fluvial landscape ecology: addressing uniqueness within the river discontinuum. *Freshwater Biology* 47(4), 641–660.

Rice, S. (1998) Which tributaries disrupt downstream fining along gravel-bed rivers? *Geomorphology* 22(1), 39–56.

Sklar, L. and Dietrich, W.E. (1998) River longitudinal profiles and bedrock incision models: stream power and the influence of sediment supply. In Tinkler, K. and Wohl, E.E. (eds) *Rivers over Rock: Fluvial Processes in Bedrock Channels*. Geophysical Monograph Series vol. 107. American Geophysical Union, Washington, DC, pp. 237–260.

Strahler, A.N. (1952) Hypsometric (area–altitude) analysis of erosional topography. *Bulletin of the Geological Society of America* 63, 1117–1142.

Whipple, K.X. (2004) Bedrock rivers and the geomorphology of active orogens. *Annual Review of Earth and Planetary Sciences* 32, 151–185.

Chapter 4 Catchment hydrology

Andrews, E.D. (1980) Effective and bankfull discharges of streams in the Yampa River basin, Colorado and Wyoming. *Journal of Hydrology* 46(3–4), 311–330.

Beven, K.J. (2001) *Rainfall–Runoff Modelling: The Primer*. John Wiley and Sons, Ltd, Chichester. 360 pp.

Beven, K.J. and Kirkby, M.J. (1979) A physically-based, variable contributing area model of basin hydrology. *Hydrological Sciences Bulletin* 24(1), 43–69.

Emmett, W.W. and Wolman, G.M. (2001) Effective discharge and gravel-bed rivers. *Earth Surface Processes and Landforms* 26(13), 1369–1380.

Gordon, N.D., McMahon, T.A., Finlyason, B.L., Gippel, C.J. and Nathan, R.J. (2004) *Stream Hydrology: An Introduction for Ecologists*. John Wiley and Sons, Ltd, Chichester. 429 pp.

Junk, W.J., Bayley, P.B. and Sparks, R.E. (1989) The flood pulse concept in river-floodplain systems. In: Dodge, D.P. (ed.) *Proceedings of the International Large River Symposium (LARS)*. Special Publication of the Canadian Journal of Fisheries and Aquatic Sciences 106, NRC Research Press, Ottawa, pp. 110–127.

Ladson, A. (2008) *Hydrology: An Australian Introduction*. Oxford University Press, Melbourne.

Pickup, G. and Warner, R.F. (1976) Effects of hydrologic regime on magnitude and frequency of dominant discharge. *Journal of Hydrology* 29, 51–75.

Reinfelds, I. and Nanson, G.C. (2004) Torrents of terror: the August 1998 storm and the magnitude, frequency and impact of major floods in the Illawarra region of New South Wales. *Australian Geographical Studies* 39(3), 335–352.

Chapter 5 Impelling and resisting forces in river systems

Baker, V.R. and Costa, J.E. (1987) Flood power. In Mayer, L. and Nash, D. (eds) *Catastrophic Flooding*. Allen and Unwin, Boston, MA, pp. 1–21.

Bull, W.B. (1979) Thresholds of critical power in streams. *Geological Society of America Bulletin* 90, 45–464.

Corenblit, D., Tabacchi, E., Steiger, J. and Gurnell, A.M. (2007) Reciprocal interactions and adjustments between fluvial landforms and vegetation dynamics in river corridors: a review of complementary approaches. *Earth Science Reviews* 84, 57–86.

Costa, J.E. and O'Connor, J.E. (1995) Geomorphically effective floods. In: Costa, J.E., Miller, A.J., Potter, K.W. and Wilcock, P.R. (eds) *Natural and Anthropogenic Influences in Fluvial Geomorphology*, Geophysical Monograph 89. American Geophysical Union, Washington, DC, pp. 45–56.

Darby, S.E. (1999) Effect of riparian vegetation on flow resistance and flood potential. *Journal of Hydraulic Engineering* 125(5), 443–454.

Gurnell, A.M., Piégay, H., Gregory, S.V. and Swanson, F.J. (2002) Large wood and fluvial processes. *Freshwater Biology* 47, 601–619.

Hey, R.D. (1979) Flow resistance in gravel bed rivers. *ASCE Journal of the Hydraulics Division* 105(4), 365–379.

Hickin, E.J. (1984) Vegetation and river channel dynamics. *Canadian Geographer* 28(2), 111–126.

Knighton, A.D. (1999) Downstream variation in stream power. *Geomorphology* 29(3–4), 293–306.

Magilligan, F.J. (1992) Thresholds and the spatial variability of flood power during extreme floods. *Geomorphology* 5, 373–390.

Miller, A.J. (1990) Flood hydrology and geomorphic effectiveness in the Central Appalachians. *Earth Surface Processes and Landforms* 15, 119–135.

Miller, R.G. (1999) Grain and form resistance in gravel-bed rivers. *Journal of Hydraulic Research* 37(3), 303–312.

Montgomery, D.R. and Piégay, H. (2003) Wood in rivers: interactions with channel morphology and processes. *Geomorphology* 51(1–3), 1–5.

Osterkamp, W.R. and Hupp, C.R. (2010) Fluvial processes and vegetation: glimpses of the past, the present, and perhaps the future. *Geomorphology* 116, 274–285.

Osterkamp, W.R., Hupp, C.R. and Stoffel, M. (2012) The interactions between vegetation and erosion: new directions for research at the interface of ecology and geomorphology. *Earth Surface Processes and Landforms* 37(1), 23–36. DOI: 10.1002/esp.2173.

Pie'gay, H. and Gurnell, A.M. (1997) Large woody debris and river geomorphological pattern: examples from S.E. France and S. England. *Geomorphology* 19(1–2), 99–116.

Piégay, H., Thévenet, A. and Citterio, A. (1999) Input, storage and distribution of large woody debris along a mountain river continuum, the Drome River, France. *Catena* 35(1), 19–39.

Wilcock, P.R. (1993) Critical shear stress of natural sediments. *Journal of Hydraulic Engineering* 119(4), 491–505.

Chapter 6 Sediment movement and deposition in river systems

Andrews, E.D. (1983) Entrainment of gravel from naturally sorted riverbed material. *Geological Society of America Bulletin* 94(10), 1225–1231.

Andrews, E.D. (1984) Bed-material entrainment and hydraulic geometry of gravel-bed rivers in Colorado. *Geological Society of America Bulletin* 95(3), 371–378.

Bagnold, R.A. (1966) An approach to the sediment transport problem from general physics. *US Geological Survey Professional Paper*, 422 J.

Brasington, J., Langham, J. and Rumsby, B. (2003) Methodological sensitivity of morphometric estimates of coarse fluvial sediment transport. *Geomorphology* 53(3–4), 299–316.

Brierley, G.J. (1996) Channel morphology and element assemblages: a constructivist approach to facies modelling. In: Carling, P. and Dawson, M. (eds) *Advances in Fluvial Dynamics and Stratigraphy*. Wiley Interscience, Chichester, pp. 263–298.

Church, M.A., McLean, D.G. and Wolcott, J.F. (1987) River bed gravels: sampling and analysis. In: Thorne, C.R., Bathurst, J.C. and Hey, R.D. (eds) *Sediment Transport in Gravel Bed Rivers*. John Wiley and Sons, Inc., New York, pp. 43–88.

Gomez, B. (2011) Sediment I: properties. *Geography Compass* 5–6, 390–411.

Gomez, B. (2011) Sediment II: modification and provenance. *Geography Compass* 5–7, 494–516.

Gomez, B. and Church, M. (1989) An assessment of bed load sediment transport formulae for gravel bed rivers. *Water Resources Research* 25(6), 1161–1186.

Hassan, M.A. and Reid, I. 1992. The influence of microform bed roughness elements on flow and sediment transport in gravel bed rivers. *Earth Surface Processes and Landforms* 15(8), 739–750.

Miall, A.D. (1985) Architectural element analysis: a new method of facies analysis applied to fluvial deposits. *Earth Surface Reviews* 22, 261–308.

Miall, A.D. (2000) *Principles of Sedimentary Analysis*, third edition. Springer, New York. 616 pp.

Parker, G. (2008) Transport of gravel and sediment mixtures. In: Garcia, M.H. (ed.) *Sedimentation Engineering: Processes, Measurements Modelling and Practice*. ASCE Manuals and Reports on Engineering Practice No. 110. American Society of Civil Engineers, Virginia, USA, pp. 165–243.

Pizzuto, J.E. (1995). Downstream fining in a network of gravel-bedded rivers. *Water Resources Research* 31(3), 753–759.

Prosser, I.P., Rutherfurd, I.D., Olley, J.M., Young, W.J., Wallbrink, P.J. and Moran, C.J. (2001) Large-scale patterns of erosion and sediment transport in river networks, with examples from Australia. *Marine and Freshwater Research* 52(1), 81–99.

Rice S. (1999) The nature and controls on downstream fining within sedimentary links. *Journal of Sedimentary Research* 69(1), 32–39.

Rice, S. and Church, M. (1996) Sampling surficial fluvial gravels: the precision of size distribution percentile estimates. *Journal of Sedimentary Research* 66(3), 654–665.

Rust, B.R. and Nanson, G.C. (1989) Bedload transport of mud as pedogenic aggregates in modern and ancient rivers. *Sedimentology* 36 (2), 291–306.

Whipple, K.X., Hancock, G.S. and Anderson, R.S. (2000) River incision into bedrock: mechanics and relative efficacy of plucking, abrasion, and cavitation. *Bulletin of the Geological Society of America* 112(3) 490–503.

Wilcock, P.R. and Crowe, J.C. (2003) Surface-based transport model for mixed-size sediment. *Journal of Hydraulic Engineering* 129(2), 120–128.

Wilcock, P.R. and McArdell, B.W. (1993) Surface-based fractional transport rates: mobilization thresholds and partial transport of a sand–gravel sediment. *Water Resources Research* 29(4), 1297–1312.

Wilcock, P.R. and McArdell, B.W. (1997) Partial transport of a sand/gravel sediment. *Water Resources Research* 33(1), 235–245.

Wohl, E.E., Anthony, D.J., Madsen, S.W. and Thompson, D.M. (1996) A comparison of surface sampling methods for coarse fluvial sediments. *Water Resources Research* 32(10), 3219–3226.

Wolman, M.G. (1954) A method of sampling coarse river-bed material. *Transactions of the American Geophysical Union* 35, 951–956.

Chapter 7 Channel geometry

Darby, S.E. and Thorne, C.R. (1996) Development and testing of riverbank-stability analysis. *Journal of Hydraulic Engineering* 122(8), 443–454.

Eaton, B.C., Church, M. and Millar, R.G. (2004) Rational regime model of alluvial channel morphology and response. *Earth Surface Processes and Landforms* 29(4), 511–529.

Ferguson, R.I. (1986) Hydraulics and hydraulic geometry. *Progress in Physical Geography* 10(1), 1–31.

Florsheim, J.L., Mount, J.F. and Chin, A. (2008) Bank erosion as a desirable attribute of rivers. *BioScience* 58(6), 519–529.

Hession, W.C., Pizzuto, J.E., Johnson, T.E. and Horwitz, R.J. (2003) Influence of bank vegetation on channel morphology in rural and urban watersheds. *Geology* 31(2), 147–150.

Hey, R.D. (1986) Stable channels with mobile gravel beds. *Journal of Hydraulic Engineering – ASCE* 112(8), 671–689.

Hooke, J.M. (1979) An analysis of the processes of river bank erosion. *Journal of Hydrology* 42(1–2), 39–62.

Hooke, J.M. (1980) Magnitude and distribution of rates of river bank erosion. *Earth Surface Processes* 5(2), 143–157.

Knighton, A.D. (1974) Variation in width–depth discharge relations and some implications for hydraulic geometry. *Bulletin of the Geological Society of America* 85, 416–426.

Leopold, L.B. and Maddock Jr, T. (1953) The hydraulic geometry of stream channels and some physiographic implications. *United States Geological Survey Professional Paper*, P 0252, USGS, Washington, DC, 57 pp.

Millar, R.G. and Quick, M.C. (1993) Effect of bank stability on geometry of gravel rivers. *Journal of Hydraulic Engineering – ASCE* 119(12), 1343–1363.

Park, C.C. (1977) World-wide variations in hydraulic geometry exponents of stream channels: an analysis and some observations. *Journal of Hydrology* 33(1–2), 133–146.

Rhodes, D.D. (1977) The b-f-m diagram; graphical representation and interpretation of at-a-station hydraulic geometry. *American Journal of Science* 277, 73–96.

Rhodes, D.D. (1987) The *b–f–m* diagram for downstream hydraulic geometry. *Geografiska Annaler A* 69(1), 147–161.

Schumm, S.A., Harvey, M.D. and Watson, C.C. (1984) *Incised Channels: Morphology, Dynamics and Control*. Water Resources Publication, Littleton, CO.

Simon, A. and Rinaldi, M. (2006) Disturbance, stream incision, and channel evolution: the roles of excess transport capacity and boundary materials in controlling channel response. *Geomorphology* 79(3–4), 361–383.

Simon, A. and Thomas, R.E. (2002) Processes and forms of an unstable alluvial system with resistant, cohesive streambeds. *Earth Surface Processes and Landforms* 27(7), 699–718.

Simon, A., Curini, A., Darby, S.E. and Langendoen, E.J. (2000) Bank and near-bank processes in an incised channel. *Geomorphology* 35(3–4), 193–217.

Thorne, C.R. (1982) Processes and mechanisms of river bank erosion. In: Hey, R.D., Bathurst, J.C. and Thorne, C.R. (eds) *Gravel-bed Rivers: Fluvial Processes, Engineering and Management*. John Wiley and Sons, Ltd, Chichester, pp. 227–271.

Chapter 8 Instream geomorphic units

Chin, A. (1998) On the stability of step–pool mountain streams. *Journal of Geology* 106(1), 59–69.

Church, M. and Jones, D. (1982) Channel bars in gravel-bed rivers. In: Hey, R.D., Bathurst, J.C. and Thorne, C.R. (eds) *Gravel-bed Rivers: Fluvial Processes, Engineering and Management*. John Wiley and Sons, Ltd, Chichester, pp. 291–338.

Clifford, N.J. (1993) Formation of riffle–pool sequences: field evidence for an autogenetic process. *Sedimentary Geology* 85(1–4), 39–51.

Corenblit, D., Steiger, J., Gurnell, A. and Tabacchi, E. (2007) Darwinian origin of landforms. *Earth Surface Processes and Landforms* 32(13), 2070–2073.

Friedman, J.M., Osterkamp, W.R. and Lewis, W.M. (1996) The role of vegetation and bed-level fluctuations in the process of channel narrowing. *Geomorphology* 14, 341–351.

Hupp, C.R. and Simon, A. (1991) Bank accretion and the development of vegetated depositional surfaces along modified alluvial channels. *Geomorphology* 4(2), 111–124.

Montgomery, D.R. and Buffington, J.M. (1997) Channel-reach morphology in mountain drainage basins. *Geological Society of America Bulletin* 109(5), 596–611.

Page, K.J. and Nanson, G.C. (1982) Concave bank benches and associated floodplain formation. *Earth Surface Processes and Landforms* 7, 529–543.

Wohl, E.E., Vincent, K.R. and Merritts, D.J. (1993) Pool and riffle characteristics in relation to channel gradient. *Geomorphology* 6(2), 99–110.

Chapter 9 Floodplain forms and processes

Brice, J.C. (1960) Index for description of channel braiding. *Bulletin of the Geological Society of America* 71, 1833.

Brice, J.C. (1974) Evolution of meander loops. *Bulletin of the Geological Society of America* 85, 581–586.

Brierley, G.J., Ferguson, R.J. and Woolfe, K.J. (1997) What is a fluvial levee? *Sedimentary Geology* 114, 1–9.

Hickin, E.J. and Nanson, G.C. (1984) Lateral migration rates of river bends. *Journal of Hydraulic Engineering* 110(11), 1157–1167.

Hooke, J. (2003) River meander behaviour and instability: a framework for analysis. *Transactions of the Institute of British Geographers* 28(2), 238–253.

Moody, J.A., Pizzuto, J.E. and Meade, R.H. (1999) Ontogeny of a flood plain. *Bulletin of the Geological Society of America* 111(2), 291–303.

Nanson, G.C. (1980) Point bar and floodplain formation of the meandering Beatton River, northeastern British Columbia, Canada. *Sedimentology* 27(1), 3–29.

Nanson, G.C. (1986) Episodes of vertical accretion and catastrophic stripping: a model of disequilibrium floodplain development. *Geological Society of America Bulletin* 97, 1467–1475.

Nanson, G.C. and Croke, J.C. (1992) A genetic classification of floodplains. *Geomorphology* 4, 459–486.

Pizzuto, J.E. (1987) Sediment diffusion during overbank flows. *Sedimentology* 34(2), 301–317.

Reinfelds, I. and Nanson, G. (1993) Formation of braided river floodplains, Waimakariri River, New Zealand. *Sedimentology* 40(6), 1113–1127.

Chapter 10 River diversity

Brierley G.J. and Fryirs, K. (2000) River Styles, a geomorphic approach to catchment characterisation: implications for river rehabilitation in Bega catchment, New South Wales, Australia. *Environmental Management* 25(6), 661–679.

Church, M. (1992) Channel morphology and typology. In: Callow, P. and Petts, G.E. (eds) *The Rivers Handbook*. Blackwell, Oxford, pp. 126–143.

Eaton, B.C., Millar, R.G. and Davidson, S. (2010) Channel patterns: Braided, anabranching, and single-thread. *Geomorphology* 120, 353–364.

Fryirs, K. and Brierley, G.J. (2010) Antecedent controls on river character and behaviour in partly-confined valley settings: upper Hunter catchment, NSW, Australia. *Geomorphology* 117, 106–120.

Knighton, A.D. and Nanson, G.C. (1993) Anastomosis and the continuum of channel pattern. *Earth Surface Processes and Landforms* 18, 613–625.

Millar, R.G. (2000) Influence of bank vegetation on alluvial channel patterns. *Water Resources Research* 36(4), 1109–1118.

Montgomery, D.R. (1999) Process domains and the river continuum. *Journal of the American Water Resources Association* 35(2), 397–410.

Montgomery, D.R., Abbe, T.B., Buffington, J.M., Peterson, N.P., Schmidt, K.M. and Stock, J.D. (1996) Distribution of bedrock and alluvial channels in forested mountain drainage basins. *Nature* 381, 597–589.

Newson, M.D., Clark, M.J., Sear, D.A. and Brookes, A. (1998) The geomorphological basis for classifying rivers. *Aquatic Conservation: Marine and Freshwater Ecosystems* 8(4), 415–430.

Nanson, G.C. and Knighton, D.A. (1996) Anabranching rivers: their cause, character and classification. *Earth Surface Processes and Landforms* 21(3), 217–239.

Schumm, S.A. (1985) Patterns of alluvial rivers. *Annual Review of Earth and Planetary Sciences* 13, 5–27.

Thorne, C.R. (1997) Channel types and morphological classification. In: Thorne, C.R., Hey, R.D. and Newson, M.D. (eds) *Applied Fluvial Geomorphology for River Engineering and Management*. John Wiley and Sons, Ltd, Chichester.

Tooth, S. (2000) Process, form and change in dryland rivers: a review of recent research. *Earth Science Reviews* 51(1–4), 67–107.

Tooth, S. and McCarthy, T.S. (2004) Anabranching in mixed bedrock-alluvial rivers: the example of the Orange River above Augrabies Falls, Northern Cape Province, South Africa. *Geomorphology* 57(3–4), 235–262.

Van den Berg, J.H. (1995) Prediction of alluvial channel pattern of perennial rivers. *Geomorphology* 12(4), 259–279.

Ward, J.V., Tockner, K., Arscott, D.B. and Claret, C. (2002) Riverine landscape diversity. *Freshwater Biology* 47, 517–539.

Wohl, E.E. and Merritt, D.M. (2001) Bedrock channel morphology. *Bulletin of the Geological Society of America* 113(9), 1205–1212.

Chapter 11 Interpreting river behaviour

Brierley, G.J. and Fryirs, K.A. (2005) *Geomorphology and River Management: Applications of the River Styles Framework*. Blackwell Publications, Oxford. 398 pp.

Downs, P.W. (1995) Estimating the probability of river channel adjustment. *Earth Surface Processes and Landforms* 20(7), 687–705.

Downs, P.W. and Gregory, K.J. (1995) Approaches to river channel sensitivity. *Professional Geographer* 47(2), 168–175.

Fryirs, K., Spink, A. and Brierley, G. (2009) Post-European settlement response gradients of river sensitivity and recovery across the upper Hunter catchment, Australia. *Earth Surface Processes and Landforms* 34, 897–918.

Ward, J.V., Tockner, K., Edwards, P.J., Kollmann, J., Bretschko, G., Gurnell, A.M., Petts, G.E. and Rossaro, B. (1999) A reference river system for the Alps: the 'Fiume Tagliamento'. *Regulated Rivers: Research and Management* 15(1–3), 63–75.

Wohl, E. and Merritts D.J. (2007) What is a natural river? *Geography Compass* 1(4), 871–900.

Chapter 12 River evolution

Blum, M.D. and Tornqvist, T.E. (2000) Fluvial responses to climate and sea-level change: a review and look forward. *Sedimentology* 47(Suppl 1), 2–48.

Brierley, G.J. (2009) Landscape memory: the imprint of the past in contemporary landscape forms and processes. *Area* 42(1), 76–85.

Brooks, A.P. and Brierley, G.J. (2002) Mediated equilibrium: the influence of riparian vegetation and wood on the long-term evolution and behaviour of a near pristine river. *Earth Surface Processes and Landforms* 27(4), 343–367.

Brunsden, D. and Thornes, J.B. (1979) Landscape sensitivity and change. *Transactions of the Institute of British Geographers* NS4, 463–484.

Church, M. and Slaymaker, O. (1989) Disequilibrium of Holocene sediment yield in glaciated British Columbia. *Nature* 337(6206), 452–454.

Knox, J.C. (1995) Fluvial systems since 20,000 years BP. In: Gregory, K.J., Starkel, L. and Baker, V.R. (eds) *Global Continental Palaeohydrology*. John Wiley and Sons, Inc., New York. pp. 87–108.

Lewin, J. and Macklin, M.G. (2003) Preservation potential for late Quaternary river alluvium. *Journal of Quaternary Science* 18(2), 107–120.

Merritts, D.J., Vincent, K.R. and Wohl, E.E. (1994) Long river profiles, tectonism, and eustasy: a guide to interpreting fluvial terraces. *Journal of Geophysical Research* 99(B7), 14,031–14,050.

Nott, J., Young, R. and McDougall, I. (1996) Wearing down, wearing back, and gorge extension in the long-term denudation of a highland mass: quantitative evidence from the Shoalhaven catchment, southeast Australia. *Journal of Geology* 104(2), 224–232.

Paine, D.M. (1985) 'Ergodic' reasoning in geomorphology: time for a review of the term? *Progress in Physical Geography* 9(1), 1–15.

Simon, A. (1992) Energy, time, and channel evolution in catastrophically disturbed fluvial systems. *Geomorphology* 5(3–5), 345–372.

Whipple, K.X., Kirby, E. and Brocklehurst, S.H. (1999) Geomorphic limits to climate-induced increases in topographic relief. *Nature* 401(6748), 39–43.

Chapter 13 Human impacts on river systems

Brandt, S.A. (2000) Classification of geomorphological effects downstream of dams. *Catena* 40(4), 375–401.

Brierley, G.J., Cohen, T.J., Fryirs, K. and Brooks, A.P. (1999) Post-European changes to the fluvial geomorphology of Bega catchment, Australia: implications for river ecology. *Freshwater Biology* 41, 1–10.

Brierley, G.J., Brooks, A.P., Fryirs, K. and Taylor, M.P. (2005) Did humid-temperate rivers in the Old and New Worlds respond differently to clearance of riparian vegetation and removal of woody debris? *Progress in Physical Geography* 29(1), 27–49.

Brooks, A.P., Brierley, G.J. and Millar, R.G. (2003) The long-term control of vegetation and woody debris on channel and floodplain evolution: insights from a paired catchment study in southeastern Australia. *Geomorphology* 51, 7–29.

Chin, A. (2006) Urban transformation of river landscapes in a global context. *Geomorphology* 79(3–4), 460–487.

Costa, J.E., Miller, A.J., Potter, K.W. and Wilcock, P.R. (1995) *Natural and Anthropogenic Influences in Fluvial Geomorphology*. Geophysical Monograph, vol. 89. American Geophysical Union, Washington, DC. 239 pp.

Doyle, M.W., Stanley, E.H. and Harbor, J.M. (2003) Channel adjustments following two dam removals in Wisconsin. *Water Resources Research* 39(1), ESG21–ESG215.

Fryirs, K. and Brierley, G.J. (2000) A geomorphic approach for the identification of river recovery potential. *Physical Geography* 21(3), 244–277.

Graf, W.L. (2006) Downstream hydrologic and geomorphic effects of large dams on American rivers. *Geomorphology* 79, 336–360.

Gregory, K.J. (2006) The human role in changing river channels. *Geomorphology* 79(3–4), 172–191.

Gregory, K.J., Davis, R.J. and Downs, P.W. (1992) Identification of river channel change to due to urbanization. *Applied Geography* 12(4), 299–318.

Hooke, R. (2000) On the history of humans as geomorphic agents. *Geology (Boulder)* 28(9), 843–846.

Kondolf, G.M. (1994) Geomorphic and environmental effects of instream gravel mining. *Landscape and Urban Planning* 28, 225–243.

Kondolf, G.M. (1997) Hungry water: effects of dams and gravel mining on river channels. *Environmental Management* 21(4), 533–551.

Kondolf, G.M., Piégay, H. and Landon, N. (2002) Channel response to increased and decreased bedload supply from land use change: contrasts between two catchments. *Geomorphology* 45(1–2), 35–51.

Marston, R.A., Girel, J., Pautou, G., Piegay, H., Bravard, J.-P. and Arneson, C. (1995) Channel metamorphosis, floodplain disturbance, and vegetation development: Ain River, France. *Geomorphology* 13(1–4), 121–131.

Starkel, L. (1987) Man as a cause of sedimentological changes in the Holocene. *Striae* 26, 5–12.

Walter, R.C. and Merritts, D.J. (2008) Natural streams and the legacy of water-powered mills. *Science* 319(5861), 299–304.

Williams, G.P. and Wolman, M.G. (1984) Downstream effects of dams on alluvial rivers. *US Geological Survey Professional Paper* 1286. USGS, Washington, DC.

Chapter 14 Sediment flux at the catchment-scale: source-to-sink relationships

Benda, L. and Dunne, T. (1997) Stochastic forcing of sediment routing and storage in channel networks. *Water Resources Research* 33(12), 2865–2880.

Fryirs, K. and Brierley, G.J. (2001) Variability in sediment delivery and storage along river courses in Bega catchment, NSW, Australia: implications for geomorphic river recovery. *Geomorphology* 38, 237–265.

Fryirs, K., Brierley, G.J., Preston, N.J. and Kasai, M. (2007a) Buffers, barriers and blankets: the (dis)connectivity of catchment-scale sediment cascades. *Catena* 70, 49–67.

Fryirs, K., Brierley, G.J., Preston, N.J. and Spencer, J. (2007b) Catchment-scale (dis)connectivity in sediment flux in the upper Hunter catchment, New South Wales, Australia. *Geomorphology* 84, 297–316.

Harvey, A.M. (2001) Coupling between hillslopes and channels in upland fluvial systems: implications for landscape sensitivity illustrated from the Howgill Fells, northwest England. *Catena* 42, 225–250.

Harvey, A.M. (2002) Effective timescales of coupling within fluvial systems. *Geomorphology* 44, 175–201.

Harvey, A.M. (2012) The coupling status of alluvial fans and debris cones: a review and synthesis. *Earth Surface Processes and Landforms* 37(1), 64–76. DOI: 10.1002/esp.2213.

Hooke, J. (2003) Coarse sediment connectivity in river channel systems: a conceptual framework and methodology. *Geomorphology* 56(1–2), 79–94.

Howard, A.D., Dietrich, W.E. and Seidl, M.A. (1994) Modeling fluvial erosion on regional to continental scales. *Journal of Geophysical Research* 99(B7), 13,971–13,986.

James, L.A. (2010) Secular sediment waves, channel bed waves and legacy sediment. *Geography Compass* 4–6, 576–598.

Korup, O., McSaveney, M.J. and Davies, T.R.H. (2004) Sediment generation and delivery from large historic landslides in the Southern Alps, New Zealand. *Geomorphology* 61(1–2), 189–207.

Madej, M.A. and Ozaki, V. (1996) Channel response to sediment wave propagation and movement, Redwood Creek, California, USA. *Earth Surface Processes and Landforms* 21(10), 911–927.

Phillips, J.D. (1993) Pre- and post-colonial sediment sources and storage in the lower Neuse Basin, North Carolina. *Physical Geography* 14(3), 272–284.

Reid, L.M. and Dunne, T. (1996) *Rapid Evaluation of Sediment Budgets*. Catena Verlag, Reiskirchen.

Sear, D.A., Newson, M.D. and Brookes, A. (1995) Sediment-related river maintenance: the role of fluvial geomorphology. *Earth Surface Processes and Landforms* 20(7), 629–647.

Syvitski, J.P.M. and Milliman, J.D. (2007) Geology, geography and humans battle for dominance over the delivery of fluvial sediment to the coastal ocean. *The Journal of Geology* 115, 1–19.

Trimble, S.W. (2009) Fluvial processes, morphology and sediment budgets in the Coon Creek Basin, WI, USA, 1975–1993. *Geomorphology* 108, 8–23.

Walling, D.E. (1983) The sediment delivery problem. *Journal of Hydrology* 65, 209–237.

Walling, D.E. (1988) Erosion and sediment yield research – some recent perspectives. *Journal of Hydrology* 100(1–3), 113–141.

Walling, D.E. (2005) Tracing suspended sediment sources in catchments and river systems. *Science of the Total Environment* 344(1–3 Spec. Iss.), 159–184.

Walling, D.E. and Fang, D. (2003) Recent trends in the suspended sediment loads of the world's rivers. *Global and Planetary Change* 39(1–2), 111–126.

Wilkinson, B.H. and McElroy, B.J. (2007) The impact of humans on continental erosion and sedimentation. *Geological Society of America Bulletin* 119, 140–158.

Chapter 15 The usefulness of river geomorphology: reading the landscape in practice

Brierley, G.J. and Fryirs, K. (2009) Don't fight the site: geomorphic considerations in catchment-scale river rehabilitation planning. *Environmental Management* 43(6), 1201–1218.

Brierley, G., Fryirs, K. and Hillman, M. (2008) River futures. In: Brierley, G.J. and Fryirs, K.A. (eds) *River Futures: An Integrative Scientific Approach to River Repair*. Island Press, Washington, DC, pp. 275–284.

Fryirs, K. and Brierley, G.J. (2009) Naturalness and place in river rehabilitation. *Ecology and Society* 14(1): 20. [online] URL: http://www.ecologyandsociety.org/vol14/iss1/art20/.

Gilvear, D.J. (1999) Fluvial geomorphology and river engineering: future roles utilising a fluvial hydrosystem framework. *Geomorphology* 31, 229–245.

Index

Note: Page numbers in *italic* refer to figures; those in **bold** to tables.

Geomorphic Analysis of River Systems: An Approach to Reading the Landscape, First Edition. Kirstie A. Fryirs and Gary J. Brierley.
© 2013 Kirstie A. Fryirs and Gary J. Brierley. Published 2013 by Blackwell Publishing Ltd.

www.ingramcontent.com/pod-product-compliance
Lightning Source LLC
Chambersburg PA
CBHW080553270125
20834CB00020B/249